Thin-Section Petrography of Stone and Ceramic Cultural Materials

Thin-Section Petrography of Stone and Ceramic Cultural Materials

Chandra L. Reedy

Archetype Publications

First published 2008 by Archetype Publications Ltd.

Archetype Publications Ltd.
6 Fitzroy Square
London W1T 5HJ
www.archetype.co.uk

Tel: 44(207) 380 0800
Fax: 44(207) 380 0500

ISBN: 978-1-904982-33-3

British Library Cataloguing in Publication Data

A catalogue record for this book is available from the British Library.

Printed on acid-free paper
Typeset by Kate Williams, Swansea
Printed and bound in Great Britain by Latimer Trend & Co. Ltd., Plymouth

Contents

Acknowledgments

My interest in thin-section petrography of cultural materials began when I was a graduate student at the University of California, Los Angeles (UCLA). My first acknowledgment and thanks must go to Jonathon E. Ericson (now a Professor at University of California, Irvine) who first introduced me to the potential of this technique of analysis. Two years of course work in UCLA's Department of Earth and Space Sciences, all including labs in thin-section petrography, gave me enough background to begin working with my own samples. Although I do not claim to be as expert in petrography as a professional geologist would be, I have spent many years peering through the microscope at the oddities often encountered with human-made objects.

I began the initial work on this book under a summer General University Research grant from the University of Delaware, when I explored the issue of method and theory in provenance studies. That fellowship allowed me to begin to collect and read the vast literature and to think about how one might organize an overview book. However, it was a Samuel H. Kress Publication Fellowship from the American Institute for Conservation of Historic and Artistic Works that inspired me to focus specifically on thin-section petrography. That fellowship provided some leave time to begin writing the book and taking the many photomicrographs needed to illustrate it.

The Hugh M. Morris Library at the University of Delaware gave me the use of a dedicated faculty study in which to work. That quiet space in the library was very helpful as I needed to peruse hundreds of books and articles for this project. Many of these materials were not available online. I made heavy use of the services of the Interlibrary Loan Department throughout my work on this book. I thank them for locating and obtaining so many materials for me so quickly.

Two grants from the National Center for Preservation Technology and Training (NCPTT) were crucial in helping me to finish the writing and the photography. One grant provided funds to purchase image analysis software and time to explore its usefulness to quantitative petrography. I credit Elizabeth Goins (now a professor at the Rochester Institute of Technology, Departments of Fine Arts and Material Culture Sciences) with introducing me to the potential of digital image analysis in thin-section petrography. The other NCPTT grant provided funds to develop the CD-ROM that accompanies the book, and to add many more photomicrographs. Without NCPTT support, this project would most likely never have been completed.

The photomicrographs were taken using a Nikon Labophot transmitted light polarizing microscope with a Polaroid digital microscope camera system. All images were taken as three separate photographs, each focused on a different area of the thin section (usually at the center, at one corner, and mid-way between center and corner). Then, using Image-Pro Plus image analysis software, a depth-of-field correction was applied, and the three photographs were merged into one photograph with improved focus.

Although many photographs are included in the book, in order to illustrate each type of material discussed, the CD-ROM organizes them all together and permits the viewer to compare plane polarized and crossed polarized light images. Becky Kinney of Moonlight Multimedia (http://moonlightmultimedia.org/) did all the Flash programming and designed the xml file and the overall layout and workings of the CD-ROM. Jan Broske, University Museums (University of Delaware) provided objects to be photographed by Terry J. Reedy for the chapter menus on the CD-ROM.

To augment my own thin-section collections, I needed a variety of materials beyond those which I have worked on myself in past projects. Thin sections, specimens, and/or objects for sampling were kindly made available to me by Pamela B. Vandiver, University of Arizona, Materials Science and Engineering Department; Paul Jett, Janet Douglas, and Blythe McCarthy of the Freer Gallery of Art; Lu Ann DeCunzo, Keith Doms, and Nedda Moqtaderi of the University of Delaware Anthropology Department; and Tom Meierding, University of Delaware Geography Department. I am very grateful to all of them for their generosity.

Accompanying CD-ROM

This book is accompanied by a CD-ROM that illustrates all the photomicrographs available in the book, but in addition allows the viewer to study each image in both plane polarized and crossed polarized light. If the CD is inserted into a PC with Autorun activated, then the program will immediately begin running. If Autorun is not activated, or if using a Mac, then click on the icon titled "Petrography" to initiate the program.

The chapter, section, and figure numbers of the CD all correspond to chapter, section, and figure numbers in the book. The first screen to appear is the Chapter Menu. To begin viewing images, click on a chapter title.

Within each chapter use the section and figure drop-down menus to navigate between sections and figures. To page through the figures in sequence, use the next and previous arrows provided at the bottom of the screen. All figures can be viewed in both plane polarized and crossed polarized light.

Within a chapter, click on the top left image to return to the Chapter Menu. Clicking on the x in the upper right will close the program. There are a few figures in the book that do not appear in the CD, because those figures are histograms or image analysis products rather than views through the microscope. Otherwise, all thin section views in the book appear on the CD, but in both plane polarized and crossed polarized views while only one of these was chosen for the print image in the book. Inclusion of captions on the CD-ROM means that it can be used as a self-contained training vehicle for gaining familiarity with thin-section images of cultural materials.

The content of the CD-ROM can also be accessed at http://www.udel.edu/CHAD/petrography so this tool will be available in the event of damage or loss of the CD.

1

Introduction

Thin-section petrography involves the examination under transmitted polarized light of a sample taken from a mineral-containing material. The technique was developed by geologists to describe and classify rocks, soils, and sand. Archaeologists and art or architectural conservation scientists also use thin-section studies to characterize and interpret many inorganic materials used in the production of cultural objects or structures. Purposes of such analyses in cultural material studies may include making correct material identifications, grouping specimens according to similar features, comparing materials-based groupings to archaeological or art-historical data, identifying possible geological sources for the object or some of its components, identifying and studying workshops, reconstructing production technology, examining technological changes and variations over time and space, and explaining technological choices. For some art materials, structural and mineralogical changes on weathered surfaces in comparison to unaltered interior sections of a sample may provide information concerning the authenticity of a piece. Thin sections are also used to investigate the type and degree of deterioration of objects and structural materials and to check the effects of conservation treatments or preservation efforts on those materials.

This book does not attempt to teach the basics of mineralogy or thin-section petrography; many textbooks and visual resources exist for that purpose (Heinrich 1956; Phillips 1971; Allman and Lawrence 1972; Folk 1974; Kerr 1977; Scholle 1979; MacKenzie and Guilford 1981; Phillips and Griffen 1981; MacKenzie et al. 1982; Williams et al. 1982; Gribble and Hall 1985; Ehlers 1987; Yardley et al. 1990; Carozzi 1993; MacKenzie and Adams 1994; Adams et al. 1997; Scholle and Ulmer-Scholle 2003). Instead, the focus is on the specific applications of thin-section petrography to the study of stone and ceramic cultural materials. The petrography of cementitious materials such as mortars, plasters, and cements is not included within the scope of this work as published resources for that specialized research are already available (Erlin and Stark 1990; DeHayes and Stark 1994). The following sections give a brief overview of materials preparation and the basic methodology of thin-section petrography, sampling issues and other problems unique to work with cultural materials, and a summary of the main goals of each of the next eight chapters.

1.1 Methodology of Thin-Section Petrography

In thin-section petrography, samples of rocks, ceramics, or other mineral-containing materials are ground down to a standard thickness, usually 30 μm. The standard thickness allows a variety of optical features to be observed under polarized light, facilitating comparison of different samples and the use of reference tables of mineral optics for identifying unknowns. The use of a specific, known thickness for all standard petrographic samples differs from cross-sections or particle microscopy often used in the study of cultural materials.

One advantage of thin-section petrography is that the primary piece of equipment needed, a polarizing microscope, is inexpensive relative to many analytical instruments, making the technique potentially available to almost any laboratory as a routine method of analysis. Although a significant investment in training (such as introductory courses in optical mineralogy provided by a university department in geology or earth sciences) is required to use this technique to full advantage, thin-section analysis is still a very efficient way to obtain crucial information about many inorganic materials. Thin sections are often the most useful starting point for a study of stone and ceramic materials, even if they need to be supplemented by other approaches. They can provide crucial information not available by other analytical techniques (Reedy 1994).

Thin-section petrography does require the availability of a small sample of the material. A variety of sampling techniques suitable for removing very small solid samples from an artifact or art object has been developed specifically for dealing with cultural materials where sample size must be kept to a minimum (Kempe and Templeman 1983). When larger specimens are available, a diamond wafering saw is generally used to remove a manageable size piece from the larger one. In order to prevent damage or deformation of the specimen, the saw and blade selected must be appropriate for the type of material being cut or sectioned.

The direction of cutting is often important. The cutting procedure must produce a specimen in which all the features of interest can be seen. For example, for stone samples, the transition from surface to interior would be important in deterioration studies. Ceramic sherds are usually cut transverse to the plane of the sherd within a few millimeters of an edge, giving views of both inner and outer surfaces as well as surface decorations, although multiple sections each cut in different directions may be required for some research questions. For example, Woods (1985b) discussed how one section cut perpendicular to the base and rim of the vessel (a vertical section) is usual for provenance studies or technological studies; and that some researchers have advised that distinguishing between wheel-thrown and coil/ring-built vessels is facilitated by the use of a horizontal section (taken parallel to the base and rim). Her research indicated that both vertical and horizontal sections are insufficient for distinguishing between these fabrication methods; a tangential section (taken tangent to the surface of the vessel) is more helpful for addressing this particular research question.

Specimens for thin sections must have a flat surface to attach to a glass slide with the aid of an epoxy resin or other mounting medium. Often, the flat surface is obtained by slicing with a diamond saw (if not already done when removing the sample from a larger specimen). This step is followed by grinding (often on a cast-iron wheel with silicon carbide or diamond abrasive powders of various grit sizes, or with abrasive papers). Friable specimens, such as many ceramic materials, or specimens with pores and cracks, are often vacuum impregnated with epoxy as a preliminary step to stabilize the sample enough to permit cutting and grinding processes.

Once one flat surface has been prepared, the specimen is then is mounted in epoxy (or sometimes Canada balsam) onto a glass slide (the usual size is 46 × 27 mm although other sizes may be used for specific research protocols). Roughening the surface of the glass slide through grinding, prior to mounting the specimen, ensures that the slide is of uniform thickness, and the roughened surface helps in the bonding process (Ahmed 2000). The epoxy or other mounting medium is usually selected to have a refractive index similar to that of quartz, the most common mineral seen in thin sections. The refractive index of other minerals can then be analyzed relative to quartz (using the Becke line test) (Kerr 1977). Dyed epoxy is useful if quantitative measurements of the amount of porosity are of interest, or if other measurements are to be taken for the pore fraction (such as average size and shape). With clear epoxy, pores can be difficult to discern (they are clear like quartz and feldspar grains in plane polarized light; and dark, like opaque particles, minerals in extinction, and many clay minerals in crossed polarized light). In contrast, with a dyed epoxy the pores are easily observed in plane polarized light (Figs. 1.1.1 and 1.1.2). Blue or red dyes are most common.

After the epoxy has cured, the top of the specimen is trimmed with a diamond wafering saw, if needed, parallel to the slide, to a thickness of 1–2 mm to reduce the grinding time needed. Thin specimens or grain mounts will need only grinding. Abrasives in a liquid slurry are used for grinding on the wheel, using progressively finer abrasive powers, to a finished thickness of 30 μm. The standard procedure is to use deonized or distilled water for the grinding powders. However, if water-sensitive materials are anticipated, then grinding in an oil rather than water slurry is required. For example, gypsum can dissolve in water, so studies of dete-

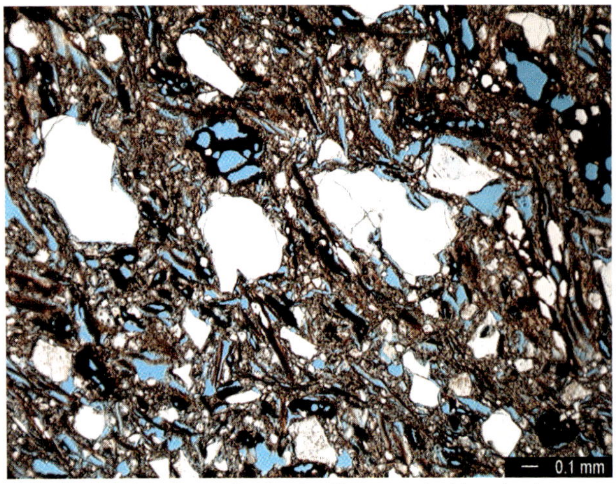

Figure 1.1.1 Thin section (historic brick) mounted in blue-dyed epoxy, plane polarized light. This historic brick sample (excavated at the Read House, Newcastle, Delaware) is porous, sandy, and rich in organic material. In plane polarized light the pores stand out when the thin section is mounted in blue-dyed epoxy.

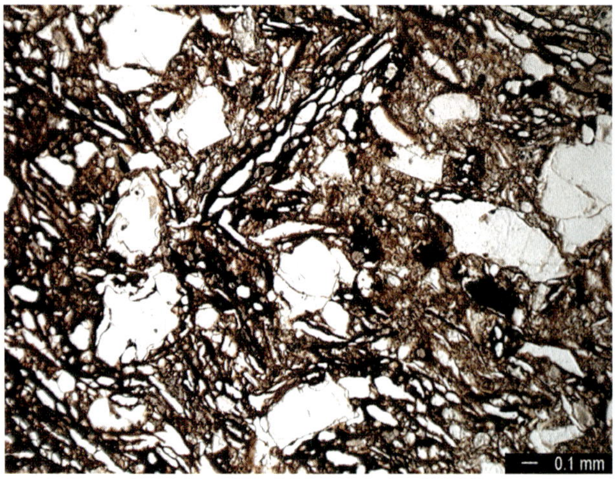

Figure 1.1.2 Thin section (historic brick) mounted in clear epoxy, plane polarized light. A sample from the same historic brick sample shown in Figure 1.1.1 is mounted here in clear epoxy. In contrast to the sample in blue-dyed epoxy, it is difficult to discern pores since in plane polarized light quartz and feldspars are also clear.

rioration may require oil-grinding procedures to preserve gypsum weathering products. Clay minerals will swell upon exposure to water, so to prevent changes in ceramic materials that are simply artifacts of preparation procedures, again grinding in oil is advised.

Grain mounts are necessary for materials that are not solid (such as clay core material from inside a bronze sculpture). In that case, the loose-grained material is sprinkled onto the epoxy-covered glass slide. Usually only a small amount of careful grinding will be needed to bring a loose-grained specimen down to 30 μm. Vacuum impregnation of the material is usually required to prevent plucking out of grains or other loss of material during the grinding procedure, and to prevent inclusion of grinding grit into the specimen.

A glass cover slip is often mounted on top of the finished specimen to protect the thin section and to reduce the scattering of light. Alternatively, the surface can be polished, allowing the thin section to be used for other purposes such as reflected light microscopy, hardness tests, or surface chemical tests. In one example, in a study on the effects of pollutants on individual components of stone (Jones and Thompson 1996), polished thin sections of freshly quarried stones (limestone, granite, and sandstone) were used as the experimental units. Since the pores in the vacuum-impregnated specimens were sealed by resin, the effect of porosity was removed as a parameter. The thin sections were exposed to pollutants by dry and wet deposition, and the effects on individual mineral components were analyzed (by light microscopy, cathodoluminescence, and scanning electron microscopy with elemental mapping).

Details of thin-section preparation methods can be found in Moreland (1968), Hutchison (1974), Kerr (1977), Elssner et al. (1999), Ahmed (2000), Chandler (2001), and Chinn (2002). There is also a variety of commercial companies who will prepare thin sections to specification. Expertise is required to ensure that the surface of the specimen is ground very flat and evenly, to avoid some areas being too thick (in which case the minerals would appear with nonstandard interference colors) and/or too thin (which would result in some of the material being plucked or ground completely away). Relying on professional thin-section preparation services has the advantage of allowing the researcher to concentrate on analysis and interpretation, rather than on time-consuming and often tricky preparation procedures. Alternatively, if a laboratory is preparing many thin sections, some of the procedures can be semi-automated using one of the various thin-sectioning systems available on the market. Protocols have also been suggested for setting up a field laboratory for making thin sections at archaeological field sites, so that fieldwork can benefit from new insights provided by immediate thin-section petrography (Hunt and Griffiths 1989; Chandler 2001).

Staining of uncovered thin sections, or staining prior to covering, can aid the identification of minerals that are otherwise difficult to differentiate in thin section (Figs. 1.1.3 and 1.1.4). Examples of the use of stains include distinguishing between alkali and plagioclase feldspars or between either of these and quartz, or highlighting the presence of calcite (Allman and Lawrence 1972; Hutchison 1974). For example, calcite can be stained with alizarin red S, potassium feldspars with sodium cobaltinitrite, and plagioclase with barium chloride plus rhodizonate.

When examined under a transmitted polarizing microscope, minerals mounted in thin sections of standard thickness can be identified through a variety of optical properties, usually at magnifications ranging from about 10× to about 500×. In plane polarized light, useful properties include transparency versus opaqueness, color, pleochroism, refractive index, relief, morphology, and cleavage. Between crossed polarizers, important properties include isotropism versus anisotropism, birefringence, extinction angle, and the presence or absence of other features such as zoning, twinning, undulose extinction, and anomalous polarization

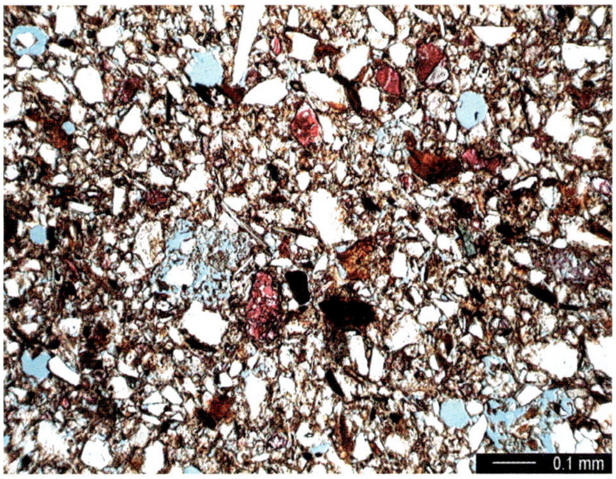

Figure 1.1.3 Thin section (fired loess) with calcite stain, plane polarized light. This sample is fired (at 700 °C) loess from the site of Dolni Věstonice, Czechoslovakia. The material is primarily quartz and micas (clear muscovite and brown biotite), with a small amount of chlorite and clay. Calcite is a minor constituent, and is easier to see (such as the grain in the center of the field of view) and quantify when stained red with alizarin red S.

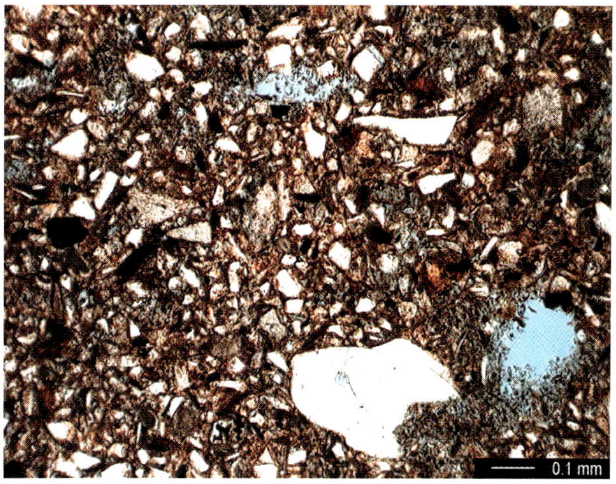

Figure 1.1.4 Thin section (fired loess) unstained, plane polarized light. This thin section is made with the same material as seen in Figure 1.1.3. In contrast to that previous slide, however, the calcite is not easily discernable. Without the red stain, calcite is colorless, and does not stand out well when present in small quantities of fine grains.

colors. This book is not intended to serve as an introductory text in polarized light microscopy, or as a reference to methods of mineral identification: for information on these subjects see Heinrich (1956), Phillips (1971), Allman and Lawrence (1972), Folk (1974), Kerr (1977), Scholle (1979), Tröger (1979), Deer et al. (1980), MacKenzie and Guilford (1981), Phillips and Griffen (1981), MacKenzie et al. (1982), Williams et al. (1982), Gribble and Hall (1985), Ehlers (1987), Yardley et al. (1990), Carozzi (1993), MacKenzie and Adams (1994), Adams et al. (1997), and Scholle and Ulmer-Scholle (2003).

In addition to the identification of specific minerals, thin-section petrography is also used to identify rock types and potential sources, characterize textures (such as grain size and morphology and the relationship between grains), and the relative or quantitative percentage of various constituents. Features relevant to specific stone or ceramic types are discussed in each of the following chapters.

Although much of the information collected in thin-section petrography is qualitative in nature (such as mineral identification, mineral textures, rock textures, etc.), quantitative information can also be collected (e.g. the relative or quantitative percentage of each component present, particle size and morphology characteristics, etc.). Even qualitative information can be used for statistical analysis by converting binary and categorical information to arithmetic data, or using statistical methods designed for binary and categorical data (Chayes 1954a, 1954b; Keith and Cooper 1974; Stoltman 1989; Reedy 1991; Agresti 2002; Lawal 2003; van der Ark et al. 2005). A variety of statistical analyses using categorical or quantitative data appears in the literature, and will be seen in the relevant chapter sections. Quantitative data, especially for ceramic thin sections, is most often obtained through point-counting procedures, discussed in detail in the ceramics chapters (6 through 9). However, newer methods of digital image analysis are also beginning to result in collection of quantitative data for a variety of research questions involving both stone and ceramic materials (Reedy 2006a). Unfortunately, multivariate statistical analysis, although commonly used for elemental analysis of ceramics, is still only rarely used in the interpretation of stone or ceramic thin-section data. Some of the most innovative uses of multivariate statistical analysis of thin-section data are highlighted in the following chapters.

Avoidance of bias is especially important when qualitative or subjective information is being recorded. Some of the best studies discussed in this book maintain a rigorous research design that includes "blind" analysis, so that information such as archaeological site, stylistic group, presumed provenance, etc. is masked from the analyst, to prevent any impact, even subconsciously, on what the analyst thinks he may be seeing in thin section. A simple numbering system, keyed into a reference table where information on the object sampled is available for later comparison, will suffice. In addition, a research design can incorporate randomization in the order of analysis, so that any predefined groups of objects will not result in measurement of all objects in one group together and separately from all objects of another group. Randomization is required for some statistical analyses to be valid, since a randomized order of analysis is an underlying assumption of many statistical tests. In any case, randomization combined with the use of blind analysis increases the chances that results can be duplicated by other researchers.

Finally, many studies include fieldwork or laboratory research in conjunction with thin-section analysis. If raw materials from potential quarry locations are available, and if the regional geology is known, then a larger number of research questions can be addressed than is possible if only thin sections of objects are available. Other research questions can be considered if laboratory experiments are conducted on raw materials or objects, and comparative thin sections are available from those experiments.

1.2 Problems Unique to Cultural Materials

Random sampling from a representative collection, which would ensure that the results are generalizable to the population of interest, is usually not possible when working with archaeological artifacts and art objects. Instead, the sampling must be pragmatic. Given the objects available, which will only be a subset of those that were made and used originally, the decision has to be made as to how many can be used for destructive sampling. This sampling strategy also affects the type of statistical analyses that can be performed (some approaches assume that a random sampling strategy was applied). Often, we want to be sure to include some of each of a number of subcategories (such as some of each vessel type or decorative technique). In that case, a random sample within each subset (a stratified random sample) may be possible.

The amount to be taken for an analysis is often quite small compared with typical geological specimens. If sampling is done from a whole object rather than a sherd or fragment, then the specific location where the sample is taken often cannot be controlled—it must be taken without damaging the object, and often from an area not visible to, for example, museum viewers.

There are also significant differences between natural and artificial materials that can complicate thin-section interpretations. Ceramic materials, in particular, differ from natural materials as to how the petrographer can interpret information obtained. For example, as ceramics are heated in the kiln, various effects are observed as certain temperatures are reached and sustained for a period of time. Natural organic components or additives will burn out in an oxidizing atmosphere, leaving behind voids; they will partially or completely char in place in reducing atmospheres. Mineral transformations may take place, with higher temperature minerals replacing the original ones. For example, quartz may begin to alter to cristobalite and tridymite, and as temperatures get even higher, may alter to a silica glass. Although some geological references may provide helpful comparisons, such as works on pyrometamorphism (Grapes 2006), it may not always be easy to determine whether or not certain minerals are present because of kiln conditions or because of the suite of naturally occurring minerals in the region of production. Knowledge of the geology of the relevant region(s), collection of comparative samples, and laboratory experiments may be necessary for clear interpretations of thin-section data.

Textural features of ceramic materials are also strongly affected by production techniques. For example, homogeneity may be affected by processes such as wedging to remove air pockets, mixing together more than one clay type, and the amount of mixing or kneading done after temper additives are incorporated into the clay. When new studies are begun, therefore, it is good practice to conduct some initial experimentation to assess the degree of homogeneity within and between thin sections. This information can be used to develop an experimental design that includes an appropriate number of thin sections and fields of view within a thin section, to ensure that results are representative of the material under study and that the questions of interest can be adequately addressed.

Even where natural materials have been used, such as with sculptural and building stones, the petrographer of cultural materials may encounter situations not experienced by the geologist-petrographer. For example, some stone objects may have undergone heat treatment, while others may have had surface treatments as part of patination or other decorative procedures, cleaning, use wear, repairs, or preservation efforts. Aggressive surface treatments might result from attempted forgery endeavors. And, as with ceramics, the number, size, and location of sampling may be severely constrained.

Geologists usually work with samples from known geographic locations, and are interested in reconstructing past geological processes. Petrographers of cultural materials often do not know the geological source of the material they are analyzing—at best, they may know the findspot of artifacts or the location of a structure under study. For art objects, often no confirmed findspot is known. Even for excavated archaeological materials, the findspot is frequently not the location where the raw material was obtained or where the object was made. At least one goal of analysis is usually to identify the region or site where an object may have been made and/or where its raw material originated, so that various questions regarding past relationships and communication networks can be explored. However, thin-section petrography of cultural materials can address a wide range of potential research questions, as the following chapters will demonstrate.

1.3 Organization of Chapters

This book consists of eight topical chapters, divided about evenly into stone and ceramics. Chapters 2 through 5 cover the stone topics, divided into volcanic igneous, plutonic igneous, sedimentary, and metamorphic stone materials. Chapters 6 through 9 cover the ceramics chapters, divided into three chapters concerned specifically with pottery (including product types and inclusions, provenance studies, and additional research questions concerning pottery), and one chapter on nonpottery ceramic and clay materials. No attempt is made to cover the material in each chapter in exactly the same way or format, or to the same depth. Instead, discussions are tailored to the research questions that tend to be posed for each type of material, the extant literature, and the specific issues in thin-section petrography relevant to each type of material.

Chapter 2 discusses volcanic igneous stone materials. Twelve volcanic rocks are reviewed in relation to their use as a cultural material (obsidian, pumice, scoria, tuff, rhyolite, dacite, trachyte, andesite, basalt, diabase, phonolite, and tephrite). The literature on tuff is particularly extensive, as it has been used for a variety of cultural materials including tools, architecture, and sculpture. Research on deterioration mechanisms of tuff, and tests of conservation treatments, includes many studies in which thin sections were utilized. Another volcanic rock that figures prominently as a cultural material is trachyte. Many thin-section analyses exist of trachytes used as building materials and for utilitarian objects. As a reflection of its prominence as a cultural material, many studies have been made of the deterioration mechanisms of trachyte that include thin-section analysis.

Chapter 3 reviews plutonic igneous stone materials that have found use as cultural materials. Among the plutonic rocks, granite is the one most often exploited, for both decorative and structural uses, and thus the literature involving thin-section studies of that material is the most abundant. Much of that literature focuses on identifying deterioration mechanisms, the effects of weathering, and the effects of conservation treatments. Granodiorite is also relatively common as a cultural material, used for sculpture, architectural components, and small utilitarian objects. In a provenance study on granodiorite used in buildings of the Roman Forum, Peacock et al. (1994) made the significant observation that problems can arise in investigations of rocks that are coarse grained and exhibit considerable variation within a single block of stone. Given the small samples that must be taken from many cultural materials, it can be difficult to extract a large enough sample in any one location to be representative of the mineralogy of the stone as a whole. This problem highlights the need to carefully examine the cultural object or structure from which samples are taken, and where necessary and possible to take multiple small samples to ensure that a representative sample is obtained if needed to answer the specific research questions posed.

Gabbro is another plutonic rock that is encountered with some frequency in architecture, sculpture, and tools, particularly for Neolithic stone axes; many characterization and provenance studies have focused on thin-section petrography of those objects. Other igneous rock types that are occasionally found with cultural materials (but not as frequently as granite, granodioriate, and gabbro), include tonalite, syenite, monzonite, and anorthosite.

Chapter 4 concerns sedimentary stone materials. A subset of these holds the greatest importance in the history of cultural materials: sandstone, limestone, and dolomite. Sandstone is used in architecture and sculpture, and is also found in a variety of other contexts (as whetstones, grinding tools, commemorative pillars, tombstones, and ornamental stones). Limestone and dolomite are both primarily employed for architecture and sculpture. Because of the ubiquity of sandstone, limestone, and dolomite as cultural materials, there are also many studies that employ thin-section petrography to investigate deterioration mechanisms of these materials, and to test the effects of conservation treatments. Other sedimentary stones encountered as cultural materials include shale, siltstone, conglomerate and breccia, sedimentary carbonates beyond limestone and dolomite (chalk, travertine, and tufa), sedimentary silicates, and evaporites.

Chapter 5 focuses on metamorphic stone materials. Marble has received the most attention in art-historical, archaeological, and analytical studies, and is represented by the greatest abundance of published literature that incorporates thin-section petrography. As a result, much of this chapter is devoted to marble. Both white and red marbles have found use in architecture, sculpture, decorative stones, and tools. Much of the marble literature that includes thin-section analysis focuses on characterization and provenance studies (source identification). Another large set of published work deals with thin-section investigations of marble deterioration, for authenticity and preservation studies, and to test conservation treatments for deteriorating marble. The next most popular metamorphic rock used as a cultural material is schist—as a building stone, for sculpture, and for tools. This chapter includes separate discussions for talc schist (and a related material, steatite/soapstone), greenschist (and the related materials greenstone and epidioriorite), and blueschist. Other metamorphic rocks encountered in studies of cultural materials include slate, meta-argillite, metasiltone, phyllite, gneiss, amphibolite, hornfels, quartzite, and serpentinite.

Chapter 6 begins the section of the book devoted to thin-section petrography of ceramics. The first part of the chapter discusses basic pottery types, including earthenware, stoneware, porcelain (both hard paste and soft paste), and stonepaste (also known as fritware). An overview is given of the types of information that can be seen in thin sections of each of these pottery products. Since nonplastic inclusions are usually the most crucial focus of thin-section studies of pottery, the second part of the chapter discusses these in detail. First, methods that are used for describing, characterizing, and interpreting nonplastics are reviewed, ranging from the traditional (qualitative estimation of parameters and various quantitative approaches to point counting) to newly emerging technologies of digital image analysis. Then, major types of nonplastics (which may be natural inclusions in clay or deliberate additions, or temper) are each discussed in detail. These include lithics, sand, calcium carbonates, organic material, grog, and others.

Chapter 7 looks at pottery provenance studies in which thin-section petrography has been employed. A wide variety of research questions has been posed within the overarching rubric of provenance studies. Questions vary according to the type of pottery under study, type and location of archaeological sites, the regional geology, and the cultural interactions of interest. The chapter begins with a discussion of research aimed at classifying pottery by nonplastic inclusions. An initial classification is often of importance to archaeologists. The more precise information provided by thin-section analysis allows criteria to be identified that can be used to develop a set of standards that can subsequently be used for classifying a larger number of objects in the field using only low-power binocular microscopy. Including thin-section data also helps to avoid mistakes regarding temper identification that can have ramifications in archaeological interpretations. There are many ways in which thin-section petrography can inform archaeological classification schemes and their interpretations. The second section of Chapter 7 focuses on characterization studies that aim to identify production groups or workshops. The third section reviews work using thin sections to examine temper changes and variation over time and space. A necessary first step toward eventual provenance determination is determining whether sherds from one site or region exhibit mineralogical homogeneity, while also being different from those of other locales. Only then can one begin to examine possible exchange relationships or identify specific geological sources for nonplastics.

The chapter concludes with the topic of identifying specific geological sources of nonplastics. Rock and mineral inclusions in clay provide information about the geology of the area in which the clay and temper materials were collected. If those inclusions are distinctive enough, it may be possible to deduce the geology of the source, and from this use information about regional geology to locate and pinpoint a probable source. As Freestone (1995) pointed out, in some cases a single thin section will suffice. However, each region and ceramic under study has its own particular set of problems and constraints, so a variety of approaches has been taken. Some of the most fruitful studies involve collection of geological samples in the field, as well as utilization of geological maps. Others use ethnographic studies to help identify possible clay or temper sources. Strangely, multivariate statistical analysis has been underutilized. Although many statistical approaches exist for taking advantage of categorical or presence/absence data, qualitative thin-section studies of ceramics rarely use them. Point counting provides quantitative data amenable to statistical analysis, but even when thin-section analysis appears as an adjunct to elemental analysis that incorporates sophisticated data treatment, multivariate statistical analysis is rarely utilized for interpretation of thin-section data. Some important exceptions are highlighted in this chapter. For example, Heidke and Miksa (2000) provide a model for sophisticated provenance studies of sand-tempered ceramics, combining geological fieldwork with quantitative point counting and a sequential approach to multivariate statistical analysis.

Chapter 8 looks at how thin-section petrography can be used to elucidate issues regarding pottery fabrication, use, and deterioration. Thin sections can provide clues that are helpful in identifying clay processing choices, reconstructing forming methods, inferring firing conditions, and deducing intended function. Employing thin sections for interpreting voids and porosity data can lead to new ideas about how a vessel was made and used. Thin sections are an excellent tool for studying decoration and surface/clay body interfaces, providing information about slips, paint and enamel, organic and salt coatings and ash or salt "glazing," or about true glazes, transfer print processes, or incised, impressed, and applied decoration. The chapter ends with examples of how thin-section analyses have informed efforts to explain technological choice, variation, and change, or to study deterioration products (usually those resulting from burial).

Finally, Chapter 9 explores thin-section analysis of nonpottery ceramic and clay materials. These include terracotta sculptures, larger scale clay sculptures, and small molded or stamped objects such as cuneiform tablets and Tibetan *tsha-tsha*s (votive clay tablets). Thin-section studies of these materials have examined issues such as characterization of materials, provenance, deterioration, and conservation or preservation efforts. Similar aims are found with two other important classes of nonpottery ceramics: bricks and tiles. The final material, clay cores from inside cast bronzes, has mainly been of interest for provenance or attribution studies.

No attempt is made to survey every publication devoted to thin-section petrography. Instead, there is a mix of landmark publications from the past that are still relevant today, and more recent scholarship. Similarly, the stone and ceramic materials included here are not exhaustive of all types of materials studied in thin section. However, they are representative of most of the cultural materials for which thin-section petrography has been used as a major research tool. Although newer methods of analysis exist, and new analytical techniques are constantly emerging, thin-section petrography still has a place in the toolkit of archaeological science and conservation research. It can provide data not easily obtained by other methods, and, as traditional petrographic methods continue to be updated with sophisticated digital image analysis approaches, it has the potential to contribute even more today than it has in the past.

2

Volcanic Igneous Stone Materials

Volcanic igneous rocks solidify from magma that has erupted onto the surface of the earth. If the magma cools rapidly, it forms fine-grained rocks, as with rhyolite, andesite, basalt, and trachyte. Some larger crystals called phenocrysts, formed before eruption, may be present. If cooled abruptly, the magma may quench and form glass, as with obsidian and pumice. The magma may flow as a liquid (lava) or be ejected from the earth as separate fragments known as pyroclasts. Loose consolidation of such fragments is termed tephra; when cemented or welded, they are called pyroclastic rocks, as with tuff. Such rocks may also be called volcaniclastic (Dietrich and Skinner 1979; Best 1982).

Classification of pyroclastic rocks is based mainly on the size of the fragments of which they are comprised. Pebble-sized fragments (2–64 mm in diameter) are called lapilli; smaller fragments (less than 2 mm) are called ash and larger ones (more than 64 mm), bombs (if partially or wholly molten during formation of the rock, identified by a more rounded or amorphous shape) or blocks (if chunks of solid rock during formation, identified by their angularity). Rocks formed mainly from the consolidation of bombs are called agglomerates; those formed from consolidation of blocks are known as volcanic breccia. Tuffs are formed mainly from consolidation of lapilli and/or ash-sized fragments (Williams et al. 1982). Since they are more friable with less mechanical strength, agglomerates and volcanic breccia are rarely used in art or architecture and thus are not discussed here.

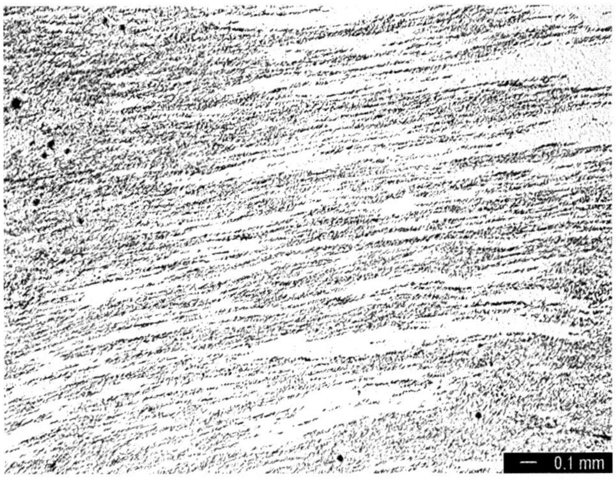

Figure 2.1.1 Obsidian, plane polarized light. This specimen shows black banding resulting from parallel striations formed by carbon particles. The glass component is transparent.

2.1 Obsidian, Pumice, and Scoria

2.1.1 Obsidian

Obsidian is volcanic glass in massive form (with conchoidal fracture), often dark in color, formed from the very rapid cooling (quenching) of silica-rich magmas. As with manmade glass, the magma is so viscous that if cooled very rapidly crystal growth is inhibited and an amorphous, noncrystalline material forms. Therefore, when a sample is viewed through crossed or uncrossed polarizers, no image is visible, as with any glass specimen. Over time, some of the glass may devitrify, however, so small crystals of cristobalite, tridymite, microcrystalline feldspar, or cryptocrystalline feldspar may be visible (Williams et al. 1982; Le Maitre 1989). There may also be some primary crystalline materials or other inclusions present. For example, the obsidian shown in Figure 2.1.1 has parallel striations of carbon particles, resulting in a banded appearance.

Although some minerals or other inclusions or devitrification products may be present, these are generally not enough to be useful in completely characterizing an obsidian specimen or identifying its quarry source. Instead, elemental analysis is the standard procedure used for characterizing and sourcing obsidian.

2.1.2 Pumice

Pumice is a fine-grained and glassy cooled lava with a frothy or foamy texture. It has essentially the same chemical composition as obsidian (silica rich), and is also wholly or substantially glass, but in a vesicular form. Much material ejected through volcanic activity bubbles and forms vesicles, and fragments by the bursting of expanding gases from magma; extreme vesiculation of the viscous silica-rich magma produces pumice, often accompanied by violent eruptions and a large quantity of ash, lapilli, and bombs. The vesicles of pumice may be either rounded or elongated (Fig. 2.1.2). Phenocrysts from crystals formed earlier and suspended in the magma before eruption may also be present (Fig. 2.1.3) (Williams et al. 1982; Le Maitre 1989).

Since pumice usually appears in relatively small fragments and is rather friable, it is rarely used as a sculptural or building material. Unbonded, loose pumice has been used as a filler for concrete, but is not suitable for carving in blocks; more densely welded pumice can be difficult to quarry, as it is brittle and has a high fracture density (Heiken 1979). The primary artifact use of pumice has been as an abrasive or polishing stone; for example, it was used in the Cyclades for polishing marble, for sharpening pens, and as a component of cosmetics (Renfrew 1971; Penny 1993). Pumice has also been used as an insulating material in buildings in Britain (Dimes 1990a).

Petrographic studies of pumice have been used to differentiate deposits originating from different volcanic eruptions. For example, a variety of volcanic eruptions in the Aegean region, such as those of the islands of Melos, Santorini, and others, has produced pumice of differing mineralogical characteristics (Fornaseri et al. 1975; Bond 1976). Major differences include the percentage of phenocrysts (versus glassy matrix), and relative percentages of quartz, feldspar, orthopyroxene, clinopyroxene, and biotite.

Helmi (1994) studied the deterioration of volcanic rocks from Egypt. Thin sections showed the usual texture of vesicles in a glassy matrix. After artificial aging, deterioration products of a brownish iron oxide appeared. Helmi found that the glass began to devitrify and iron, calcium, sodium, potassium, and magnesium were leached from the glass and redeposited in pits on the surface of the material, causing a change in color. One result of his study was the conclusion that glassy rocks such as obsidian and pumice deteriorate more rapidly than crystalline rocks.

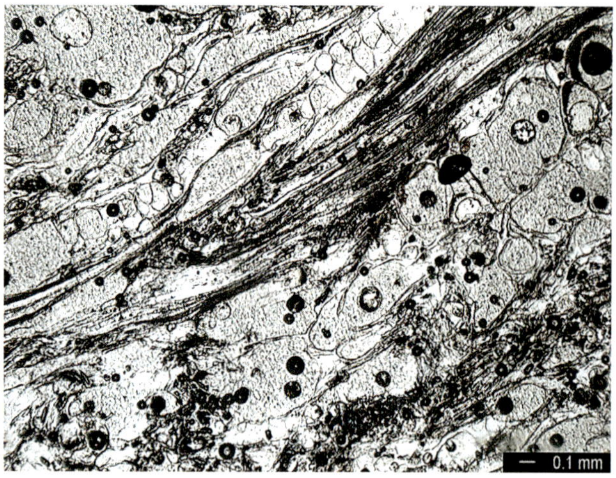

Figure 2.1.2 Pumice, plane polarized light. There are both rounded and elongated vesicles; some black magnetite is also visible within the foamy, glassy, colorless matrix.

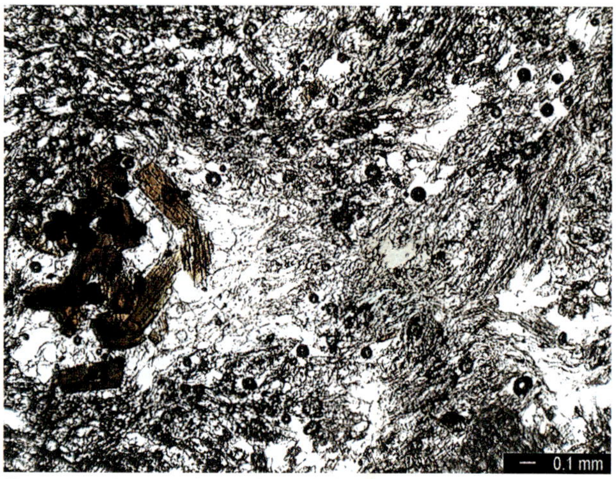

Figure 2.1.3 Pumice, plane polarized light. There are scattered phenocrysts of brown flakes of biotite and black magnetite within a foamy, glassy matrix.

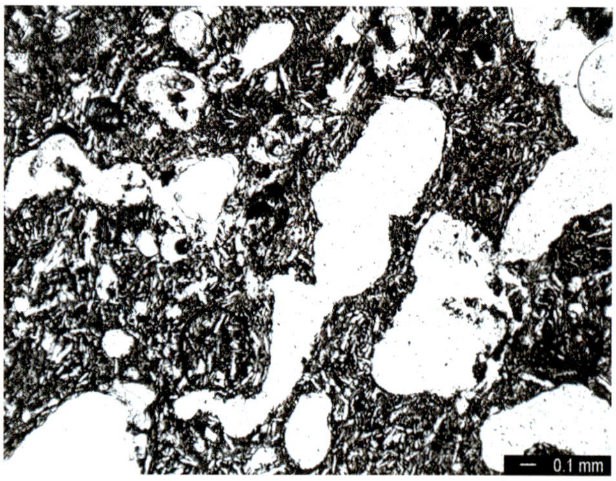

Figure 2.1.4 Scoria, plane polarized light. Many vesicles are visible, here set within a matrix of glass and microlites of feldspar.

2.1.3 Scoria

Scoria is also highly vesicular (Fig. 2.1.4), of basaltic composition (less silica rich than pumice). It too may have crystalline phenocrysts embedded within the vesicles (Fig. 2.1.5). Scoria, like pumice, tends to occur in particles rather than larger consolidated rocks, and has therefore been of somewhat limited use as an artifact material, apart from products such as road-building material and aggregate for concrete mixtures (Perath 1984). One exception is a complex of rock-hewn churches in Ethiopia that was carved from a deposit of scoria (Grissom 1994). Another famous exception can be found on Easter Island, where scoria was used to form headdresses (*pukao*) atop large stone images (*moai*) (Baker 1993); red scoria was also used to carve full-scale statues although only about a dozen remain intact (Charola et al. 1994; Charola 1997).

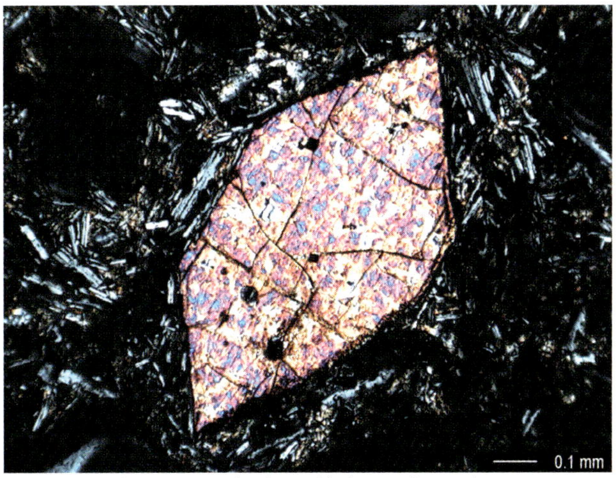

Figure 2.1.5 Scoria, crossed polarized light. An olivine phenocryst embedded within the microlites and vesicles shows a characteristic shape, high birefringence, and a fractured, altering surface.

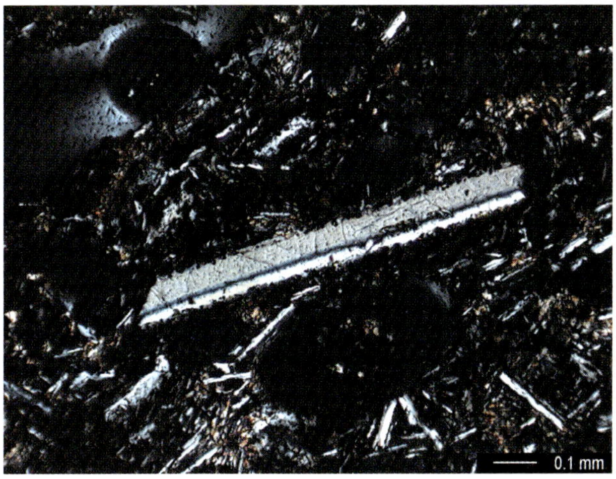

Figure 2.1.6 Scoria, crossed polarized light. The matrix consists of elongated feldspar microlites, isotropic glass, and patches of devitrified material. There are many large vesicles, dark in crossed polarized light, and some larger feldspar phenocrysts such as this one, much larger than the feldspars within the matrix. All the feldspars show weak birefringence.

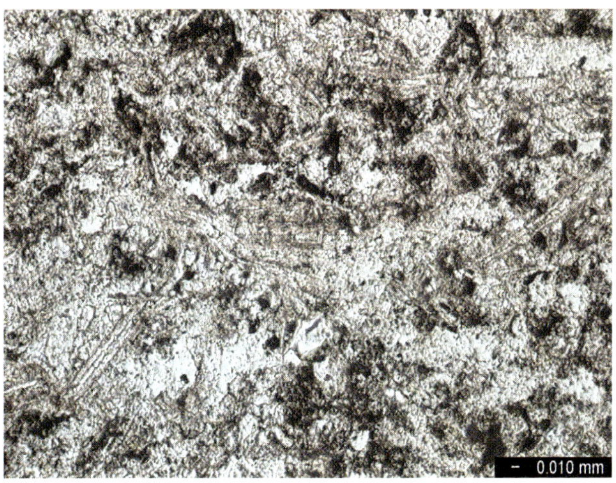

Figure 2.2.1 Tuff, plane polarized light. This welded tuff has glass shards that have been flattened and/or welded (flowing or fusing into adjacent areas).

Lazzarini et al. (1996) illustrated and described thin sections of Easter Island scoria used to produce the *pukao* atop the heads of some *moai*. They described this material as vesicular, and as having a glassy matrix rich in unaltered labradoritic plagioclase feldspars which appear as thin elongated lamellae (Fig. 2.1.6). These feldspars were associated with lesser amounts of clinopyroxene, altered olivine, and iron oxides.

Scoria was also used in ancient Mexico as a material for hand tools. Outwater (1957) described scoria pieces about 20 cm long and 10 cm in diameter found near the site of Malinalco where there is a monumental temple, much of it hewn from solid rock. He noted that the scoria tools were very effective, with sharp edges that easily cut and remove wet stone. The porosity of the scoria helped to prevent clogging during the cutting operations.

2.2 Tuff

Tuff, as a volcanic pyroclastic rock, is formed of fragments that may include crystals, lithics, glass, and ash. The relative proportion of these materials helps determine the full name of the tuff. Lapilli tuffs have the majority of fragments with mean diameters ranging between 2 and 64 mm; ash tuffs have most fragments less than 2 mm in diameter. The unmodified term "tuff" is understood to refer to an ash tuff. Tuffs with more glass than mineral or rock particles are called vitric tuffs (or vitroclastic texture); those with minerals as the predominant constituent are termed crystal tuffs; and those with rock fragments as the dominant constituent are lithic tuffs. Rocks formed of blocks mixed with abundant ash are called tuff-breccia.

Welded tuffs are formed when the fragments are so hot they immediately fuse into solid rock upon contact with each other or with other materials. Rhyolitic tuffs are welded tuffs of an acid composition similar to that of rhyolites and obsidians (silica rich) and are sometimes called ignimbrites. They form when very hot pyroclastic flows contain a mixture of magma and gases that flow and form into rocks rapidly (Williams et al. 1982).

In thin section, welded tuffs may have glass shards that have been flattened and/or welded (flowing or fusing into adjacent areas) (Fig. 2.2.1). Welded tuffs formed from ash flows may have a more uniform texture than those formed from ash falls. As welding occurs, cooling vapors rich in silica and alkalis precipitate within pore spaces, and alkali feldspars, trydimite, and cristobalite may form there. In addition, glass material may devitrify in certain zones of an ash flow, forming quartz and feldspar. Sometimes the crystalline material replaces the glass while maintaining the original texture (Dietrich and Skinner 1979; Best 1982).

2.2.1 Tuff as a cultural material

Tuffs have been used for tools (axes, whetstones, and touchstones), building materials, and sculpture. A large literature exists on the locations of quarries, characterization of stones, and studies of the deterioration and conservation of tuff. Heiken (1979) discusses how the mineralogy and structure of tuff give it strength and ease of carving and quarrying. He states that these properties are due to the devitrification process and presence of fibrous crystalline silica (cristobalite) and small alkali feldspar grains; and to void spaces being partially or completely filled with somewhat coarser grained silica (trydimite) and alkali feldspar. The intergrowth of minerals bonds the pyroclastic material together, yet tuff is still porous enough to be easily cut, carved, or scraped with simple tools. Tuff is light but has enough strength to be suitable for building construction. Heiken also noted that devitrified tuffs are relatively resistant to weathering, especially after they develop a slight patina of microcrystalline carbonate and sulfate.

Tools

In a large project on identification of the materials of prehistoric stone axes, undertaken by the Council for British Archaeology (Bunch and Fell 1949; Keen and Radley 1971), 350 implements were examined by thin-section petrography. One group included an axe production center at Great Langdale in northern England where tuff was used for over 100 tools. The tuff was described as an ash, of epidotized intermediate or basic tuff. Thus some alteration to form epidote crystals has occurred. (Intermediate and basic refer to the silica concentration of a rock, derived from elemental analysis, with intermediate being 52–63% silica and basic 45–52% silica.) This tuff is described as a dull green color, mainly glass with some lapilli, quartz, and microlites of alkali feldspar, with "nests" of epidote.

An additional 30 implements not from the above group (29 axes and one perforated hammer) were made from a variety of other tuffs. These included ones described as epidotized and epidotized feldspathic silicified tuffs. Again, some alteration to form epidote crystals had occurred; the second type also had small feldspar crystals visible, as well as silicification (impregnation and/or replacement of some components by a form of silica such as quartz, chalcedony, trydimite, cristobalite, or opal), either as vapor-phase crystallization during cooling or through devitrification of glass material.

Volcanic rocks frequently undergo alteration that may drastically change original minerals and textures. Some alteration may occur during original cooling, as gases and fluids continue to move and cause changes in areas that had already begun to solidify. Hydrothermal exposure (percolation of hot aqueous solutions from the environment) can also cause alterations. Under certain conditions of temperature and water pressure, primary silicates often recrystallize into fine-grain aggregates of the hydrous mineral epidote. Along with quartz, these epidote aggregates can precipitate in open cavities. The resulting rocks resemble greenstone, a weakly metamorphosed igneous rock that often includes epidote (Best 1982). Greenstones are frequently used for stone axe production, so the resemblance of epidotized tuff to greenstone may account for its selection for implement production in the cases identified by Keen and Radley.

Highly detailed thin-section information was reported in a study of the petrography and origin of two whetstones from a Middle Bronze Age tomb at the site of 'Enan in northern Israel (Shimron 1985). The study was instigated because the artifacts appear to be imported rather than made of local material. One of the whetstones was found to be made of tuff. The matrix was described as finely crystalline, containing small particles of quartz, feldspar, and white mica, with a minor amount of green chlorite. About 70% of the rock consists of larger sand- or silt-sized angular fragments of quartz, plagioclase feldspar, lithics, micas, and opaque crystals. A "flow structure" or preferred orientation, especially of micas, from metamorphic stress was clearly visible. Conclusions were that the absence of coarser lithic fragments and the fine and uniform grain size indicate that the deposit formed somewhat distant from the source volcano, and that the presence of abundant small micas, along with the preferred orientation, suggests low-grade regional metamorphism that caused recystallization of clays. The resulting rock was hard and dense, quite suitable for use as a whetstone for grinding and sharpening. Two possible regions of origin were pinpointed.

Touchstones are fine-grained black stones used for assaying gold or silver. They function by comparing the color of streaks made on the surface of alloys of known composition with the colors made on those of unknown composition. A thin-section study of touchstones showed that many are made of tuff (Moore and Oddy 1985). Examples included a tuff with a fine-grained crystalline or devitrified glass matrix containing quartz, secondary silica, and often mica (as with the whetstone discussed previously, indicating some degree of metamorphism had occurred). Banded tuffs were also found, with coarser quartz, secondary silica, and bands of opaque material, possibly carbon.

Welded tuff (also known as ignimbrite) was found in thin-section studies of Roman millstones (Williams-Thorpe 1988). Such material has a hard and relatively rough surface that would be effective for grinding of grain. Cutting and grinding areas of the hard glassy material would be enhanced by projecting quartz grains or by cavities of vesicles; additional ones would be produced as crystals are pulled out of the rock surface through repeated grinding action. Thin sections of this material showed glass with scattered quartz and feldspar phenocrysts (Fig. 2.2.2).

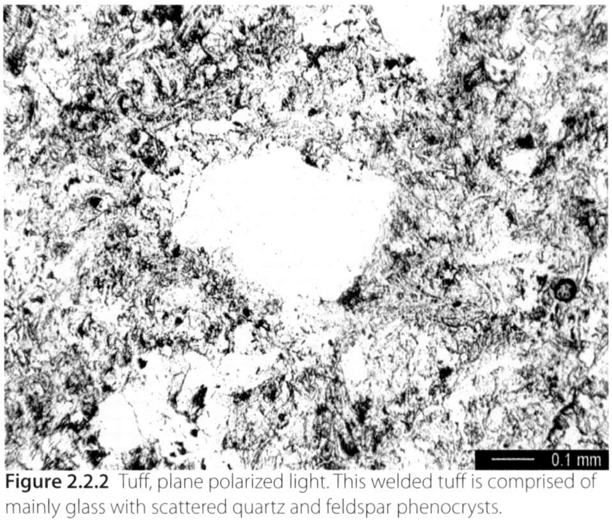

Figure 2.2.2 Tuff, plane polarized light. This welded tuff is comprised of mainly glass with scattered quartz and feldspar phenocrysts.

Architectural materials

Ignimbrite (welded tuff) was used as an architectural material in ancient Mexico. Williams and Heizer (1965b) studied samples from the valleys of Oaxaca and Mitla, looking both at architectural samples and at samples from quarry sites. In the quarries, the material available varied in the degree of welding and in the amount of crystallization of the glass. Williams and Heizer noted that these variations meant that a wide variety of uses could be made of the rock (and, in fact, some hammer stones and picks are indeed made from the ignimbrite).

Underground chapels, cities and tunnels have sometimes been carved out of tuff. One example is the city of Vardzia in Soviet Georgia (Heiken 1979). Heiken noted that tuff blocks have also been used widely for construction materials in the former USSR, Mexico, and South America. It was valued because of its resistance to weathering and its low thermal conductivity.

Blocks of tuff were used to construct pre-Columbian pyramids and platforms at the site of Cochasquì in Ecuador. The tuff was characterized in thin section by Rossi-Manaresi and Chiari (1980). Thin sections revealed that about 50% of the rock was comprised of minerals and lithic fragments set into a glassy matrix (Fig. 2.2.3). Minerals present (0.01–3 mm in size) were primarily plagioclase feldspar and hornblende, with smaller amounts of pyroxene. Oxidation of iron in the rock created an intensely colored stone (Fig. 2.2.4). Tuff was also used at many other pre-Columbian sites; for example, in central Mexico unwelded and partially welded tuffs were used, and continue to be used, in architecture, for paving stones, and for sculpture (Heiken 1979).

Bianchetti et al. (1994) included petrographic analysis, along with other techniques, to examine tuffs used in construction at the Forum and Palatine areas in Rome. They illustrated an example of a thin section of a densely compacted tuff (an ignimbrite). Their example contained crystals of pyroxene, biotite, sanidine and analcimized leucite set into a very fine ash matrix (Fig. 2.2.5).

Another Roman building, the Tabularium, located in Rome on the Capitol, was studied in thin section by De Casa et al. (1994); they also used a variety of ancillary analytical techniques. When the building was used for storing salt in the Middle Ages, severe degradation of the stone occurred. The 1994 study of the tuff blocks and related ancient quarries was intended to provide a materials characterization and to evaluate restoration efforts made 20 years previously. The tuff consisted of abundant lapilli within a matrix of ash and glass, with the glass extensively altered to zeolites (Fig. 2.2.6). Minerals were rare within the matrix, but included leucite, pyroxenes, biotite, and olivine. Lithic fragments were volcanic and sedimentary, with leucitic rocks common. A variety of carbonates was also present, some containing fossils, and some microcrystalline quartz. De Casa et al. published thin sections which illustrated leucite (often extensively replaced by zeolites) (Fig. 2.2.7), zoned pyroxene, and zeolites (phillipsite and chabazite).

Tuff is abundant in the Latium region of Italy. Since it can be quarried relatively easily and has good compression strength, it was used extensively as a construction material

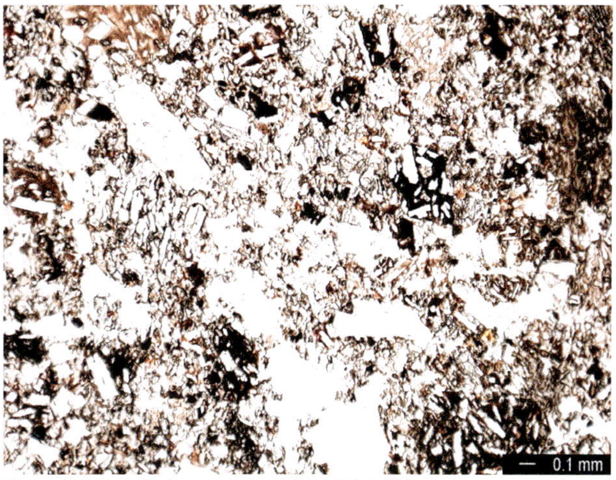

Figure 2.2.3 Tuff, plane polarized light. This tuff has a large proportion of minerals and lithic fragments set into the glassy matrix.

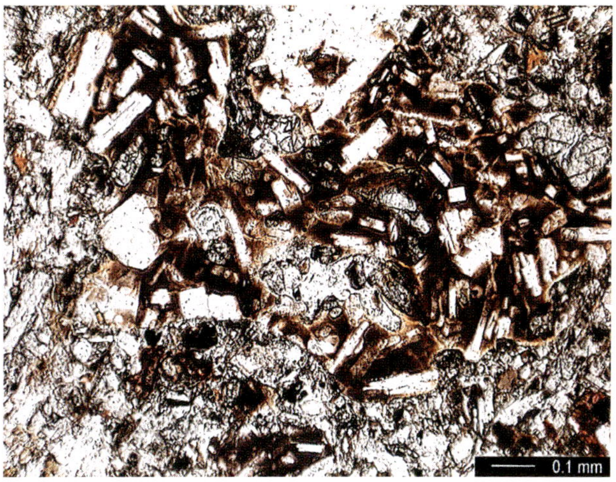

Figure 2.2.4 Tuff, plane polarized light. Oxidation of iron in this tuff specimen creates an intensely colored stone. Here the reddish iron oxidation products stand out from the lighter glass matrix and mineral crystals.

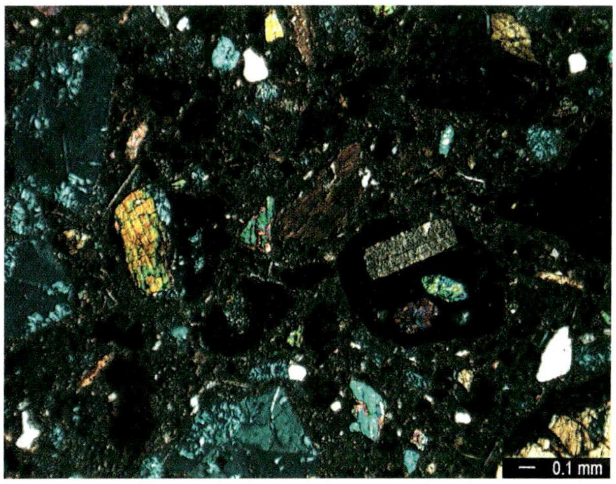

Figure 2.2.5 Tuff, crossed polarized light. This is a densely compacted tuff with a variety of minerals set into a very fine ash matrix (which is dark in crossed polarized light). Dark gray leucite grains have secondary analcime zeolite alterations (with light gray interference colors). Higher birefringent grains include biotite, feldspars, and pyroxenes.

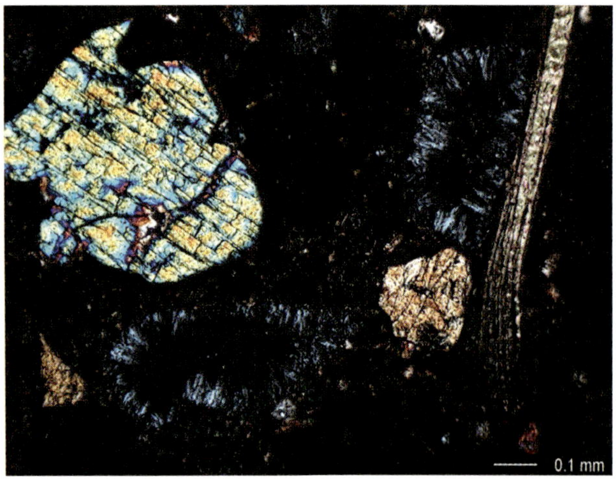

Figure 2.2.6 Tuff, crossed polarized light. Fragments (lapilli) of minerals and lithics in this tuff specimen are surrounded by a matrix of ash and glass. The glass is extensively altered to zeolites. These secondary zeolite minerals can be seen as tiny, oriented, weakly birefringent grains surrounding and lining some of the shapeless, black, isotropic glass areas.

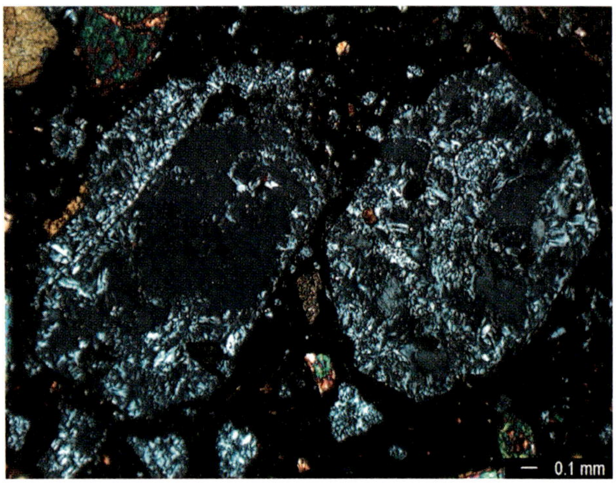

Figure 2.2.7 Tuff, crossed polarized light. This tuff contains leucite grains partially replaced by zeolites. Leucite typically shows an unequally developed trapazohedral form.

in ancient Rome. The mechanical characteristics of this material have been studied in detail, including thin-section examinations (Sappa et al. 1995; Jackson et al. 2005). In addition to the lithified tuffs, a great deal of loose granular volcanic material is found in this area, and has been used as an additive (pozzalana) to hydraulic mortars.

Jackson et al. (2005) characterized Roman tuffs in thin section. They plotted point count data on a ternary diagram, with the abundance of lava and sedimentary rock fragments plus crystals and crystal fragments on the upper apex, glass plus authigenic clay on the lower left apex, and authigenic zeolite plus calcite on the lower right apex. The resulting plot illustrated a diversity of tuff building stones with a fairly wide range of petrographic variability even within a single lithology.

Sculpture

Some of the most famous tuff sculptures are those of Easter Island. Stone images (*moai*) include ones carved from a lapilli tuff (Roa 1994). This tuff ranges from brown to yellowish or greenish in color, is moderately lithified, and is a relatively soft stone that would have been easy to carve when wet (Charola 1997). Charola has described it petrographically as having a glassy, vesicular matrix with lapilli consisting mainly of olivine and augite, with some plagioclase feldspar and lithic fragments of mainly basalt; the size and coherence of inclusions varies. Secondary accessory minerals such as zeolites are also present, along with many vacuoles (trapped air bubbles) in the glassy matrix. There are also crystals of ilmenite together with some palagonite as an alteration product (Baker 1993).

Lazzarini et al. (1996) also carried out petrographic studies of the tuff used to produce Easter Island *moai*. They described the tuff as having a clastic structure, with a slightly altered (through incipient devitrification) light brown glassy micropumicieous-ashy mix. They noted that the main components of the clastic portion are scoriaceous lapilli, with both rounded and elongated vacuoles. There are abundant (unaltered) plagioclase feldspars (andesine and labradorite), pyroxene (augite), olivine, and iron oxides, all of which are found in both the clasts and in the glassy matrix, as well as lithic fragments (basalt and andesine). Different tuff layers were found to be essentially the same compositionally, differing mainly in grain size and coherence (and thus in porosity).

2.2.2 Tuff deterioration mechanisms and conservation treatments

Beurdeley (1989) reported on an authenticity study of Indonesian stone sculpture made of tuff. He noted that over time, the feldspar-rich surface of tuff begins to alter to form zeolites (fibrous anhydrous silicates) and the phosphate mineral monazite. The idea is that if these alteration products are found only on the surface and not deep within the stone, then the alterations occurred after the piece was sculpted. The presence of a gradual, long-term secondary mineral formation on the surface of a stone piece could occur after initial carving through exposure to the atmosphere or to groundwater during burial. This phenomena and its utility in dating or authenticity studies of tuff sculptures is still under study and requires more experimental work in order to assess the length of time it might take for the alteration products to form. Although much of his work involved the use of scanning electron microscopy (SEM) to better visualize the small zeolites and monazite crystals, coupled with elemental analysis of crystals to ensure correct identifications, some of these alteration products are visible in thin sections under high magnification.

In a study of architectural materials, Martinez and Martinez (1991) used thin-section petrography to characterize a deteriorated building stone from the facade of the Metropolitan Cathedral in Mexico, as part of an examination into deterioration in urban environments. Goals were to characterize the stone type prior to selecting conservation treatments, and to study the degree of decay in relation to other characteristics of the stone such as porosity. In thin section, the stone appeared to consist almost entirely of tridymite, with patches of clay, a small fraction of sodium plagioclase feldspar, and many vesicles. Martinez and Martinez hypothesized that the porosity resulting from the presence of these vesicles might increase susceptibility to deterioration from environmental effects, yet that same porosity should also allow for good penetration of consolidating and water-repelling agents.

In another study of architectural materials, thin sections were used to assess the degree of deterioration of historic chapels carved into volcanic tuff in the Cappadocia region of central Anatolia (Türkmenoğlu et al. 1991). Samples were taken systematically to give a continuous view of the rock from the surface to a depth of about 10 cm. The rock had earlier been identified as a dacitic lapilli tuff (De Witte et al. 1988). A variety of mineralogical alterations (such as the breakdown of plagioclase feldspar into clays and alteration of biotite into iron oxide) could be detected, depending upon the depth. Microbiological activity had drastically altered the surface mineralogy and texture and penetrated at least 1 cm into the rock. Other rock alterations were observed as deep as 8 cm.

Additional studies of the Göreme valley tuff of central Anatolia, undertaken to better understand the types and depth of alteration, were aimed at developing potential conservation treatment approaches (Caner-Saltik et al. 1995). Excavated chapels date back to the seventh century, Byzantine churches, chapels and monasteries were carved into cliffs during the eighth and ninth centuries, and during Arab invasions underground cities connected by tunnels provided protection (Heiken 1979). Caner-Saltik et al. studied thin sections from the church of Santa Barbara, which showed quartz, plagioclase feldspar, and biotite phenocrysts in a clay-rich matrix; there were also lithic fragments present, with pumice being the most common. Grain size was generally less than 1 mm. The rock was sampled down to a depth of 10 cm from the surface, and the geological formation from which the church was hewn was also sampled for comparison. Thin sections from the upper 3 cm have extensively altered feldspars, with alteration greatly decreasing by 6 cm. At 10 cm, few feldspar grains showed any alteration.

Topal and Doyuran (1996) also studied the weathered tuffs of the Cappadocia region, with the goal of contributing to conservation treatment approaches. Thin sections were prepared from block samples in order to gain a continuous view from the surface down to a depth of 23 cm. In the upper 2 cm, they found slight fracture in the plagioclase grains, while biotite crystals and lithic fragments created iron oxide staining that ranged from slight to severe. In the specimens with severe iron oxide staining, biotite grains were highly discolored, with the outer boundary of the crystal difficult to identify, and with cleavages visibly widened. As depth into the stone increased, the effects of mechanical weathering decreased (essentially ceasing at about 8–10 cm) while chemical weathering was not seen further than 2 cm into the rock.

Volcanic glass and lithic fragments had also altered to produce a smectite-type of clay mineral, to a greater degree next to joints and to a lesser degree where lichen cover existed on the surface of the rock. This situation created a differential strain and swelling pressure around joints, and the formation of tension cracks around altered rock fragments. Topal and Doyuran concluded that oxidation of biotites and lithic fragments caused the tuff to discolor, and that there was potential for further oxidation of biotites and rock fragments around joints. They therefore advised avoiding the use of iron-containing materials during conservation efforts.

Boulders at an ignimbrite outcrop in central Italy were carved into monumental sculptures in the sixteenth century. Fratini et al. (1994) studied the material in thin section as part of efforts to characterize the stone and its degree and mechanisms of deterioration. A wide variety of minerals and lithics was identified, with few changes found as they went from the surface to deeper within the stone (weathered to unweathered). The main weathering changes were apparent in the glassy matrix, which underwent some devitrification into chlorite and serpentine. Some pyroxenes on the outer weathered surface also underwent serpentinization, illustrated in a micrograph (Frantini et al. 1994).

Salvadori et al. (1994) also studied the deterioration of volcanic tuffs identified as ignimbrite, from the Vulsini Volcanic Complex in central Italy. Their goal was to better understand the effects of lichen-induced deterioration on tuff, to identify which species of lichen are the most harmful to the stone, and to confirm or disprove the hypothesis that in some cases the lichen can form a protective coating that actually prevents deterioration of the stone surface. The ignimbrite in this study was located between Lake Bolsena and the Tiber valley, and was covered with lichen. Samples were taken from natural outcrops and from an active quarry. The tuff was described as having a vitro-clastic texture, slightly cemented, formed of scoria, blackish pumice, xenoliths of lava, and phenocrysts embedded in a microgranular groundmass formed of glassy fragments. The phenocrysts included sanidine, clinopyroxenes, leucite, biotite, and labradoritic plaioclase feldspar. Iron oxides and hydroxides were present as secondary minerals, along with zeolites, another product of weathering processes.

In thin section, the stone showed evidence of mechanical erosion and leaching along surfaces of clasts and pumices, probably due to rainwater. There were deep fractures and cracks, sometimes also with leaching, reducing the cohesion between clasts and groundmass. Zeolites were abundant at the surfaces, probably forming as a deterioration product of glass, pumices, and siliceous fragments. Thin sections also revealed that for some of the lichen species, fungal hyphae can penetrate deeply into the tuff (by as much as 4–5 mm), causing detachment of several millimeters of stone. Large amounts of weddellite (calcium oxalate) were seen in some of the thin sections, accumulated in the thallus of some lichen, probably due to removal of calcium from the substrate by the lichen. Other lichen were only loosely adhered to the rock substrate, and weddellite was present in small amounts.

The authors concluded that the high porosity of tuff, as well as its heterogeneity, contributed to its susceptibility to lichen penetration. Phenocrysts were surrounded while the vitreous pumices and microgranular groundmass were broken into. Some initial protection from degrading effects of rainwater would be provided, but deep penetration of the rock leads to detachment of stone, and the production of calcium oxalate is problematic.

Jackson et al. (2005) described deterioration evidence seen in thin sections of tuff building stones from ancient Rome. They noted that the tuff is often shielded by stucco or decorative stone cladding, but despite this, water sorption led to decay in Roman tuff building stones. The surface layer of the tuff swelled and corroded as water remained in contact with hydrophilic minerals. Zeolite and calcite grain cements sometimes dissolved, and microcracking and disaggregation of the tuff could occur. As sun and wind evaporated moisture from the surface, dissolved components sometimes created surface efflorescence and crusts. Exfoliation and alveolar erosion may cause disaggregation and erosion along the surface.

2.3 Rhyolite

Rhyolite is an acid rock (> 63% silica), containing phenocrysts of quartz and alkali feldspar (and often minor amounts of plagioclase feldspar and biotite mica) within a microcrystalline or glassy matrix. It has the same composition as a plutonic igneous rock, granite, but unlike coarse-grained granite, rhyolite tends to be aphanitic (constituents are too small to be visible without the aid of microscopy) because of its rapid cooling regime.

The most distinctive constituent of rhyolites is usually glass, along with very small crystal grains. The glass is often partially devitrified to microcrystalline or cryptocrystalline feldspar appearing as slender fibers (Fig. 2.3.1), and silica forms such as tridymite, cristobalite, opal, or chalcedony, in feathery or fibrous aggregates. Somewhat larger quartz grains may also be present, as well as separate small feldspar grains in a glassy matrix; or microgranular aggregates of interlocking crystals of feldspar and quartz (Williams et al. 1982; Le Maitre 1989) (Fig. 2.3.2).

Rhyolite has a long history as a cultural material—it was one of the stones used at Stonehenge in England (Thomas 1923). Thomas described that material as flinty and dark gray, and very fine grained. The feldspars were mainly albite and albite-oligoclase, although quartz was the primary component.

Rhyolite was also one of the volcanic rocks used as a building material in ancient Egypt. Thin-section analysis showed quartz and plagioclase feldspar with a spherulitic texture (Helmi 1994). Fibrous spherulites and spherulitic aggregates were often produced through devitrification of glass in acidic volcanic rocks.

It was also identified as a material used in the building of the Holy Trinity Memorial in Czechoslovakia in the eighteenth century. Petrographic studies of this material were conducted as part of studies into its state of preservation and conservation treatment needs (Krauze et al. 1991). Described as a rhyolite porphyry that was pink-gray in color, thin sections showed mainly recrystallized glass with quartz inclusions, and calcium feldspars in the matrix. However, rhyolite is generally considered to have too rough an appearance and texture (combined with a lack of ability to take any polish) to be suitable for most construction uses or for interior decorative work (Merrill 1891).

Rhyolite was used to produce stone tools in pre-Columbian North America. Excavations at a rhyolite source in the White Mountains of New Hampshire showed that the source was exploited over a period of 7,000 years (Gramly 1984). The source provided mainly locally used materials, primarily flaked tools.

2.4 Dacite

Dacite is also an acid rock (> 63% silica), with less than 35% total feldspar. It has a fine-grained matrix texture similar to that of rhyolite. The groundmass is sometimes glassy,

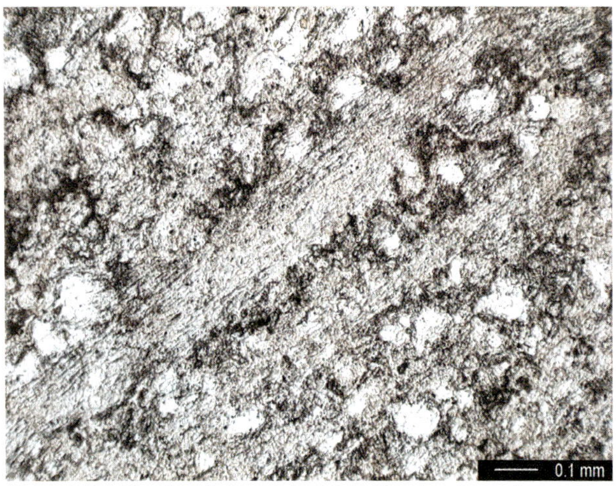

Figure 2.3.1 Rhyolite, plane polarized light. The glass in rhyolite tends to be partially devitrified to microcrystalline or cryptocrystalline feldspar, appearing as aggregates of aligned slender fibers, and to silica forms in microcrystalline or fibrous aggregates.

Figure 2.3.2 Rhyolite, crossed polarized light. Larger quartz grains can also be present in rhyolite, set into the glassy matrix along with tiny rods or fibers of feldspar, or within microgranular aggregates of interlocking crystals of feldspar and quartz. In crossed polarized light the larger, birefringent quartz grains stand out from the dark, isotropic or fibrous matrix.

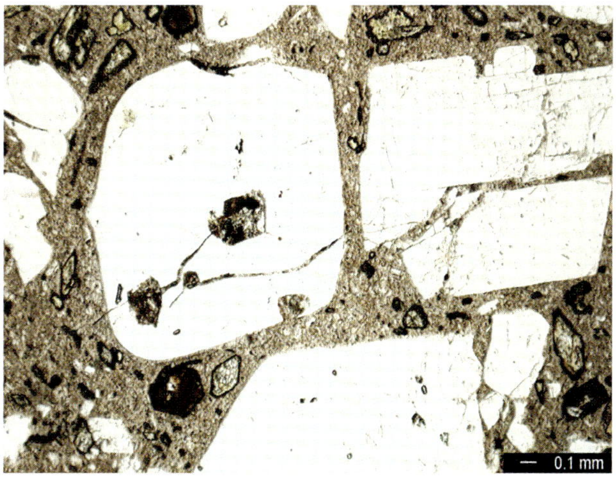

Figure 2.4.1 Dacite, plane polarized light. The fine-grained matrix texture is similar to that of rhyolite, with a groundmass of fine interlocking crystals of quartz and feldspar microlites surrounded by glass. There are many large phenocrysts of plagioclase feldspar with a few subrounded grains of sanidine; there are also smaller microphenocrysts of biotite and hornblende.

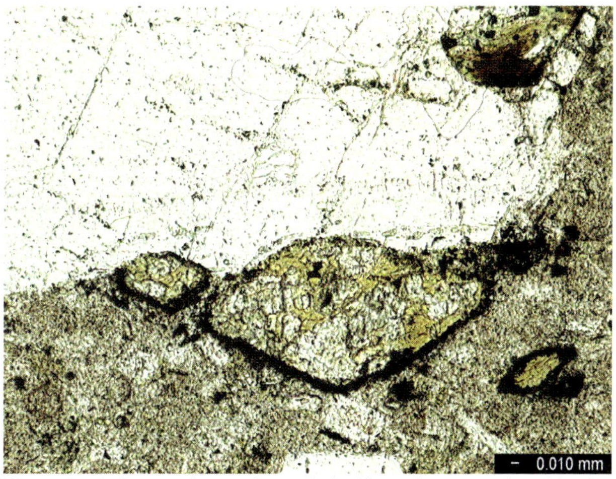

Figure 2.4.2 Dacite, plane polarized light. The yellow-green hornblende microphenocrysts have the characteristic rhombic and pseudohexagonal cross sections of amphiboles and are smaller than the colorless phenocryst of feldspar above them.

sometimes composed of fine interlocking crystals of quartz and feldspar, or of small feldspar microlites surrounded by glass. However, phenocrysts of quartz and plagioclase feldspar are common (Fig. 2.4.1), and there may also be some phenocrysts of mafic minerals (such as pyroxenes, amphiboles, and biotite) (Fig. 2.4.2) and other accessory minerals (Williams et al. 1982; Le Maitre 1989). It is reported only rarely in the literature as a cultural material, but is known to have been used, for example, at the temple of Karnout in Armenia (Grissom 1994).

Brown and Harrell (1995) used thin-section petrography to classify as dacite some of the rock samples collected at ancient Egyptian quarries. These specimens were described as porphyritic, with phenocrysts of plagioclase feldspar and hornblende. The groundmass was plagioclase, hornblende and biotite; and accessory minerals were magnetite and apatite.

2.5 Trachyte

Trachytes have a 52–63% SiO_2 content, with alkali feldspars being the predominant or only feldspar present. Generally light in color, they tend to be porphyritic rocks, with large crystals of feldspar within a finely crystalline groundmass of small feldspar microlites (Fig. 2.5.1). Minor constituents may include olivine, pyroxene, amphibole, biotite, and accessories such as iron oxides (Fig. 2.5.2) and apatite (Williams et al. 1982).

2.5.1 Trachyte as a cultural material

Trachyte has often been used as a building material and for utilitarian objects (Hull 1872). For example, it was used for churches and other buildings dating back to the medieval period in central France, for Roman buildings in Italy, and for street pavement in Pompeii. Trachyte was also used in the construction of some of the monumental stone temples in pre-Columbian Mexico (Outwater 1957).

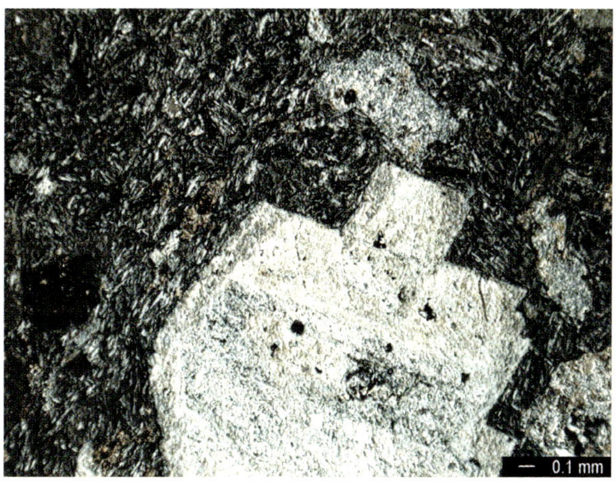

Figure 2.5.1 Trachyte, crossed polarized light. Trachytes tend to be porphyritic rocks. Within the matrix of fine-grained aggregates of feldspar are large phenocrysts of feldspar, strongly altered.

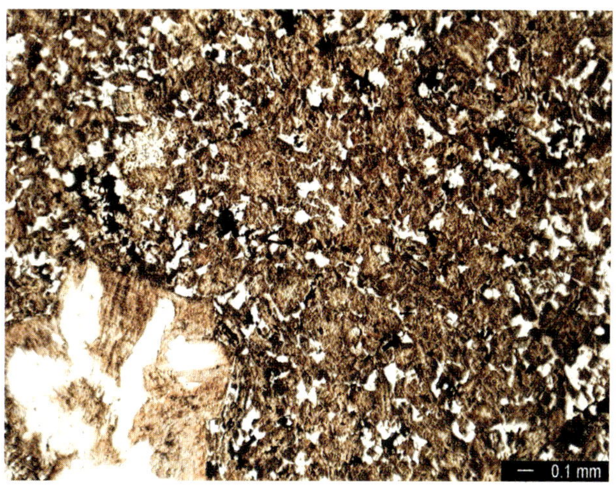

Figure 2.5.2 Trachyte, plane polarized light. Accessory constituents of this trachyte include black magnetite partially altered to hematite. There is a matrix of small tabular feldspar, and highly altered phenocrysts of feldspar.

Trachyte was identified through thin-section studies as one of the materials used for Roman millstones (Williams-Thorpe 1988). Samples were described as having phenocrysts of iron-titanium oxides along with heavily altered olivine and pyroxene crystals. Elsewhere, trachyte was sometimes used for sculpture production on Easter Island where seven intact statues made of this material have been documented (Charola 1997). These statues range in size from 1 m to over 2 m high. Trachyte was also used for carving small household images (Baker 1993).

Zantedeschi and Zanco (1993) conducted a detailed study of trachytes quarried from the southwestern Venetian region. This material was extensively used during the Roman period for buildings, bridges, and pavements. Trachytes from different quarries were clearly distinctive petrographically. A wide variety of attributes seen in thin section was useful in characterizing and distinguishing the trachyte samples. Distinguishing characteristics included the presence or absence of certain minerals as phenocrysts, accessories, or groundmass components (alkali feldspars, plagioclase feldspars, quartz, hornblende and other amphiboles, pyroxene, rutile or other opaques, apatite, zircon, and biotite) (Fig. 2.5.3).

The authors also found that textural features were extremely important in characterizing the trachytes. For example, they described thin sections of one type, the Monselice trachyte, as being "holocrystalline glomeroporphyritic, rarely cumuloporphyric texture produced by the presence of monomineralic or polymineralic aggregates that are immersed in a microcrystalline groundmass of trachytic texture." This means that the rock is composed wholly of crystals rather than glass (holocrystalline), with phenocrysts (the larger crystals) mostly being found in distinct clusters (glomeroporphyritic) (Fig. 2.5.4), rarely with the clusters of phenocrysts resembling cumulus clouds, this overall texture being produced by aggregates of either single or multiple mineral types set into an overall microcrystalline groundmass of trachytic texture (crowded microlites of feldspar arranged in a flow pattern that is somewhat parallel in orientation, with the interstices filled by micro- or cryptocrystalline material).

Zantedeschi and Zanco also listed very specific and characteristic textures for many individual minerals. For example, the same Monselice trachyte was described as having quartz that appears as crystals that are subroundish, anhedral (often of irregular shape) and sometimes elongate, and usually found clustered together. The most abundant minerals, alkali feldspars, were specifically identified as anorthoclase and sanidine. These were usually subhedral (only partially bounded by crystal faces) and often formed clusters with biotite and hornblende; it was also reported that they sometimes show weak exsolution phases (referring to the process of unmixing of the solid solution of the feldspar phases).

Zantedeschi and Zanco's paper is an outstanding example of the utility of being very specific in mineral and texture descriptions. It included many more examples of such descriptions, and its specificity permitted clear distinctions to be made between trachytes of different quarries. This work showed that with detailed petrographic descriptions it is possible to identify the provenance of many building materials down to the specific quarry where the rock originated.

Capedri et al. (2003) also conducted provenance studies of trachyte, in this case working with flagstones used in the paving of Roman roads in the Po plain of Italy. Trachyte was useful as a paving material because it is resistant to both abrasion and to surface alteration in the outdoor environment. Although the trachyte used for this purpose came mainly from the Euganean Hills Complex near Padua in northern Italy, more than 70 quarries were in use at various times. Capedri et al. sampled 269 flagstones, carrying out both thin-section and chemical analyses and comparing those data to results of analyses on samples from the 35 most exploited quarries. They were able to conclude that

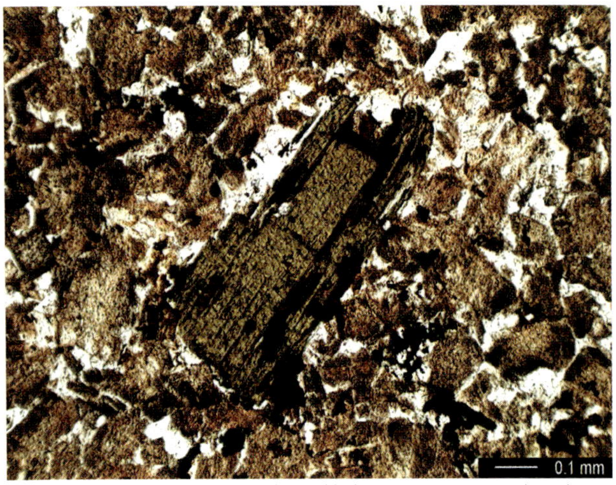

Figure 2.5.3 Trachyte, plane polarized light. Accessory minerals in this trachyte include biotite, altering to iron oxides, and smaller chunks of magnetite, some altering to hematite. These accessory minerals are set within a matrix of small tablets of feldspar that appear brown due to alteration products, and small interstitial accessory grains of clear, colorless anhedral quartz.

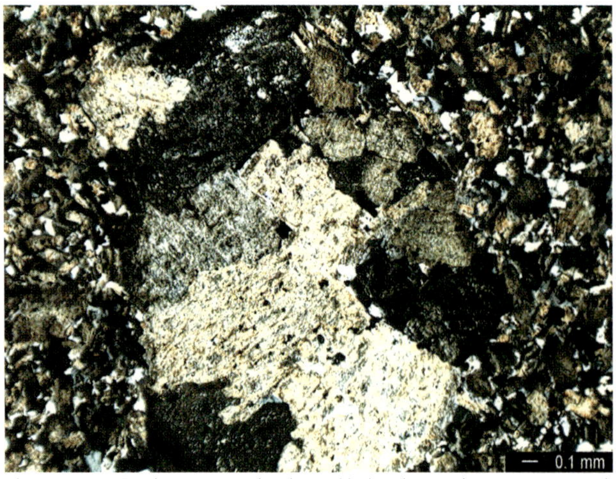

Figure 2.5.4 Trachyte, crossed polarized light. This trachyte is composed wholly of crystals rather than glass, with a matrix of mainly small tablets of feldspar. The phenocrysts (larger crystals) of altered feldspar are most often gathered in distinct clusters; this texture is termed glomeroporphyritic.

the flagstones came from along the eastern border of the Euganean hills, probably because those sources were closer to transportation routes including canals that connected these quarries to the Adige River.

In thin section, Capedri et al. found that the trachytes were clearly porphyritic, with phenocrysts of feldspars (anorthoclase and plagioclase) and lesser amounts of mafics (mainly biotite with some hornblende and clinopyroxenes). Accessory minerals included quartz, opaques (ilmenite and magnetite), zircon, apatite, and titanite. Variation especially appeared in proportions of feldspar phenocrysts, the proportion of anorthoclase to plagioclase, groundmass texture, geometric aspects of anorthoclase phenocrysts, and in the distribution of clinopyroxene and hornblende.

Using these data, six petrographically distinct types were identified among the flagstone trachytes, and the provenance of each group could be inferred through comparison with trachyte samples from quarry localities. Detailed petrographic descriptions were provided for each of the six groups, supplemented with photomicrographs (Capedri et al. 2003).

Antonelli et al. (2004) conducted thin-section petrography of trachytes used for protohistoric grinding tools found in the Karst plateau on the northeastern border of Italy and Slovenia and in the Istria peninsula of Croatia. The trachytes were found to share a number of petrographic features, including being mildly vesiculated and porphyritic with phenocrysts of euhedral plagioclase, anhedral to subhedral anorthoclase and skeletal biotite, along with zircon and apatite accessories. The matrix was described as being of hyalopilitic texture, so composed of microlites of sanidine-plagioclase-opaques plus or minus interstitial quartz and minute interstitial brownish glass. Most of the artifact materials could be traced to sources in the Euganean hills near Padua. Previously, it was known that these sources were exploited in northern Italy during protohistoric times, but the new data indicated that use of these materials for grinding tools extended all the way to Istria.

2.5.2 Trachyte deterioration mechanisms

A building in Germany constructed partially of trachyte, Cologne Cathedral, was studied petrographically as part of a deterioration study (Mirwald et al. 1988). Trachyte was used mainly in the early stages of construction, during the thirteenth–sixteenth centuries. Thin sections of this material showed a fine-grained groundmass with both large phenocrysts and smaller mineral crystals. Sanidine and plagioclase feldspar were the major components, with minor amounts of quartz and accessory levels of diopside, biotite, rutile, apatite, and ilmenite. The trachyte was deteriorating by extensive foliation, and sanidine phenocrysts were being partially destroyed, leaving holes in the rock reflecting the original morphology of the sanidine crystals. Chemical incompatibilities with other rocks with which the trachyte is in contact, especially an alkalic limestone, were accelerating deterioration.

Laue et al. (1996) identified trachyte as one of three building materials used for construction of the crypt of St. Maria in Kapitol (near Cologne, Germany), under study because of its severe salt deterioration problems. The trachyte, characterized as having sanidine feldspar phenocrysts set into a fine-grained groundmass, exhibited scaling in 1–20 mm-thick fragments. Small cracks could be seen starting at the edges of sanidine crystals, eventually developing into major fissures. Small fragments of groundmass near sanidine crystals often fell out, leaving the phenocrysts protruding out of the surface. Salt patches also formed along the trachyte surface during dry periods, dissolving and creating a moist residue on the rock surface during wetter periods. This wet area stays on the surface rather than being pulled into the rock by capillary forces.

As part of their study, Laue et al. used thin sections to examine the pore structure of the trachyte, looking at size and distribution of pores in relation to distribution and formation of deterioration products. The trachyte contains a heterogeneous pore structure, built up of irregularly distributed fissures and cracks which are not well interconnected. The groundmass material and the phenocrysts do not contribute any free porosity, except in the area of weathered sanidine phenocrysts, where there is only a slight amount of free porosity—most of the porosity of the stone consists of trapped pores, thus the pore system is not capable of transporting solutions by capillary force. With diffusion as the main transport mechanism of water, solutions can penetrate into the fissures and cracks near the surface. Salts in those solutions then alternate between crystallizing and dissolving, processes that promote scaling problems along the surface of the trachyte.

2.6 Andesite

Andesite also falls into the category of intermediate volcanic rock (52–63% SiO_2), but with plagioclase the predominant or only feldspar present, rather than alkali feldspar. Most andesites are porphyritic, with large phenocrysts of plagioclase (Fig. 2.6.1) which are often zoned, and mafic minerals such as olivine, pyroxenes, hornblende, and/or biotite. The minerals may be highly altered (Fig. 2.6.2). The matrix or groundmass often has a pilotaxitic texture, in which there are many small microlites of feldspar arranged in subparallel order as a result of magma flow, with the interstices filled in with microcrystalline or cryptocrystalline material; sometimes the interstitial material may be glass (Williams et al. 1982; Le Maitre 1989) (Fig. 2.6.3).

Curtis (1959) identified andesite as one of the building materials at the Mexican site of La Venta. He described one type as a hornblende andesite, with phenocrysts of plagioclase, black hornblende, and green augite in a fine-grained groundmass. Another type is described as a pyroxene andesite, composed of phenocrysts of plagioclase, augite, and hypersthene in a glassy matrix. Weathering was found to extend to a depth of 1.2 mm.

Andesite was widely used as a sculptural and building material in ancient Mexico. Williams and Heizer (1965a) examined thin sections from a variety of Olmec objects

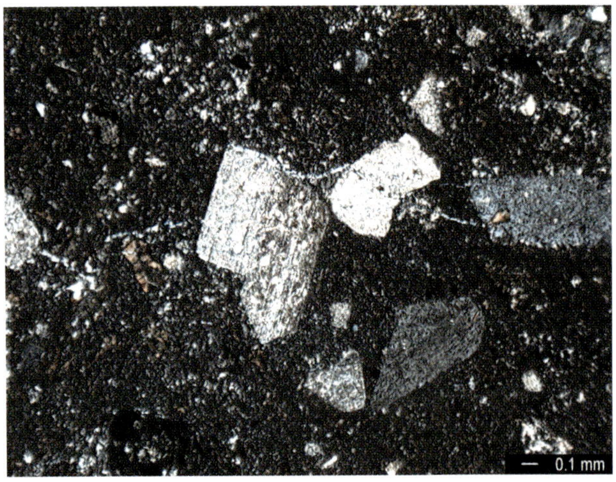

Figure 2.6.1 Andesite, crossed polarized light. Most andesites are porphyritic, with large phenocrysts of plagioclase feldspar. In crossed polarized light, the highly altered feldspars are visible, while the fine-grained matrix of feldspars and glass is mostly dark.

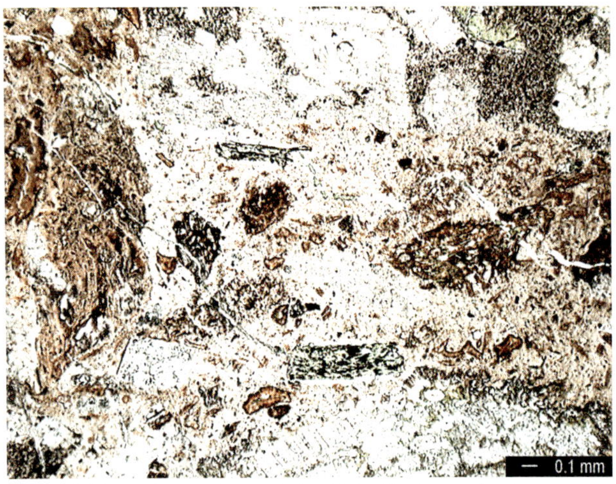

Figure 2.6.2 Andesite, plane polarized light. Most andesites have many mafic minerals such as olivine, pyroxenes, hornblende, and/or biotite, along with phenocrysts of plagioclase feldspar. All these minerals may be highly altered. A volcanic flow texture is visible.

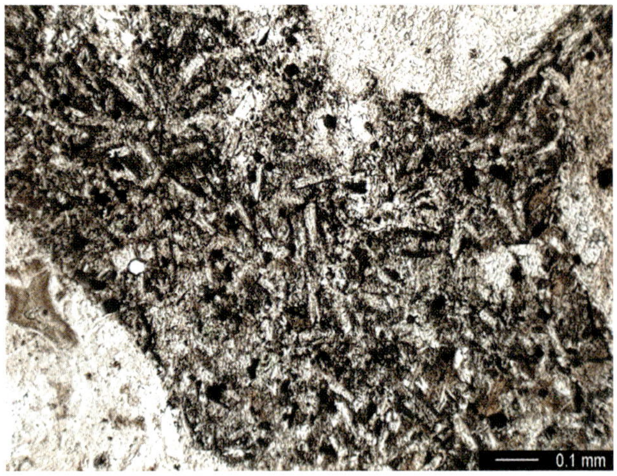

Figure 2.6.3 Andesite, plane polarized light. Many andesites have a groundmass with a "pilotaxitic" texture, i.e. there are many small microlites of feldspar arranged in subparallel order as a result of magma flow, with interstices filled in with microcrystalline or cryptocrystalline material, or glass. The tiny rods of feldspar are visible here.

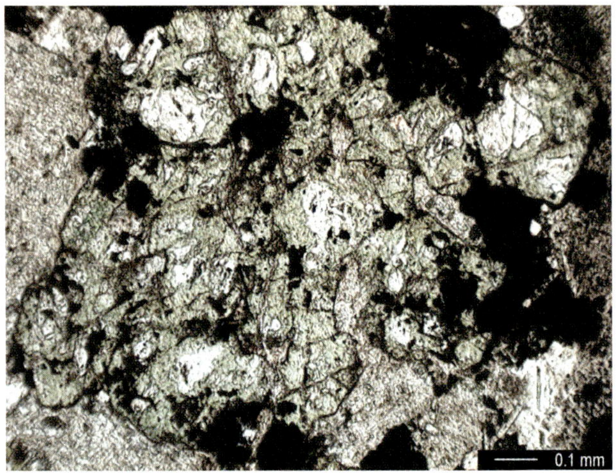

Figure 2.6.4 Andesite, plane polarized light. The green hornblende in this andesite is being replaced by black iron oxides as well as by brown chlorite and colorless calcite, in an alteration process known as propylitization, to which andesites are especially susceptible.

made of andesite. They described one type as a vitrophyric (having a glassy groundmass into which large phenocrysts are set) hornblende-augite andesite. About 25% of the rock consisted of large euhedral phenocrysts of hornblende, pleochroic in thin section, with edges rimmed by magnetite. About 15% of the material consisted of euhedral phenocrysts of augite, and 30% of small crystals of labradorite plagioclase. There were also some grains of magnetite, a matrix of glass and cryptocrystalline/microcrystalline feldspar, and many vesicles lined with small spheroids of cristobalite.

Williams and Heizer reported finding other types of andesite, easily distinguishable in thin section. One consisted of about 40% phenocrysts of feldspar with oscillatory zoning; and one-third hornblende fringed with magnetite. Hypersthene made up about 5%, and the remainder of the stone was comprised of microlites of feldspar along with cryptocrystalline/microcrystalline feldspar and cristobalite lining vesicles. Another andesite type is especially distinctive because it contains tridymite. The illustrations of andesite samples in thin section provided by these authors are especially helpful.

Williams and Heizer (1965c) also examined colossal sculpture at and near the site of Teotihuacan, again identifying andesite. Thin sections were described as showing a porphyritic, pilotaxitic (a subparallel flow pattern to the microlites of feldspar with interstices filled by microcrystalline or cryptocrystalline material) pyroxene andesite with resorbed hornblende (being replaced by magnetite or magnetite-augite) (Fig. 2.6.4). Plagioclase phenocrysts had intense oscillatory zoning. Accessories included hypersthene, iron oxides, and cristobalite.

Andesite (described as a vesicular andesite) was used in the construction of the Indonesian religious monument of Borobodur. Thin sections of this material were studied in an attempt to understand deterioration mechanisms resulting from mosses, algae, and lichens, which have adversely affected the aesthetic appreciation of the carved bas reliefs at the site (Picot and Ragot 1972). A hard white deposit seems to be associated with the appearance of the bioorganisms. The deposit was identified as amorphous silica (opal) incorporating calcite, hydrated iron oxide, and rock debris. Thin sections illustrated a glassy rock with many vesicles and few microlites. As the rock becomes more and more devitrified, it becomes less porous and microlites are more abundant. Other thin sections showed lines of iron oxides in glassy zones, along with iron hydroxides. Abundant cristobalite had developed in the pores of a devitrified sample.

Andesite was also one of the building materials used in the construction of the eighteenth-century Holy Trinity Memorial in Czechoslovakia, although it was mainly utilized for sculptural decoration. This material was described (Krauze et al. 1991) as having a porous texture that gives a lightness and baroque appearance to sculptures carved from it, and therefore Krauze et al. hypothesized that it was exploited mainly for aesthetic purposes (most of the construction of the building itself was done using rhyolite instead of the andesite). In thin section the andesite was found to consist largely of plagioclase feldspar, pyroxene (diopside-augite), amphibole, and highly altered biotite. Plagioclase grains were often zoned, with bytownite found at the cores and labradorite along the edges. The groundmass consisted of microlites of plagioclase along with iron oxides and a vesicular glass. The authors were careful to point out the great variability that would be expected in this type of pyroclastic material, therefore caution must be exercised in the analysis of small samples to characterize the rock.

However, because of its porous structure and lack of ability to take a polish, andesite was rarely used for building construction or for sculpture. Most of the andesite sources in North America were not quarried, or were only lightly used. A few exceptions have been noted (Merrill 1891); it was sometimes used for building and paving work in Greece and Turkey (Kempe and Harvey 1983).

2.7 Basalt and Diabase

2.7.1 Basalt

Basalt is a basic volcanic rock (45–52% SiO_2), often dark in color as it is rich in ferro-magnesian minerals (such as pyroxenes and olivine); the groundmass tends to be fine grained, even glassy (Fig. 2.7.1). The main constituents are usually calcic plagioclase and pyroxene (Fig. 2.7.2). Olivine and small amounts of interstitial quartz are also frequently present, along with glass. Basalt is often porphyritic with phenocrysts of mafic minerals or plagioclase feldspar (Figs. 2.7.3 and 2.7.4). The most silicic type of basalt is called tholeiitic basalt (Williams et al. 1982; Le Maitre 1989). By far the most common volcanic rock, basalt has been used for tools, sculpture, and building materials.

Curtis (1959) found basalt used as a building material at La Venta (Mexico). One type, from a plaque, was an olivine basalt with large phenocrysts of augite and abundant phenocrysts of plagioclase feldspar within a dark microcrystalline groundmass. The olivine is altering along the edges to iddingsite (a magnesium-iron silicate that only occurs as partial or complete pseudomorphs after olivine) (Kerr 1977).

Pre-Columbian uses of basalt in Mexico were very prominent, for sculptures and for building material. Olmec heads were carved in a coarse-grained porphyritic basalt, rich in olivine and augite. Williams and Heizer (1965a) reported on thin-section studies of several large-scale monuments, including a colossal head and a carved stela. They described one type of material as being distinctive by an unusual abundance of large crystals of olivine and augite (each comprising about 25% of the sample), with only a few large crystals of plagioclase feldspar. In thin section the olivine crystals were colorless except for rims of brown iddingsite (a common alteration product of olivine). The augite was a pale yellowish green, often with oscillatory zoning between colorless and green. They noted that plagioclase feldspar is usually abun-

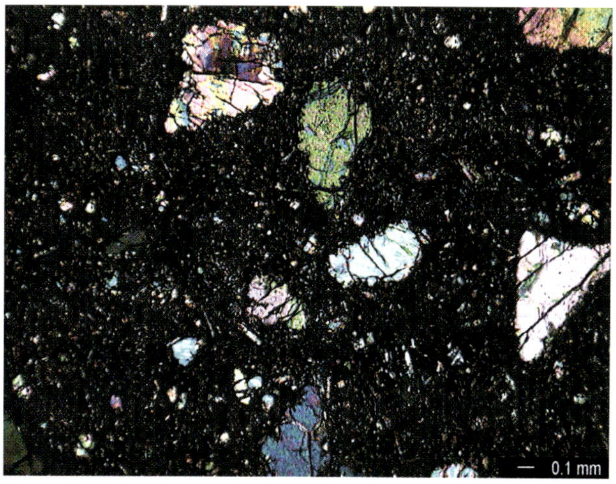

Figure 2.7.1 Basalt, crossed polarized light. Basalt tends to have a fine-grained matrix with phenocrysts. Here the groundmass includes some poorly oriented aggregates of small plagioclase feldspar laths, as well as tiny grains of augite, olivine, and black magnetite. The phenocrysts are augite and olivine, with moderate to strong interference colors.

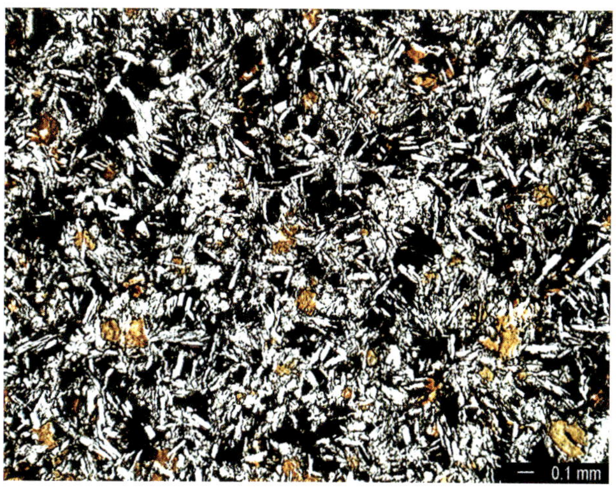

Figure 2.7.2 Basalt, plane polarized light. In this basalt the matrix consists of thin colorless laths of plagioclase feldspar (labrodorite) and granular clinopyroxene, black magnetite, and some interstitial glass which is altering to a yellow-brown iron-rich chlorite.

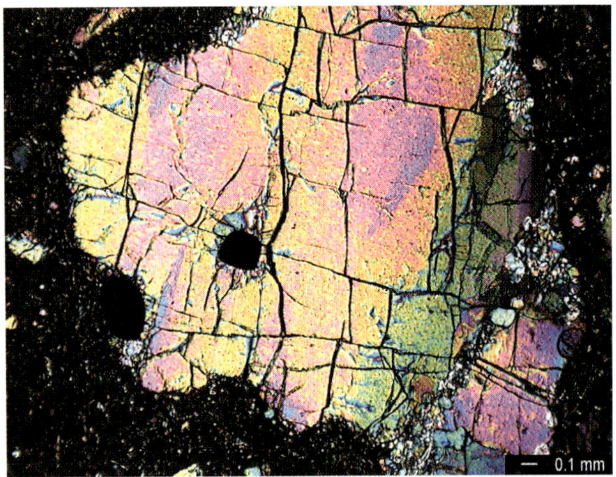

Figure 2.7.3 Basalt, crossed polarized light. Basalt is often porphyritic. This phenocryst of pyroxene (augite) in a basalt shows moderate interference colors in crossed polarized light. Note the pyroxene cleavage with two lines intersecting at 90 degrees. Two grains of black magnetite are also visible.

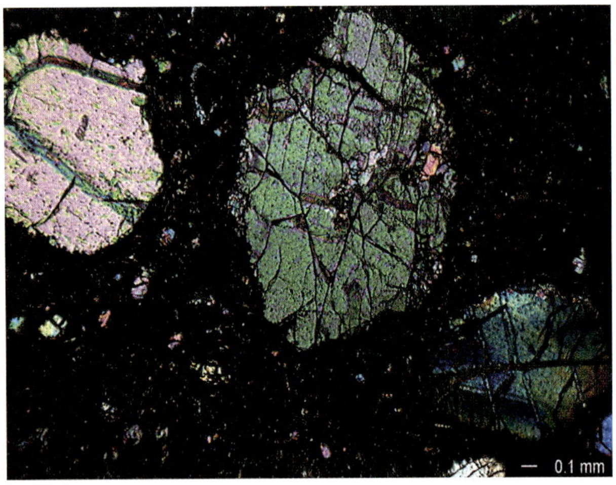

Figure 2.7.4 Basalt, crossed polarized light. These olivine phenocrysts in a basalt show very high birefringence in crossed polarized light, so have an almost pastel appearance. They also have typical irregular cracks or fractures sometimes filled with alteration products. Here the olivine is partially altered in fractures and along the edges to iddingsite, a magnesium iron silicate.

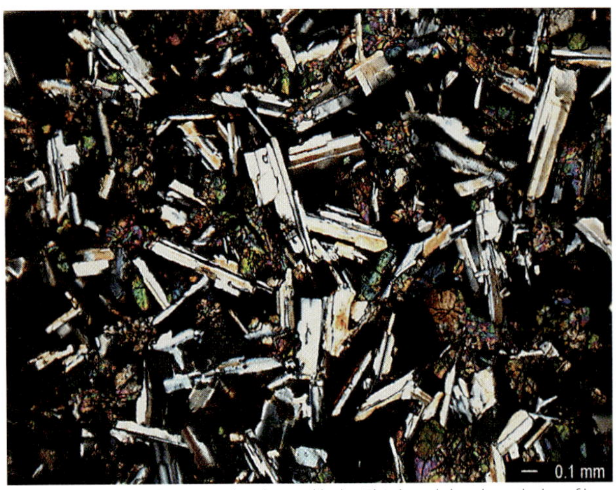

Figure 2.7.5 Basalt, crossed polarized light. This basalt has long laths of low birefringent plagioclase feldspar set amid clusters of augite with moderate interference colors. The feldspars exhibit both simple and polysynthetic twinning. Reddish-brown iron oxides fill some of the interstices.

dant in basalts, but not here. Accessory minerals included apatite needles and hematite flakes.

Williams and Heizer also found a very different basalt—this one with many small phenocrysts of olivine (about 15% of the specimen) but none of augite or plagioclase feldspar. The remainder of the material consisted of small laths of plagioclase and augite, along with iron oxides. Yet other basalts found include one with many large phenocrysts of feldspar, with fewer and smaller ones of augite and olivine (with the olivine being extensively altered to iddingsite), and one distinctive by its scarcity of olivine and abundance of augite, along with many large phenocrysts of plagioclase feldspar. The combination of mineralogical and textural differences identified allow the authors to hypothesize about which stones came from a common source, and through some geological work to try to identify sources for some of the stones. Their paper includes illustrations of materials studied, which is very useful for comparing with thin sections from other similar sculptures, as an adjunct to the verbal descriptions of stone types.

Basalt was also used as an architectural material in historic America. For example, Steward et al. (1981) discussed a fine-grained basalt used in nineteenth-century New York. It was quarried along the Hudson River just north of the city, and was used mainly for interior lintels of windows and in interior windowsills.

In Richard Newman's petrographic study of stone sculptures of India (1984) he characterized in detail a sculpture made of basalt. The thin section showed long laths of plagioclase feldspar with polysynthetic and albite twinning (Fig. 2.7.5). There was also augite, slightly pleochroic and formed in clusters. Also present were iron oxides, as crystals and as minute granules, with a few yellow-orange iron-rich zones in interstitial areas. Apatite was present as an accessory mineral.

Perath (1984) surveyed historical building materials of Israel, including basalt deposits. He noted that most Israeli basalts can be classified as calkalkaline (associated with a basalt-andesite-dacite suite) olivine basalts. Although they are somewhat abundant, they were only sparsely used as an architectural material. Perath suggested that this may be because the deposits are somewhat isolated from major settlement areas, and tend to co-occur with calcareous rocks that were preferred for architectural uses. Still, basalt was used here for a wide variety of applications including storehouses, monumental public buildings, religious buildings, and relief sculptures.

Many of the early Buddhist and Hindu rock-cut cave sites in India (Elephanta, Karle, Bhaja, Kanheri, Ajanta, Ellora, and Pitalkhora) were hewn out of basalt. These basalts were characterized petrographically by Lal (1985), who also carried out extensive studies of their deterioration.

A study of millstones recovered from the wreck of an ancient Greek ship off the coast of Spain showed basalt as the material, the source being identified as the island of Pantelleria off North Africa in the Sicilian channel (Williams-Thorpe and Thorpe 1990). Thin sections of the material showed a fine matrix of plagioclase feldspar, pyroxene, and iron-titanium oxides, along with phenocrysts of feldspar, clinopyroxene, and sometimes olivine.

Basalt was also used for grinding stones in ancient Egypt. Petrographic analysis of such grinding stones in Upper Egypt was employed to characterize the rock types used and to identify their sources (Mahmoud and Bard 1993). Since the rocks were not local to the artifact findspots, the geological studies are helpful for obtaining information about regional and long-distance trade and exchange relationships.

Basalt was first used in ancient Egypt for small vessels; later, it became an important material for temple pavements, for the construction of royal sarcophagi, for statuary, and for architectural elements (Harrell and Brown 1995). Thin-section studies have included samples from both artifacts and ancient quarries. For example, one quarry was described as containing slightly porphyritic basalt comprising mainly labradorite plagioclase, augite, and volcanic glass (partially devitrified) with minor amounts of magnetite, and accessory hornblende, olivine, quartz and apatite (Harrell and Brown 1995). The basalt from this quarry was found to match some artifact basalts.

Schneider (2002) found basalt as one of the materials quarried for use as milling tools by indigenous peoples of the American West. Vesicular basalt was selectively quarried in the Maya Highlands, as well as by the Hohokam of the Phoenix basin in Arizona. Ethnographic studies in the Guatemala Highlands showed that metates were used for many household purposes, but those that were reserved for use in grinding maize were almost exclusively made of vesicular basalt. This material was preferred because it did not require frequent resharpening and the production of grit was minimal. The Hohokam were prehistoric maize agriculturalists who imported coarse, nonlocal basalt because the vesicular and coarse-grained basalt made maize processing more efficient. This material provided a good cutting/shearing texture for processing maize kernels. Finer grained or nonvesicular stones were more often used for processing wild seeds.

2.7.2 Diabase

Diabase is coarser grained than basalt, due to slower cooling; it typically has a medium-grained texture, often described as being ophitic (having laths of plagioclase in a matrix of coarse pyroxene, typically clinopyroxene, so that the smaller feldspars appear to be enclosed by pyroxene) (Fig. 2.7.6) or subophitic (where the average length of the plagioclase is larger than that of the pyroxene so that only some feldspars are partly enclosed by pyroxene) (Williams et al. 1982).

There is a long history of interest by British archaeologists and geologists in characterizing the stones found at Stonehenge, and in identifying the localities of their sources. In the nineteenth century, small samples were taken for microscopic study, and one of the four stone types was determined to be diabase (Maskelyne 1878), under the British definition of the term (with heavy chlorite content). Both the feldspar and augite pyroxene are relatively coarse, giving the rock a porphyritic texture, and both show visible alteration. Accessories include opaque metallic oxides and quartz.

Thomas (1923) also studied these Stonehenge diabases. He noted that visually these stones are moderately crystalline, with blue-green to greenish-gray color. He also noted that although Maskelyne's work had shown there are minor variations in textures and mineral proportions among the diabases found at Stonehenge, those variations are similar in range to what could be expected within a single source of diabase. Thomas reported that there were white or pink spots distributed throughout the stone, formed by groups of oligoclase-albite feldspar, which distinctly characterize the rock. However, these crystal spots are often so widely spaced that they might not be represented in a thin section; yet they serve to characterize and identify this particular diabase. Thomas's observations highlight the importance of examining the full object from which a thin section is taken, to ensure that samples are representative of the variation present in the rock.

In historic North America, diabase was quarried mainly for use as paving blocks, although sometimes it was used for building purposes (often marketed as "black granite"); it was also quarried near the Gettysburg battlefield where it was used for headstones (Merrill 1891). The term often used in Britain to refer to this rock type is "dolerite," with diabase referring to altered dolerites (ophitic or subophitic texture, with visibly altered feldspars and pyroxenes being replaced by chlorite and fibrous amphibole) (Fig. 2.7.7).

The term picrite is another British name for a diabase variant, one that is very rich in olivine (usually greater than 60%) and pyroxene. This rock was identified as material used to produce ancient stone axes in Britain (Evens et al. 1962), and was examined in thin section. The matrix was comprised of an almost isotropic chlorite. Within that matrix were found rounded or irregular masses of colorless hornblende needles and blades, and irregular patches of cracked olivine. No feldspars were seen.

One of the basalt types reported by Curtis (1959) at La Venta, Mexico, used in the construction of a column, was described as picrite, an olivine basalt with abundant phen-

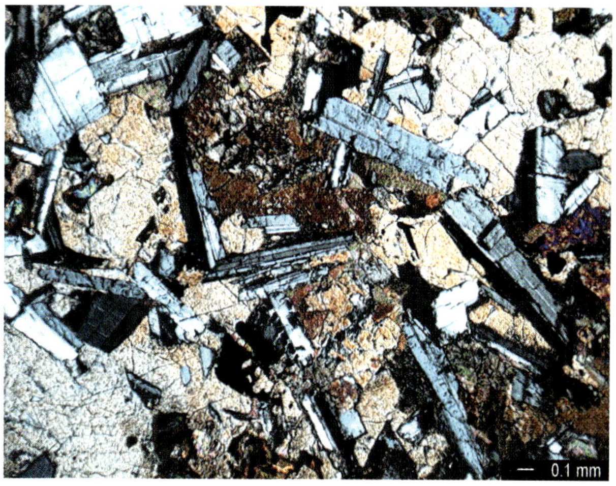

Figure 2.7.6 Diabase, crossed polarized light. Diabase tends to be more coarse grained than basalt due to slower cooling, with a medium-grained texture that is often ophitic (having laths of plagioclase feldspar within a matrix of coarse pyroxene so that the feldspar appears to be enclosed by pyroxene).

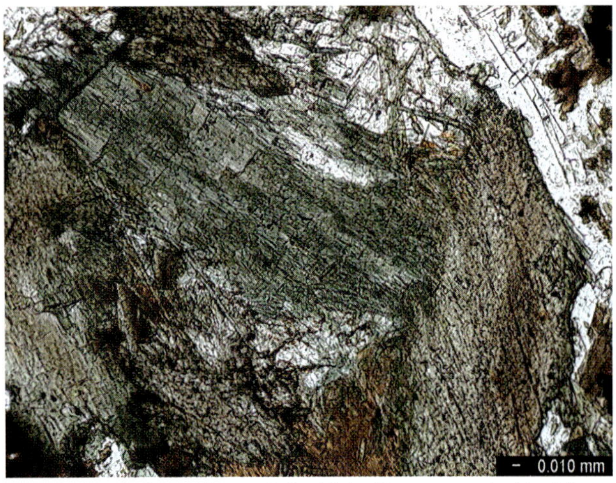

Figure 2.7.7 Diabase, plane polarized light. The pyroxene in this diabase (dolerite) is being replaced by green and brown chlorite and by green fibrous amphibole.

ocrysts of olivine in a fine-grained ground mass of pyroxene and plagioclase feldspar. Picrite is sometimes, as in this case, a term used to refer to a basalt with abnormally high phenocrystic olivine, augite, or both (Williams et al. 1982).

Diabase has been identified as a sculptural material in eastern India, particularly in Bengal (Asher 1987). It is fine grained, a dark bluish-gray color in hand sample, and in thin section was seen to be equigranular and comprised of 55–70% plagioclase feldspar, 20–40% clinopyroxene, plus minor amounts of olivine and opaque minerals. Altered samples included uralitized pyroxene and olivine, and sausseritized plagioclase. Uralite is a secondary tremolite-actinolite or hornblende-like amphibole formed by hydrothermal alteration of clinopyroxene, especially augite. Saussuritization refers to alteration of a calcic plagioclase to a mixture made up largely of albite and one or more epidote-family minerals plus or minus calcite, sericite and one or more zeolites (Kerr 1977; Dietrich and Skinner 1979).

2.8 Phonolite and Tephrite

2.8.1 Phonolite

Phonolite is a volcanic rock with SiO_2 at 52–60%; the major constituents are alkali feldspars and feldspathoids (such as nepheline, sodalite, and leucite) (Fig. 2.8.1). Most phonolites are porphyritic, with phenocrysts of feldspar, feldspathoids, or other minerals in a microcrystalline groundmass of alkali feldspar, feldspathoids, pyroxenes, and sometimes glass (Williams et al. 1982; Le Maitre 1989). Both nepheline phonolites and leucite phonolites (Fig. 2.8.2) were used by the Romans for millstones (Williams-Thorpe 1988).

2.8.2 Tephrite

Tephrite (Fig. 2.8.3) is a basic volcanic rock (45–52% SiO_2) rich in plagioclase feldspar (20–40%), with 10% or more feldspathoids (nepheline, leucite, etc.) and with little or no olivine present. Other minerals that may be present include pyroxenes, hornblende, biotite, opaque metallic oxides, sphene, zeolites, and apatite (Williams et al. 1982).

Petrographic studies, including thin-section analyses, of a material called "yellow tuff" used in the construction of the temple of Cibele (Rome) showed that it is actually a leucitic tephrite (Bianchetti et al. 1982). Multiple samples taken from the building allowed for a full assessment of variability within the material. Samples from four quarries that might have supplied the material were compared, and the probable quarry identified. The rock was very porous, with zeolites lining the wall of the vesicles; zeolites were also present as devitrification products of the groundmass. Other minerals present included clinopyroxenes, mica, sanidine and labrodorite plagioclase feldspars, magnetite, and apatite, all dispersed within the glassy, zeolitic groundmass along with lithic fragments.

Tephrites rich in leucite crystals as large as 2 cm across, and often equant-shaped, were used for Roman millstones

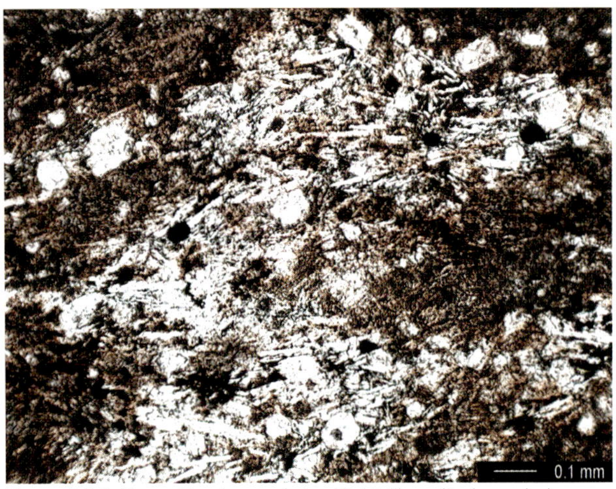

Figure 2.8.1 Phonolite, plane polarized light. Phonolite has colorless alkali feldspars and feldspathoids as major constituents. Here the matrix consists of thin elongated laths of orthoclase feldspar crystals and stubby nepheline euhedra, along with accessory black magnetite. Nepheline can be in the form of small hexagonal or rectangular crystals.

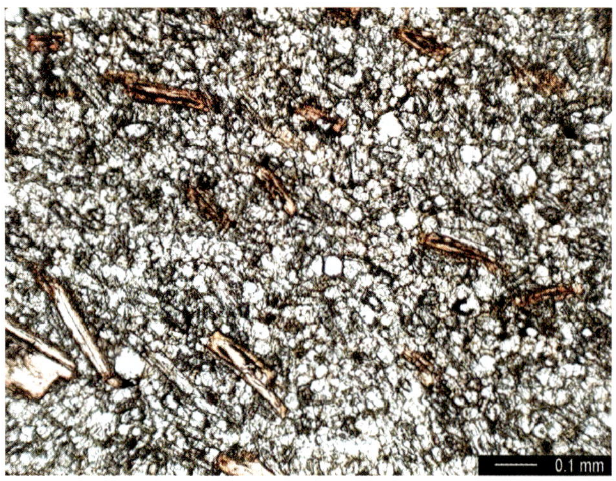

Figure 2.8.2 Phonolite, plane polarized light. This is a leucite phonolite, with small euhedral crystals of leucite forming much of the groundmass, along with some small colorless alkali feldspar laths. In plane polarized light leucite is also colorless and shows a trapezohedral outline with octogonal sections. The larger brown flakes are phlogopite mica.

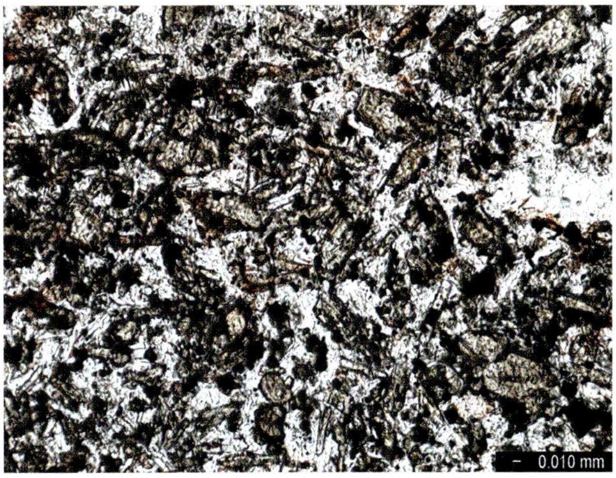

Figure 2.8.3 Tephrite, plane polarized light. Tephrite has both plagioclase feldspar and feldspathoids. This leucite nepheline tephrite has small laths of plagioclase, leucite showing the trapezohedral outline with octogonal sections, and nepheline in the form of small hexagonal or rectangular crystals. This tephrite is very porous, with some glass (noncrystalline) component. There are also opaque iron oxides and accessory pyroxenes and hornblende visible.

(Williams-Thorpe 1988). The leucite crystals and the cavities they left behind when ripped out of the rock through grinding operations provided good cutting and grinding edges for milling work. Thin sections show the presence of phenocrysts of pyroxenes, iron-titanium oxides, and a matrix of feldspars and iron-titanium oxides. Nepheline tephrites (with very green pyroxene and nepheline phenocrysts) were also identified by Williams-Thorpe as being used as Roman millstones.

Building material used in the construction of the 700-year-old Orvieto Cathedral in central Italy is called "basaltina," but was actually described as a leucite-bearing rock of phonolite/tephrite composition (Poli et al. 1992). Samples were taken as part of a study to identify the source of the stone, and to check the building stone against the source stone to characterize and measure the deterioration that had clearly occurred during the 700-year history of the cathedral. Petrographic studies were used to divide the stone into a fine-grained type and a coarse-grained type, referring to the size of leucite phenocrysts. Illustrations of thin sections also showed zoned clinopyroxenes and feldspars (both plagioclase and alkali feldspars). Two different quarry sites were identified as having been used for the construction of the cathedral.

However, typical deterioration mechanisms did not appear to be causing the decay problems visible with the cathedral stone; the study showed no major mineral alterations or salt crystallization. Instead, Poli et al. hypothesized that the deteriorated areas may be receiving thermal stress due to their greater exposure to sunlight than other areas of the cathedral; a contributing factor may be mechanical stress due to construction methods and load distribution.

Thin sections were used to study weathering mechanisms of tephrites used in the construction of a Romanesque cathedral built during the twelfth–fifteenth centuries in southwest Germany (Mausfield and Grassengger 1994). The material was identified as tephrite and leucite tephrite with pyroxene (augite) and plagioclase phenocrysts, along with crystals of leucite and magnetite/ilmenite in a glassy or microcrystalline matrix. Some areas of the cathedral showed visible damage, with cracking and spalling of layers of material. Since the fine-grained ash weathers more rapidly than the coarser lapilli and bombs, surface relief is intensified. Mausfield and Grassengger illustrated thin sections that show weathering of the pyroxene (augite) in the form of corrosion, cracking, and development of iron-rich halos. They also showed details of weathering effects on the glass content, which displays cracking, corrosion pitting, and development of a microporosity. These weathering effects were all much more visible near the exposed surface of the stone. Microcrystalline areas created a microporosity between crystals, which was often connected with a brown iron-hydroxide-rich crust. The plagioclase feldspar, in contrast, was not strongly affected by weathering in this particular situation.

3

Plutonic Igneous Stone Materials

Plutonic igneous rocks are formed from molten material (magma), solidifying deep within the earth. Because they solidify slowly, they tend to be relatively coarse grained, often with a phaneritic texture (where individual crystals can be distinguished with the naked eye). Quartz is often very abundant, along with feldspar. The total percentage of silicates is one way of classifying plutonic rocks: acid if more than 63%, intermediate if 52–63%, basic if 45–52%, and ultrabasic if less than 45%. The minerals commonly found in plutonic rocks are also grouped by color. Felsic minerals are light in color, and include quartz, feldspars, and feldspathoids. Mafic minerals are dark in color and include olivines, pyroxenes, amphiboles, biotite, and iron oxides. Typical minor accessory minerals often include zircon, sphene, apatite, muscovite, carbonates, and iron oxides (Williams et al. 1982; Le Maitre 1989).

Among the plutonic igneous rock types, those most frequently used for cultural materials include granite, tonalite, granodiorite, diorite, gabbro, syenite, monzonite, and anorthosite.

3.1 Granite

Granite is an acid rock that consists mainly of quartz, alkali feldspar, and plagioclase feldspar (with more than 35% of total feldspars being alkali feldspar, typically orthoclase or microcline), and often micas and/or amphiboles. Granites can include rocks with alkali feldspar being the predominant feldspar, and also rocks with approximately equal plagioclase and alkali feldspar; the latter are sometimes called quartz monzonites.

Most granites have 20–40% quartz, which tends to be anhedral (occupying irregular spaces between other constituents, rather than forming phenocrysts) (Fig. 3.1.1). The quartz grains often have many dusty and acicular inclusions (such as rutile or apatite) (MacKenzie and Guilford 1981; MacKenzie et al. 1982; Williams et al. 1982; Le Maitre 1989).

Typical of granites is that the alkali feldspar often has a considerable amount of sodium, so exsolution of the sodic phase during the slow cooling period frequently leads to the formation of a microperthitic texture. In this process, the originally homogenous mixture separates into two solid feldspar phases, one rich in soda (plagioclase), the other in potash (alkali feldspar). A typical appearance is of narrow plagioclase lamellae forming a tightly braided pattern on an alkali feldspar host (Fig. 3.1.2) (Kerr 1977; Shelley 1993; MacKenzie et al. 1997).

The most common mafic mineral in granite is a dark brown biotite, often altering to chlorite and/or granular sphene (Fig. 3.1.3), and with inclusions of smaller minerals. For example, zircon grains with pleochroic halos are frequent inclusions in biotite (Fig. 3.1.4). A typical amphibole in

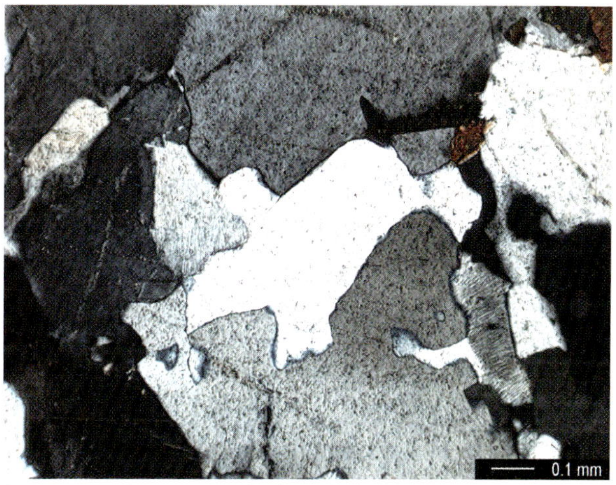

Figure 3.1.1 Granite, crossed polarized light. Most granites have 20-40% quartz, which, as seen here, tends to be anhedral (occupying irregular spaces between other constituents rather than forming phenocrysts). The quartz is interstitial to feldspars, and some flakes of biotite mica are also visible.

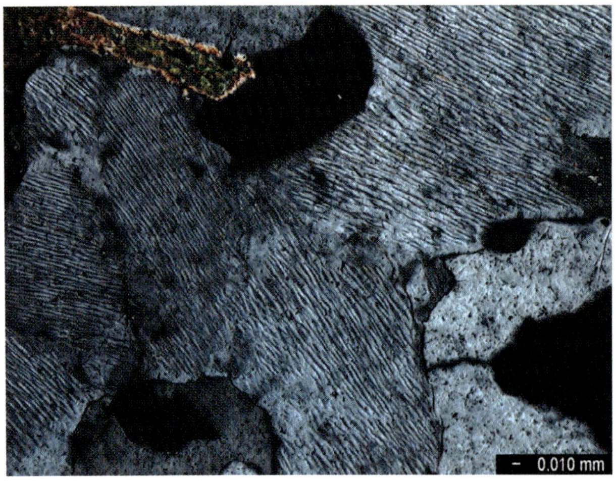

Figure 3.1.2 Granite, crossed polarized light. The feldspars show microperthitic texture, typical of granites. In granite, the alkali feldspars often have a considerable amount of sodium, so exsolution of the sodic phase during the slow cooling period often leads to formation of this texture. During this process, the originally homogeneous mixture separates into two solid feldspar phases, one rich in soda (plagioclase) the other in potash (alkali).

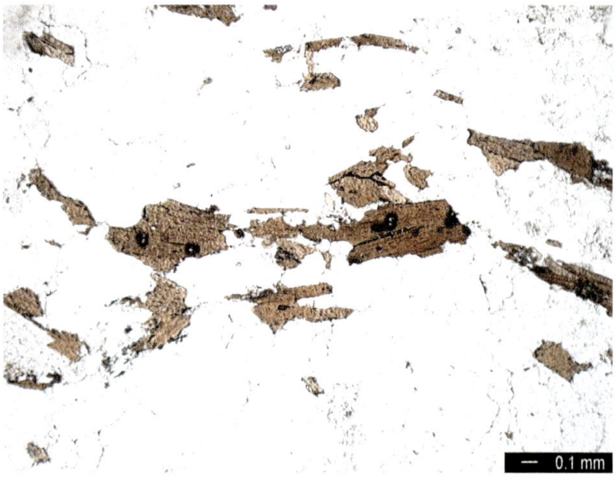

Figure 3.1.4 Granite, plane polarized light. Zircon grains are common inclusions in biotite, and are often accompanied by a feature called "pleochroic halos" that appear like dark burned-out areas around the small colorless zircon grains.

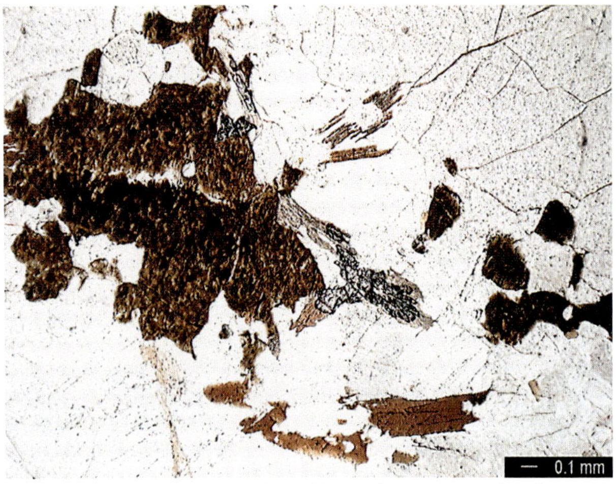

Figure 3.1.3 Granite, plane polarized light. The most typical mafic mineral in granite is a dark brown biotite, often altering to chlorite and/or granular sphene. Some unaltered tabular biotite is seen in the lower area of the slide. Most of the biotite in the upper left has been altered to a flaky chlorite, while in the center, some colorless but very high relief sphene is taking over a biotite flake.

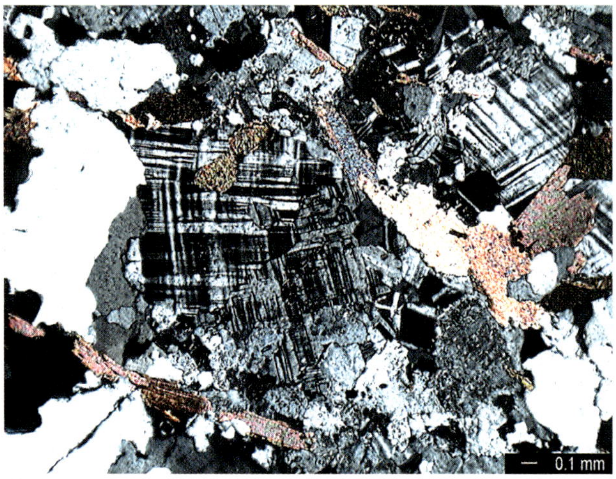

Figure 3.1.5 Granite, crossed polarized light. The microcline feldspar grains exhibit the typical "Scotch plaid" twinning of that mineral. The microcline is surrounded by quartz, that, like the microcline, exhibits low birefringence, and by highly birefringent biotite flakes.

granite is green hornblende, in the form of small prisms. A variety of other amphiboles may be found in alkali-rich granites, which may also have some pyroxenes present. Most granites have accessory minerals that often include apatite, sphene, zircon, and opaque iron oxides (Williams et al. 1982).

3.1.1 Granite as a cultural material

As a cultural material, granite has been used for a variety of purposes, especially structural and decorative. It is hard and durable, with exceptionally good weathering properties, which makes it very useful as a building material. For example, pink and gray granites from Aswan were important building materials in ancient Egypt (Clarke and Engelbach 1990); it was also identified as a material used in the construction of some of the more than 300 ancient temples at the south Indian site of Kancheepuram, near Madras (Singh 1993). Granite was used for many other monumental sculptures and temples in India, and thin sections have helped to characterize the material in detail and study its condition (Lal 1985).

For example, Lal (1985) reported that the tenth-century monumental statue of Gomateshwara in south India is a granite composed of quartz and abundant microcline feldspar (Fig. 3.1.5), with some plagioclase feldspar and biotite mica, along with apatite and magnetite as accessory minerals. The quartz and feldspars are clear and unaltered; the biotite shows only a small amount of alteration to a reddish-brown limonite. Exposed to sunlight most of the time, the rock surface does not remain wet for long periods of time, which Lal suggested helps to reduce the chemical alteration of mineral constituents.

Granite was often used as a structural material in Sri Lanka (formerly Ceylon) due to its aesthetic appearance, hardness, compression strength, and resistance to weathering. It was used, for example, for pillars and columns, as well as for sculptures, construction of the royal throne at Sigiriya, and other types of monuments and structures throughout the island (Kularatnam 1979). Many of the Sri Lanka granites are a type called charnokite, a hypersthene granite.

Granite is prized for its ornamental and decorative qualities; the relatively large feldspar grains can impart a white, pink, or red color, and the large quartz crystals and micas can produce a shiny quality, enhanced by polishing. It has been used extensively as an architectural and decorative facade material, as a veneer, as an engineering material for bridges and docks, and for street paving and flooring, columns and monumental sculpture, smaller scale sculptures, tombstones and sarcophagi, and bowls and vases (Merrill 1891; Henson 1953; Kempe and Harvey 1983; Stanier 1985; Asher 1987; Dodge 1988; Bell 1990; Penny 1993; Miller et al. 2006).

Shipley and Graham (1987) conducted petrographic studies in thin section of granite outcroppings and granites used for ground stone tools (such as manos, metates, and other implements) from Maya sites in Guatemala and Belize. Their goal was to determine whether it would be possible to pinpoint granite sources used for the tools, in order to reconstruct past trade networks. Some tools were made of a microcline-perthite coarse-grained granite. Its source was fairly extensive, so locating an exact site that supplied it was not possible, but the authors were able to define a larger trade network region. Photomicrographs of this material showed coarse to very coarse grains, with microcline, quartz, plagioclase, and minor biotite and muscovite micas. Some grain boundaries had what is called cataclastization, referring to a texture in which crushing and fragmentation of crystals has occurred as a result of post-consolidation movement.

Galetti et al. (1992) described plutonic rocks used in antiquity by the Egyptians and Romans for the production of columns, pillars, tubs, and vases. Twelve rock types were studied in thin section, and point counting provided data for modal analysis. These data were augmented by a program of elemental analysis to look at chemical composition patterns of the different rock types. Five of the rocks studied were granites.

Mahmoud and Bard (1993) studied thin sections of Predynastic grinding stones from Egypt. One group of grinding stones made of granite consisted of euhedral to subhedral and prismatic grains of orthoclase feldspar, altered microcline and plagioclase feldspars, quartz, biotite, and augite pyroxene. Based on the thin-section work, Mahmoud and Bard were also able to propose possible sources for the grinding stone material, most likely located as far as 150 km away from the grinding stone findspots, indicating the presence of a widespread exchange network.

Brown and Harrell (1995) provided detailed thin-section descriptions of two granites from ancient Roman quarries in the Eastern desert of Egypt. One (from Wadi Abu Maamel) is pink and medium to coarse grained, with primary minerals including alkali feldspar (orthoclase microperthite), quartz, plagioclase feldspar (oligoclase), and biotite. Accessory minerals include apatite, magnetite, sphene, and zircon, with some hornblende. The other granite, from Badia, is similar, but is speckled pink and black, and includes hornblende among the main components.

Hernández (1996) reported on a large-scale study of granite building materials that included thin-section petrography to characterize building stones and their quarries. A wide variety of granite mineralogies and textures is illustrated with photomicrographs of the thin sections. Included are microcracking and development of porosity networks in weathered granites with examples of granites from Spain, France, and Portugal.

3.1.2 Granite deterioration mechanisms

Dearman and Irfan (1978) used petrographic analysis as part of their assessment of the degree of weathering in granites. They started with hand specimens to characterize the appearance of samples with the unaided eye. Thin sections then provided a quantitative evaluation of mineralogical and textural changes produced by weathering processes, with a focus on changes that can affect the physical and mechanical properties of the granite. Properties analyzed included mineral types present (both primary minerals and secondary alteration products); texture (grain size and grain boundary relationships, size, location and characteristics of voids and microcracks); and structure. Quantitative data were obtained by point counting, and microfracture analysis was performed by counting the number of clean and stained microcracks present. Changes occurring at seven stages of weathering were measured and then correlated with a range of physical index tests to see how physical parameters were affected by the changes brought about by weathering.

Examples of weathering processes tracked by Dearman and Irfan included a change from Stage I, fresh granite (mineral constituents fresh and rock fabric unfractured), to Stage Iiiii, partially stained granite (plagioclases partially gritty, biotite slightly decomposed, partial staining along microcracks), to Stage III–IV, weakened granite (slight grittiness of potash feldspars, quartz microfractured, opening up of grain boundaries, and intensely microfractured throughout).

Ordaz et al. (1978) also used thin-section petrography to chart the deterioration state of granites and to correlate that state with physical and mechanical properties. Using four architectural and monumental stones from northwest Spain, two of them granites, they developed a deterioration index through illustrated thin-section studies of mineralogical and textural characteristics then compared that with properties such as porosity and compressive strength.

One of the stones studied was a pink granite. The primary minerals were quartz and orthoclase feldspar, with lesser amounts of microcline, plagioclase, and biotite. Accessories were apatite, zircon, titanite, and allanite. Secondary alteration products included chlorite, sericite, and, within biotite and chlorite, magnetite (Fig. 3.1.6). The quartz grains had undulous and mosaic extinction, and microcracking. The orthoclase had perthitic texture, but was relatively unaltered. The plagioclase altered to sericite, and had some myrmekitic intergrowths (rod-like inclusions of quartz within a plagioclase grain) (Fig. 3.1.7). The biotite is described as having bent cleavages and undulous extinction, as well as altering to chlorite. Microcracks were easily visible in thin section.

The second granite type was variegated in color, with white, gray, pink, and black areas. Feldspars (orthoclase and plagioclase) were more abundant than quartz; biotite was also a major mineral. Accessories included zircon (the most abundant), apatite, titanite, and allanite. Alteration products included chlorite, sericite, magnetite, and epidote. The quartz grains had many inclusions; orthoclase was perthitic. Plagioclase grains were extensively altered to sericite, with zonal extinction. Biotite, in contrast, was little altered. Although there were fewer microfractures than with the pink granite, as before these were most often concentrated within the quartz grains (Figs. 3.1.8, 3.1.9, and 3.1.10).

An index of deterioration was developed for quartz, feldspars, and mafics. Individual grains within a thin section were then assessed for their "grade" (ranging from 0 for unaltered to 4 for extremely altered). Point counting for each grain type within the thin section, to obtain a statistically representative sample, allowed an overall score to be developed to describe the average degradation state (based on the degree of deterioration for each mineral type taking into account its abundance within the rock).

Rodriguez-Rey and Montoto (1978) performed deterioration studies on a wide range of plutonic rocks. Fourteen rocks used for ornamental purposes in Spain were selected, studied in thin section and then exposed to heating and cooling cycles. Weight loss was used to derive a weathering index (since weathering creates mechanical microdisintegration accompanied by a weight loss), which was then corre-

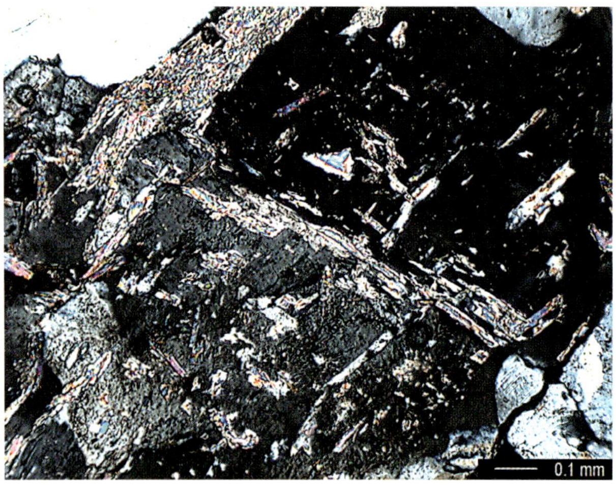

Figure 3.1.6 Granite, crossed polarized light. Secondary alteration products have formed on this feldspar grain. The abundant high birefringent scaly flakes are sericite, the minutely crystalline variety of muscovite mica. The alteration products clearly stand out in contrast to the low birefringent feldspar.

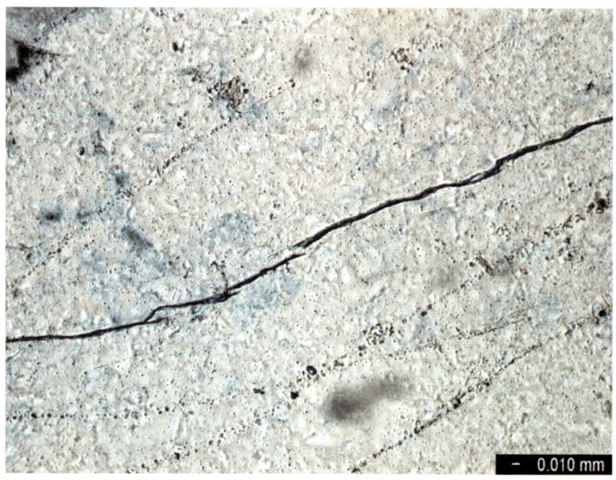

Figure 3.1.8 Granite, crossed polarized light. A microfracture within a quartz grain can be documented as one measure of deterioration of the stone.

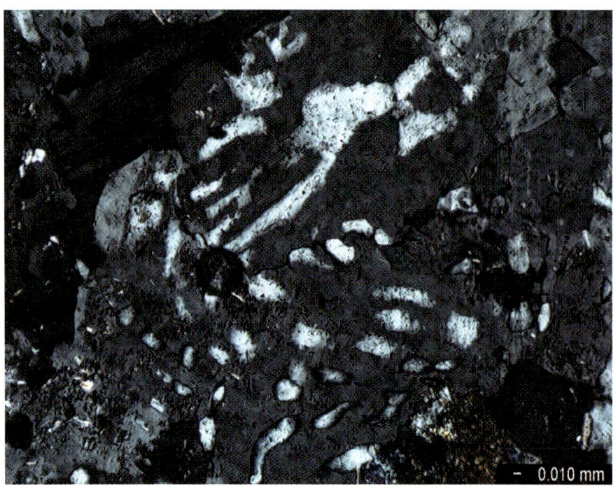

Figure 3.1.7 Granite, crossed polarized light. The plagioclase feldspar has myrmekitic intergrowths, which are rod-like inclusions of quartz within a plagioclase grain.

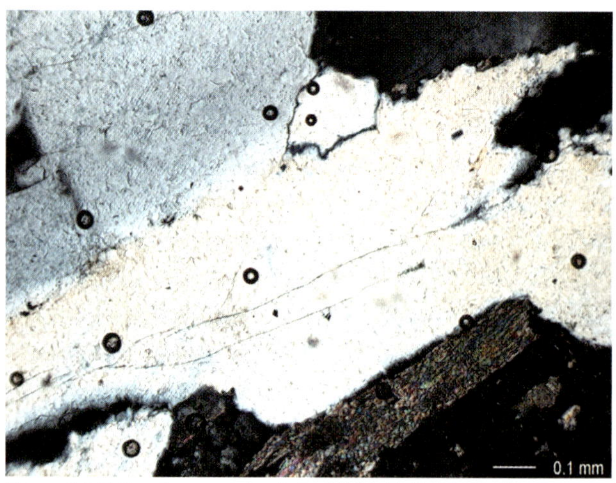

Figure 3.1.9 Granite, crossed polarized light. Multiple microfractures are found within a single quartz grain, extending for the full length of the grain. The cracks are relatively wide and are partially filled in with secondary interstitial material.

lated to tension strength. They concluded that weathering behavior is dependent on the acid or basic character of the rock. Among the 14 rocks studied, five were granites. One is described as a pearl gray porphyritic granite and three as biotite granites. One is adamellite, a type of granite in which plagioclase and alkali feldspars are present in about equal amounts (each between 35 and 65% of total feldspars). They thus have more alkali feldspars than granodiorites, and more quartz than monzonites, so are closer to granites in composition. The mafics are usually biotite and hornblende (Williams et al. 1982; Le Maitre 1989).

Kelly (1985) studied deterioration of granite building materials in Ireland, finding a tendency towards scaling and exfoliation. This was ascribed to the fact that the granites examined all had some degree of banding and foliation; effects of temperature and moisture on mechanical properties of minerals in such a foliated rock lead to the deterioration effects observed. Although there was variation, much of the stone was a light gray muscovite granite with microcline, orthoclase, and plagioclase feldspars. There was also some biotite, and accessory minerals of zircon, sphene, apatite, rutile, garnet, beryl, fluorspar, cassiterite, and tourmaline. This stone was used in the construction of several major castles, a friary, and a college. Illustrated thin sections showed features such as the fracture system in microcline, which extends in a pattern parallel to the outer surface, and the concurrence of the dominant orientation of micas and the fracture system.

Thin sections were also useful in the study of deterioration of granite used in the building of Toledo Cathedral in Spain (Lopez Azcona et al. 1991). Degradation of the stone was mainly due to climatic factors (humidity, thermal change, and frost) rather than to air pollution. Deterioration results seen in thin section included biotite altering to chlorite, and altered plagioclase feldspar. In places, deterioration is so severe that the granite has altered to sand.

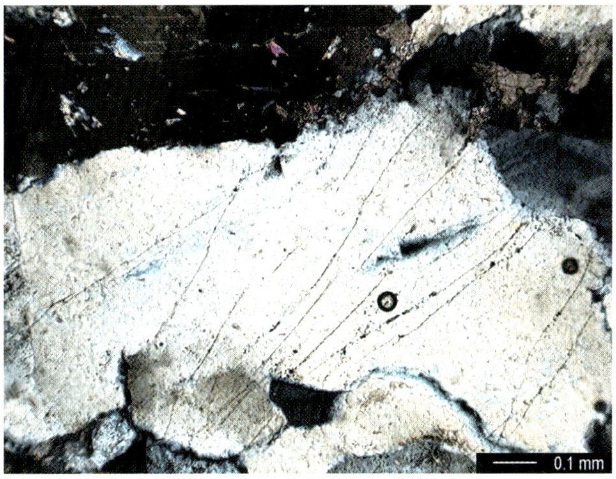

Figure 3.1.10 Granite, crossed polarized light. There is even more extensive cracking in this quartz grain, with multiple microfractures found throughout the grain, most extending for the full length of the grain and some beginning to be filled in with secondary material.

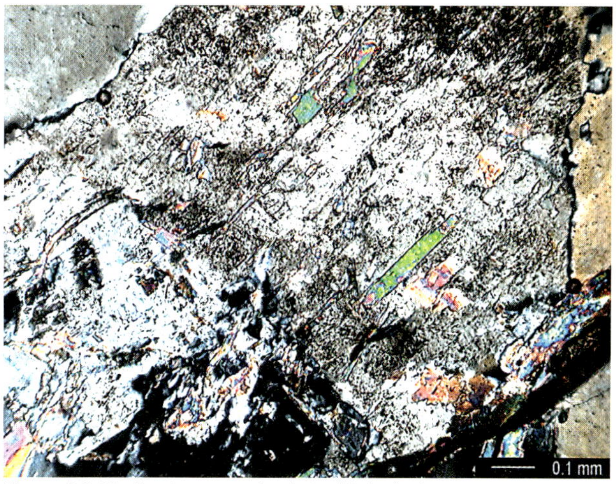

Figure 3.1.12 Granite, crossed polarized light. In this granite specimen, alteration products include stringy platelets of sericite within feldspars, forming especially along cleavage planes.

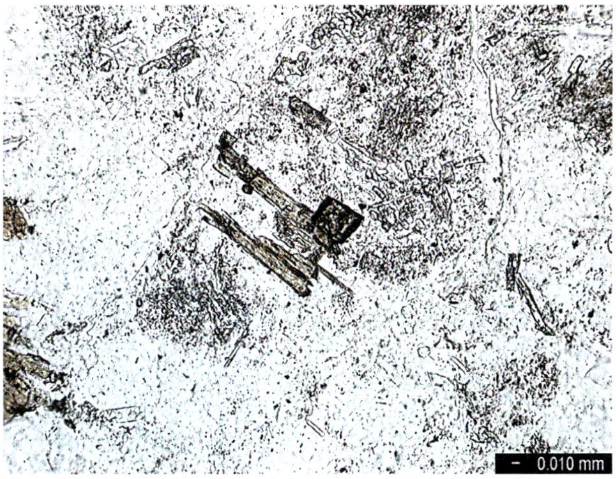

Figure 3.1.11 Granite, plane polarized light. Sphene shows high relief and a typical acute rhomic cross-section. Here it is set within a background of colorless, low-relief quartz, colorless but altering feldspars with a "grungy" appearance, and brown flakes of biotite mica.

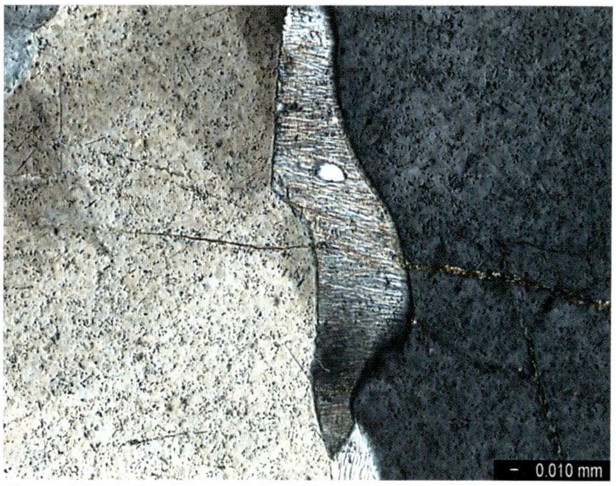

Figure 3.1.13 Granite, crossed polarized light. An intergranular fracture partially filled with alteration products is visible, starting from the quartz grain on the right and extending into the feldspar grain in the center and on into the quartz grain on the left.

Rainey and Whalley (1994) studied the role of microfractures in the decay of granite building stones from southern California (Alabama Hills, Owens Valley). Thin sections were used to characterize the stone. Two samples were of biotite granite; one sample had about 65% plagioclase feldspar and thus is classified as a granodiorite. All three samples have quartz, biotite, feldspars (alkali and plagioclase), and sphene (Fig. 3.1.11). Alteration products included sericite forming along cleavage planes of the feldspars (Fig. 3.1.12), green chlorite on the biotite, and epidote. Only a few microfractures were visible in the thin sections, mainly within quartz crystals. Some sealed intergranular fractures with alteration products were also visible (Fig. 3.1.13). Rainey and Whalley developed an index for quantitative classification of the microfractures, with seven parameters: total length of microfracturing, average length, average number, density, connectivity, width, and percentage inter- versus intragrain fractures.

Microfissuring development due to salt crystallization was examined on a granite from northwest Spain in a study by Pérez-Ortiz et al. (1994). Although backscattered scanning electron microscopy (SEM) was the primary tool used (in conjunction with digital image analysis), polarized light microscopy of the thin sections of the granite was also crucial. These thin sections clearly show cracks that appear in quartz grains, as well as the accompanying alteration of plagioclase feldspar grains to sericite and chlorite.

Fiora et al. (1996) used thin sections in studies of deterioration of granites from historic buildings in Turin, Italy. The authors provided detailed descriptions of the mineralogy and texture of the different types of granites used. A pink granite, colored by very small inclusions of hematite, contained alkali feldspar that was beginning to alter to clay, quartz, and plagioclase feldspar, and biotite that was altering to chlorite. Accessories were fluorite, monazite, zircon, and aggregates of pyrite and magnetite. A white granite consisted of plagioclase (zoned and altered) and alkali feldspars, quartz, and biotite that was altering to chlorite, with accessory zircon, monazite, apatite, and opaque iron oxides.

Some of the deterioration evidence is best seen in the thin sections, for example, microfractures that form a network of intergranular and intragranular cracks developing just below a black patina. Thin sections showed that the black layer, usually less than 1 mm thick, included opaque microspheres and grains of minerals such as phyllosilicates and quartz. Fiora et al. hypothesized that this layer was formed by dust and smog settling on the granite surface. Other thin sections illustrated how biotite grains under the black layer have started altering to chlorite, causing swelling along the mica cleavage planes; intergranular and intragranular fractures were also clearly visible under the black layer. Mineral alteration and microcracks were clearly worse in areas where the black layer had formed.

Thin-section petrography was one of a suite of techniques employed in a study of the paleoweathering profiles by depth of granites used in the construction and restoration of Avila Cathedral in Spain (Hernández 1996; Molina et al. 1996). Three weathering levels were identified. A lower gray level consisted of anhedral quartz, microcline, slightly altered plagioclase, biotite partially altered to epidote, and chlorite; accessories were apatite, zircon, rutile, and opaque iron oxides. Some areas of this level of the granite were coarse grained, while others were fine grained. A middle level had undergone deep weathering, resulting in changes in color and structure; resistant minerals such as quartz were found in a matrix of clays and areas of opal (cryptocrystalline silica). The upper level had strong color variations from red to ochre yellow to white. Here, opal was more abundant, along with fibrous silica in cavities; both were formed by leaching of silica from above. All three levels have been quarried for building materials in the past, but because of their differing mineralogies, textures and degrees of weathering, they vary greatly in physical and mechanical properties.

Esbert et al. (1996) included thin-section examinations in their research into the use of a water repellent to protect ornamental granite from the humid and saline atmosphere of a seaside location. The granite under study (from quarries near Madrid, Spain) was carefully characterized mineralogically. Many of the feldspars were altering to clay minerals, and an abundance of microcavities created a material with high open porosity, contributing to the need to find ways of protecting the granite from excess humidity and the aggressive saline environment.

Thin sections were used in a study of granite deterioration caused by lichens at a castle in Spain (Ariño et al. 1996). At this site, the granite blocks of which the castle is constructed were darkened by a dense covering of lichens, which in turn initiated mechanical rupture. The interface between the lichens and the rock was examined with optical and scanning electron microscopy. The thin sections clearly showed the interface area, with fissures where the lichen is penetrating into the rock, and disintegration occurring in those areas. In addition to many fractures at this interface area, feldspars were undergoing alteration, and some of the fissures were filled with a carbonate material. Lichen penetration goes as deep as 5 mm into the rock surface. As existing pores and figures were filled, surrounded mineral fragments were eventually pulled out of the rock matrix, leaving new cavities to be colonized by the lichens.

Rodrigues (1996) also used thin sections as part of a study of lichens and their effects on granite. The thin sections were useful in showing the mechanisms by which the lichens attack and penetrate into the granite, the depth of penetration (as much as 45 mm), and the ways in which different minerals are affected (biotite and feldspar crystals penetrated much more than quartz crystals). Thin sections also showed the development of secondary alteration minerals (such as clays) on the biotite and feldspars that accompany the fracturing of those grains, and they facilitated examination of surface patinas; Rodrigues illustrated an example showing a gypsum-rich patina with the granite substrate beneath it severely cracking and splitting, with minerals from the substrate pulled out and deposited into the patina layer.

Sterflinger et al. (1996) included thin sections in studies of biodeterioration of granite. They focused on three granite monuments in the Mediterranean area (Spain and Greece), looking at the relationship of environmental factors and biodeterioration processes. They examined visual appearance, mineralogy, rock fabric, chemistry, and biological activity on and below the rock surface. Thin sections revealed that in one case fungi had penetrated into the granite rather than being just on the surface, and their cells could be seen several millimeters under the surface. It appeared that fungi used the weaker places of the granite for this attack (the micas and feldspars which could easily be leached or even removed mechanically leaving empty spaces that could be occupied by a fungal colony). Cracks between micas and quartz grains could also be penetrated more easily. The consequences appeared to be disintegration of minerals sometimes leading to larger areas of exfoliation, and changing porosity and permeability characteristics.

Sterflinger et al. also used thin sections to study surface stromatolite patinas that developed on these granites. The biodeterioration process involves mineral precipitation as a result of microbial biogeochemistry, and the biofilms are also able to trap and consolidate mineral particles or other substances. The patinas in their examples consisted of calcite, weddellite and/or whewellite, and apatite, with an uppermost layer of gypsum. They illustrated two forms in which these patinas were found: a compact micritic layer with only some remnants of lamination, and a stromatolitic form with the shape closely following irregularities of the rock surface and with distinguishable layers.

3.1.3 Granite conservation treatments

Conservation treatments of granite were the focus of a study by O'Brien and Cooper (1994). Two commonly used methods for cleaning granite were compared: hydrofluoric (HF) acid and grit blasting. Thin sections examined before and after cleaning revealed that quartz was fractured by both conservation treatments. Gravimetric measurements were also used to measure loss of material, and runoff was collected and analyzed. The grit-blasted stone suffered as much as three times more material loss as occurred with the HF treatment.

Thin sections were taken from the surface of uncleaned stone, from the cleaned stone, and from fresh unexposed areas deeper within the stone. The rock is medium grained, with an orange discoloration extending down as far as 6 cm. The major minerals were found to be quartz, microcline, and plagioclase feldspar, each consisting of about 30% of the rock. Minor constituents were biotite and muscovite, with accessories of rutile and zircon. In the section taken from deeper into the stone (greater than 6 cm), natural feldspar weathering was seen, along with fracturing of quartz grains; hematite was also present. On the uncleaned surface, fractures were covered with iron oxides, while feldspars were completely clouded with alteration products, broken down, and extremely friable.

After HF cleaning, the first 4 mm showed greatly increased fracturing, with significant increases as far down as 15 mm. In the grit-blasted samples, the first 10 mm showed the greatest increase in fracturing. Photomicrographs of thin sections were digitized for the calculation of fracture statistics to allow for a quantitative analysis. Image analysis was based on the fracture length per unit area of quartz. Although the fracture lengths were somewhat lower for the HF than for the grit blasting, both cleaning methods caused an increase in fractures of a factor of three or four compared to the uncleaned surface and the stone interior.

A similar approach was taken in studies by Pavía Santamaría et al. (1996a, 1996b), where polarized light microscopy of thin sections was instrumental in assessing the effects of cleaning regimes on granite. The granite studied is described as medium to coarse grained, composed of quartz, microcline and plagioclase feldspars, and micas, and is affected by various deterioration mechanisms. Thin sections taken before and after treatments were prepared without water in order to preserve any water-soluble salts that might be present. The cleaning methods were assessed by examining their effects on mineralogy (such as alteration of feldspars to sericite and of muscovite to chlorite, and of formation of calcite, gypsum, or anhydrite crystals) and on degree of fracturing (both intergranular and intragranular); an average alteration score for each cleaning method was calculated. The authors illustrated thin sections that show features such as fractures filled with gypsum flakes within feldspar grains, and gypsum flakes initiating breaking of biotite grains.

3.2 Granodiorite

Granodiorite (Fig. 3.2.1) is also an acid rock that consists primarily of quartz, plagioclase feldspar, and lesser amounts of alkali feldspar (10–35% of total feldspars) with minor amounts of hornblende and biotite (often about equal). The plagioclase grains are usually either euhedral or subhedral rather than anhedral; some may be partially or wholly enclosed by orthoclase, which rarely forms phenocrysts and is sometimes perthitic (but not as often as in granites) (Williams et al. 1982).

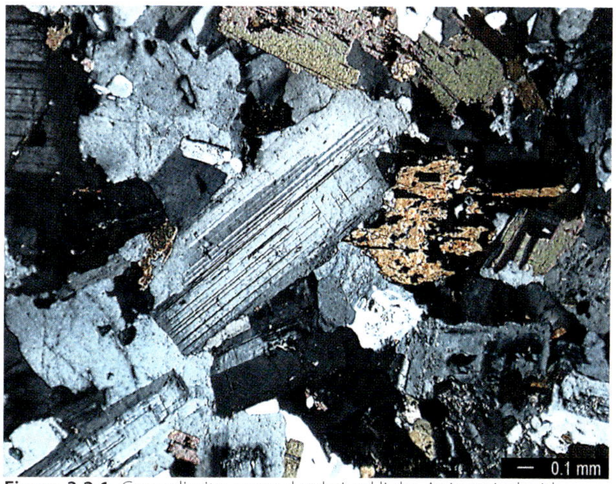

Figure 3.2.1 Granodiorite, crossed polarized light. As is typical with granodiorite, feldspars are mainly subhedral or euhedral plagioclase, surrounded by quartz and minor (but usually equal amounts) of biotite (here ranging from mid to very high interference colors) and hornblende (of lower birefringence, a yellowish orange here).

3.2.1 Granodiorite as a cultural material

Granodiorites have been used for much the same types of building materials and objects as granites. For example, Galetti et al. (1992) described plutonic rocks used in antiquity by the Egyptians and Romans for the production of columns, pillars, tubs, and vases. Twelve rock types were studied in thin section with point counting for modal analysis, augmented by elemental analysis. Two of the rock types studied were identified as granodiorite. In addition to architectural uses, granodiorite is found as a sculptural material, often called "black granite" by art historians and archaeologists (Penny 1993).

A granodiorite known as *granito del foro* was used in buildings of the Roman Forum. Peacock et al. (1994) studied this rock in detail: they characterized it, identified the source, determined where it was and was not used, and speculated about the cultural interpretations of its use patterns. They found that thin sections from some specimens of the stone show strongly zoned plagioclase feldspar, yet of the 130 quarries known for the rock, none had similar zoning, or had only weakly developed zoning. This led the authors to study gray plutonic rocks to identify distinguishing criteria, and to try to define sources. They included outcrops that might have been ancient sources used by the Romans as well as archaeological artifacts; methods included visual comparison, thin-section petrography, and elemental analysis.

Thin sections showed that the Mons Claudianus source in the Eastern desert of Egypt had little zoning of plagioclase feldspars, whereas other sources did have the zoning. Some sources were also distinguished by a bluish-green color of the amphiboles, the presence of epidote, and a distinctive reddish color of the biotites. Accessory minerals were not very characteristic, as the same ones were found widely distributed (apatite, zircon, and sphene). The authors concluded that while thin-section mineralogy gives some clues to artifact source, it is unlikely to be completely diagnostic. Part of the problem is the difficulty of taking large enough samples from objects to be representative of the mineralogy of the stone as a whole, given the coarse grain size and variation within the rock. The thin sections are thus complemented with electron microprobe analyses of individual grains of plagioclase feldspar, biotite, and amphibole. The combination proved to be very successful for distinguishing between granodiorite sources.

Brown and Harrell (1995) also characterized a granodiorite from the Eastern desert sampled from an ancient Roman quarry site at Bir Umm Fawakhir. They described it as mottled pink and gray in hand sample, and medium to coarse grained. The primary minerals were plagioclase feldspar (oligoclase), quartz, alkali feldspar (microcline with some microperthite), and biotite. Accessory minerals included apatite, magnetite, muscovite, sphene, and zircon, with a little hornblende.

3.2.2 Granodiorite deterioration mechanisms

In the study by Ordaz et al. (1978) mentioned above, they also used thin-section petrography to characterize the deterioration state of rock samples and to correlate that state with physical and mechanical properties. Focusing on two granodiorites from the four architectural and monumental stones from northwest Spain, they used the deterioration index they had developed for a comparison of properties such as porosity and compressive strength.

One of their granodiorites was described as being a gray color dotting a white background. The major minerals present were quartz, feldspars (plagioclase, orthoclase, and microcline), and muscovite mica. Accessory minerals were apatite (the most abundant), biotite, and zircon; the main secondary alteration product was sericite. The quartz component was distinctive in that it had very sutured grain boundaries (Fig. 3.2.2); mosaic extinction patterns were also common. The orthoclase was sometimes perthitic. The plagioclase was oligoclase, in prismatic crystals, with large sericite crystals within it. The muscovite was often clumped or in fibrous, radiating aggregates. Microcracks were moderately abundant, most often occurring in quartz, but sometimes also in feldspars. Cracks only rarely extended beyond grain boundaries. Micas had bent cleavages and undulous extinction.

The second granodiorite studied by the authors was described as gray in color. Primary minerals were quartz (uniformly distributed), feldspars (plagioclase, orthoclase, and microcline), biotite (in aggregates), and muscovite. Accessory minerals included zircon, apatite, and rutile. Secondary alteration products were sericite, chlorite, and epidote. The quartz grains often had inclusions, anomalous extinction, and sutured grain boundaries. Orthoclase grains tended to be microclinized; plagioclase had heavily altered to sericite (sometimes more than half of the grain) and showed zoned extinction. The biotite was tabular, and riddled with zircon pleochroic halos, along with being altered to chlorite and epidote. The less abundant muscovite sometimes formed feathered aggregates. There were few microcracks in this rock, but where present, they were found in the quartz and feldspar grains.

In the study by Rodriguez-Rey and Montoto (1978) (see above) on 14 rocks used for ornamental purposes in Spain, two were granodiorites, the only two rocks that were not extremely coarse grained.

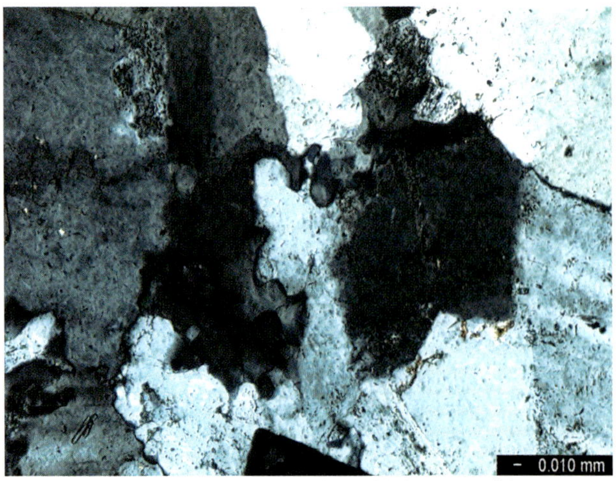

Figure 3.2.2 Granodiorite, crossed polarized light. Some of the quartz component exhibits sutured grain boundaries versus the straighter quartz boundaries more often seen in igneous rocks.

Christaras (1990) focused on weathering and resulting mechanical behavior changes of plutonic rocks of northern Greece. Included in the study were three granodiorite deposits. The first was coarse grained with a grayish-black color due to abundant biotite. Many small cracks were visible under the microscope, along with alteration of feldspars to kaolinite and sericite. The biotites were sometimes partially altered to chlorite. The second granodiorite was similar, but the minerals showed strain effects such as bending and twinning. The third example was medium grained, lighter in color, and contained hornblende.

Weathering was assessed through point counting of thin-section features under the petrographic microscope. The weathered minerals were counted (mainly feldspars and micas), as were microfissures. A weathering index developed by Christaras was then calculated as a percentage of altered minerals plus microfissures divided by the percentage of unaltered minerals. After statistically comparing the weathering index with data from tests of mechanical and physical properties of the rocks, he concluded that even slight increases in weathering can cause significant decreases in physico-mechanical resistance.

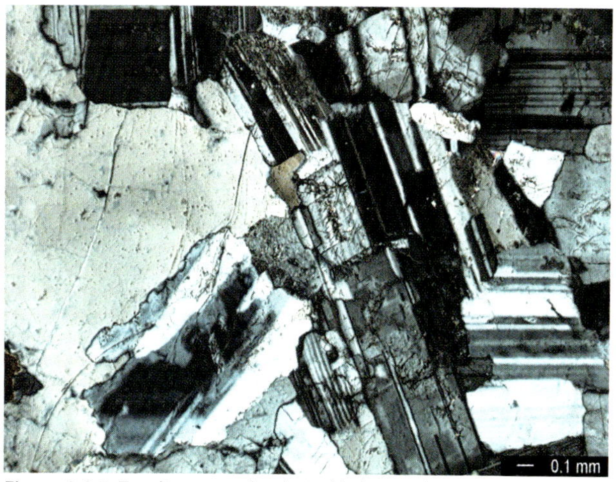

Figure 3.3.1 Tonalite, crossed polarized light. Tonalite consists mainly of plagioclase feldspar and quartz; it is similar to granodiorite except that it has less alkali feldspar and also tends to have lower levels of quartz.

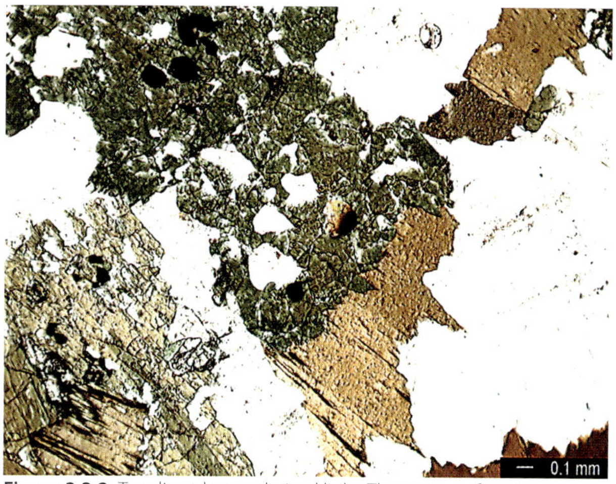

Figure 3.3.2 Tonalite, plane polarized light. The main mafic minerals that darken tonalite are hornblende and biotite. The quartz and feldspar components are colorless, the brown biotite appears as tabular plates, and the green hornblende is more broken up with the amphibole cleavage lines visible. There are also black opaque iron oxides present as accessories.

3.3 Tonalite

Tonalite is a plutonic rock consisting mainly of sodic plagioclase feldspar and quartz (Fig. 3.3.1), usually with some biotite and amphibole present as well (Fig. 3.3.2). It greatly resembles granodiorite but has much less alkali feldspar and somewhat lower levels of quartz. It tends to be a bit darker than granodiorite, with the main mafic minerals being hornblende and biotite, sometimes with augite that is rimmed with one of these other two minerals. Accessory minerals may include apatite, sphene, and opaque iron oxides (Williams et al. 1982; Le Maitre 1989).

A study of petrographic thin sections taken from sculpture materials in India showed that tonalite was used for sculpture (Asher 1987). An image fragment from a monastery at the site of Nalanda was found to be tonalite. That example was described as medium grained with plagioclase feldspar, biotite, chlorite, muscovite, and quartz.

3.4 Diorite

Diorite consists mainly of sodic plagioclase feldspar (andesine or oligoclase), as the greatly predominant or only feldspar; and often with some hornblende as the primary mafic mineral (sometimes also including biotite, and occasionally a pyroxene) (Fig. 3.4.1). Diorites can be either medium or coarse grained, and there is a variety of textures (for example, some may be porphyritic, others may have a segregation of light-colored and dark-colored minerals in irregular dots or layers). Quartz is always interstitial, and may be intergrown with orthoclase feldspars. The dominant plagioclase grains in diorites are often zoned, which can be highlighted by alteration of the grains, as some zones become altered while others remain clear (Fig. 3.4.2).

Inclusions are often found in the plagioclase grains (such as mafic minerals, apatite, and opaque iron oxides). Microcline is sometimes present, although rarely; more often the alkali feldspar is orthoclase. The perthite frequently found in granites is rarely seen in diorites, although myrmekite is sometimes present.

Hornblende is the most common mafic mineral in diorites. Biotite is often present as well, typically altering to chlorite or granular sphene. If pyroxene is found, it is usually a colorless or pale-colored augite, often with reaction rims of hornblende. Hypersthene may also be found in diorites. Accessory minerals may include apatite, zircon, sphene, or opaque iron oxides (Williams et al. 1982).

Diorite has been used for architectural and ornamental purposes, as well as for sculpture, bowls, and vases (Hankey 1976). For example, Galetti et al. (1992) described a variety

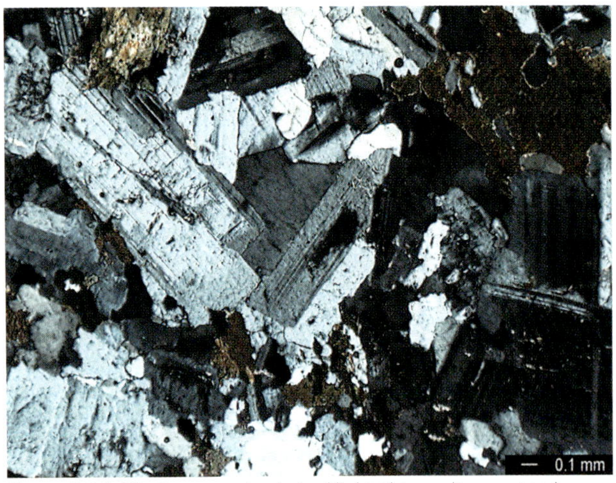

Figure 3.4.1 Diorite, crossed polarized light. This medium-grained diorite is dominated by laths of plagioclase feldspar. The quartz component is interstitial. The mafic minerals are hornblende and biotite.

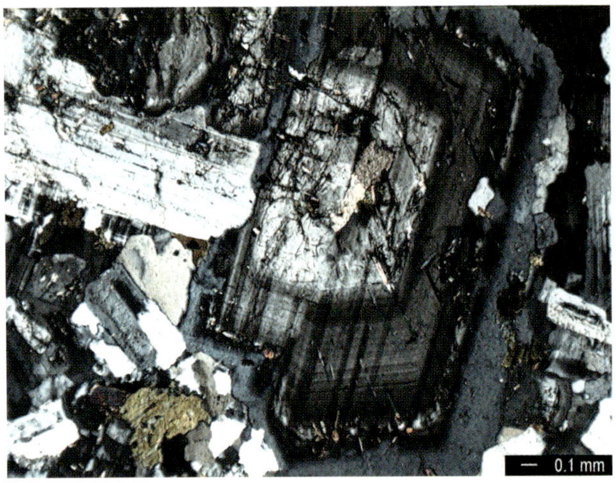

Figure 3.4.2 Diorite, crossed polarized light. The plagioclase feldspar grain in the center is zoned. The zoning appears as interference color gradations ranging from light gray in the center of the grain to dark or black along the edges. The zoning is highlighted by differential alteration of the grain. Some areas, particularly in the center of the grain, are heavily altered while the zones outside of the center are relatively clear of alterations.

of plutonic rocks used in antiquity by the Egyptians and Romans. One of the rock types they characterized is a chloritized diorite. Röder (1992) discussed a diorite source that was utilized by the Romans during the reign of Emperor Claudius, and is thus called "Claudianus granite." The stone is actually, as described by Röder, a leuco-quartz diorite. It was used for a wide variety of purposes, from the building of monumental pillars to the production of bowls, fountains, and tubs. Röder also described a similar material that was quarried by the Romans in the Odenwald region along the Rhine River north of the Alps, and which was used for similar purposes. Although called Felsberg granite, it too is actually diorite, although somewhat lighter in color than the leuco-quartz diorite.

Brown and Harrell (1995) reported on thin-section analyses of Egyptian diorites, from four ancient Roman quarry locales. Three were described as quartz diorites, ranging from fine to coarse grained, and colored white or yellowish white with dark speckles. Primary minerals included plagioclase, biotite, and quartz; and, in some locales, hornblende and/or alkali feldspar (orthoclase). Accessory minerals included apatite, magnetite, sphene, and zircon. The fourth quarry discussed was a source of diorite, rather than quartz diorite. It was described as a mottled light gray in color, and coarse to very coarse grained. The primary minerals here were plagioclase feldspar, hornblende, and magnetite. No quartz was present.

3.5 Gabbro

Gabbros are of basic composition (45–52% silicates). They consist mainly of calcic plagioclase feldspar (more than diorites), pyroxene, and iron oxides (Fig. 3.5.1). Mafic minerals (usually pyroxene, most often diopsidic augite and sometimes hypersthene) generally make up more than 40% of the rock, giving it an overall darker appearance than diorite. The pyroxenes frequently show exsolution phenomena due to slow cooling, visible as thin parallel lamellae, or the pyroxene crystals may be rimmed with hornblende (Fig. 3.5.2) or partially replaced with fibrous amphibole. Biotite, alkali feldspar (orthoclase or microcline) and quartz may be present as minor constituents; sometimes olivine or hornblende occur. Common accessory minerals include opaque iron oxides and apatite. Gabbros can vary from coarse to medium grained, with a variety of textures (Williams et al. 1982).

3.5.1 Gabbro as a cultural material

Gabbro has been used for architectural purposes, sculpture, and tools. A long-term British study of stone axes and their sources found that one group was made of gabbro, described as uralized gabbro (Shotton 1969). Uralite is a

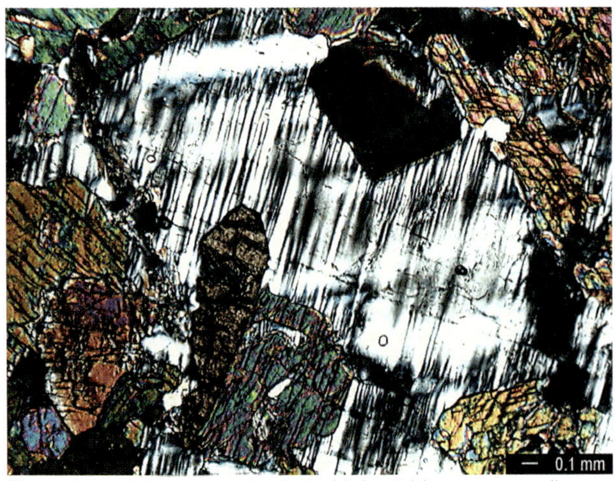

Figure 3.5.1 Gabbro, crossed polarized light. Gabbros are generally comprised of more than 40% mafic minerals in addition to plagioclase feldspar. Here the feldspar is in the center, surrounded by hornblende, orthopyroxene, and clinopyroxene. Hypersthene has rather weak birefringence up to about 0.016, whereas hornblende has moderate and clinopyroxenes have higher interference colors.

secondary alteration product of augite, which forms by hydrothermal alteration (Kerr 1977). The augite in this material was colorless to pale brown in thin section, had no clear crystal outlines, and was altering to a green, fibrous hornblende along cleavage cracks and outside edges of the grain (Wallis 1963). Wallis noted that this hornblende creates an extremely tough stone axe. The feldspars, too, were quite altered, and filled with small actinolite needles, which was diagnostic for this stone group. The main accessory mineral was ilmenite, which was sometimes altering to leucoxene (a common alteration product of titanium minerals, appearing as an opaque white substance along the borders of the titanium).

Thin-section studies of Mesolithic-Neolithic stone axes from Sweden, along with analyses of stone outcroppings, indicated that a fine-grained gabbro (dolerite) of local origin was used in the production of Swedish stone tools (Kars et al. 1991). The unusual shapes of these tools were thought to be due to properties of the material. For example, thin-section studies showed features such as a relatively coarse grain size, veining, and small cracks, which may have affected the size and shape of tools made from that material. This was an important finding because it demonstrated that the material available may have exerted more of an influence on technology and form than could be attributed to cultural factors or craft skill.

Mandal (1997) also reported on Neolithic stone axe studies, with a focus on Irish materials. Thin sections were examined, in conjunction with elemental analyses, for axes studied along with known or proposed rock sources. Mandal reported specifically on a group of 45 axes found to be made of gabbro. Two gabbro types were identified in thin section. One type had pyroxene (altered mainly along cleavage and grain edges to a green fibrous amphibole) and fibrous blue-green amphibole radiating from the rims of coarse opaque clusters, with no primary amphibole present. Opaque iron oxides included magnetite and ilmenite, with only a minority showing alteration. The second gabbro type demonstrated much more variation in mineralogy and texture, the main difference between this and the first gabbro group being the presence of primary amphibole (Fig. 3.5.3). The oxide phases were more altered; for example, remnant ilmenite oxidation exsolution lamellae were visible in sphene or void spaces. Some axes, however, were made of a gabbro that did not fit within either of the two major type categories.

Geochemical analyses confirmed the groupings identified by the thin-section analysis, and did not discern any distinct types not already clarified through the thin-section study. Comparison with existing data on British axes indicated that one of the two Irish gabbro groups was identical to a British group, and that those axes probably originated from the Cornwall region in southwest England.

Petrographic analysis of igneous deposits in Britain was the focus of another study, which aimed to compare new data with data from earlier stone axe studies (Williams-Thorpe et al. 1999). This analysis included some fine-grained gabbros (called dolerites, although this term can also refer to a volcanic rock, diabase). Both petrographic and chemical features were used to characterize source areas.

Ixer (1997) summarized some of the extensive work done on tracing the provenance of the dolerite "bluestones" used at Stonehenge in England. He demonstrated that combining the usual thin-section petrographic transmitted light methods with examination of thin sections in reflected light can provide more detailed information, an approach he termed "total petrography." The identification and observation of detailed characteristics were enhanced by the use of reflected light. Ixer compared Stonehenge samples with those from a number of outcrops considered likely sources. The outcrop dolerites could be grouped by the following characteristics:

- the presence/absence, thickness, and density of ilmenite oxidation/exsolution lamellae in titanomagnetite
- the degree and type of ilmenite alteration
- the presence or absence of chromite, and coarse-grained sulfides (especially pyrrhotite and pyrite).

He found that this technique of "total petrography" was more sensitive than geochemical analysis at detecting small-scale inhomogeneities within a single outcrop, making it more likely that the area of extraction for the Stonehenge

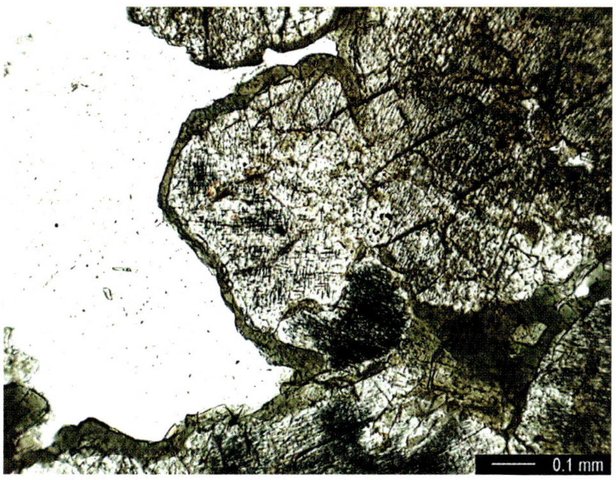

Figure 3.5.2 Gabbro, plane polarized light. The pale pyroxene crystals are rimmed with greenish-brown hornblende.

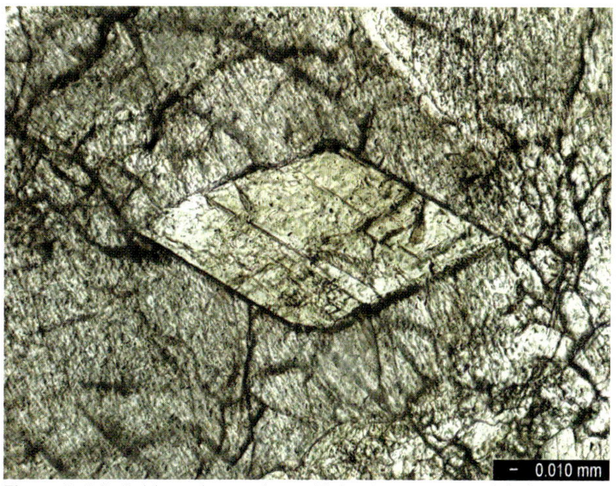

Figure 3.5.3 Gabbro, plane polarized light. A primary amphibole grain is visible among the pyroxenes, with a typical amphibole cross-section.

monoliths can be pinpointed. In some cases, outcrops for monoliths for which samples were available were matched to outcrops and areas within those outcrops.

Finally, some other occurrences of gabbro as a cultural material are known. Galetti et al. (1992) found two gabbros in their study of plutonic rocks used in antiquity by the Egyptians and Romans. Brown and Harrell (1995) also identified gabbro as one of the rocks exploited in ancient Roman stone quarries of the Eastern desert of Egypt.

3.5.2 Gabbro deterioration mechanisms

Two gabbros were included in the previously mentioned study by Rodriguez-Rey and Montoto (1978), and Leavengood et al. (1996) identified gabbro as the material used for a monumental sculpture by Isamu Noguchi. It was carved from a single 30-ton block of stone quarried in Brazil; the finished piece weighs 12 tons. The surface, of very uniform grain size, was highly polished by the sculptor. A very dark color resulted from the presence of extensive biotite and ilmenite, magnetite, and other opaque iron oxides. Hairline cracks in a radial pattern began to appear on the sculpture, and samples were thus analyzed petrographically to determine whether the cracks were due to weathering or mineral transformations. A thin section was taken from a plane perpendicular to a crack to study the form of fracturing and locate any incipient mineral transformations, as well as to characterize the rock type. The authors concluded that the cracking most likely relates to stresses in the stone, as the cracks pass through minerals of all types rather than following the boundaries between them. No secondary weathering products were found on the minerals along the crack edges, even at the surface, and there were no grain losses at crack locations. The solution for prevention of further cracking of the surface was to closely monitor temperature changes, sheltering the sculpture in the winter. The cracks were filled with Paraloid B-72 bulked with microballoons, to prevent water penetration and development of secondary alteration products.

3.6 Syenite

Syenite consists mainly of alkali feldspar as the predominant or only feldspar, with lesser amounts of sodic plagioclase feldspar along with biotite, pyroxene, amphibole, and sometimes fayalite olivine. Minor amounts of quartz or feldspathoid (usually nepheline) may also be present; the rock is then generally called nepheline-syenite (Fig. 3.6.1). Quartz is usually in the form of interstitial anhedral grains making up less than 10% of the rock. Very distinctive of syenites is the presence of perthitic intergrowths between orthoclase or microcline and albite or sodic oligoclase (Fig. 3.6.2). Sometimes the intergrowths are extremely minute (Williams et al. 1982). Accessory minerals include sphene, apatite, zircon, and iron oxides.

Syenite is not frequently encountered in the literature on cultural materials, partly because it is often mistakenly called "granite," and vice versa (for example, the locality that gives this stone type its name, Syene, Egypt, is now known to actually consist mainly of granites not syenites). Syenite is known, however, to have been used for ornamental work, street paving, and architectural purposes (Merrill 1891).

A variety of deeply colored minerals that may be present, especially in nepheline-syenite, contributes to the aesthetic appearance of this rock. For example, syenites often contain a deep blue sodalite, yellow or pink cancrinite, green aegerine or aegerine-augite, or black ferromagnesian minerals, all of which contrast with the light-colored groundmass of nepheline and alkali feldspar. Although found in relatively small pockets, and therefore not a widely used material, it has been employed for ornamental stones in particular. Today this material is also used in the production of a wide variety of products such as glass, ceramics, cements, and roofing

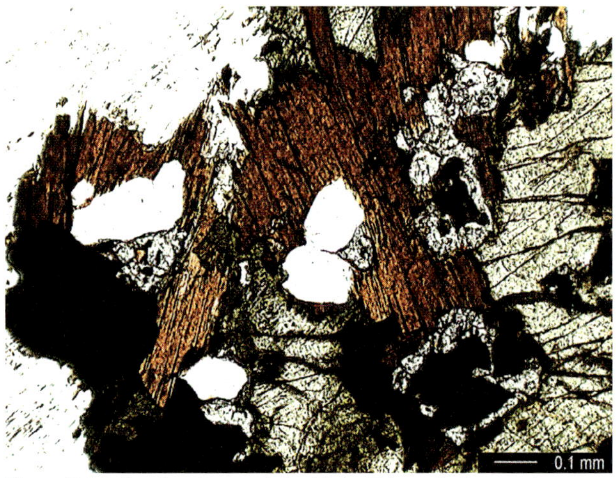

Figure 3.6.1 Syenite, plane polarized light. This nepheline-syenite consists of large areas of alkali feldspar as on the upper and lower left sides of this image, colorless in plane polarized light, brown flakes of biotite, black opaque iron oxides, pale yellow-green fayalite olivine with many fracture lines, small amounts of colorless, grungy pyroxene with typical cleavage, and colorless nepheline crystals with typical hexagonal and rectangular outlines.

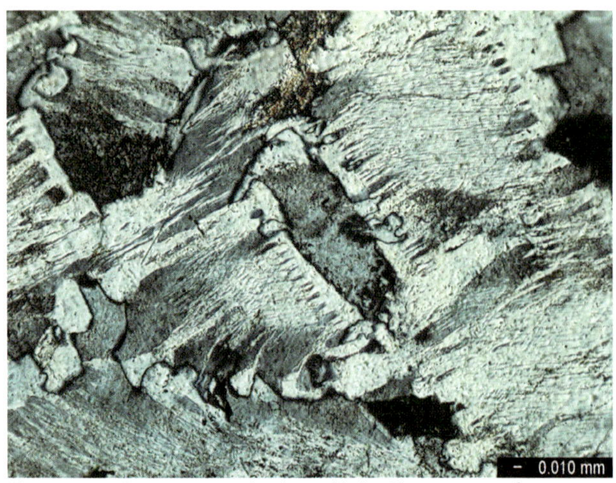

Figure 3.6.2 Syenite, crossed polarized light. Characteristic of syenites is the presence of perthitic intergrowths between alkali feldspar and plagioclase feldspar. The intergrowths can range from minute braids to larger patterns.

materials; it is an extender in paint, and a filler in plastics, rubber, and insecticides (Allen and Charsley 1968).

3.7 Monzonite

Quartz is only a minor component of this rock type, usually less than 5% of the mineral fraction. Feldspars make up between one-third and two-thirds of the rock, with approximately equal amounts of alkali and plagioclase feldspars (Fig. 3.7.1). Mafic minerals may include biotite, hornblende, augite, hypersthene and olivine. Myrmekite (quartz rods intergrown in feldspar) is sometimes seen, and augite is often altered to fibrous amphibole (uralite) (Williams et al. 1982).

Monzonite and quartz monzonite (similar in composition, with equal amounts of alkali and plagioclase feldspars and a bit more quartz, but not enough to be termed a granite) are found used as sculptural and architectural materials. For example, Asher (1987) identified quartz monzonite as the material used to carve a superstructure molding found lying beside a temple at the eastern Indian site of Nalanda. And, Galetti et al. (1992) found one quartz monzonite in their study of plutonic rocks used in antiquity by the Egyptians and Romans.

In the study by Shipley and Graham (1987) already mentioned, petrographic studies of thin sections of rock outcroppings along with materials used for ground stone tools (manos, metates, etc.) from Maya sites in Guatemala and Belize were carried out to determine whether it would be possible to pinpoint sources used for the tools and to reconstruct past trade networks. Some of the tools were found to be made of a quartz monzonite. This material matched two potential sources, both with the same mineralogy and coarse-grained texture.

In the deterioration studies of Spanish ornamental rocks performed by Rodriguez-Rey and Montoto (1978), three of the rocks were monzonites. Of these, two are classified as "larvikite," a Norwegian monzonite. This is a pyroxene (augite) monzonite of close to basic composition (52–55% silicates); it is sometimes also called an augite syenite. The primary minerals present are rhomb-shaped feldspars, both alkali feldspar (perthitic orthoclase) and plagioclase (oligoclase or labradorite), and augite, with small amounts of biotite, quartz, iron-titanium oxides, and apatite; sometimes other pyroxenes may also be present, along with olivine, nepheline, sphene, or other minerals (Williams et al. 1982; Le Maitre 1989). Larvikite is famous as a decorative facing stone. Minute inclusions develop along cleavage planes within the abundant feldspars and impart a prized iridescent quality to the stone (Henson 1953; Kempe and Harvey 1983).

3.8 Anorthosite

Anorthosites consist largely, or even entirely, of coarse-grained plagioclase feldspars (bytownite, labradorite, or andesine) (Fig. 3.8.1). There may also be minor amounts

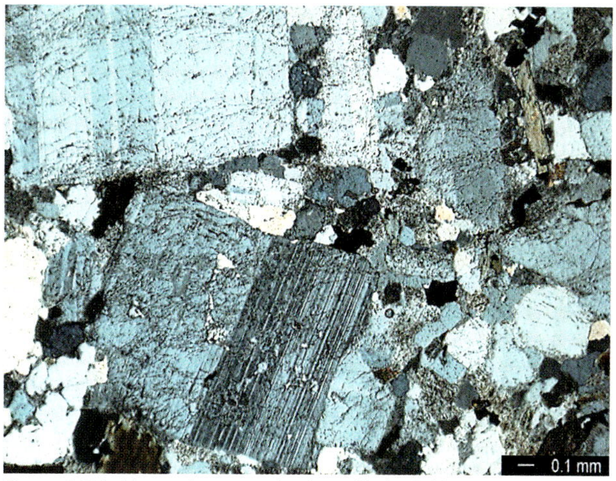

Figure 3.7.1 Monzonite, crossed polarized light. Feldspars make up much of this rock. Both alkali feldspars—with Carlsbad twinning (simple twinning with a grain appearing as two individuals) and plagioclase feldspar with both polysynthetic (multiple) and Carlsbad twinning—are present and comprise most of the rock. Minor amounts of quartz are found as smaller interstitial grains. Mafic minerals include biotite, hornblende, and black iron oxides.

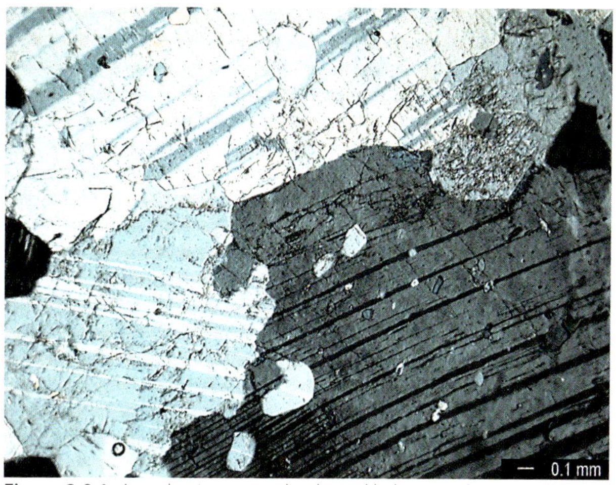

Figure 3.8.1 Anorthosite, crossed polarized light. Anorthosites consist largely of coarse-grained plagioclase feldspars. Plagioclase here exhibits the usual polysynthetic twinning.

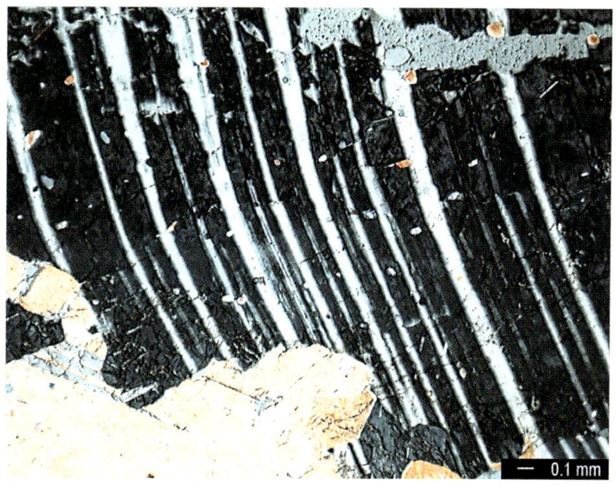

Figure 3.8.2 Anorthosite, crossed polarized light. Strain is sometimes evident in anorthosites, with some plagioclase feldspar grains showing undulous extinction and curving twin lamellae.

of mafics, such as pyroxenes, hornblende, or biotite; accessories may include a wide variety of iron oxides. Strain is often evident with undulatory extinction and curving twin lamellae (Fig. 3.8.2). Anorthosite is sometimes called "plagioclasite" (Williams et al. 1982; Le Maitre 1989).

Rodriguez-Rey and Montoto (1978) identified anorthosite among Spanish building stones they studied. However, this stone appears to be used only rarely in the production of cultural materials.

4

Sedimentary Stone Materials

Sedimentary rocks are formed by deposition of material at or near the earth's surface, under relatively low temperatures and pressures. As the layering progresses, the earlier layers can end up buried rather deeply; postdepositional (diagenetic) processes tend to create firmly lithified rocks out of the initial loose sediments. If the depth of burial increases enough, with accompanying increases in temperature and pressure, sedimentary rocks can grade into metamorphic rocks.

The two deposition processes that form sedimentary rocks are: (1) detrital—the mechanical accumulation of sediments, minerals, and rock fragments produced from the disintegration, erosion, and transportation of older rocks (also called "terrigenous" or "clastic"; and (2) chemical precipitation of minerals that become deposited into rocks. Examples of detrital sedimentary rocks are shale, siltstone, sandstone, conglomerate, and breccia. Examples of chemically precipitated sedimentary rocks include the calcium carbonates (such as chalk, limestone, dolomite, travertine and tufa), silicates (such as chert, flint, jasper, and chalcedony), and evaporites (gypsum and anhydrite) (Folk 1974; Williams et al. 1982).

For sedimentary rocks, grains are classified by size according to a standard system, although several different systems exist, one of the most common being the Udden-Wentworth scale, which classifies grains on the basis of maximum size as clay, silt, coarse silt, very fine sand, fine sand, medium sand, coarse sand, very coarse sand, and gravel. Krumbein's phi scale is a logarithmic transformation of the Udden-Wentworth scale, ranging from 8 for clay to –2 for gravel (Folk 1974).

4.1 Shale

If a consolidated rock is comprised mainly of clay-sized grains (less than 0.005 mm in size), and has a tendency to split parallel to the bedding direction into thin plates or flakes, it is called a shale. Because of the small size of clay minerals, thin-section petrography of shales is often augmented by other techniques such as electron microscopy. However, there are many features that can be observed with optical microscopy, for example, strong parallel orientation of the clay flakes can be seen (Fig. 4.1.1), as can inclusions such as silt (usually quartz, feldspars, and micas), organic materials, carbonates, or metallic oxides (Williams et al. 1982). Detrital clay matrix can sometimes be distinguished from authigenically formed clays; a pisolitic texture may be visible (globules with a pea-like structure), often showing lamination and dessication cracking. This texture is typical for many soil-type deposits such as bauxites, caliche crusts, and laterite hardpan that may form part of a shale (Scholle 1979).

Although the clay-sized particles often cannot be distinguished in thin section, larger inclusions such as quartz, feldspars, chalcedony clusters, calcite, chlorite, muscovite, biotite, and iron oxides may be distinct. Quartz may vary in

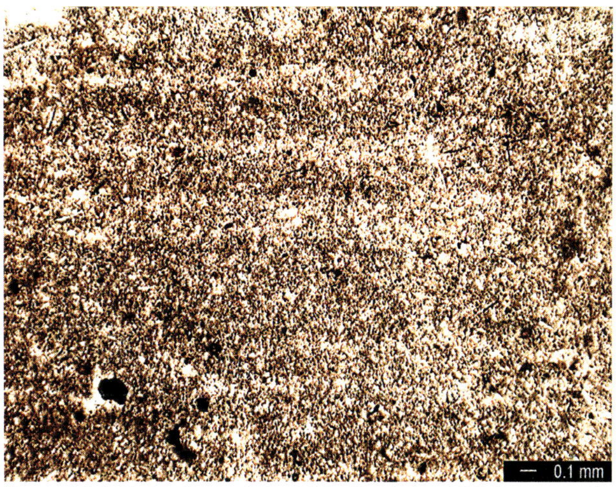

Figure 4.1.1 Shale, plane polarized light. The dark reddish-brown matrix of this argillaceous shale is comprised of illitic clay minerals. Strong parallel orientation of the clay flakes can be seen.

degree of rounding and other textural characteristics, and may range from silt up to sand-sized particles (Fig. 4.1.2). The feldspar grains may vary in type and degree of alteration (Carozzi 1993).

A variety of lamination types can be seen in thin-section studies of shales. For example, shales may be finely laminated, thickly laminated, wavy laminated, or lenticular laminated. Other variations include the amount of bitumen, carbon, and/or iron content; presence of silt, algae spores, fossils (Fig. 4.1.3), and other nonclay materials (Fig. 4.1.4) (O'Brien and Slatt 1990).

4.1.1 Shale as a cultural material

Shale was used as an early stone tool material (Glover 1975; Kempe and Harvey 1983); and occasionally for carved-stone artifacts (Ritchie 1973; Green et al. 1980; Page and Greenwood 1986). In Roman Britain, Kimmeridge shale, obtained at several locations in the British Isles, was carved to make small decorative items such as jewelry (mainly bracelets) and elements for furniture (Ritchie 1973). Other Roman shale artifacts include spindle whorls and vessels on which deterioration consisting of cracking and delamination can be seen (Page and Greenwood 1986).

Shale has sometimes been used as a building stone. However, since it is not strongly lithified it is rather friable and does not resist abrasion well (Fig. 4.1.5). If permeated by water, shale tends to disintegrate almost completely (National Materials Advisory Board 1982). These properties limit its usefulness as a cultural material. Conservation efforts thus far have been mainly directed at geological and paleontological collection specimens. Such specimens tend to deteriorate through physical damage from inappropriate handling and storage, delamination due to environmental conditions that lead to dehydration and shrinkage of clay minerals, or decay of the pyrite component (Dollery 1994; Collins 1995).

Glover (1975) included thin-section petrography in his studies of Aboriginal artifacts (chips and flakes) from southwestern Australia. One illustrated photomicrograph showed a siliceous ferruginous shale. The shale had many small bodies with a mixture of straight and curved walls of

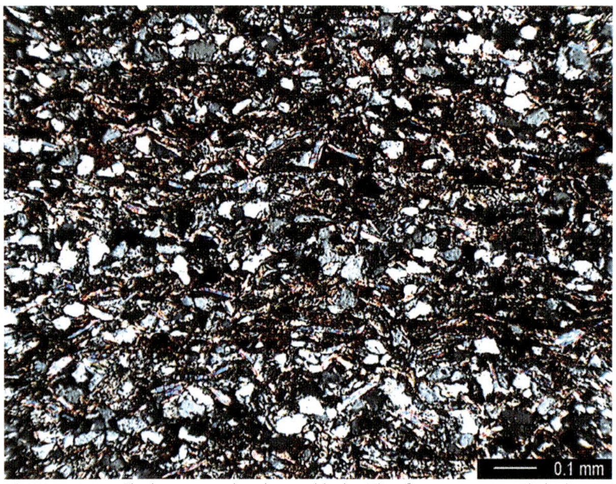

Figure 4.1.2 Shale, crossed polarized light. This ferruginous red shale has a large fraction of silt-sized particles of quartz, along with abundant oriented flakes of high birefringent muscovite and biotite mica. The iron oxides are dark red, brown, and black.

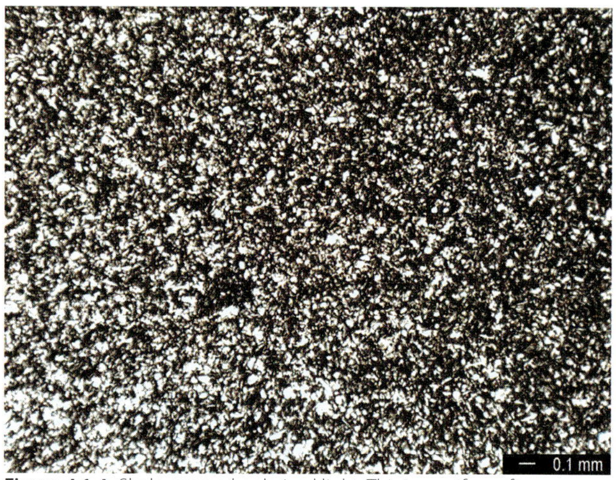

Figure 4.1.4 Shale, crossed polarized light. This is a uniform, fine-grained, quartz-rich shale. Abundant black carbonaceous material obscures much of the thin section.

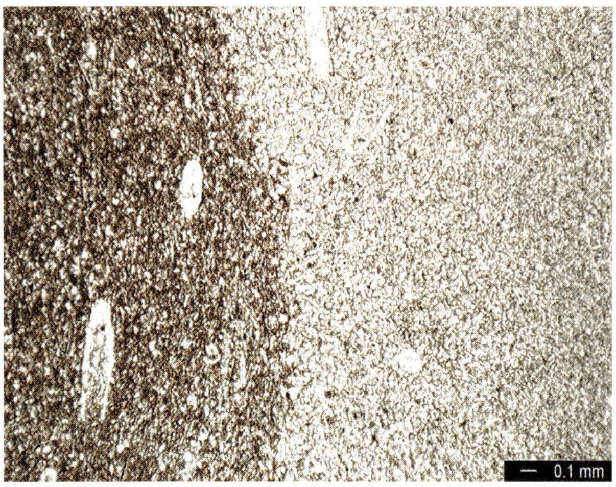

Figure 4.1.3 Shale, plane polarized light. This is a fossiliferous shale, comprised of a mixture of clay and abundant calcium carbonates, with clear layering. A number of small microfossils are scattered throughout the shale.

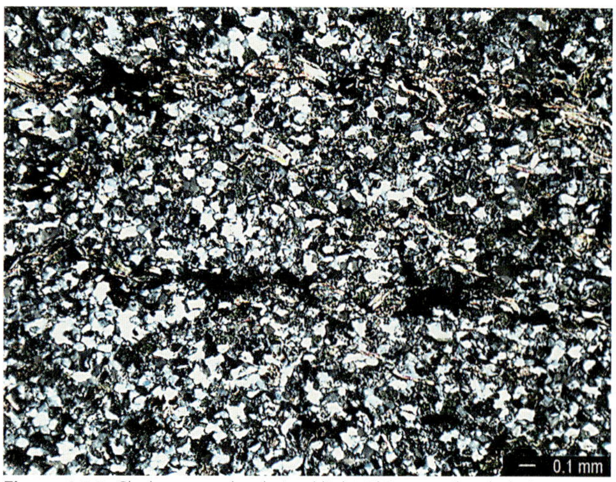

Figure 4.1.5 Shale, crossed polarized light. This very silty shale is not strongly lithified, so is rather friable and is traversed with cracks. There are many grains of rounded, low birefringent quartz, along with some platy muscovite mica and the clay.

microcrystalline quartz, ranging from 0.01 to 0.05 mm thick. Glover thought these bodies may be algal, comprising up to 15% of some of the flakes. The cement consisted of hematite and goethite. He described the shale flakes as being tough, laminated, and with conchoidal fracture, while ranging in color from grayish red to brown, and he was able to identify the geological formation from which this shale derived, showing that it was obtained locally.

In England, shale was mixed with sand, clay, straw and water to form a building material called "cob." Typically, cob would consist of about 25% shale in gravel-sized pieces, about 20% each clay and coarse sand, and a little over 30% fine sand, mixed with a small amount of straw and water (Williams-Ellis 1919).

4.2 Siltstone

If a clastic rock is comprised mainly of silt-sized grains (0.004–0.0625 mm, larger than clay but smaller than sand), it is called a siltstone. Unlike larger grained sandstones, which contain a mixture of polycrystalline lithic grains and single-crystal mineral grains, siltstones tend to have very few of the polycrystalline lithics. Instead, they generally have abundant clay, along with single crystals of quartz and feldspar, all cemented and lithified (Fig. 4.2.1). Cements include the same ones found with sandstones, such as carbonates, gypsum, iron oxides, silica, etc. Other constituents of a siltstone may include fossils, glauconite, phosphatic pellets, and other detrital material. Siltstones may be either massive or fissile (having a tendency to split along a relatively smooth surface parallel to the bedding plane) (Fig. 4.2.2) (Pettijohn 1975; Williams et al. 1982; Pettijohn et al. 1987; O'Brien and Slatt 1990; Carozzi 1993).

Although it is often possible with thin-section petrography to identify the most abundant clay mineral present in a siltstone, those clays present in small amounts will generally require other techniques for identification, such as x-ray diffraction analysis (XRD). However, it may be possible with thin sections to find clay minerals present in small patches that are difficult to locate in bulk analysis techniques like XRD; these small patches of clay minerals may help in identifying the geological source of a siltstone (Folk 1974).

4.2.1 Siltstone as a cultural material

Although not commonly used for the production of cultural materials, siltstone does occasionally appear as an artifact material. For example, touchstones—fine-grained dark stones used for assaying gold by comparing the color of streaks made on the surface to the color made by alloys of known composition—were often made of siltstone. Thin-section petrography of some of these siltstone touchstones showed that inclusions that may pinpoint sources of the material are sometimes present: one touchstone was found to contain tourmaline, a fact that allowed researchers to suggest a provenance in the southwest of England (Moore and Oddy 1985).

Siltstone was also used occasionally in the production of stone axes. Keen and Radley (1971) reported on petrographic analyses that identified a highly siliceous siltstone and a micaceous chlorite siltstone used in the production of prehistoric axes in Britain.

Newman (1997) reported on thin-section analysis of a siltstone fragment of a relief depicting the Buddha with two attendants, from eighth-century central Thailand. He found that about 10% of the rock consisted of detrital silt-sized quartz grains, with the remainder consisting of a fine-grained matrix that included smaller quartz grains, mica, chlorite, and calcite.

Some of the ancient monuments at Petra in Jordan were carved from siltstone, which occurs there along with the more abundant sandstone (Fitzner and Heinrichs 1994). Thin sections were used to characterize the rock and assess the degree and mechanisms of deterioration. The mineralogy and porosity were examined, and both were found to be related to susceptibility to weathering.

A geological survey (Miller et al. 2006) of 172 early and late medieval sculpted stones from central and western

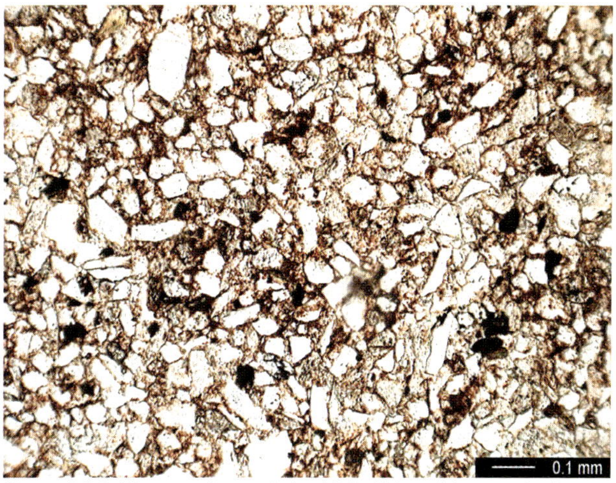

Figure 4.2.1 Siltstone, plane polarized light. This rock is comprised mainly of a reddish clay along with single crystals of silt-sized quartz and feldspar, all cemented and lithified.

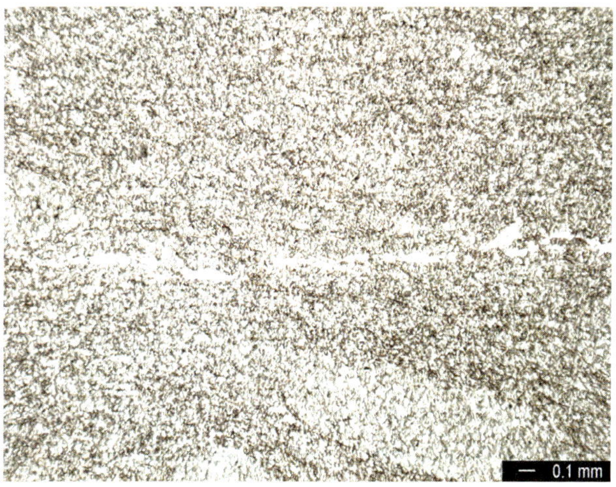

Figure 4.2.2 Siltstone, plane polarized light. This fine-grained siltstone is fissile, having a tendency to split along a smooth surface parallel to the bedding plane. This fissility is visible in thin section as well as in hand sample.

Scotland revealed that all but two of the early ones are sandstone; a wider variety of stones appears with the later ones. One of the two nonsandstone early medieval objects was siltstone (the remaining one was granite). The standing stones were carved with various designs and symbols, both abstract and pictorial. The sole siltstone example was identified as being a calcareous siltstone. The grains were very angular and well sorted, and included monocrystalline quartz, weathered feldspar, kinked chlorite, biotite and other micas. There were also minor opaque grains, along with both clay and calcic cements.

4.3 Sandstone

If a detrital rock is comprised mainly of sand-sized grains (0.0625–2.0 mm), it is called a sandstone. The grain particles pack together to form the framework of the rock. Sometimes smaller clay or silt particles lodge between the framework grains or in pores, and are called the matrix component. The pore space between grains often becomes filled with a cement component, comprised of authigenic minerals precipitated from solutions in the pores between grains (Fig. 4.3.1). Quartz is usually the dominant mineral in the sand component of the framework grains, followed by feldspars. Small fragments of rock, called lithic grains, are also common. Thus, one way of classifying sandstones is by the relative proportions of these three components: a quartz sandstone if 90% of grains are quartz (Fig. 4.3.2); feldspathic sandstone if less than 90% of grains are quartz and feldspars exceed lithic grains (Fig. 4.3.3); and lithic sandstone if lithic grains predominate over feldspars (Fig. 4.3.4) (Williams et al. 1982; Greensmith 1989; Carozzi 1993).

For example, early petrographic studies to characterize and identify the sources of the large stones of Stonehenge in England showed that some were quartz sandstone (also termed "quartzose sandstone") (Thomas 1923). The stone is

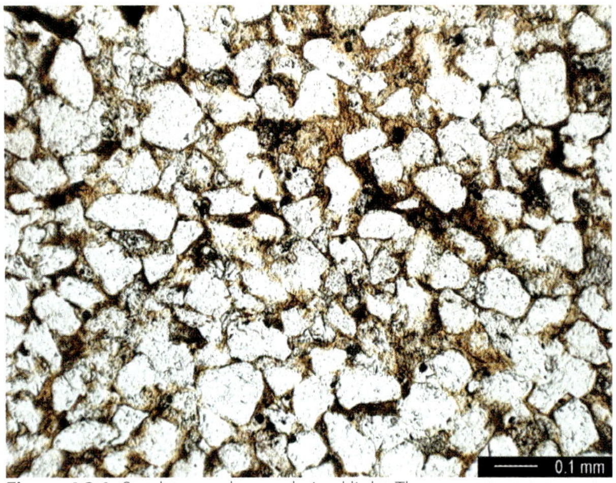

Figure 4.3.1 Sandstone, plane polarized light. The cement component found in many sandstones is comprised of authigenic minerals (in this case hematite) precipitated from solution into the pores between grains.

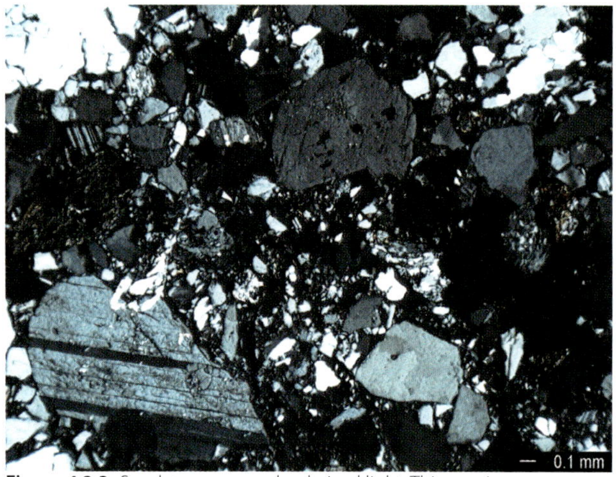

Figure 4.3.3 Sandstone, crossed polarized light. This specimen can be classified as a feldspathic sandstone, since less than 90% of the framework grains are quartz and the feldspar fraction exceeds the lithic fraction. Many of the feldspars visible here are plagioclase, with polysynthetic twinning.

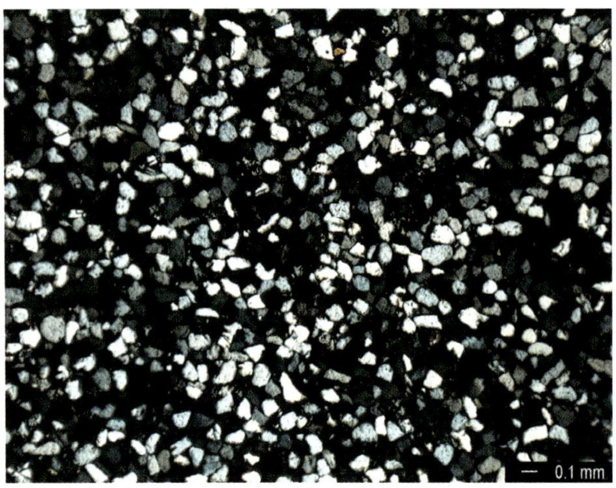

Figure 4.3.2 Sandstone, crossed polarized light. This specimen can be classified as a quartz sandstone, with at least 90% of the framework grains being quartz.

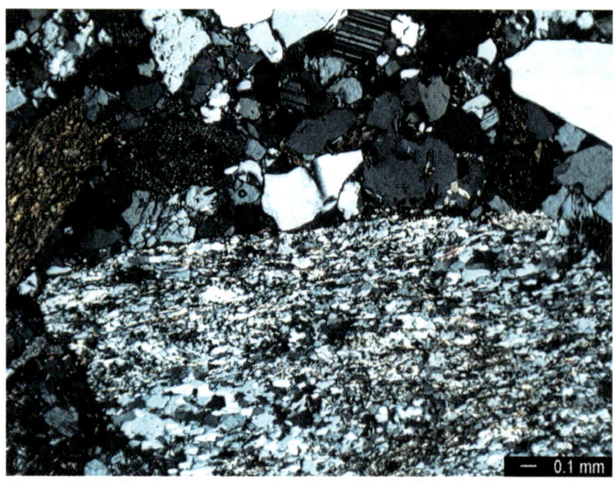

Figure 4.3.4 Sandstone, crossed polarized light. This specimen can be classified as a lithic sandstone, because less than 90% of the framework grains are quartz and the lithic fraction exceeds the feldspar fraction. This sandstone contains mainly metamorphic lithic fragments, such as the large schist piece in the lower part of the image.

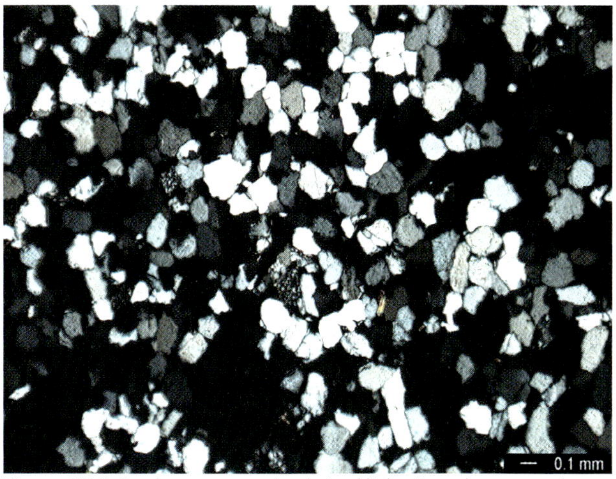

Figure 4.3.5 Sandstone, crossed polarized light. A sandstone may be classified as an arenite if it is relatively free of clay and fine silt, thus a "clean" sandstone that usually has less than 15% matrix material. Most of the material here is sand-sized quartz, with a few sedimentary lithics and an occasional sericite flake. There is no real authigenic cement, only some concentrations of hematite along some of the grain boundaries; the large black patches on this image are voids, showing that this is a very porous stone with the pores being clear and relatively clean.

described as being a fine-grained micaceous sandstone with a cement that is partly calcareous and partly siliceous. In thin section, fine-grained angular quartz, muscovite flakes, abundant greenish chlorite, and another green mineral that may be glauconite were all visible within a fine-textured calcareous-siliceous matrix. Accessory minerals included garnet, zircon, rutile, tourmaline, and anatase. Several possible source areas for the stones were identified.

In contrast, feldspathic sandstone was used in the construction of the Kailasanatha Temple, built in the eighth century in Kancheepuram, near Madras in southern India (Singh 1993). Singh's petrographic analyses showed that feldspars present were plagioclase (altered to sericite and kaolinite), orthoclase, and microcline; also found were quartz, chlorite, chloritized bioite, and lithic fragments. The cementing material was mainly calcite, with some ferruginous cement. Deterioration due to atmospheric exposure and capillary action of groundwater was a serious problem, with both the cement and the feldspars being greatly affected, and in turn affecting the rest of the stone.

Additional subdivisions within each of these categories include use of the term arenite if the stone is relatively free of clay and fine silt ("clean" sandstones, usually with less than 15% matrix) (Fig. 4.3.5), or wacke if the stone is argillaceous (filled with clay and silt between sand grains) (also called "dirty" sandstones, with more than 15% matrix) (Fig. 4.3.6). A graywacke (Fig. 4.3.7) is a sandstone that has been buried and developed into a hard, dark-colored rock with a much lithified matrix that obscures boundaries of individual sand grains; sometimes these are further subdivided into feldspathic graywackes and lithic graywackes depending on the predominant particle type (Williams et al. 1982; Pettijohn et al. 1987; Greensmith 1989; Carozzi 1993).

Dolomite-rich sedimentary rocks are sometimes called dolostones, and as they are harder than similar calcite-rich rocks and have greater fragility, are relatively uncommon as cultural materials. However, a dolomitic-calcareous arenite, comprised of dolomite, calcite, and quartz, was used as a building material in Italy (Spadea et al. 1996). Thin-section studies of this arenite showed that its use was restricted to the Middle Ages. Thin sections augmented with XRD analysis were used to try to confirm the possible local source for the rock, by comparing building stone samples with possible quarry materials. About 40 samples were studied for mineralogical and textural characterization. Although the source could not be identified, the building stones were similar enough to suggest they did all come from a single source, which appeared to best match local deposits. The illustrated thin sections showed an arenite texture (relatively clean) with clasts of dolomite as rhombohedric crystals, micritic limestone, and mono- and polycrystalline quartz or chert, with some altered opaque minerals. Very scarce calcite occurred as cement.

Graywacke, or "dirty" sandstone contrasts with arenite, or "clean" sandstone. Graywacke has been used in the produc-

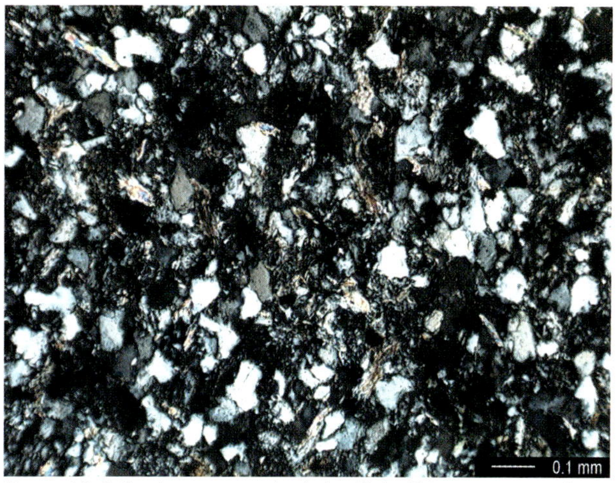

Figure 4.3.6 Sandstone, crossed polarized light. This specimen can be identified as a wacke, or an argillaceous sandstone, filled with clay and silt between sand grains (a "dirty" sandstone), and having more than 15% matrix. The framework of fine-grained quartz and feldspars (orthoclase and plagioclase) is cemented by a mixture of sericite, chlorite, clay, carbonates, and some iron oxides.

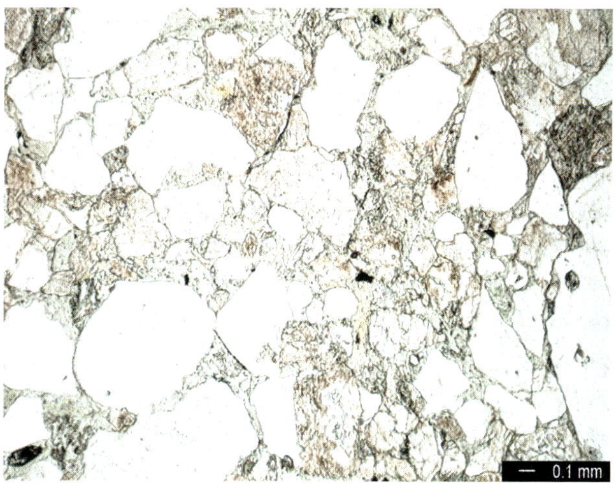

Figure 4.3.7 Sandstone, plane polarized light. This sandstone is categorized as a graywacke, with a dark, lithified matrix. Many of the framework grains are very angular, and the green and reddish-brown areas of the matrix indicate the presence of chlorite.

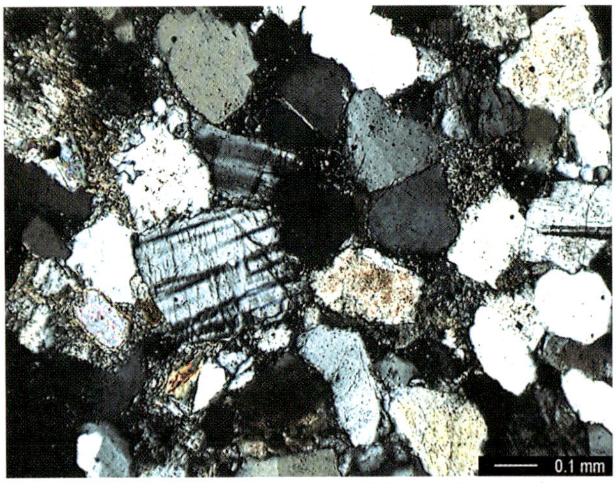

Figure 4.3.8 Sandstone, crossed polarized light. An arkose, or arkosic sandstone, such as this example, is formed principally by mechanical disintegration of a granitic rock. Feldspars account for 25% or more of the framework grains, and quartz is abundant. Accessories include flakes of mica (muscovite, biotite, and chlorite), along with some black iron oxides.

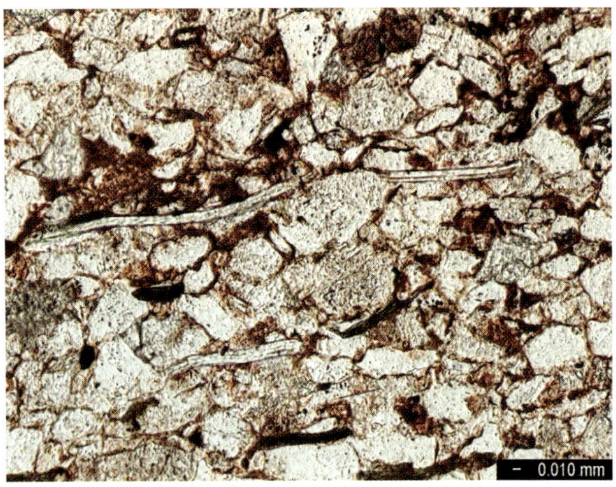

Figure 4.3.9 Sandstone, plane polarized light. Here the framework grains include a large amount of mica (both colorless muscovite and brown-green pleochroic biotite flakes). Colored red by hematite, the sandstone is termed a red micaceous subgraywacke.

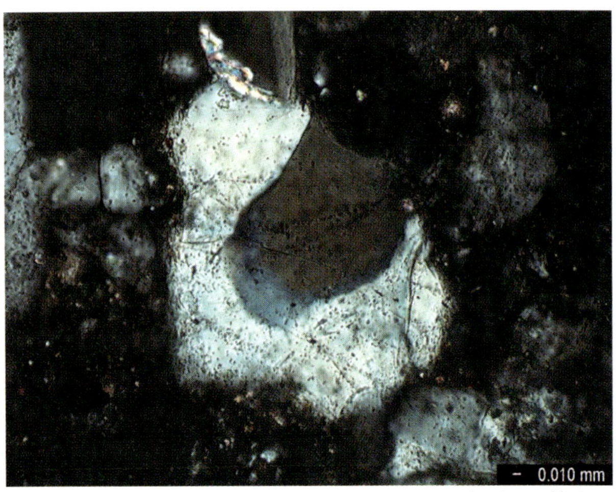

Figure 4.3.10 Polygonized quartz in sandstone, crossed polarized light. When two or three subcrystals are present in a single quartz grain, each large enough to be identified as quartz, the term polygonized quartz is used.

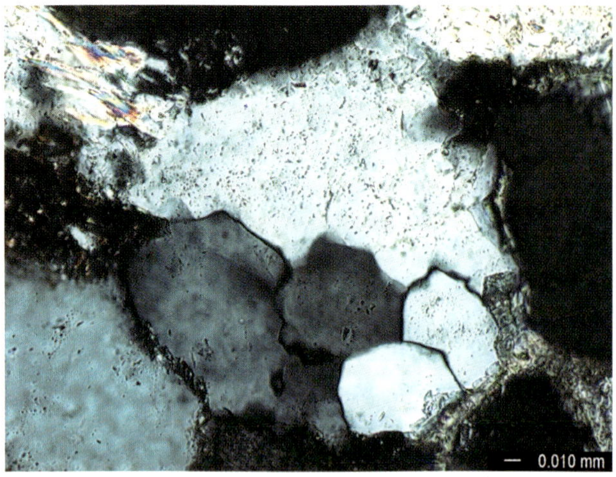

Figure 4.3.11 Polycrystalline quartz in sandstone, crossed polarized light. When there are more than three subcrystals present in a single quartz grain, each large enough to be identified as quartz, the term polycrystalline quartz is applied.

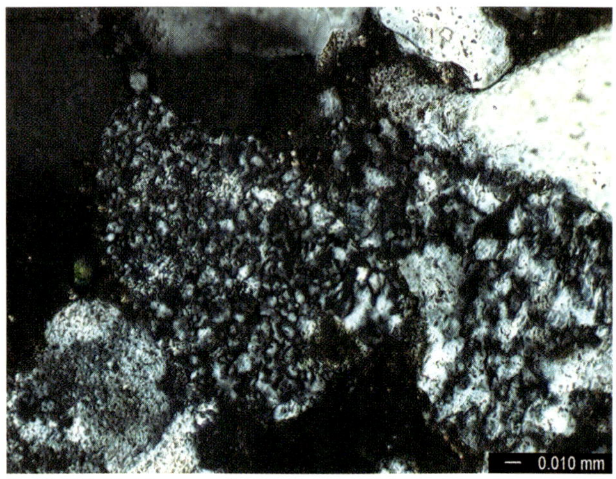

Figure 4.3.12 Microcrystalline quartz in sandstone, crossed polarized light. This term is used when there is a quartz aggregate in which individual grains are less than 0.03 mm in length, as in chert.

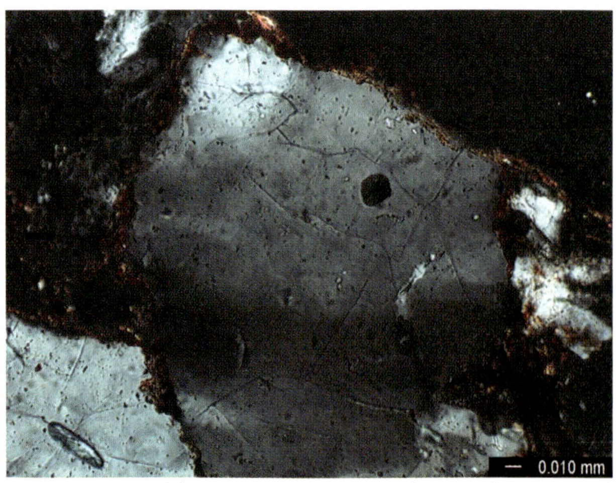

Figure 4.3.13 Undulous extinction in quartz within a sandstone, crossed polarized light. As the microscope stage is rotated between crossed polarizing filters, this quartz grain is going extinct slowly over more than five degrees of stage rotation, in wave fashion, rather than going dark all at once.

tion of stone tools. The hardness and lithified matrix developed during burial make this material appropriate for tool use, unlike most sandstones. Thin-section analyses of stone axes made of graywacke found in England (Shotton 1959; Wallis 1963) showed one group made of unevenly graded, subangular, and unstrained quartz grains in parallel formation. There were some grains of metamorphic quartzite present, with clear interlocking sutures. Also found were microcrystalline quartz and accessories of devitrified glass, feldspars, lithics (trachyte, andesite, schist, pegmatite, and chert), mica, epidote, chlorite, and chalcedony. The cement was a mixture of silica and fine mica. Other groups of graywacke axes looked very different in thin section; for example, they had a finer texture, showed less alteration of feldspars, and had a heavy mineral component (such as sphene).

Other categories of classification include subdivisions within feldspathic sandstones. When quartz and feldspars are both abundant, as when the sandstone is formed principally by mechanical disintegration of a granitic rock, the terms arkose or arkosic sandstone are often used, generally when feldspar is 25% or more of the framework grains and quartz is abundant (Fig. 4.3.8). When quartz is even more abundant, with less than 25% feldspar, the term subarkose is used (Williams et al. 1982; Pettijohn et al. 1987).

In addition to quartz, feldspar, and lithics, framework grains may include micas, zircon, tourmaline, rutile, and a wide variety of other minerals depending upon the age and degree of weathering that the sands underwent prior to cementation (Fig. 4.3.9). If carbonate rock fragments predominate (detritus from limestones or dolomites), the stone is called a calclithite, although most often such stones are conglomerates (made of larger pebbles). Calclithites often include noncarbonate grains such as quartz and lithics, and are usually cemented with calcite (Folk 1974).

Various systems have been developed to use the composition of sandstones to identify provenance. For example, point counting in thin section can obtain a quantitative "detrital mode," in which grains are divided into basic types. In one system (Dickinson 1985) grains are classified as quartzose, feldspar, or lithics. Polycrystalline quartz is sometimes counted as quartzose and sometimes as a lithic fragment (chalcedony).

The most abundant cementing materials include carbonates (calcite, dolomite, ankerite, and siderite), silica (quartz, chert, chalcedony, and opal), clay minerals (smectite, chlorite, illite and kaolinite), feldspars, iron oxides (hematite, goethite, and limonite), anhydrite and gypsum, barite, and apatite. However, many other minerals may crystallize in a sand after deposition and form more unusual cements (Williams et al. 1982; Pettijohn et al.1987; Greensmith 1989).

In addition to overall texture and identification of cements and detrital constituents, textural characteristics of certain detrital minerals, such as quartz, may be very helpful in characterizing and sourcing a sandstone. The following quartz textural features may be useful (Scholle 1979).

- Polygonized: two or three subcrystals are present in a single grain, each large enough to be identified as quartz (Fig. 4.3.10).
- Polycrystalline: more than three subcrystals are present in a single grain, each large enough to be identified as quartz (Fig. 4.3.11). The amount of polycrystalline quartz present has been shown to be related to grain size (the amount increases with increase in grain size) (Conolly 1965). Both polygonized and polycrystalline quartz can be described further according to degree of uniformity and size of the grains, presence of elongation, type of contact between grains (such as straight or sutured), and presence of undulatory extinction (Pettijohn et al. 1987).
- Microcrystalline: an aggregate in which individual grains are less than 0.03 mm in size, as in chert (Fig. 4.3.12).
- Undulatory extinction: when the crystal is rotated between crossed polarizing filters, the grain goes to complete extinction slowly over more than five degrees of microscope stage rotation in a wave fashion, rather than all at once (Fig. 4.3.13). The amount of undulatory extinction found in quartz grains of a sandstone is related

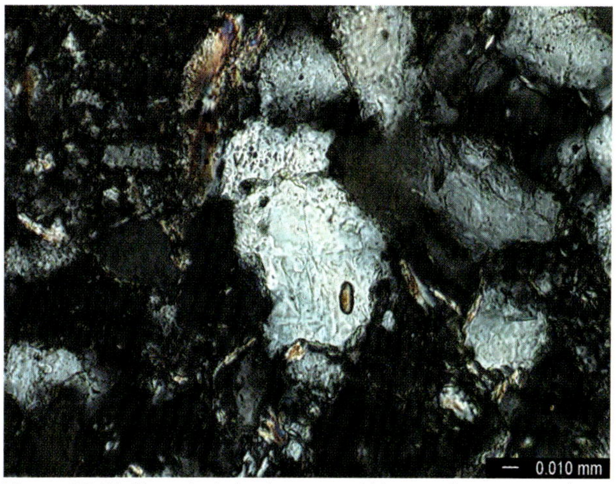

Figure 4.3.14 Zircon inclusion in quartz within a sandstone, crossed polarized light. The very high relief and higher interference color of zircon stand out against the low relief and low birefringence of quartz.

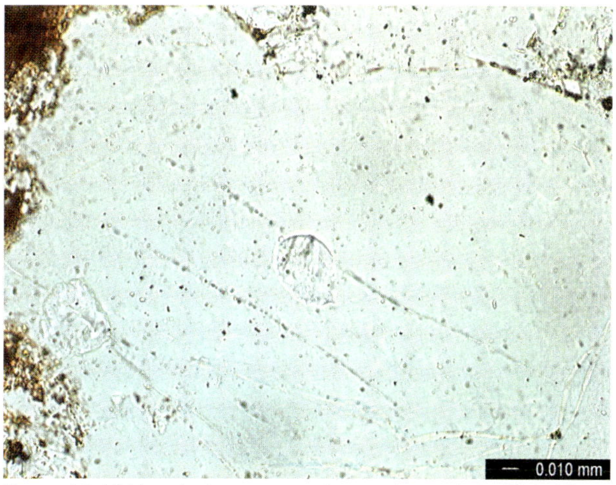

Figure 4.3.15 Vacuoles in quartz, sandstone, plane polarized light. These appear as tiny inclusions, always dark in crossed polarized light and almost opaque-looking under low magnification in plane polarized light. However, under high magnification in plane polarized light they are seen to be transparent bubbles having strong negative relief and often filled with liquid.

to post-depositional processes of folding and faulting, and this feature also increases with increasing grain size (Conolly 1965).

- Inclusions: other minerals are present in the quartz grain (Fig. 4.3.14).
- Vacuoles: these appear as tiny inclusions (black in crossed polarized light) in the grain but under high magnification are seen to be transparent bubbles having strong negative relief and often filled with liquid (Fig. 4.3.15).

Other quartz characteristics that may be helpful in characterizing and sourcing a sandstone sample include the average size and range of sizes of grains, and the grain shape, usually defined as angular/subangular (Fig. 4.3.16) or rounded/subrounded (Fig. 4.3.17).

Sandstones cemented with hematite are often red in color (see Fig. 4.3.1). Red sandstones were used in the production of medieval querns in England (Wallis 1963). Thin-section studies of medieval hones, or whetstones (Morey and Dunham 1953; Wallis 1963), showed that several types of sandstones were used, including micaceous (see Fig. 4.3.9) and glauconitic sandstones (Fig. 4.3.18), all composed of small angular quartz grains set into a soft matrix of mica, clay minerals, or calcite. Some of these could be sourced as to location of collection, which provided information about trade connections between England and Scotland during the medieval period.

In addition to mineral composition, sandstones are described petrographically using a wide variety of characteristics of the framework grains, especially quartz. These characteristics can include grain size, morphology (the shape and degree of roundness), surface textures, the type of grain-to-grain contacts, degree of sorting (uniformity in size distribution), and quartz and lithic types (plutonic, volcanic, or metamorphic) (Fig. 4.3.19) (Folk 1974; Pettijohn et al. 1987).

Various features of feldspar grains in feldspathic sandstones can also be useful in characterization and sourcing studies. Some of these features are observable in thin section, including zoning and twinning. Helmold (1985) discussed these characteristics, and also the ways in which feldspars may show weathering and deterioration.

Types of porosity in sandstones can be useful in identifying provenance. Shanmugam (1985) illustrated very clearly

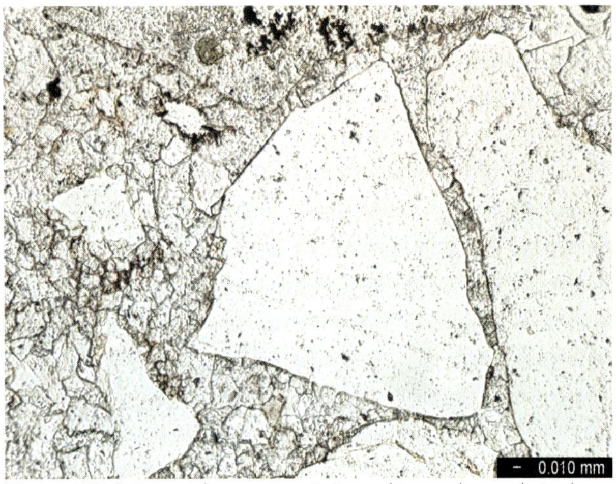

Figure 4.3.16 An angular quartz grain in sandstone, plane polarized light. Here the colorless quartz is surrounded by a thick calcium carbonate cement.

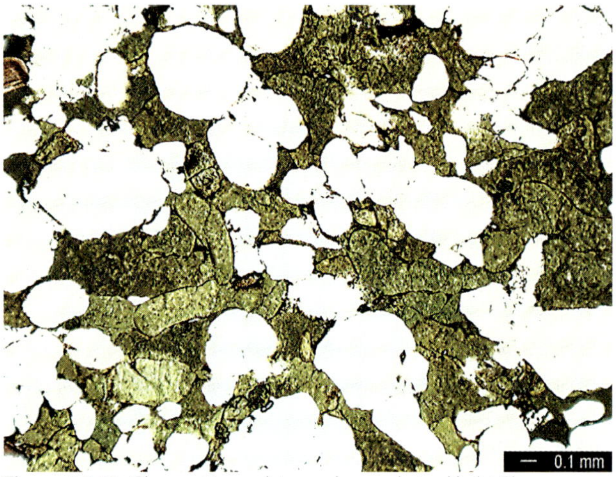

Figure 4.3.18 Glauconitic sandstone, plane polarized light. The abundant glauconite here is a deep yellow-green color and surrounds the colorless quartz grains.

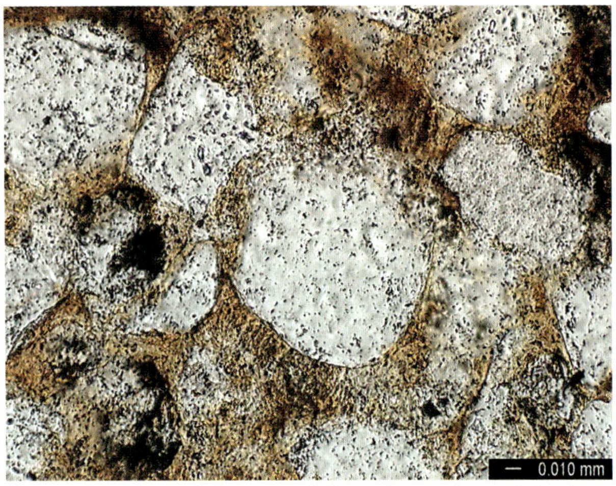

Figure 4.3.17 A rounded quartz grain in sandstone, plane polarized light. A reddish hematite cement surrounds the rounded quartz framework grains in this sandstone.

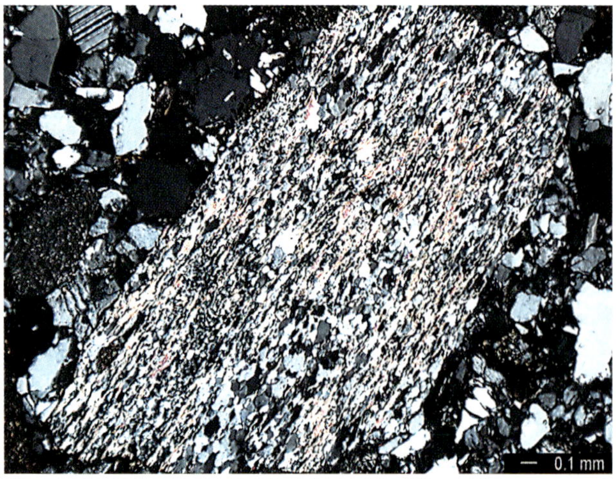

Figure 4.3.19 A metamorphic lithic fragment within sandstone, crossed polarized light. This schist fragment shows alternating layers of fine-grained quartz (rounded interlocking grains with weak birefringence) and elongated stringy mica (biotite flakes with strong birefringence).

the major types of primary porosity (indicating depositional framework) and secondary porosity (arising from postdepositional processes). These porosity types include primary and secondary intergranular and intragranular, grain fracture, and rock fracture. Other illustrations of interest in Shanmugam's work, because they show phenomena often observed with deteriorated cultural materials, are the progressive stages of feldspar dissolution.

Other weathering effects frequently seen with sandstone and which can affect provenance interpretations are discussed by McBride (1985). Some of these processes occur deep within the subsurface, and almost grade into metamorphism; others occur in upper zones of weathering. Many such changes have already taken place when the sandstone is quarried for use as a cultural material. These latter processes can also continue after the stone has been used for architectural or monumental purposes; the former processes may reoccur if an artifact has become deeply buried. Some of the changes noted by McBride include the dissolution of nonquartz grains under severe weathering; replacement of detrital grains by authigenic carbonates, clays, zeolites, and other minerals; albitization of plagioclase and alkali feldspar when deep burial conditions exist; detrital ferromagnesian minerals such as magnetite and rutile may undergo oxidation to hematite; and primary textures of carbonate and other rock fragments may be lost through recrystallization or compaction. McBride lists the changes each component of a sandstone may undergo, and illustrates many of these with thin sections.

4.3.1 Sandstone as a cultural material

Architectural uses

Sandstone is a very common building material. For example, blocks of quartzose sandstone were used for some of the larger stones that form the main circle (Sarsens) at the ancient site of Stonehenge (Storey-Maskelyre 1878; Thomas 1923). Sandstone was also the building material used for another very early structure, the temple of Karnak in Luxor, Egypt (Saleh et al. 1992). Thin sections showed quartz with fine crack patterns, orthoclase feldspars, plagioclase feldspars decomposing into sericite and kaolinite (Fig. 4.3.20), muscovite and rutile. And, sandstone was one of the building materials used in construction of the ca. 1300 B.C. Seti-1 Temple at Abydous, Upper Egypt (Helmi and Attia 1996).

Thin-section work showed that among the 27 different types of stones used in the construction of Cologne Cathedral, two sandstones were among the four major materials, the Schlaitdorfer sandstone and the Obernkirchener sandstone (Efes and Luckat 1976; Mirwald et al. 1988). The former displayed more problems with deterioration than the latter. The illustrated thin sections showed that the Schlaitdorfer sandstone was relatively coarse grained, with large quartz crystals cemented by kaolinite. There were also minor amounts of dolomite, feldspars, and muscovite. The Obernkirchener sandstone, in contrast, was much finer grained, with small, more interlocking grains of quartz cemented by kaolinite, and a number of accessory minerals.

The researchers reported that the Schlaitdorfer sandstone has problems with gypsum formation resulting from the reaction of SO_2 in the polluted air of the Cologne region with moisture and with dolomite in the sandstone. As a result, large areas of this stone were covered by a black crust. Exfoliation was also a problem. In contrast, the Obernkirchener sandstone was reported to be in relatively good condition.

According to Richard Newman (1984), sandstone was one of the most important materials used in central India for construction as well as for sculpture. A hard, feldspathic gray sandstone called "bluestone" was often used as a flooring material. However, the most common architectural use of sandstone was as the primary construction material in masonry buildings.

De Geyter (1985) characterized and illustrated thin sections of a dark brown iron-rich sandstone that was described as being an important building stone for Romanesque and Gothic monuments in several regions northeast of Leuven. He called the stone an "iron sandstone" and described it as having angular sand-sized quartz grains, with well-rounded glauconite pellets which tended to be weathered, especially along cracks. The cement was iron oxide, mainly goethite, with minor amounts of lepidocrocite and hematite.

De Geyter also illustrated and characterized another sandstone building material used in Belgium, with a greenish color, called "veldsteen." He described it as a glauconitic siliceous quartz sandstone. The thin sections showed it was comprised primarily of fine sand-sized quartz grains (subangular), with more than 10% coarse (up to 1 mm diameter) well-rounded glauconite pellets. The cement was opal-CT, with a fibrous chalcedony filling irregular intergranular spaces. Siliceous skeletons of sponges and diatoms were sometimes present, while in some varieties the remains of bivalve, echinoderm, and foraminifera skeletons were almost completely dissolved and replaced by opal-CT.

Calcarenite was used for the excavation of Mycenaean tombs in Cephalonia, Greece (Koumantakis and

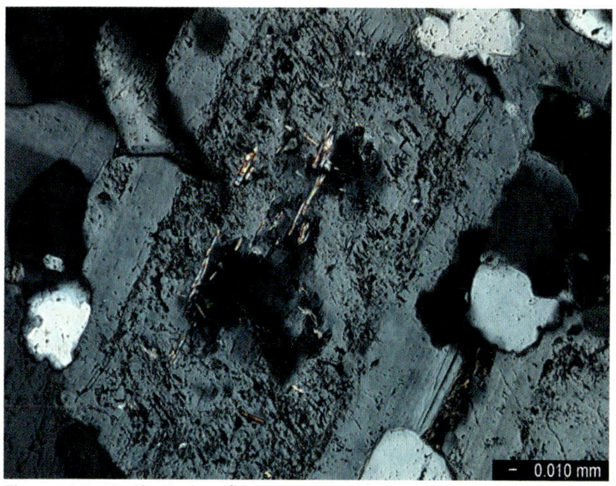

Figure 4.3.20 Plagioclase feldspar grain within a sandstone, crossed polarized light. The plagioclase is altering into clay (fine dark particles within the grain) and sericite (tiny birefringent flakes aligned along cleavage traces).

Orfanoudakis 1988). The rock was described as consisting primarily of micrite (microcrystalline calcite), and only a small amount of quartz, with a calcite cement. Petrographic analysis was one of the examination techniques that showed how erosion and weathering post-excavation had become a major problem for preservation of the sites.

Sandstones used for architectural and ornamental purposes at ancient cities of Aegean Thrace were shown in thin section to be very heterogeneous (Xidakis and Samaras 1989), ranging in grain size from a very fine sand up to pebble-sized, and varying in quartz shape from angular to round. Veins or nodules of hematite were often apparent, and kaolinite appeared as a breakdown product of feldspars. The cement was usually silica and hematite.

Rossi-Manaresi and Tucci (1989) demonstrated with thin-section work that deteriorating stone used in the construction of the Greek temples of Agrigento was a highly porous type of sandstone described as biocalcarenite. It consisted primarily of fossil debris with calcareous shells and fragments of micritic limestone cemented together with calcium carbonate. Glauconite and limonite grains imparted a brown-yellow color to the stone.

Quarrying sites and structures dating to the Bronze (Minoan) Age on Crete showed that sandstone of a type called "eolian sandstone," or "eolianite," was used as a building material in the eastern coastal zone of the island (Papageorgakis et al. 1992). This variety of sandstone originated from marine sand that had drifted and been deposited by wind along the shores (sand dunes). It was consolidated by calcareous material (mainly calcium carbonate ranging from micrite to sparite) precipitated from aqueous solutions in the pores of the sand. It was very porous—with the cement often incompletely filling spaces between grains—and had a high calcite composition. Thus the rock was also called a biocalcarenite, as the main constituents were calcareous clastic grains (fossilized shell fragments), with secondary amounts of quartz, and more rarely, feldspar, muscovite, chlorite, clays, and iron oxides.

A similar stone described as a calcarenite (calcareous arenite) was used in an interesting way in Italy. At Sassi di Matera in Basilicata, some houses and churches were hollowed out of the formation (the Calcarenite di Gravina), while others were built with stones cut from it. Laviano (1992) reported on a study of 11 thin sections taken from two deteriorated churches at the site. All the samples show the stone being composed of closely packed carbonatic lithoclasts, ranging from a few microns to hundreds of microns in size, cemented by sparitic calcite. Clay, organic matter, and iron hydroxides were also seen. Between and within the carbonatic clasts were very rare quartz, feldspars, pyroxenes, and amphiboles. Rare bioclasts could also be seen (gastropods, lamellibranchs, echinides, and foraminifers). Small pores were visible everywhere, although they were more evident in the outer layer. Most of these pores appeared to be a result of dissolution of cement. In some areas, incipient recrystallization of calcite in voids made a very compact external crust a few centimeters thick. This crust had a high content of organic matter and iron hydroxides, and tended to scale off. In thin section, the transition from crust to underlying rock was marked by a change of color and by an increased number of voids and coarser clasts along the detachment surface.

American brownstone buildings, very popular in the eastern United States, were made of a very brown or reddish-brown colored sandstone with a very fine feldspathic cement. Unfortunately, this cement is prone to weathering by conversion to clay minerals, especially in humid environments (Winkler 1994) (Fig. 4.3.21). Nonetheless, many important buildings throughout the United States are constructed of brownstone masonry (Behre 1933) (Fig. 4.3.22).

The 800 monuments of the ancient city of Petra in Jordan were carved from sandstone bedrock. Thin sections have been used to quantify the mineral composition of this rock, and to examine its structural properties (Fitzner and Heinrichs 1994). Porosity studies were done by thin-

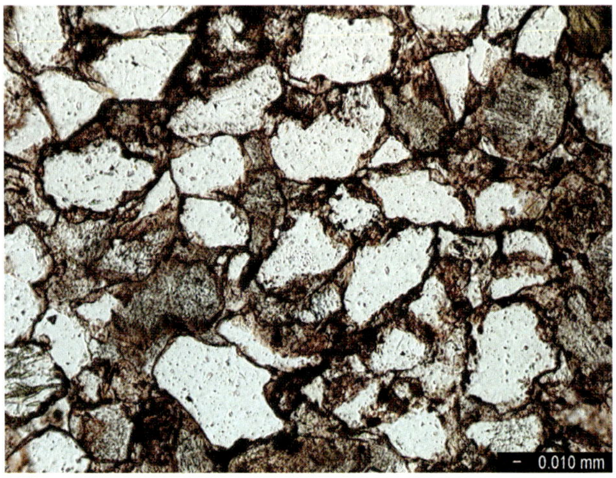

Figure 4.3.21 Sandstone, plane polarized light. This "redstone" from East Longmeadow, Massachusetts, is a popular building stone material in the eastern United States. It is a reddish-brown colored sandstone, rich in hematite, with fine feldspathic cement that is prone to weathering by conversion to clay minerals, especially in humid environments. Clay minerals are forming both in the cement material and as alteration products on larger feldspar framework grains.

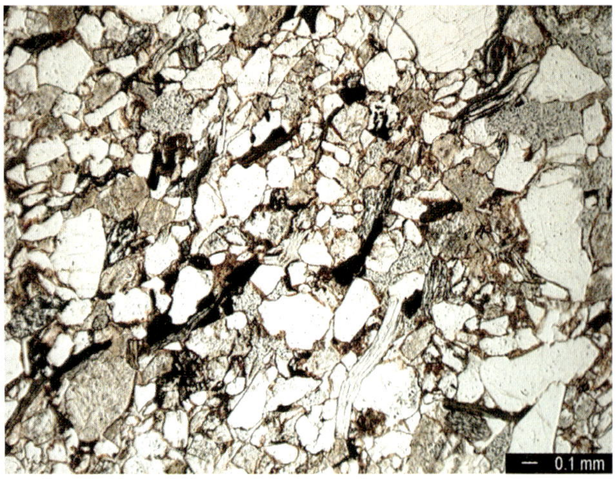

Figure 4.3.22 Sandstone, plane polarized light. Many historic buildings in the United States are constructed of brownstone masonry. This example is called Portland Brownstone, from Portland, Connecticut. As in the redstone of Figure 4.3.21, feldspars are converting to clay minerals; this example also has abundant micas (muscovite and biotite) along with black iron oxides, in addition to the quartz and hematite.

section analysis, mercury porosimetry, and nitrogen sorption method. Weathering appeared to be mainly through granular disintegration. The stone was susceptible to this process because of its structure that includes many large pores where water can infiltrate easily.

Two sandstones used in the construction of historic buildings in Florence were characterized and their source quarries identified (Banchelli et al.1997). Locating quarries is considered important in permitting restoration work to utilize the same stone material as the original building. One stone type was identified as feldspathic graywacke with clay matrix, the other as lithic arenite with carbonate cement. The extensive study included 220 samples taken from 11 different quarries; 129 samples were taken from buildings either under restoration or about to be restored. Unweathered samples were obtained by coring. The thin-section studies were supplemented by XRD analysis so that the clay minerals could be identified, considered crucial to distinguishing between quarry sites.

Even where sandstones consist of more than 99% quartz, they have properties that enable them to be distinguished. Götze and Siedel (2004) subdivided such sandstones (from the Elbe Zone and used as an ancient building material in Saxony): one type was medium to coarse grained, without pore-filling cement but with authigenic quartz overgrowths; the other type was fine to medium grained, with pore-filling cements consisting mainly of clay minerals (especially kaolinite) and/or iron oxides. Cathodoluminesence microscopy helped to subdivide the materials even further.

Sculpture

Sandstone is also an important sculptural material. For example, it was the most common material used for sculpture in western and north-central India. Richard Newman (1984) noted that the sandstones chosen were fine grained (with grain size about 0.1–0.2 mm) and well cemented (often with silica of low porosity), making them suitable for carving into intricate, three-dimensional sculptures. Indian sandstone sculptures can vary greatly in color according to the iron oxide content, from nearly white to purplish red. A cream-colored sandstone was used to construct the Mauryan-period edict columns in central India. Sculptures from Mathura, the southern capitol of the Kushan empire and an important artistic center throughout the Gupta period, are famous for their unique appearance, a very red stone spotted with white areas. It was quarried from outcrops located in northern Madhya Pradesh. This stone is described as a quartz arenite. Newman found that it often contained 5–10% detrital potassium feldspar (usually microcline). The detrital grains were coated with dusty red iron oxide, coloring the rock; gaps in the coating provided the white splotches.

The sculptures from the Buddhist stupas of Bharhut and Sanchi are sandstone (Newman 1984). Some pieces are quartz arenite (having less than 15% fine-grained matrix and more than 75% quartz among the detrital grains); and some are a quartz wacke (with more than 15% fine-grained matrix along with more than 75% quartz making up the detrital grains). The Bharhut pieces sampled had only a few percent matrix, and included a rare phyllite rock fragment as part of the detrital component. The Sanchi samples had a calcareous matrix that ranged from 5% up to about 20%.

Newman found that many other sculptures from central India of the fourth through twelfth centuries were also made of quartz arenite, quartz wacke, or argillaceous quartz arenite. For example, Gupta-period sculptures from Sarnath (the site where the Buddha gave his first sermon) were sandstone, with one sampled piece being a quartz arenite and another a quartz wacke containing 30–40% matrix. In these rocks, feldspars generally accounted for less than 10% of the detrital grains, and lithic fragments were rare (usually chert or fine-grained metamorphic rock). Feldspars were most often microcline, although sometimes plagioclase was present, and/or perthitic feldspar. The accessory minerals tended to be those that are relatively stable and resistant to weathering, such as tourmaline (usually well rounded), zircon, and rutile (in clusters of prisms). Opaque iron oxides included mainly magnetite or hematite, rarely ilmenite; these often occurred in patches filling spaces between detrital grains. Cement was generally silica, which tended to form overgrowths on the detrital grains. Sericite or other micas were sometimes found as matrix material.

In the quartz wackes, the matrix was often sericitic, creating some foliation in thin section; sometimes, however, the matrix was argillaceous. They were silica cemented, with more than 90% of the detrital grains, separated by matrix, comprised of quartz. As with the quartz arenites, the accessory minerals were limited to rutile, tourmaline, zircon, and iron oxide.

Newman was unable to identify specific locations where the materials for the sculptures were quarried, mainly because detailed petrographic descriptions of outcrops were not available, although the fact that some of the outcrops were quite extensive (continuing for as much as 400 km in some cases) would probably make specificity difficult.

He also identified an argillaceous quartz arenite as the material used for a sandstone sculpture from Orissa in eastern India. Most of the detrital grains were found to be quartz, about 0.1 mm in diameter, cemented by about 10% silica matrix (Newman 1984). Some fine-grained iron oxide trapped between quartz grains gave the stone a pinkish-brown color. Accessory minerals among the detrital grains included rutile, magnetite, zircon, and ilmenite.

Newman also noted that sandstone was widely used at the site of Konarak, an important Orissan temple devoted to the Hindu sun god, Surya, quartz arenite was used for sculptures in Rajasthan (western India), and, further north, sandstone appeared as a sculptural material in the Kangra district of Himachal Pradesh (northwest India) (Ohri 1975). He analyzed one sandstone sculpture from Himachal Pradesh, and described it as being a gray-brown quartz wacke with between 25 and 50% reddish-brown, iron-oxide stained matrix containing strings of mica. Some of the matrix appeared to be crushed phyllitic rock fragments, and there were a few chlorite-rich areas. Detrital grains were mainly angular to subangular quartz, most between 0.6 and 0.12 mm in size.

In the geological survey mentioned above of 172 early and late medieval sculpted stones from central and western

Scotland (Miller et al. 2006), identified types among the sandstones included litharenite, sub-litharenite, arkose, subarkose, and mudstone. Potential sources for the rocks were identified.

Douglas and Sorensen (2007) conducted detailed petrographic analyses of Southeast Asian Khmer sandstone sculptures in Bayon style of the Angkor period (twelfth–thirteenth century). One goal was to link distinctive mineralogical criteria to the types of sandstone used by the royal workshop. Long-term goals include identifying geological sources of distinctive stone types used in the Angkor period, and providing some criteria for classifying or interpreting sculptures that lack secure archaeological provenance. Thin-section petrography was augmented by scanning electron microscopy (SEM) and electron probe microanalysis.

The samples are described as terrigenous clastic sandstones with low carbonate content, medium to very coarse grained, and ranging from submature to immature texture. The clastic grains consist of about one-third each of quartz, feldspar, and lithic fragments, so are classified as lithic arkoses or feldspathic litharenites. The matrix consists of iron oxides and hydroxides, and clay.

The Bayon sculptures were further divided into two groups, based on several criteria. First was the presence or absence of finely divided acicular feldspar laths, which indicates that the sandstones formed from relatively young and unaltered volcanic rocks; secondly, the presence or absence of mottled albitized feldspar grains, where a calcium, sodium, and potassium-bearing feldspar in the anorthite-orthoclase-albite solid solution is transformed into the sodium end member, albite. Douglas and Sorensen proposed that all the samples containing partially albitized feldspars may have derived from the same, or a similar, geological source. Finally, amphibole grains were studied. Electron microscope analysis identified specific compositions. The data suggested that the same igneous source provided the amphibole detritus for those sandstones.

Ultimately, the authors concluded that many of the sculptures were mineralogically similar enough that they may have derived from the same quarry, and perhaps even from a single sandstone layer within that quarry. A small group of the samples differs from the main group; however, there is no evidence to suggest this second group derives from a single sandstone source. Characterization of a royal workshop stone type will be useful in identifying other sculptures that may have been produced in that workshop, and should facilitate locating the geological source.

Other uses

Shimron (1985) provided a thin-section analysis of a perforated stone tool, probably a whetstone, dating to the Middle Bronze Age and excavated in Israel. He confirmed his mineral identifications with SEM. The sandstone was gray, with alternating bands of gray and black. The black bands were formed by dark, opaque minerals while the gray band consisted of colorless light-transmitting minerals. About 65% of the black area was ilmenite, with minor amounts of rutile and zircon and a wide variety of additional accessory minerals. The main constituent of the lighter bands was white mica, with lesser amounts of quartz, plagioclase feldspar, and the dark minerals found in the black bands. The laminated banded structure was due to very weak metamorphism, leaving the rock harder and denser than it was originally, and thus quite suitable for use as a grinding and sharpening tool.

Sandstone was also used for grinding tools at Maya sites. Shipley and Graham (1987) studied thin-section samples taken from 18 *manos* (grinding tools) excavated at Maya sites in Guatemala and Belize. Of these, two were identified as being graywackes (both described as a porphyry volcanic crystal-lithic graywacke), and one was described as a semi-schistose metaquartzo-feldspathic sandstone. The two graywackes had crystal and lithic fragments that appeared to derive from a siliceous porphyry volcanic source, with a chlorite-rich matrix. The two samples examined were very similar. The sandstone material showed faint banding in thin section due to variations in grain size and concentration of inclusions. Through similar thin-section studies of nearby rock outcroppings, Shipley and Graham were able to deduce that, for this region, characterization of the sedimentary stones may not be very useful in pinpointing specific sources of raw material as the variation was not as great as with some other types of rock such as granite.

Sandstone was used in the construction of one of the most famous group of monuments from ancient India, the Ashokan Pillars. These were erected by the Mauryan emperor Ashoka, to commemorate the teachings of the Buddha (Upadhyay 1991). The pillars are made of a very porous sandstone, highly polished, and at one time were topped by carved capitals.

Visser et al. (1996) discussed thin-section studies of sandstones used in Germany for the construction of buildings, for ornamental purposes, and for Renaissance tombstones. They noted that in major trade centers such as Hamburg, sandstones from many distant locales were trade items in the medieval period and later. The authors hope to prepare a register for sandstones used in German monuments, including information about provenance, characteristics that could be used to identify materials from specific sources, and weathering phenomena associated with each sandstone. Their first two case studies were of dated tombstones: one in Hamburg dated 1610 using a stone quarried in the Elbe Sandstone Mountains in eastern Germany, and one in Hanover dated 1918 using Wealden sandstone from the south of Hanover. The primary method of analysis used was thin-section petrography combined with a heavy mineral analysis.

The Hamburg tombstone was shown in thin section to be fine grained and quartz rich, with lithic fragments (siltstone and schist) (Fig. 4.3.23), kaolinite and magnetite. The Hanover tombstone was made of an argillaceous sandstone (see Fig. 4.3.6) covered by a weathering crust of palmierite ($K_2Pb(SO_4)_2$, hypothesized to have formed by the reaction of a weathering of a paint layer that may have once covered the tombstone.

Dingelstadt et al. (2000) developed a petrographic atlas to use in selecting replacement and substitution materials for ornamental sandstones found in historic buildings and

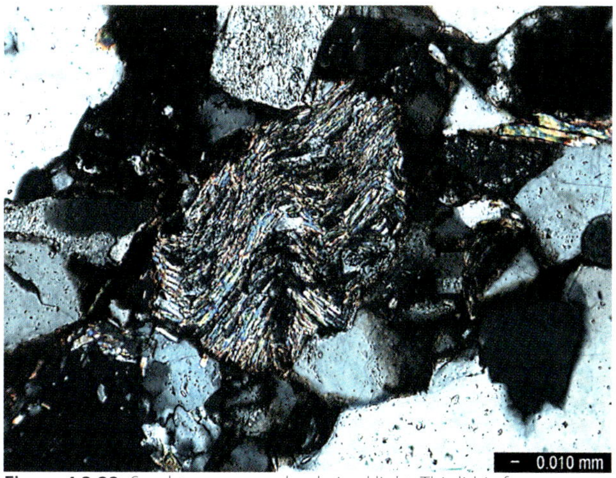

Figure 4.3.23 Sandstone, crossed polarized light. This lithic fragment is schist (muscovite mica, graphite, and some quartz) exhibiting a kink-folded texture.

monuments, mainly in southern Belgium. The atlas is intended to help not just in selecting matching rock types for restorations, but also in identifying the weathering potential of such rocks. Often the original ornamental stone is no longer available, or it may be too prone to deterioration. Samples from about 300 stones were taken from abandoned quarries in Belgium, Germany, Luxemburg, France, and the British Isles. Thin sections confirmed that many of the specimens are sandstones. The atlas first illustrates general macroscopic features, then microscopic features, followed by details of those microscopic features (such as the cement and fine minerals). For example, one sandstone shows fine and well-sorted detrital grains of quartz, plagioclase, glauconite, and iron oxides. The detail illustrates three layers of cement surrounding the detrital grains, composed of opal, lightly colored chalcedony, and fibrous chalcedony.

4.3.2 Sandstone deterioration mechanisms and conservation treatments

Thin-section analysis of sandstones has aided many studies of deterioration and tests of conservation treatment approaches. For example, Rossi-Manaresi (1976) conducted deterioration studies and experiments on consolidation of exfoliating sandstone buildings in Bologna. Samples were taken from the exfoliating crust, from the layer underneath that was lacking cohesion, and from the lesser altered core at depths of 6–10 cm into the stone. Thin-section analyses were conducted to compare the deteriorated surface areas with the less altered interiors. One sandstone type, for example, had a core area with good porosity, poor cohesion, and a large amount of calcitic cement. The main cause of decay appeared to be the dissolution of the calcitic cement by acid water near the surface areas. Cohesion becomes worse and porosity is increased as cement is lost, and a hard crust forms through recrystallization of calcite and gypsum. When the sandstone's quarry was identified, thin-section analysis showed that the same phenomenon had occurred there, along with a color change as iron minerals were oxidized and transported towards the surface.

Ten consolidation treatments were tried with the aim of finding a way of filling pores and fissures and consolidating the friable material to impart strength and prevent disintegration, and to avoid further deterioration by reducing the movement of water into the stone and thus preventing dissolution of the calcitic cement. A variety of tests was used to assess effectiveness before and after accelerated aging, such as porosity measurements, compressive strength, ultrasonic test, water absorption by capillarity, and yellowing. Thin-section petrography was not used in this phase of the research, although it was the primary tool used to characterize the degree and cause of deterioration.

Gruber and Sternad (1981) combined thin-section petrography with the use of ultraviolet fluorescence microscopy to develop a new means of identifying and mapping the presence of newly formed gypsum on deteriorating sandstone. They used Titan yellow stain on thin sections viewed in ultraviolet light. The stain produces an intense orange-yellow fluorescence on gypsum. This method was used to study deterioration of a recently built cathedral in Linz, Austria, undergoing severe pollutant attack.

Ordaz and Esbert (1985) illustrated thin sections from a study that compared petrographic characteristics of sandstones used in the construction of historic Spanish monuments to factors affecting their deterioration susceptibility (porosity and capillary suction). They noted that deterioration processes in monumental stones are closely related to movement of fluids (mainly water) through their interior, and that the movement of water is dependent on the quantity of pores (Fig. 4.3.24), pore size, and degree of connectivity between pores. However, they found that differences in clay mineral content, along with other textural and mineralogical factors, were often the most significant contributory factor—where clay content was more abundant, the stone was more inclined to disintegrate.

Ordaz and Esbert found that in sandstones, the nature of clay minerals, their abundance, disposition, and crystallographic habit tended to control the geometry of the pore system. Their illustrated thin sections showed areas of concentration of phyllosilicates as matrix material, as well as

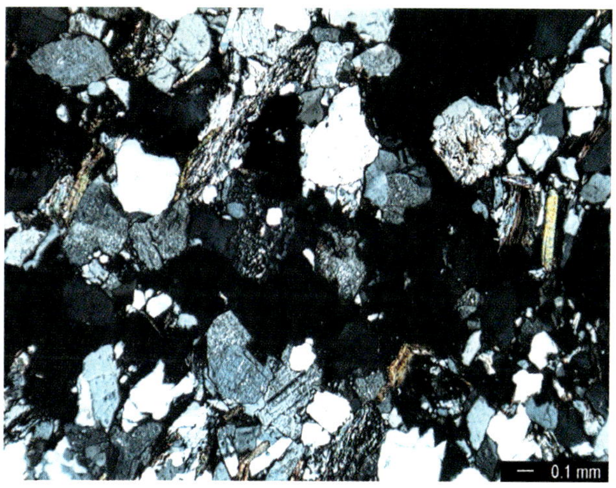

Figure 4.3.24 Sandstone, crossed polarized light. This brownstone is very porous. In crossed polarized light the pores appear black.

areas where clay minerals coat detritic grains. These concentrations of clay minerals were shown to affect a variety of pore parameters, overall porosity, and susceptibility to further alteration. Other textural aspects also seemed to be related to porosity (such as grain size and sorting). One implication of their study was that it may not be possible to generalize about susceptibility of sandstones to deterioration, and that assessing that susceptibility is a complex enterprise that must be done on a case-by-case basis.

Twilley and Podany (1986) employed thin sections to characterize Arizona red sandstone used on a variety of buildings in California, and to study its deterioration. The stone was identified as being an immature, poorly sorted calcareous feldspathic sub-orthoquartzite. Particle size was intermediate between sand and silt, and grains were angular. The striking violet-red color was due to coarse diagenic hematite and to iron-bearing red clays. Residual gypsum, not completely replaced by interstitial calcite, was the primary cause of deterioration of the stone, as the movement of water through the stone has redeposited gypsum in the form of dark gypsum crusts on the surface.

Casi et al. (1988) used thin sections of samples from the fourteenth-century Ducci-Del Rosso Mansion in Sansepolcro, Italy, to characterize the stone and identify signs of deterioration (such as the presence of newly formed minerals, formation of secondary porosity, loss of cohesion, disaggregation, and presence of microcracks), by comparing unaltered cores with weathered areas. The stone was a feldspathic graywacke with calcitic cement from different geological formations. One type had about 20% calcite, while another had about 15–20% dolomite, and a finer grain size. Nearer the surface, water has caused a loss of calcite cement, which has reprecipitated to form a surface crust. Degradation of the cement increases overall porosity and creates disaggregation of clastic grains.

Kirchner et al. (1991) studied the weathering characteristics of Baumberger Kalksandstein, a lime sandstone widely used as a building material in northwestern Germany. Unfortunately, this material is strongly affected by air pollution. Samples taken from quarries and from buildings were studied in thin section. It was described as having a calcite and micritic matrix with fossil remains, subangular grains of quartz, and some glauconite. Because of the presence of calcite and a high porosity (Fig. 4.3.25), the stone is easily affected by environmental changes and pollutants. Thin sections clearly illustrated the role of porosity in the development of damage.

In northeastern Belgium, a goethite-cemented ferruginous sandstone of local origin was used in the construction of ancient monuments, and thin-section analyses, in conjunction with XRD analysis, were crucial in understanding the deterioration mechanisms of that stone (Fobe et al. 1993). Quartz and glauconite are the principal components, both of about the same size (coarse), cemented by iron oxides. The pores between detrital grains may be filled only partly with cement. Black crusts are visible, and in thin section gypsum can be seen in the surficial zones, creating damage as it crystallizes in existing pore spaces created by incomplete cementation, applying pressure to the stone. In addition, many glauconite grains were fractured by the crystallizing gypsum, and the goethite cement was disrupted, fracturing the stone structure itself. In some buildings the sandstone was mixed with limestone, providing a carbonate source for gypsum. In other buildings where only the sandstone was used, lime mortars may be the major source of calcium ions.

Santamaría et al. (1994) included thin-section study among a variety of analytical techniques used to identify the origins of gypsum crusts on historic calcareous sandstone buildings in Spain. The thin sections showed that the crusts were comprised of rock minerals bound by a sulfate and calcite cement, which was heavily cracked. Sulfate cement material and cement cracking were not visible on fresh stone at the quarry sites. Although quarry rock also had calcite cement, the weathered gypsum crust showed a calcite cement that was more recrystallized; feldspars in the deteriorated samples were also more weathered than those from fresh rock at the quarry site. In the crusts, sparitic calcite replaced the original texture of the stone, and fine-grained and fibrous gypsum was found filling cracks and as a binding material replacing the original micritic cement of the stone. Gypsum was sometimes replaced by anhydrite. The authors hypothesized that the observed crusts resulted from sulfation of the rock so that micritic calcite cement was converted to gypsum, which crystallized as a cement replacing the micritic cement, and also filled cracks. Some of the micrite also changed to sparite, and the white color of the crust was due to gypsum and recrystallized calcite. The sandstone itself was the source of the calcium in the gypsum, while sulfur came either from pollution or from microorganisms dispersed in the rock.

Thin-section petrography was combined with other laboratory techniques to study deteriorating historical monuments in Thailand (Riederer and Prasartset 1995). The thin sections showed that a local sandstone was used as the building material. A very broad range of grain sizes was found, with grains ranging from as small as 0.05 mm up to 1 mm. The grains were not intergrown, but were cemented

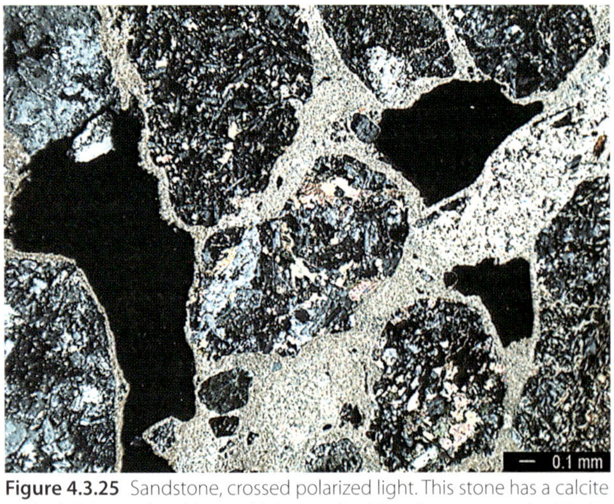

Figure 4.3.25 Sandstone, crossed polarized light. This stone has a calcite and micritic matrix, with very high porosity. Here the pores are the black areas, and the extreme birefringence of the matrix helps identify it as calcite and very fine-grained recrystallized micrite.

by a siliceous matrix. In the illustrations of the thin sections, one can easily see that feldspars between quartz grains have become completely altered to micas and clay minerals, with plagioclase greatly altered to sericite, and the orthoclase and microcline altering to clay and mica (although less completely). Micas themselves were quite altered: muscovite to sericite and biotite to chlorites. Hematite and magnetite were also altered, to iron hydrates. These alteration processes created a loosening in adherence of primary silicates to the matrix, as the silicates developed small gaps and areas of separation that appeared as open pores and fissures. The increased porosity permitted water to penetrate deeper into the stone, attacking more and more silicate minerals and accelerating the alteration processes.

Riederer and Prasartset deduced that alteration of silicates in the tropical climate was already underway in the rock outcrops before quarrying had even begun; exposure to high temperature and high humidity on the surface of buildings accelerated this natural chemical weathering process. The authors undertook laboratory and field tests to identify an effective protective system for sandstone buildings in Thailand.

Alaimo et al. (1996) reported on a fourth-century B.C. theater built at a Greek colony in Sicily and excavated in the 1950s. The builders used a local biocalcarenite cemented by gypsum. In spite of efforts to protect the structure, the stone was deteriorating. Thin-section examinations, combined with other methods of analyses, aimed to more thoroughly characterize the stone and identify the causes of deterioration. Thin sections made from samples taken from the building were examined, along with samples taken from rock outcroppings on the hills close to the theater. The authors hoped to find the origin of the stone used for the building which could then provide a supply of stone blocks for further laboratory tests of possible protective treatments.

They concluded that the stone could be classified as gypsous-biocalcarenite, of a type found in central-southern Sicily called "Serie Gessoso-Solfifera." The stone for the theater was taken from the lower part of that series, where there are several beds of a biocalcarenite with some clay and a gypsum cement. The fine grain and workability would have made it a good choice. Thin sections showed good matches in samples taken from the theater and the outcroppings.

The illustrated thin sections demonstrated that the calcareous fraction comprised mainly small microfossils and bioclasts. Identified were various microphoraminiphera, echinoids, calcareous algae, bryozoans, and mollusks. A twinned gypsum cement comprised 20–30% of the rock. In addition to calcite and gypsum, there were accessory occurrences of glauconite, quartz, feldspars, volcanic glass, and pores. Water appeared to be producing the major deterioration problems by dissolving the gypsum in the stone and redepositing it as a soluble salt. Thin sections showed that the bioclastic fabric had developed a high porosity because of dissolution of the gypsum cement. The surface ends of the thin sections showed lichen penetration, along with calcium oxalate deposits.

Another Spanish monument constructed of sandstone, the church of Betlem in Barcelona, was analyzed with thin sections, combined with other analytical techniques, in order to characterize the stone, identify the degree and mechanisms of deterioration, and evaluate a cleaning treatment to be used on the building (Garcia-Vallès et al. 1997). The thin sections were used primarily to identify the main components of the rock and the proportions in which they are present, and to describe their microtexture. Samples for thin sections were taken before, during, and after treatment.

Sample sites were selected to permit identification of all rock types and all preservation problems. These samples were then used to identify the range of materials present and their state of deterioration, to help in planning a treatment strategy, and to monitor results. An illustrated thin section taken perpendicular to the surface of a sandstone comprised mainly of quartz and feldspar in a silicic cement showed adjacent quartz and potassium feldspar grains with different levels of deterioration. The potassium feldspar was much more deteriorated, with an accumulation of dark particles forming a film on the surface. Garcia-Vallès et al. noted that as quartz is quite stable in the face of chemical and physical weathering agents, it had not altered within the historic time scale. In contrast, the feldspars were affected by hydrolysis, and released alkaline and alkaline-earth cations, producing new mineral structures such as kaolinite, which was itself subject to mechanical erosion. Different weathering patterns found with two different sandstones on the building were partly related to differing amounts of feldspars each contains. A higher proportion of feldspar to quartz would have made the stone softer and easier to work, but more prone to weathering. The degree of weathering was also related to whether or not the geometry of a particular location permitted the accumulation of water.

Other problematic deterioration processes stemmed from the presence of iron oxides, usually pyrite, within the stone. Exposure to water led to the production of limonite (rust, a mixture of iron oxides and hydroxides) and the mineral jarosite ($(K, Fe)SO_4.2H_2O$). Structural iron elements have also rusted, and their washing has produced stripes and other areas of color change on the stone. Other visual problems included a black film of fly ash over the entire surface, a result of a polluted environment. Finally, a fire in the early twentieth century caused thermal effects on some of the quartz leading to areas of severe disintegration of the stone. However, the lack of carbonates in any of the stones meant that this building stone was relatively immune to gypsum deposition, even in the polluted environment of Barcelona.

Thin sections showed how the cleaning regime (using brushes and spraying water with a neutral detergent) had different effects on different sandstone types. On the sandstone with siliceous cement, the grains were strongly linked and did not become detached during cleaning. However, a black film could only be removed from the most exposed points—it remained in deeper points. But, if there was clay in the matrix, the link between grains was weaker and the cleaning efforts caused some grains to be removed along with the black film. Visually, the cleaning process appeared to be better, but has actually resulted in the loss of original material.

Thomachot and Jeannette (2000) studied deteriorating sandstones used for building Strasbourg Cathedral. One type of sandstone used was yellow, with oxide and hydroxide brown aureoles, and now covered with a glossy black crust. The crust appeared different from typical weathering crusts in that it was very thin whereas more usually, weathering crusts become thick and spongy, leading to surface detachment and friability due to gypsum intrusions. The thin crusts, instead, seem to give the sandstone impermeability and even hydrophobia. Thin sections of the yellow sandstone covered with the glossy black crust showed that on the surface, grains were covered with an iron oxide and carbonaceous matter layer of 10–20 μm; along 1 mm under the surface, porosity was reduced and the oxides were present in large quantities. This produced a compact zone on the surface filled with oxides and with decreased porosity, which seems to have protective qualities preventing the interior of the stone from deterioration.

Dreesen and Dusar (2004) reported that a petrographic atlas developed for stones used in historical monuments in Limburg (northern Belgium) listed over 35 different stone types. The atlas included detailed stone descriptions, and is serving a useful purpose as an identification and decision-making tool for building restoration efforts. Most of the building stones are sedimentary rocks, and most are no longer available, so the inventoried historical monuments serve as evidence for a previous geological richness in quarry materials. The authors found thin-section petrography helpful in discriminating between different varieties of ferruginous, quartzarenitic, litharentic, and subarkosic sandstones, as well as in assessing provenance, weathering, and bioreceptivity potential. They used fluorescent epoxy-impregnated thin sections to aid in assessing porosity.

4.4 Conglomerate and Breccia

If a consolidated rock is comprised of mainly gravel or rubble (fragments larger than 2 mm) it is called a conglomerate (if the grains are mainly rounded) (Fig. 4.4.1) or a breccia (if the grains are mainly angular) (Fig. 4.4.2). These rocks are characterized by identifying the major rock types from which the fragments originated (volcanic, plutonic, sedimentary, or metamorphic) and the type and amount of interstitial material cementing the fragments together (Carozzi 1993). Conglomerate and breccia are only rarely used in the production of cultural materials, as their physical properties are not very amenable to uses for which stone is generally desired (detailed carving, architectural or tool applications, etc.).

However, they have sometimes been employed for architectural purposes. Thin-section analysis was helpful in research to identify building stones used in ancient monuments of Greece. Breccia and conglomerate were identified at several sites, including Athens, for use in retaining walls, foundations, and support of floor slabs (Dermentzopoulos et al. 1988). A light-colored conglomerate and a breccia were also found among the many stones used in the construction of the Minoan palace of Knossos (Papageorgakia and Mposkos 1988). A pink and yellow breccia was used for some Roman monuments and houses (Lazzarini 2002).

More recently, conglomerate was quarried in Pennsylvania for construction of railroad bridges and other heavy masonry uses; it was also utilized for building construction. It has been noted that the freshly quarried conglomerate has variegated colors in gray, tan, and pink shades, but within 50 years the colors fade to an even gray; in a city atmosphere it tends to darken even more and become stained almost black (Behre 1933).

In the eastern United States, a conglomerate composed of cemented small seashells (also called coquina) was quarried in Florida and used for architectural purposes in the seventeenth and eighteenth centuries (McKee 1971; Fenn et al. 1987). To make the material smoother and more water resistant, the blocks of coquina were usually covered with stucco. However, the porous nature of coquina makes it particularly susceptible to deterioration upon exposure to water.

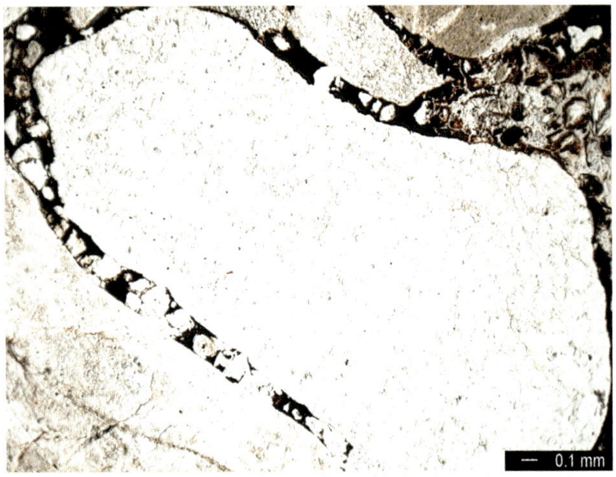

Figure 4.4.1 Conglomerate, plane polarized light. A conglomerate is defined as a consolidated sedimentary rock comprised mostly of gravel or rubble (fragments larger than 2 mm) with the grains being mainly rounded, as with this large colorless quartz grain.

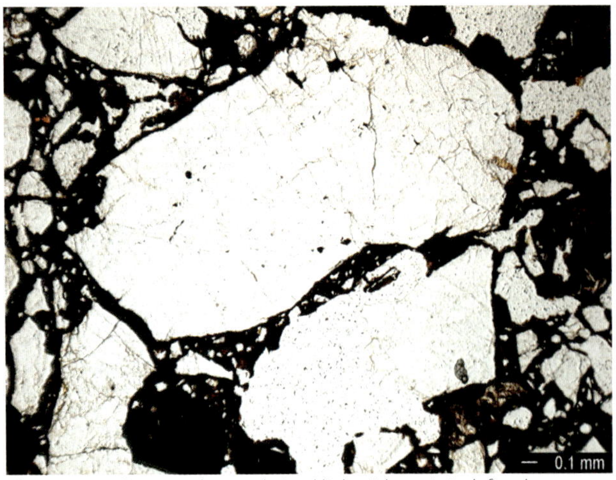

Figure 4.4.2 Breccia, plane polarized light. A breccia is defined as a consolidated sedimentary rock comprised mostly of gravel or rubble (fragments larger than 2 mm) with the grains being mainly angular, as with these colorless quartz grains.

Conglomerate and breccia are sometimes used for decorative stones. Conglomerates may have an interesting appearance because the worn pebbles (representing long transport distances) of the detrital component often include a mix of fragments of varying composition and color. Breccias are comprised of more angular pebbles deposited closer to their source, but may also include a variety of colors and fragment sizes (Winkler 1994).

One example is found in the Marmara region of Turkey. Quarries at Hereke (ancient Herakleia) were worked for a conglomerate of gray, brown, and black pebbles cemented by a red, limonitic calcium carbonate. The quarries are known to date back to at least the fifth century A.D., where they were used to provide stone for a doorframe of the second Haghia Sophia in Istanbul. Subsequently, the stone was used for many other buildings in Istanbul, mainly for window frames, courtyard columns, and exterior elements. A similar conglomerate with a yellow limonitic calcium carbonate cement, from quarries in Mihaliç (Karacabey), was used locally for architectural decorations such as window frames and door sills. Similarly, breccia quarries in Turkey have been worked for decorative material (Erguvanli et al. 1988).

A conglomerate from local river deposits was also exploited as one of the building materials (for foundations) at monuments of the ancient Olympia area (Varti-Matarangas et al. 1995). It consists of pebbles of various sizes, comprised mainly of limestone, chert, and quartz.

The Mogao grottoes in China—a site famous for its many Buddhist wall paintings and clay sculptures and for Buddhist texts found there—were carved out of a conglomerate cliff. The conglomerate is described as argillaceous, of low strength, and only loosely cemented, thus easily eroded by windblown sand, with pebbles coming loose from their surrounding matrix. The conglomerate has high porosity and high water absorption, thus undergoes additional deterioration through freeze–thaw cycles (Zuixiong et al. 1997).

In the geological survey already discussed by Miller et al. (2006) of early and late medieval sculpted stones from central and western Scotland, one of the late medieval objects was identified as being made of conglomerate, with clasts of igneous pebbles (volcanic) that were angular to subrounded, and poorly sorted. The matrix was predominantly clay. In addition to the pebbles, minerals present included polycrystalline and monocrystalline quartz, weathered feldspar, very minor biotite, chlorite, and other micas.

4.5 Sedimentary Carbonates

Chemically or biochemically deposited calcium carbonates are a primary form of nondetrital sedimentary rock. The most common minerals include calcite, dolomite, and aragonite; the rocks formed include chalk, limestone, dolomite, and, more rarely travertine or tufa. Carbonate rocks are defined as having more than 50% carbonate minerals.

A wide variety of microfossils can be expected in thin sections of carbonate rocks such as foraminifera, ostrocods, radiolaria, diatoms, sponges, corals, mollusks, worms, and algae. Many well-illustrated reference manuals permit thin-section identification, and the microfossil mix may prove crucial to characterizing and sourcing carbonate rock samples. If possible, it is helpful to have two thin sections for each sample, one cut parallel and the other perpendicular to the stratification. Thin sections of fossil-containing carbonate rocks are often made thicker than ordinary thin sections (up to 50–70 μm in thickness) because details of microfossils can be obscured if the section is ground too thin. However, this then changes the polarization colors of the minerals present (for example, quartz is yellow, orange, or red instead of gray) (Horowitz and Potter 1971; Flügel 1982; Bignot 1985; Adams et al. 1997; Adams and MacKenzie 1998; Scholle and Ulmer-Scholle 2003).

Much effort has also gone into development of methods for grain-size analysis in the study of limestones. Flügel (1982) reviewed methodology of textural analysis in the microfacies analysis of limestones. He discussed how methods originally developed for the study of siliceous clastic rocks have also been applied to carbonates and carbonate rocks in order to obtain more information about their depositional environments.

4.5.1 Chalk

Carbonate mud forming from the deposition of the remains of calcareous plankton on relatively shallow ocean floors results in chalk; there are a few freshwater deposits as well. Since chalk tends to be very porous, friable, and soft, it has had somewhat limited use as a cultural material (Fig. 4.5.1). White chalks are greater than 90% calcium carbonate (low-magnesium calcite). Most of the carbonate is in the form of very fine organic particles, such as coccoliths (lime secretions of algae), foraminifera (single-celled shellfish whose minute shells are perforated by pores), and other organisms (Fig. 4.5.2). Often it consists of foraminifera aggregates cemented by a matrix of coccoliths. Impurities can include clay minerals, glauconite, pyrite, and alkali feldspars.

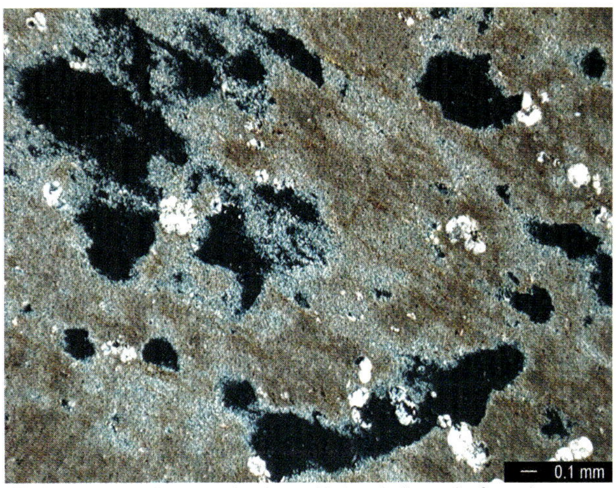

Figure 4.5.1 Chalk, crossed polarized light. Chalk results from the deposition of carbonate mud formed from the remains of calcareous plankton on relatively shallow ocean floors, and occasionally in freshwater deposits. Chalk tends to be very porous, friable, and soft. Here the many pores in this specimen appear as black holes.

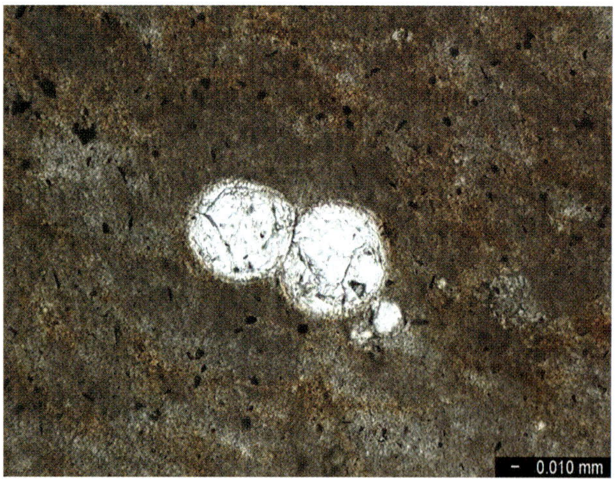

Figure 4.5.2 Chalk, plane polarized light. Larger microfossils are visible within the fine-grained microcrystalline carbonate mud.

Some chalk beds have been hardened through cementation processes (Williams et al. 1982; Greensmith 1989; Pearson 1992).

One cultural use of chalk has been as a building material, for example, one variation of "rammed earth" construction is "rammed chalk." A ramming process is used to essentially produce an artificial rock, more compressed and harder than the original material. Generally, no additives are needed, as there is enough natural moisture present to aid in compaction. Rammed chalk was used extensively in England in the production of historic period houses and other forms of vernacular architecture (Pearson 1992).

Chalk block or rubble construction was also used in England, although the chalk was sometimes faced with brick for aesthetic and protective purposes. This method, also called "chalk adobe," uses blocks made from crushed chalk mixed with water and set with a mortar of chalk and sand (Pearson 1992). Sometimes chalk is added to clay in the construction of chalk-mud walling, increasing strength and durability and decreasing setting time. Chalk has an advantage in that it is relatively easy to quarry and to work, although it is prone to moisture-related deterioration. Used alone, it is particularly good for finely carved interior elements (Wright 1991). A carvable soft chalk often called "clunch" has been used in England both as a building material and as a sculptural material (Ritchie 1973).

Thin section analysis showed that Cretaceous chalk was used as the material for construction of an eighteenth-century porch at the church of Sainte Marguerite in Belgium (Dingelstadt et al. 2000). Because of the friability and susceptibility of the material to weathering (many microcracks could be seen in thin section), the decision was made to select a different material for restoration efforts. Using a petrographic atlas, materials were chosen to be visually similar to the original chalk but more durable. Thus, a white fine-grained limestone, seen in thin section to contain few iron oxides and having a very low porosity (thus with good long-term weathering potential), was selected as a replacement for the original chalk.

4.5.2 Limestone and dolomite

Compact carbonate rocks in which calcite (calcium carbonate) predominates over dolomite (calcium magnesium carbonate) is called a limestone; one in which dolomite predominates is called a dolomite (or, occasionally, a dolostone). Limestones and dolomite rocks have some primary components, the ratios of which are often used in classification schemes to describe such rocks:

- Grains, or aggregates, of calcium carbonate (also called allochems), which are distinct particles the size of coarse silt or larger. Allochems can include fossils (Fig. 4.5.3), ooids (Fig. 4.5.4), pellets (Fig. 4.5.5), or intraclasts (fragments of eroded carbonate rock deposited as conglomerate) (Fig. 4.5.6).
- Carbonate mud (microcrystalline carbonates, or micrite) (Fig. 4.5.7).
- Detrital quartz, clay, or other noncarbonate minerals (which can include anhydrite, gypsum, chert, glauconite, feldspar, and phosphate) (Fig. 4.5.8).
- Larger carbonate crystals that often are secondary deposits filling pores (also called sparite) (Fig. 4.5.9). There can also be secondary replacement crystals of dolomite, evaporites, or other noncarbonate minerals, and pore spaces filled with air, water or hydrocarbons (Horowitz and Potter 1971; Scoffin 1987; Adams and MacKenzie 1998; Scholle and Ulmer-Scholle 2003).

The primary mineral found in limestone is calcite (rhombohedral calcium carbonate, the stable form at ordinary temperatures). Other carbonate minerals may also be present, including aragonite (orthorhomic calcium carbonate, the form usually appearing as precipitation from seawater), dolomite (rhombohedral calcium magnesium carbonate), magnesite (magnesium carbonate) and siderite (iron carbonate). Impurities may also be present, such as quartz, pyrite, clay minerals, glauconite, phosphates, albite, and calcareous or siliceous organisms (Williams et al. 1982; Greensmith 1989).

It is very difficult to distinguish between calcite, dolomite, and aragonite in thin section, so thin-section examina-

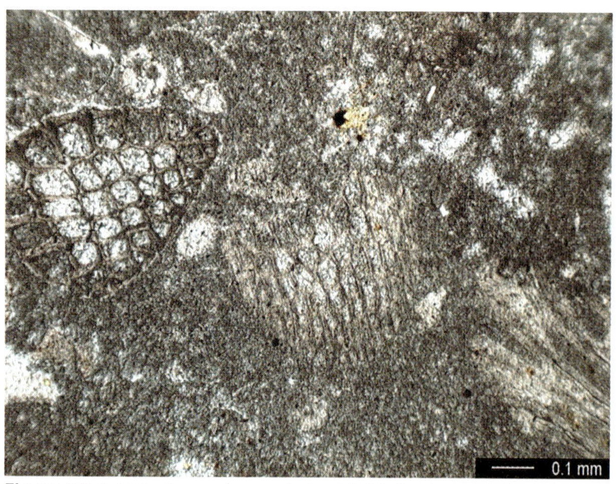

Figure 4.5.3 Limestone, plane polarized light. One type of allochem, or silt-sized or larger discrete particles or aggregates of calcium carbonate, is a fossil (fragments of the hard parts of calcareous organisms, the unbroken shells of individual organisms, or other animal or vegetable remains). These examples are bryozoans.

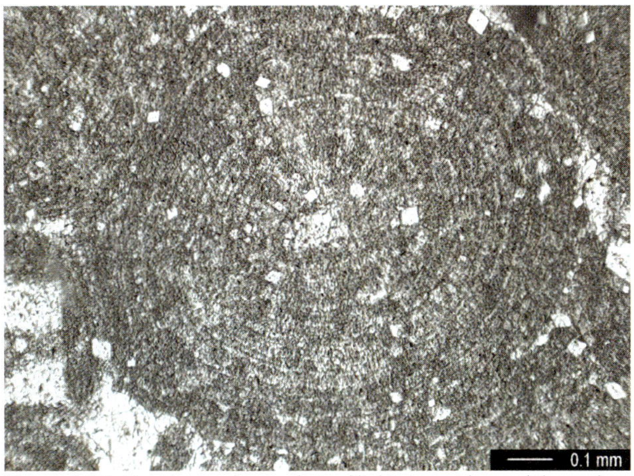

Figure 4.5.4 Limestone, plane polarized light. An ooid, or oolith, can be spherical (as here) or ellipsoidal. Ooids are sand-sized particles of calcite or aragonite formed by chemical accretion around a nuclear particle. They often form in radial or concentric layers, as a grain rolls back and forth over time in carbonate-saturated water. Here we also see some scattered rhombohedral carbonate grains.

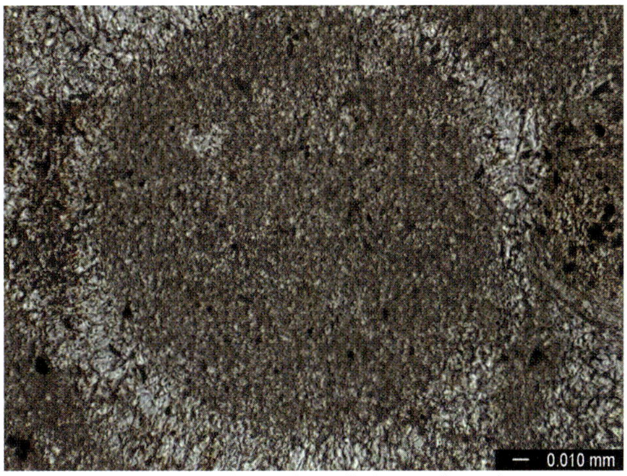

Figure 4.5.5 Limestone, plane polarized light. A pellet is also a sand-sized spherical or ellipsoidal grain, consisting of microcrystalline carbonate, but unlike ooids, pellets have no nucleus or apparent internal structure. They are usually 0.03–2.0 mm in diameter, and are often found squished together in clumps. They are thought to represent fecal pellets of marine invertebrates. This pellet is surrounded by a sparry cement.

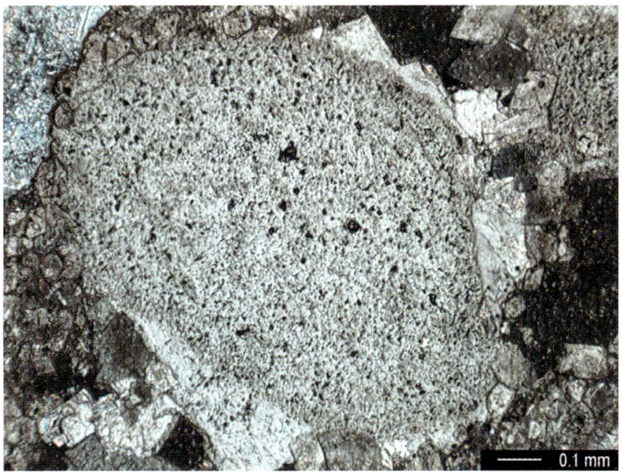

Figure 4.5.6 Limestone, crossed polarized light. An intraclast is a chunk of eroded limestone deposited as a conglomerate. Generally at least 20 µm in size, and clearly discrete grains, intraclasts are often made up of microcrystalline or cryptocrystalline carbonate, looking rather like a large pellet, as seen here.

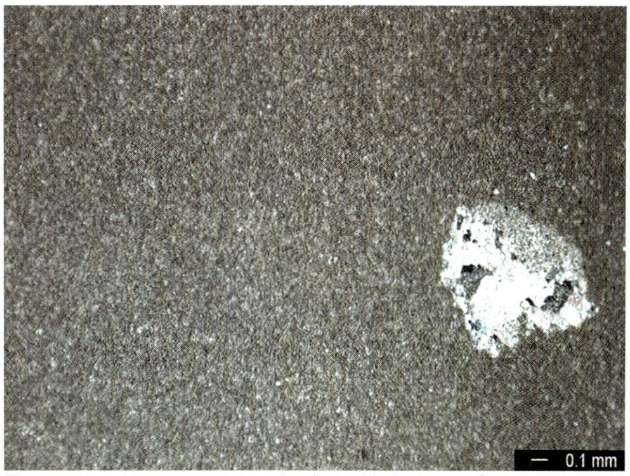

Figure 4.5.7 Limestone, crossed polarized light. Most of this specimen is comprised of carbonate mud, clay-sized calcite, generally called micrite (a word condensed from microcrystalline calcite). It is very fine grained, especially compared with the larger calcite inclusion seen on the right side of the image.

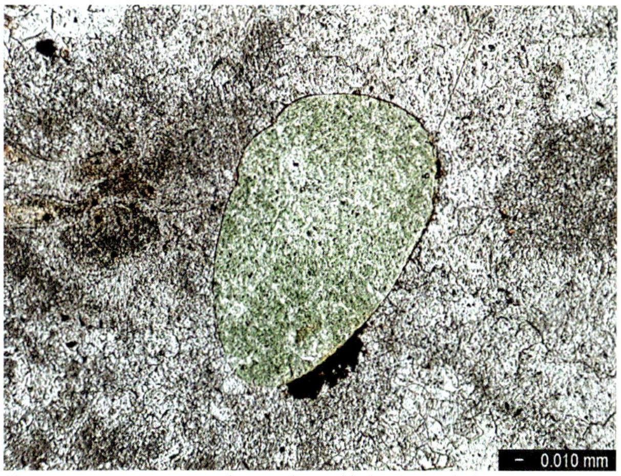

Figure 4.5.8 Limestone, plane polarized light. Detrital glauconite is often a component of limestones. In plane polarized light it is green and somewhat pleochroic. It occurs in pellets that are in part aggregates of minute crystals and in part single crystals. This pellet of glauconite is found amid a mainly micrite groundmass with some sparite.

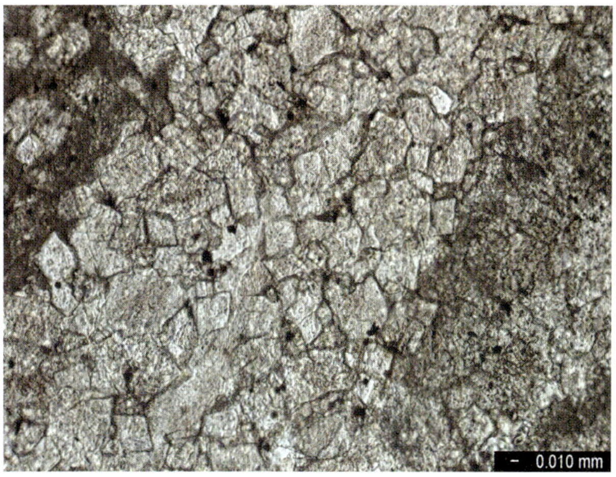

Figure 4.5.9 Limestone, plane polarized light. Sparite is crystalline carbonate, forming grains of at least 30 µm in size (0.03 mm). It is often a secondary deposit filling pores, such as this sparry cement here filling in the areas between fossils and intraclasts.

tion is often augmented by selective staining or etching, or with cathodoluminescence microscopy (Hutchison 1974; Greensmith 1989). In staining, alizarin red S is used to help distinguish calcite from dolomite, and potassium ferricyanide to distinguish iron-bearing carbonates from those with little or no iron. In cathodoluminescence, one is looking for natural materials that emit visible light when bombarded with an electron beam, and carbonate minerals are prone to this. Ordinary polished thin sections are often left uncovered to allow for staining or etching to differentiate between various carbonate minerals and to locate noncarbonate components. For example, calcites tend to show shades of yellow and orange when they luminesce, dolomites tend to be red (Flügel 1982; Adams and MacKenzie 1998).

Limestone rock consists of lithified carbonate sediments, and thus marine organisms may be present, having contributed to those sediments by precipitating calcium carbonate from seawater or secreted carbonate skeletons. Evidence of those organisms may include bioclasts, or skeletal particles and nonskeletal grains (Fig. 4.5.10); interstitial spaces of the rock may be filled by carbonate mud (micrite) or precipitated calcite (sparite) (Cruciani-Fabozzi et al. 1991; Carozzi 1993). Given the presence of such microfossil material, paleontological analysis is often an important part of characterizing a limestone. Thin sections that are thicker than normal (between 30 and 70 μm) are often used in order to best identify all the textures and structures that may be present; in contrast, very fine-grained limestones (such as micrites) may require ultra-thin sections (0.5–5 μm) (Flügel 1982).

If a rock is composed mainly of calcite or aragonite, the term "limestone" is used; if the main constituent is dolomite, it is called a "dolomite" or, sometimes, a "dolostone." A mixture may be called a "dolomitic limestone" or "calcdolomite" depending upon the dominant mineral. If additional minerals are relatively abundant, they will be included in the name, e.g. "glauconitic limestone" or "sandy limestone" (Williams et al. 1982).

One very common type of limestone is "fossiliferous limestone" (Fig. 4.5.11) formed from the calcium carbonate skeletons or shells of marine organisms. Some of this material from organisms is broken down into carbonate mud or sand, and may be incorporated in the carbonate rock. Limestones where grains (particles of calcium carbonate silt sized or larger) predominate are called "allochemical" while those in which microcrystalline calcite (micrite) predominates are termed "orthochemical." Within these categories there is a wide variety of textures and descriptive terms that can be used to characterize a specimen (Williams et al. 1982; Greensmith 1989; Gauri and Bandyopadhyay 1999; Scholle and Ulmer-Scholle 2003).

Most dolomites result from processes that replace calcium carbonates with magnesium carbonate, and the structures and textures of the original rocks are often preserved to some extent. Although the replacement mechanisms are not yet well understood, they appear to involve circulation of seawater in areas where there is a continuous flushing of fluids (Carozzi 1993). Weathering of the rock can initiate dedolomitization or reconversion to calcite. Dolomites

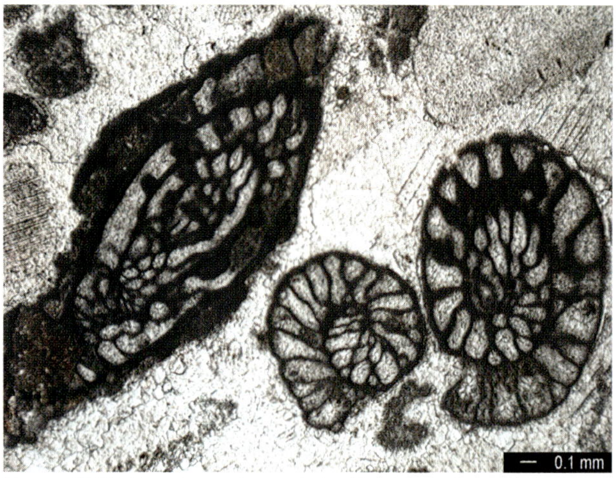

Figure 4.5.10 Limestone, plane polarized light. Limestones often include abundant bioclasts, such as these foraminifera; interstitial spaces are filled with sparite.

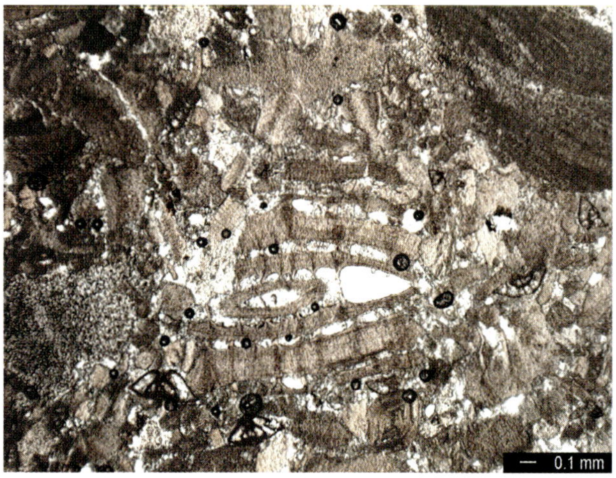

Figure 4.5.11 Fossiliferous limestone, plane polarized light. This specimen, from a carved head attributed to Egypt or Anatolia, is a biomicrite (microcrystalline calcite containing fossils). Fossils make up approximately 50% of the stone. In this view, fossils visible include a nummilite and other foraminifera, algae and algal mat, and echinoderm fragments. The stone is very porous, and many of the fossils show evidence of leaching (no internal structure, and a single dark micrite rim).

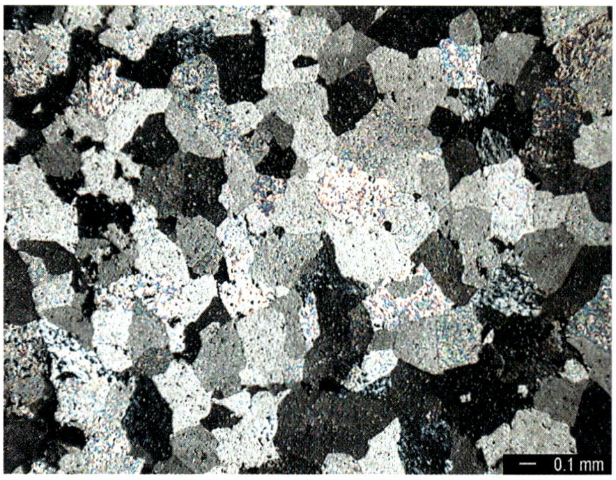

Figure 4.5.12 Dolomite, crossed polarized light. Dolomites tend to have a uniform grain size and few accessory minerals or microstructures, as in this example. Interference colors are similar to those of calcite.

tend to have a uniform grain size, and have fewer accessory minerals or microstructures than limestones (Fig. 4.5.12) (Williams et al. 1982; Greensmith 1989).

As with sandstones, one method of characterizing limestones is to determine the frequency, size, and roundness of particles in addition to identifying the types of particles present and their structures or fabrics. Grains are defined as clastic particles with well-defined boundaries. The maximum apparent grain size is measured, grain-size classifications are computed as with sandstones, and various statistical parameters are calculated such as average grain size, sorting, shape, and average degree of roundness (Flügel 1982).

4.5.3 Limestone and dolomite as cultural materials

Architectural uses

Limestone is widely used as an architectural material, for example, in the construction of medieval cathedrals in England, and for building most of the Gothic cathedrals north of the Alps, with lighter and softer limestone being used to carve intricate interior details (Penny 1993). One of the most famous limestone monuments is the monolithic Great Sphinx in Egypt, along with other limestone Pharaonic monuments dating back almost 5,000 years (Gauri and Bandyopadhyay 1999).

Hart (1988) provided thin-section descriptions of dolomites, also described as magnesian limestones, being quarried in the British Isles for building stones. In her discussion of ten different quarries, she noted that the color and texture of the stones can vary extensively along a single outcrop. The level of impurities was generally low, including minor amounts of quartz, calcite, clay, and iron oxide along with the dolomite. Exceptions included some quarry stones that were quartz rich. Most of the stones had various types of relict structures present such as ooliths, pellets, and bioclasts. Because Hart was interested in looking at issues of durability and porosity in building stone selections, in preparing her thin sections they were vacuum-impregnated with a dyed resin so that the pore structures could be observed more easily. Thin sections of samples from each of the ten quarry areas were described, and some were illustrated. Although ten were dolomite, they varied considerably. For example, the stone from one quarry (Abbey) had a rhombohedral texture, dominated by large rhombic dolomite crystals; while that from another quarry (Cadeby) had an oolitic texture, dominated by ooliths.

Thin-section analysis differentiated several limestones used for architectural and ornamental uses in the ancient cities of Aegean Thrace (Xidakis and Samaras 1989). A gray limestone was seen to be a fine-grained fossiliferous stone with mostly calcite, and 6–7% dolomite. A hard and durable material, it was used for foundations of buildings. A white limestone was described as coarse to medium grained, a biogenic rock with 30–40% dolomite; a relatively soft rock that is susceptible to weathering. In some areas fragments of nummulites, bryozoa, or other organisms were visible, and the rock was quite friable. Areas of marly material also contributed to lack of weatherability of the stone, making it more vulnerable to salt crystallization and freeze-cracking.

A famous twentieth-century building in Barcelona known as "La Pedrera" designed by Antoni Gaudi was built using two types of limestones, locally quarried (Ordaz et al. 1989). One type is gray, massive, and compact, composed mainly of micrite (microcrystalline calcite). It has a network of veins of red clay, and a low proportion of fossils. In thin section the very fine-grained micrite interrupted by many crisscrossing veins of clay is distinctive. The other type is white-yellowish, less compact, and coarser grained. It too is mainly calcite, with some quartz and clay, along with aggregates and bioclasts. In thin section the texture appears very different from the first type, with many small bioclastic aggregates held together with a micrite cement. Because of these structural differences, the second type of limestone has a coarse-grained appearance and weaker cohesion.

Dolomite, too, is a common building stone. For example, it was used for the Almería Cathedral in southeastern Spain. A published thin section of that stone showed a rather coarse texture, low compaction and cohesion, and the presence of quartz, pores, and dolomite crystals (Alonso et al. 1994).

Thin-section studies, together with other analytical methods, helped to identify the stone masonry construction at a Roman Republican sanctuary in Italy (Bugini et al. 1995). One stone type was a micritic limestone from nearby outcrops, used mainly for the masonry and for white mosaic tesserae. A black micritic limestone was used for black mosaic tesserae. A porous limestone (calcarenite) from farther away was used for slabs that comprise a podium, and for some architectural elements. A fossiliferous limestone was also used for part of a platform.

A black limestone, called "Lapis Niger" that was used as a building and pavement material at Roman monuments was the focus of an in-depth study by Fornaseri et al. (1995). The authors combined thin-section studies, XRD and isotopic analysis to characterize the stone and compare it with possible source rocks. The black color of all of the limestones comes from carbon and/or bitumen. Lapis Niger is described as having a cataclastic fabric, a micritic fabric crossed by a network of micro and macro veins filled by secondary, sometimes sparry, calcite. Stylolites were present (irregular surfaces often marked by thin seams of insoluble residue, transecting grains or fossils and indicating removal of material after deposition), frequently filled with black carbonaceous matter (which was also dispersed within the groundmass). There were many reworked bioclasts such as filaments of bivalves, radiolaria, and sponge spicules. Small detrital and authegenic quartz, potassium mica, limonite, and opaque iron oxides were also present as accessory minerals. The thin sections illustrated how illite needles and sponge spicules have a parallel orientation. Fornaseri et al. concluded that the probable source was Black Palombin limestone from Tolfa Allumiere, to the north of Rome.

Thin-section studies have also been carried out on extraction cuttings at ancient limestone quarry sites, as part of work intended to characterize the quarry material, estimate the total volume of building rock extracted at a site, and study the quarrying techniques that were used. Papageorgakis and Kolaiti (1992) conducted such studies in Greece at the ancient limestone quarry of Profitis Elias, which supplied building materials for the monuments of Delfi, and which is located about 5 km from that site. Thin sections, illustrated and described, showed that the quarry rock is a biomicrite. The only carbonate present was calcite, with quartz and ferruginous minerals as accessories.

As part of an investigation into the sources of stone used for masonry and sculptural decoration of monuments in the Paris basin, Blanc et al. (2002) collected samples from quarries in the Lutetian limestone formations of the region. Analyses included petrographic thin-section studies. All the Lutetian layers were seen to consist of pellet-formaniferal limestones, all of which were calcitic. In thin section the calcite appeared to be mainly biomicrite, with some microsparitic matrix. Calcite was the only mineral present in the bioclastic fractions of the rocks and in the matrix.

Sculpture

Limestone is a very common sculptural material. Richard Newman (1984) carried out thin-section studies on Indian limestone sculptures of the Amaravati school of the Late Andhra period (A.D. 50–320). These were described as being very fine grained (most with calcite grains of 0.01–0.03 mm in size), with a pale green or bluish color. A slight foliation was marked by a few aligned strings of mica. Neighboring crystals showed irregular polygonal grain boundaries, and there were scattered larger (0.05–1.0 mm) grains of calcite, sometimes showing twinning.

Busdraghi et al. (1991) used thin-section petrography to characterize badly deteriorating calcareous statues and ornaments in the courtyard of the Palazzo Buonaccorsi in Macerata, Italy and in the villa of Monte Santo. A variety of bioclasts was visible (such as ooliths and peloids), and these, along with mineral identifications (such as dolomite) aided the identification of probable sources of the stones.

Thin-section analysis of a sample taken from a limestone relief sculpture depicting the enticement of Nanda from Hadda, Afghanistan (near Jalalabad) and now in the British Museum, showed that it was a biomicrite (Reedy 1992b). Thus the rock was formed essentially from skeletal grains embedded in carbonate mud, cemented and lithified over time by precipitation of fine-grained carbonate. Original skeletal grains were often replaced by microcrystalline carbonates (micritization) (Williams et al. 1982).

In the sculpture from Hadda (Reedy 1992b), the carbonate was in the microcrystalline form, micrite. Bioclasts, or skeletal particles which are the hard parts of carbonate-secreting organisms, were relatively abundant (and are mainly ostracods). There were also some scattered coarser grained carbonate (sparite) patches, voids, and accessory minerals including quartz, plagioclase feldspar, hornblende, biotite, muscovite, larger calcite rhombs, hematite, and glauconite. Although microcrystalline limestone is found in many places in Afghanistan and neighboring northern Pakistan, not all of those deposits are fossiliferous. For those that are, petrographic descriptions indicate a different appearance or fossil assemblage, ruling them out as probable sources for the stone. At Hadda, stone sculptures are rare, as stucco is the preferred medium, but the limestone used extensively as construction material for the monuments was apparently obtained from the Jalalabad region (Tarzi 1985). This suggests the likelihood of a similar local, but as yet unidentified, quarry source for the sculptural sample.

Harrell (1992) employed thin-section petrography to identify the quarries of Egyptian limestone sculpture. Many ancient quarries were located in the Nile valley, and samples from them have been characterized by thin-section analysis. Because of this work it is possible to identify the provenance of many sculptures and architectural stones. Limestones from different sources can differ greatly in grain size and texture and in the types of microfossils present. They may also differ in type and ratio of carbonates present, and in the type, abundance, or ratio of various accessory components. Other analytical techniques, such as elemental analysis, have also proved quite useful in limestone provenance work (Meyers and van Zelst 1977; Holmes et al. 1986), however, thin-section analysis can often answer the question of provenance as long as the potential sources differ in one or more of the characteristics mentioned above.

The stone used for the famous Sphinx statue of Egypt has been well studied, including thin-section analyses (Kholief and Barakat 1992). The statue was carved *in situ* in about 2600 B.C., and was formed of layers that vary in composition and texture. The upper layers were a marly and ferruginous fossiliferous limestone comprised of fine sand grains in a groundmass of fine calcareous micrite streaked with clay. Because of its composition, this portion of the Sphinx is quite prone to deterioration, and has been the object of conservation and restoration efforts from Old Kingdom times up to the present. The lower layers are harder and more compact, but are composed primarily of fossiliferous limestone with micritic texture of microcrystalline calcite. There is some recrystallization into sparite, especially within the fossil remnants and along fracture planes. In the lower layer some transformation of calcite into dolomite has also occurred.

In the Gangetic plains of India, a sedimentary limestone rock known as "Kankar" was used for a variety of local purposes including sculpture and building materials. Petrographic studies of this rock (Hasan et al. 1995) showed it to be an accretionary stone that resulted from accumulation of calcium carbonate as a replacement material in weathered soil, rock, or sediment, created as water enriched in calcium carbonate percolated downward from soil. The host rock tended to be mudstone, with the Kankar stone occurring as large irregular gray masses or layers, often containing pebble- to cobble-sized clasts. Kankar is generally about 88% fine-grained calcium carbonate groundmass, with about 5% silica, 3% muscovite mica, 2% mud, and 1% feldspar. The groundmass is moderately porous with some solution cavities. Kankar stone is weak and weathers easily, nonetheless, blocks of this stone were used in many ancient construction projects, and for now heavily eroded sculptures.

4.5.4 Limestone and dolomite deterioration mechanisms and conservation treatments

Thin sections have also proved useful in studies of limestone deterioration and preservation. Van Gemert et al. (1988) employed thin-section examination as their primary analytical method in a comparison of naturally weathered limestone and samples that underwent accelerated aging. The stone selected was from a widely used quarry in Belgium, and samples from eighteenth-century buildings where the stone had been employed. Accelerated aging of quarry samples included acid attack in a humid sulfuric atmosphere, salt crystallization in an aqueous soda solution, and freeze–thaw cycles. Thin sections showed the development of microcracks in quartz grains on deteriorated surfaces, probably formed as gypsum crystallized on the surface of the stone. Erosion of calcite and calcareous cement was also observed.

Nijs et al. (1988) employed thin sections to study deterioration mechanisms of limestones used as a building stone for Flemish Gothic, Renaissance, and Baroque architecture. Two different stones were studied, both sandy limestones containing some glauconite. Deterioration mechanisms included leaching of the fine calcite matrix by rainwater, with gradual transformation into gypsum that forms crusts. The gypsum formation began in micropores around detrital quartz and inside glauconite grains. Mechanical destruction and grain fragmentation caused by recrystallization and concentration of gypsum was a major problem.

Thin-section analysis was used as one of a variety of analytical techniques to investigate possible causes for deterioration of Egyptian limestone sculptures during storage in a museum environment (Bradley and Middleton 1988). The thin sections show that the sculptures sampled were mainly fine-grained micritic limestones with only a few fossil fragments (although some were highly fossiliferous biomicrites). Some stones showed partial dolomitization with small rhombohedra of dolomite visible. No significant difference could be seen between deteriorated and undeteriorated sculptures. Although the thin sections were useful for characterizing the stone type, in this case other analytical techniques were needed to identify the probable deterioration mechanisms: chemical determination of soluble salts, gravimetric estimation and analysis by XRD of the acid insoluble fraction, SEM, and mercury porosimetry. The authors concluded that the clay content (which rendered the stone liable to structural decay), and the level and nature of soluble salts, especially soluble nitrates (making the stone more susceptible to localized powdering and pitting), were important factors in determining the nature and degree of deterioration of these sculptures.

Kozlowski et al. (1990) investigated causes for the deterioration of porous limestone building stones, the Pińczów limestone, used for many monuments in the city of Cracow (Poland). On some of these monuments, adjacent blocks of stone showed significantly different states of preservation. Petrographic studies clearly showed a correlation between the lithology of different limestone varieties and their resistance to weathering; the coarse-grained variety with large fossils was always in good condition, even where it had been exposed to an unfavorable environment. The fine-grained varieties showed weathering resistance that ranged from good to poor. The fossil components were the same for all varieties of Pińczów limestone; the main difference between the coarse- and fine-grained varieties was in the size of the species or of their fragments. Deterioration was caused by gypsum, which formed as an interaction of the limestone with the polluted atmosphere.

The coarse-grained limestones had no gypsum in their interiors, and only a few flakes on their surfaces. The well-preserved, fine-grained stones had a gypsum layer on the surface about 1 mm thick, but gypsum then decreased with depth. The surface layer was well adhered, forming a "natural patina." The weathered fine-grain stones had gypsum permeating deeply into the rock. The authors concluded that the structure of the coarse-grained limestones was less reactive to gypsum, and any water that entered could flow back out through larger pores, transporting any gypsum that had formed up to the surface, where it was washed away. In the fine-grained stones, the structure was more reactive, and water transport was hindered, facilitating formation of gypsum deep inside the stone if the protective surface coating was washed away. Thus, water-repellent treatments are important for fine-grained limestones. They also recommended that petrographic studies should be carried out when selecting placement stones to ensure that one with good weathering resistance will be chosen.

Deterioration mechanisms and characteristics of a limestone building material used in Venice were studied by Biscontin et al. (1992a). Termed "Istria stone," it was a microcrystalline limestone with microfossils and veins with spathic calcite, quartz, and limonite. Illustrated thin sections showed a variety of weathering features: a dendritic black crust with carbonaceous particles in the thick clear outer layer only, and microfissures filled with gypsum within the interior of the limestone; a more compact black crust with a very laminated dark brown layer impregnated with gypsum; a compact black crust with a very thick inner layer of gypsum; and a white crust with protruding spathic and bright calcite crystals reprecipitated on the surface of the crust. The authors concluded that the first weathering type results from surface degradation and exposure to dry atmosphere deposits in sheltered areas (a similar crust is illustrated in a thin section of a sample from inside an architectural complex in Venice, the Portale of S. Maria dei Servi, and the deterioration mechanism discussed by Bisconin et al. 1992b). The last type occurred in exposed areas where there was frequent rain-washing of the surface and wet deposition of atmospheric pollutants. The compact dark crusts resulted from a combination of wet and dry deposition of pollutants.

Garcia del Cura et al. (1992) used thin-section petrography to characterize limestone tesserae from mosaics in the Roman town of Italica, Spain. Four types of limestone rocks were identified with the tesserae: mudstones (micritic limestones with variable amounts of fossils), grainstones (biosparities with fossil fragments and oosparites), crystalline limestones (mainly red dolomitic limestones), and sandy limestone (fossiliferous). Detailed petrographic descriptions were given of variations within these overall types, with tesserae colors ranging from white to gray, yellow, and red.

Thin sections employed by Franchi and Gori (1993) to study deterioration processes on limestones used in the construction of the Ducal palace of Urbania in Italy showed that the rock was mainly calcite with traces of quartz and a small amount of argillaceous phyllosilicates. Deterioration included formation of gypsum and weddellite. With the aid of thin sections, the authors identified two variations of the stone. The first type consisted of oolitic limestone, with ooliths micritized. The ooliths were weakly cemented by sparry calcite, with a high degree of intergranular porosity. Sometimes large crystals of secondary calcite could be seen, often joined to form polygonal mosaics. Bioclasts consisted of large foraminifers and algal fragments, but were rare. Deterioration occurred mainly by dissolution of the carbonate cement, leading to disaggregation of the ooliths. Increased porosity from the dissolution of the cement could be seen down to a depth of about 1 cm. Close to the surface, large cavities had resulted from the loss of ooliths; micropitting was also evident. The second type was described as having ooliths and pellets, with micritization so complete that it was hard to identify individual structural elements. There were a few algal fragments recrystallized into a single crystal, and a few large pisoliths (larger and less regular in form than an oolith, although with the same concentric and radial internal structure) were visible on the surface. Porosity was greater than in the other type; deterioration processes were similar but the cement was in worse condition, and the dissolution created more of a problem because of porosity and because the structural elements were more oriented. Treatment of both types consisted of consolidating the internal structure.

Rodriguez-Navarro et al. (1994) used thin sections (along with mercury-intrusion porosimetry and mechanical testing) in a petrophysical and mechanical analysis of deteriorated samples from a limestone building (Jaén Cathedral in Spain) that were compared with undeteriorated quarry samples of the stone. Both micritic and biomicritic limestones were used. Digital image analysis of thin sections provided information on the macropore system of the stones. It was found that the mercury-intrusion porosimetry had severe limitations for biomicritic limestone, and that the digital image analysis of thin sections provided the most reliable data, allowing the exact dimensions of each pore type to be measured. There was a big difference in degree and distribution of macroporosity (> 5 μm in radius) between the weathered and unweathered stone samples.

In addition to micrite (microcrystalline limestone), the thin sections of micritic limestone showed small amounts of quartz and oolites, along with microcracks and fractures filled with sparite. The biomicritic limestone showed calcite, bioclasts, quartz, and glauconite aggregates. Deteriorated samples had extensive rupturing of bioclastic chambers, increasing the effective porosity of the stone.

They concluded that the building samples showed an increase in total porosity compared with quarry samples, resulting from the opening of macropores. The micritic limestone was originally almost entirely microporous with a fine lamination, and it had developed microcracks along that orientation. The biomicritic limestone, with almost 90% bioclasts, developed macropores as a result of breakage of the foraminifera chambers. This work illustrated the usefulness of incorporating computerized digital image analysis procedures into thin-section analysis.

Moropoulou et al. (1995) studied the effects of various environmental factors in the deterioration of limestone using a combination of thin-section study, XRD analysis, SEM, mercury porosimetry, and overall surface analysis using digital image processing. Working with a large-scale monument (the fortress of Rethymnon) in the marine environment and mild climate conditions of the Mediterranean, they focused on the effects of humidity sources such as capillary rise of groundwater and marine salt spray. Stone samples were taken from various locations on the monument characteristic of a variety of deterioration patterns, and examined in thin section. Weathering classifications were based on microstructure, texture, and composition.

The thin-section analysis permitted identification of the building stone. It was described as primarily a biosparitic limestone with bioclasts, sparite, micrite, and quartz. The bioclasts were either rounded fossil fragments (corals) or sharp-edged fragments with radiolaria remnants. The sparite was a binding cement, surrounding grains, and ranged from fine grained to large grained. The thin-section work also identified various weathering phenomena, including granular disintegration that left large pores behind, and intensively oxidized crusts where biodeterioration was combined with the formation of a hard carbonate crust.

Two limestone waterproofing materials were tested by Esbert et al. (1995), specifically looking at contact angle and water vapor permeability before and after experimental exposure to sulfur dioxide. The stone tested was Hontoria limestone, widely used as a material for Spanish monuments. It was described as a bioclastic limestone containing many fossil remains, voids, and a sparry cement, with some areas of recystallization. The voids and open porosity contributed to many of the physical properties of interest in studies of deterioration and preventive conservation. Thin sections were used to examine and illustrate internal differences in grain size.

In the previously mentioned study by Varti-Matarangas et al. (1995), thin sections of building stones employed in monuments of the ancient Olympia area were used to aid in identifying sources and their deterioration mechanism. Among the stones were several types of limestones. For example, one limestone had a wide variety of fossils (such as ostracods and gastropods), with a very high porosity of inter-intraparticle type, while another biomicrite had a wide array of fossils present; one of the main decay mechanisms in both was swelling of clay minerals.

Malaga-Starzec et al. (2000) conducted laboratory experiments on the weathering behavior of limestones quarried in central Sweden, comparing the differences between fresh and impregnated (colloidal silica) material. Acid rain was simulated in the laboratory, and a variety of tests conducted to compare how the materials fared. Thin sections were employed to observe optical changes. While sandstones that underwent the same tests showed no optical changes for either fresh or impregnated specimens, the limestones did show differences. On fresh limestone, weathering products (gypsum and calcium sulfate hemihydrate) were more abundant than for impregnated specimens.

Török (2004) used thin sections of limestone to compare the process of deterioration of two building materials in the polluted urban environment of Budapest (Hungary), where many public buildings were constructed of limestone. The two most common types of limestones used were a Miocene oolitic limestone (with a coarse texture) and a Pleistocene freshwater limestone. The oolitic limestone tended to show intense crust formation and blackening, with the crusts prone to scaling and/or flaking followed by rapid substrate material loss through granular disintegration. The freshwater limestone tended to have blackened surface crusts, but the crusts appeared to be stable and did not lead to scaling. In this case, the problem was aesthetic, while in the case of the ooliitic limestone, the problem was structural, with changes in strength and durability occurring. Török proposed that the differences in durability between the two stones were related to textural differences that included larger pore size, smaller effective porosity, and a higher rate of cementation and strength. Thin sections illustrated the high proportion of open pores in the coarse limestone, with a thin cement rim on rounded ooids. The black crust that developed on the coarse limestone was easily seen in thin section, along with an irregular surface to the crust and a dissolving limestone substrate. Gypsum crystals were visible within the pores deep below the crust. In contrast, the freshwater limestone showed a high proportion of closed (cemented) pores and few, unconnected open pores. The stone itself was in relatively good condition.

Quaresima et al. (2006) included thin sections in a comprehensive study of ornamental stones used in 65 medieval religious buildings in the historic city of L'Aquila (Italy). The limestones and marly limestones of the ancient quarries where these ornamental stones were carved were characterized, and correlated to the decay forms seen on the buildings. This work was part of an effort to construct a general strategy for preservation and conservation of the buildings. Thin sections were illustrated for a variety of stone types, as seen in the quarries and on the buildings. These types included fossiliferous limestone, a calcareous breccia, and biointramicrite examples; some stones are white and others are red. Differences in mechanical resistances of the red and white stones account for differences in observed decay: the red ones have more problems, due to the presence of clay and iron oxides homogeneously distributed in the matrix or concentrated in layers, as seen in the illustrated thin sections. For example, microfractures can be seen along iron-oxide veins.

4.5.5 Travertine and tufa

Most groundwater contains calcium carbonate in solution, and waters at deeper pressures and higher temperatures, such as hot springs, can become saturated since the solubility of carbon dioxide in water is increased. Solid calcium carbonate can then precipitate upon exposure to air, forming masses of travertine or tufa, often in concentric layers. Travertine tends to occur in extensive beds of hard and compact stone, and has been used as a building material. Perhaps the most extensive and famous deposit is found in the Tiber valley near Rome. It ranges from a buff color to white or gray, usually variegated, and with irregular-shaped pores (Fig. 4.5.13) created by the decaying of bacteria and other organic life forms and imparting a somewhat spongy appearance to the stone (National Materials Advisory Board 1982; Greensmith 1989; Penny 1993). Living clumps of bacteria with a continuously precipitating corona of calcite eventually decayed, leaving calcite crystals riddled with micropores. Cycling repetition under conditions of periodic bacteria growth give travertine a complex structure. Calcitic mudstone strongly cements the remains of stomatolites and bioherms, creating a durable texture (Jackson et al. 2005).

Travertine is relatively soft when fresh and during cutting, but hardens on exposure to air, making it a good building material; for example, it was used extensively in both ancient and modern Rome for buildings such as the Colosseum. A famous deposit is found near Tivoli in Italy, where water carrying volcanic carbon dioxide has created a massive deposit of travertine in the bed of a former lake. It was an important source of hard, dense building stone (Kempe and Harvey 1983).

Travertine was one of the materials used in the construction of Orvieto Cathedral, a major monument of Gothic Italian architecture. Thin-section studies by Moroni et al. (2002) of this travertine showed it to be completely produced by biogenic processes of calcium carbonate encrustation involving macrophyta and microphyta. Some samples were identified as stromatolitic travertine, derived from chemical encrustation by fibrous calcite of algal felts, with concentrical structures. Others samples were identified as phytohermal travertines, deriving from carbonatic encrustation of microphyta (hygrophyte stalks). A few samples were found

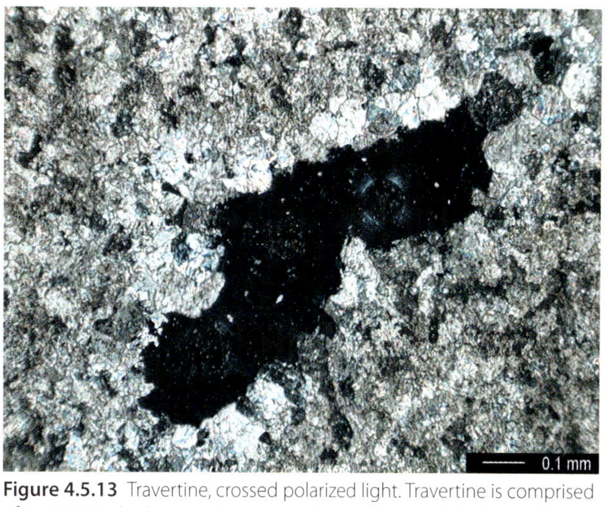

Figure 4.5.13 Travertine, crossed polarized light. Travertine is comprised of precipitated calcium carbonate, so the grains are relatively small and irregular in shape. Here is seen one of the typical irregular-shaped pores created by the decaying of organic life forms; it is the presence of many such pores that imparts a somewhat spongy appearance to travertine.

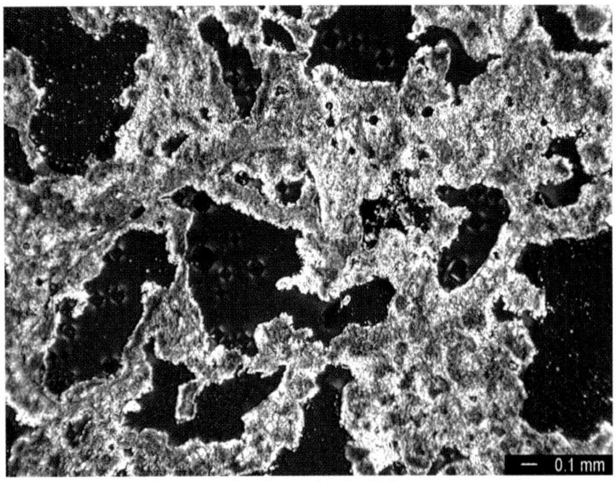

Figure 4.5.14 Tufa, crossed polarized light. Tufa is an extremely porous and spongy rock, and many irregular cavities can be seen in this example. The calcium carbonate is in the form of a very fine-grained to cryptocrystalline calcite.

to be detrital travertines. Visible components in the travertines included some fossils of Ostracoda, Gasteropoda, and Characeae, as well as small aggregates of quartz, alkali feldspars, micas, pyroxene, epidote, and chlorite grains. The thin-section evidence led the authors to suggest an origin on the outskirts of Orvieto rather than in more distant locales.

Tufa refers to extremely porous and spongy deposits (Fig. 4.5.14), often associated with abundant moss and algae growths (Scoffin 1987). These form around calcareous springs, and are too friable to be very suitable as building stones (Jackson et al. 2005). Blocks of tufa, in conjunction with brick, were used as a medieval building material in Italy, as well as for ornamental features on architecture (Penny 1993). De Geyter (1985) described and illustrated thin sections of a building material termed "tuffeau" used in Belgium for Romanesque and early Gothic religious monuments, and again in the last century for a variety of vernacular uses such as houses, farm buildings, and bread ovens. De Geyter characterized the stone as a porous, siliceous (opaline) limestone, with a small amount of detrital quartz (less than 10%). There were abundant calcareous remains of foraminifera, mollusks and echinoderms, along with about 25% cavities where siliceous sponges have dissolved. Glauconite was also present, as was garnet (mainly fine-grained spessartine crystals). The cement was opal-CT, with scattered minute calcite grains and clay flakes.

Thin sections of travertine and tufa used in the production of inscribed Etruscan stelae from Siena, Italy, are discussed by Gandin et al. (2006). Although specific geological sources have not yet been located, the thin sections of different stele showed some variety, including a cement of clear rhombohedral crystalline calcite and a cement of clear acicular crystalline calcite. One stele had a thin network of dense microcrystalline calcite with spongy texture and relicts of organic matter, which outlined irregular primary pores partially filled by crystalline calcite. Another example had rosettes of dark calcite, and yet another contained very fine clayey material thought to be derived from the vegetated zone of the ground soil and deposited within small secondary cavities.

4.6 Sedimentary Silicates

Another type of nondetrital sedimentary rock includes chemically or biochemically deposited silica. This category incorporates microcrystalline quartz, or chert, with variations such as flint and jasper, and cryptocrystalline quartz, or chalcedony and opal. However, sometimes all these materials are more broadly referred to as chert, especially by North American archaeologists (Luedtke 1992).

Water must have a very high concentration of silica for it to precipitate. This most easily occurs near the surface of hydrothermal waters through slow evaporation of relatively pure alkaline solutions, or when silica-secreting organisms extract silica from seawater and redeposit it in the amorphous form to provide rigid frameworks to support their soft tissues. Much of it redissolves after the organisms die and is then reused by other organisms, however, some is incorporated at sediments on the ocean floor, where it reprecipitates to form chert beds (Luedtke 1992; Carozzi 1993).

Chert

The name chert refers to a rock composed primarily of water-precipitated silica. It is mainly microcrystalline (Fig. 4.6.1), often defined as greater than 0.01 mm but less than 0.10 mm in crystal length. It is usually equigranular, with anhedral grains that can be separately distinguished. An undulatory extinction frequently appears in these grains because the crystals are smaller than the thickness of the section, so light may pass through several misoriented crystals.

Common constituents of chert deposits are organic remains of radiolarian tests that have been partially or completely replaced by microcrystalline quartz or chalcedony. These appear as rounded or oval forms (Fig. 4.6.2). Other recrystallized organic remains that may be seen

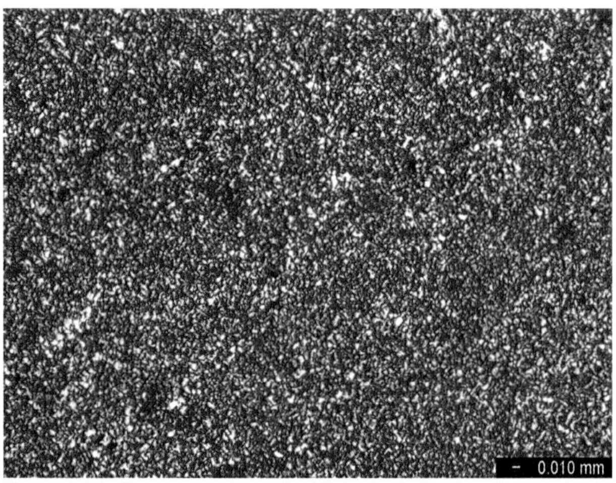

Figure 4.6.1 Chert, crossed polarized light. Typically, chert is comprised mainly of microcrystalline quartz. It is often equigranular, with anhedral grains that can be separately distinguished, as in this chert artifact from Belize (a primary core reduction flake from the Chetumal Bay coast).

include sponge spicules, foraminifera, diatoms, and other fossils; other impurities may include clay minerals, silt, carbonates, pyrite, and organic matter (Folk 1974).

Flint

Flint and jasper, terms most often used by archaeologists, are specific varieties of chert. Flint is generally used to refer to a black chert (the black is due to organic matter inclusions) (Fig. 4.6.3), but it may be brown or gray, on freshly broken surfaces, and is found as nodules or beds within chalk deposits. This term is frequently used by British archaeologists to refer to dark, nodular cherty material found in the Cretaceous chalk formations of southern England. It consists of a mixture of microquartz and chalcedony, biochemically deposited by reprecipitation in pore waters within the chalk deposits of silica sediments originally dissolved from siliceous microorganisms (mainly sponges) replacing calcium carbonate. Flint has often been used for stone implements and for cutting and trajectory blades; it has also had a variety of other uses such as fire-starting implements, flintlocks in firearms, building material (laid with a lime mortar), as a component of concrete, road gravel, and an additive to pottery and glass (Sheperd 1972; Folk 1974; Clayton 1983; Wright 1991; Carozzi 1993).

De Geyter (1985) carried out thin-section studies of flint used as a building material in Belgium during Roman and medieval periods. He found a typical flint comprised mainly of microcrystalline quartz, chalcedony, and a small amount of opal. The chalcedony was usually associated with organic remains, and occurred as a lining of the walls of foraminifera and bryozoa, or as a replacement filling for radiolaria and sponges (Fig. 4.6.4).

Jasper

Jasper is often used to refer to red or brown chert colored by ferric oxide (Fig. 4.6.5); sometimes "green jasper" is a term used to refer to chert that has been colored green from chlorite inclusions (Folk 1974). However, because all cherts share essentially the same mineralogy, geologists generally do not give specific names to identify different colors of

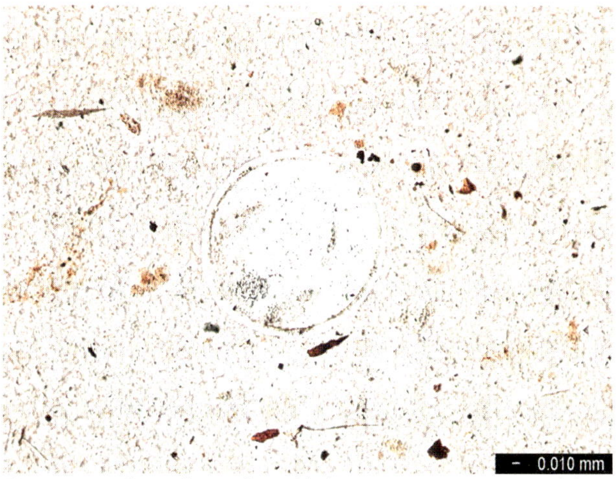

Figure 4.6.2 Chert, plane polarized light. Common constituents of chert deposits are radiolarian tests (siliceous skeletons of marine protozoa). This one shows a characteristic round shape, in a sample from a chert artifact from Belize (from the Chetumal Bay coast).

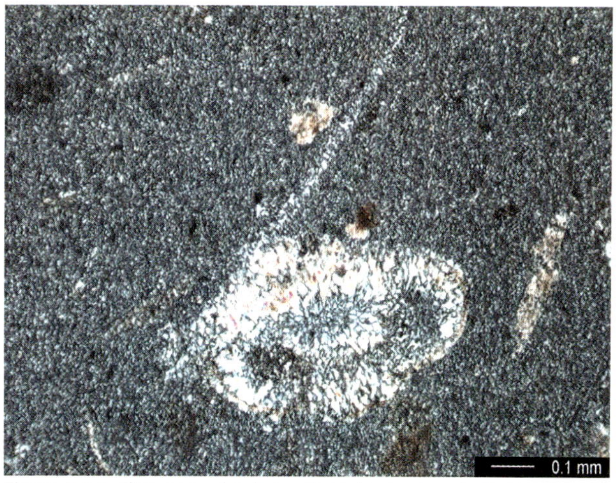

Figure 4.6.4 Flint, crossed polarized light. Chalcedony in flint is often associated with organic remains, such as a replacement filling for radiolaria (circular) and sponges (as with this long slender curving sponge spicule).

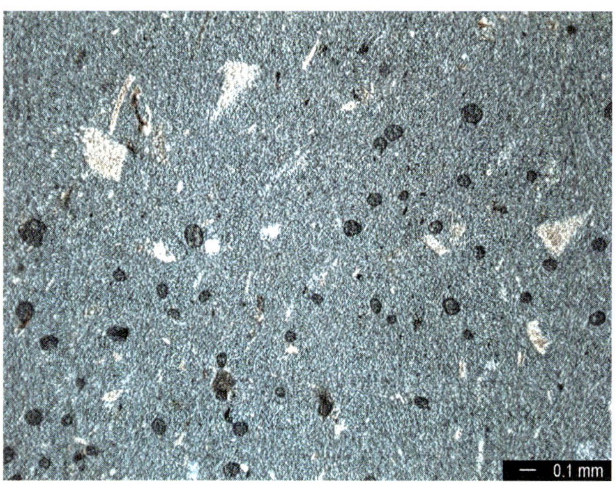

Figure 4.6.3 Flint, crossed polarized light. Flint is the term generally used to refer to black chert (mainly microcrystalline quartz colored by organic matter inclusions).

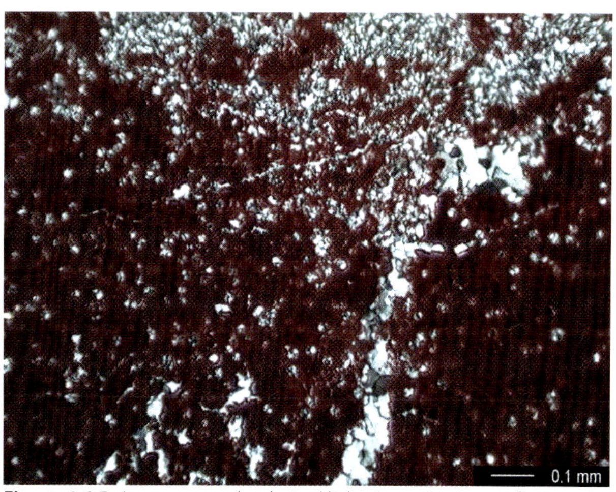

Figure 4.6.5 Jasper, crossed polarized light. Jasper is a term often used to refer to red or brown chert deeply colored by ferric oxide. The iron oxide coloring fills most of the field of view, with low-birefringent grains delineating the areas of microcrystalline quartz.

chert. Hematite dust, carbonaceous materials, and clay impurities can all cause local variations in color.

Jasper is sometimes defined as being of varied colors produced by iron oxides, but always opaque (whereas "agate" is a term that may be used for more transparent specimens, especially if they are banded). Under this definition jasper may be red, blue, green, or purple, and may contain banding. In addition to stone tool use, for example by Native Americans (Luedtke 1992), jasper has been carved into very detailed vases, bowls, and other vessels in Czarist Russia (Hull 1872), and used as a stone drilling material for perforating beads in ancient eastern Iran (Kenoyer and Vidale 1992).

Jasper and some other microcrystalline and cryptocrystalline materials used as gemstones are often stained before marketing, so many of the colors identified with them are either not natural or are natural but artificially accentuated. Confusion is introduced by this practice of artificial coloring, as it has caused some commonly used terms for colored stones to conflict with geological terms attached to naturally colored stones (Weinstein 1958).

Carnelian

Carnelian is a term often applied to translucent reddish (unbanded) chert or chalcedony. It is sometimes differentiated from jasper by the fact that jasper may contain many impurities in addition to microcrystalline quartz (such as ferric oxides, iron silicates, and clay) visible in thin section, whereas carnelian has fewer impurities. It has often been used as a bead material or for other jewelry items. Natural, unaltered carnelian is usually a pale or brownish red. Very deep red shades are often the result of artificial color enhancement treatments. For example, they may be heated and exposed to ferrous sulfate (Weinstein 1958).

Opal

Opal (or opal-A) is hydrous silica that has an almost amorphous (glassy) structure. As it is rarer than chert or chalcedony, it is categorized as a precious rather than an ornamental stone (Hull 1872). It has been used most often for stone settings within rings or other items of jewelry. Opal may also occur as small inclusions within other forms of precipitated silica.

Porcellanite

Porcellanite is defined as an argillaceous (clay rich, with 25–50% clay minerals) porous siliceous rock. It is not as vitreous as chert, and is not as hard or dense, with an appearance somewhat like that of unglazed porcelain. Sometimes the silica portion consists of a microquartz groundmass mixed with clay; sometimes it is mainly opal-CT (hydrous silica that consists of layers of cristobalite and tridymite, the high-temperature forms of silica, or quartz). The clay portion is often mullite, and other minerals present may include cordierite or corundum (Greensmith 1989; Kenoyer and Vidale 1992; Carozzi 1993).

Chalcedony

Chalcedony, composed of minute fibrous silica, is cryptocrystalline (usually less than 0.01 mm in crystal length). It often has a radiating form (Fig. 4.6.6), and is commonly found filling cracks and cavities within chert. Sometimes the sheaf-like bundles of silica fibers may be longer than usual, but still extremely thin. In addition to its association with chert deposits, chalcedony may also appear as a secondary material in cavities of igneous rocks, as nodules or bands within secondary limestone deposits, and as a replacement of calcareous fossils. It appears brown in thin section (in plane polarized light) due to the presence of many tiny water-filled pores, which give the surface a spongy appearance (Folk 1974; Greensmith 1989). In addition to being used for stone tools, the hardness, high polish, and strength of chalcedony has led to uses as vases, brooches, and ornaments (Hull 1872).

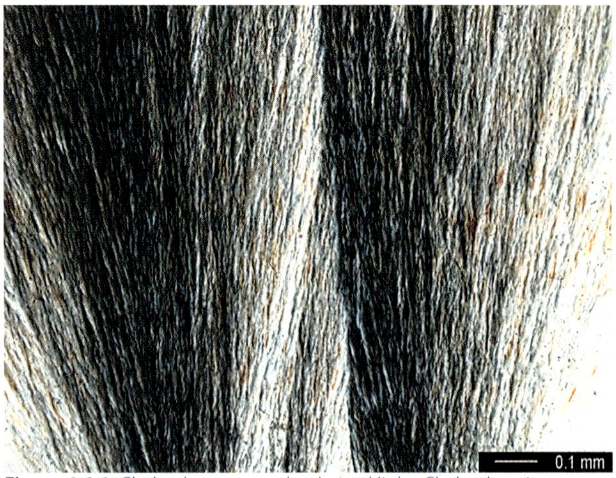

Figure 4.6.6 Chalcedony, crossed polarized light. Chalcedony is cryptocrystalline quartz (minute fibrous silica). It often has a radiating, fan-like form, as seen here.

Agate

Agate refers to striped or banded chalcedony or chert. The stripes may vary in width and color, and may be very obvious or visible only under a microscope. Colors may vary, but they are usually pale white, yellow, red, and brown. Deep colors and blue and green shades seen in some jewelry are generally the result of artificial staining. A wide variety of names is given to various color combinations of agate: for example, onyx for black and white bands and sardonyx for brown and white (Weinstein 1958). However, all these materials appear similar in thin section, highlighting the importance of using characteristics visible by eye as well as those visible under the microscope in describing a cultural material.

Chrysoprase

Chrysoprase refers to chalcedony or chert with a natural apple-green color resembling jade. It usually has many imperfections, and is seldom a deep green color except when artificially stained. The natural color is unstable, as exposure to heat tends to bleach it. The heat of polishing can affect the color, and the material is also susceptible to cracking therefore it is not often used for jewelry today, but in the past it was used as a gemstone, for mosaics, and for inlaid decorative work. Modern stones marketed as chrysoprase are often simply chalcedony that has been artificially stained a deep green, so real chrysoprase is now called "natural chrysoprase" (Weinstein 1958).

4.6.1 Sedimentary silicates as cultural materials

Chert and related materials are widely employed as cultural materials, primarily for tools, as the physical properties are ideally suited for that use. Although most analytical studies of chert artifacts focus on elemental analysis, thin-section petrography is also used to characterize chert objects. Although there is limited variation in the basic mineralogical and textural features of chert, it can contain many impurities that may allow samples from different sources to be distinguished. Thus, in addition, differences in proportions of various forms of silica (such as microcrystalline quartz, fibrous quartz, and megaquartz), and major impurities such as clays, carbonates, iron oxides, organic matter, detrital sand grains, and micas are visible in thin section, as well as fossils and other replacement artifacts. The spatial distribution of constituents and the relationships between grains (microfabric characteristics) may also be important in characterizing a chert (Luedtke 1992).

Biggs (1957) studied thin sections of nodular chert from 18 formations in Illinois, sampling from 35 separate outcroppings that occur as nodules within limestone and dolomite. The accuracy of thin-section identifications was confirmed by XRD and infrared absorption analyses. In addition to variation in the amount and type of silica forms, Biggs also found carbonate minerals (calcite and dolomite), of which some were fragments of the original host rock and others were authigenic crystals differing in composition from the host formation. Also present were fluorite crystals, iron-oxide specks, carbonaceous matter, fossils, oolites, and coquina.

In another thin-section study of chert, Folk and Pittman (1971) were able to determine when silica has replaced sulfates and evaporites by the presence of rare length-slow chalcedony. Most chalcedony is length-fast (the c-axis, or slow ray, is perpendicular to the fibers). But a length-slow chalcedony, identified with the use of the gypsum plate, was found to be characteristic of evaporitic environments. Similarly, Purdy and Blanchard's (1973) analysis of chert samples from the eastern United States and England led them to the tentative conclusion that petrographic criteria may be useful in distinguishing different source areas.

Dimitrijević et al. (1974) illustrated thin sections of siliceous stone tools (axes and adzes) made from chalcedony from the Neolithic period in Serbia. Impurities included carbonates (magnesite and dolomite) and megaquartz.

Purdy (1981) examined thin sections of chert (flint) from Florida to determine whether it would be possible to identify significant differences within a geologic source, or even within different portions of the same nodule. If so, she felt it might be possible to identify differential selection of raw material for chipping qualities. She was also interested in finding out if different locations showed significant variations that could be used to identify source quarries of stone tools and chipping debris from archaeological sites. In the illustrated photomicrographs of cherts from various locations in Florida, all were seen to be broadly similar to each other, but dissimilar to cherts from other regions; thus, although identifying specific quarries in Florida may not be possible in thin section, it may be feasible to identify and source imported materials. Within-source samples from Florida showed little differentiation, and samples taken from different portions of one nodule also evidenced little difference petrographically.

Thin sections, together with other analytical techniques, were used by Takács-Biró (1983) to study sources of raw materials of chipped stone implements commonly found in Hungary. Major differences were seen, for example, with radiolarian cherts versus those rich in spiculites. In the study by C. L. Williams (1983), thin sections were employed to examine chert that formed as a replacement of calc-arenite in Devon (England). Chalcedony filling in voids was visible, as was replacement of calcite by microquartz, and secondary quartz growths around quartz detrital grains. Thin-section studies of Pennsylvania jasper showed that a wide variety of characteristic features was visible (Prothero 1983). Examples included isolated quartz grains, fine particles of iron oxide including those that outlined ghosts of replaced quartz grains, replacement quartz in the form of relic plagioclase feldspars and small patches or streaks of unaltered quartzite.

Thin sections of chert artifacts from northern Belize were employed by Boxt and Reedy (1985) to accurately describe the nature of chert selected for artifact use, delineate the mineralogical range of variation found with chert tools from northern Belize, determine whether distinct groupings of chert can be made through thin section observations, and deduce which criteria are most useful for identifying groups. They found that criteria useful for distinguishing chert types were (1) abundant versus rare radiolaria; (2) whether chalcedony was more or less common than microcrystalline quartz; and (3) the presence or absence of carbonates, hematite, and sponge spicules. They were able to propose provisional groups by mineralogical composition.

They also found that the mineralogy, texture, and type of fossils present within chert, which appeared to indicate a source location, were not related to the color of the chert. Although several chert samples may be very similar in color, the microtexture and mineralogy may differ greatly. And, some samples which were very similar mineralogically and texturally had no color relationship. For example, samples from Colha which looked identical in thin section were red, golden, brownish golden, and white in color; and one mineralogically similar group from Sarteneja included specimens of clear, black, and red cherts. The authors therefore cautioned against using color alone as a criterion for sourcing chert (a common archaeological practice).

Morgenstein (2006) studied chert (including types identified as flint, jasper, and agate) from two archaeological sites in the United States (one in northern Nevada and one east of Seattle, Washington) in which a wide range of geological source environments was represented. A variety of approaches to provenance studies was used including thin-section petrography. Specimens taken from a Nevada quarry (the Elko Hills quarry) were illustrated in plane polarized and cross polarized light, showing an ostracod-bearing chert with opalized ostracod shell fragments. The matrix was comprised of allogenic clay and calcite ghosts, with an authigenic overprint of chert (composed of microcrystalline quartz and fibrous chalcedony). Sometimes calcite was replaced with megaquartz, chalcedony, or iron oxyhydroxides. Most of the samples from the associated archaeological site in Elko, Nevada were similar, so could have come from the Elko Hills quarry chert. However, one sample appeared to have originated from an unknown source (a chert-cemented siltstone).

Morgenstein found that the potential sources (Pleistocene drift sediments at the site itself) for the site in Washington (Tolt) are rather different. They included: petrified wood pebbles (preservation of cellular structure varies); hydrothermal metavolcanic replacement jasperoid chert (with vesicles filled in with chalcedony and microcrystalline quartz, and containing iron oxyhydroxides); and joint-fault and fracture-fill hydrothermal chert (including some with vugs filled by chalcedony, and iron and manganese oxyhydroxides in oribicular or radiating microstructures). Archaeological samples from the associated site represented all three types of chert. In addition, a volcanic (agate) chert appeared to come from outside the region.

4.7 Evaporites

Evaporite rocks are deposited by precipitation and crystallization from aqueous solutions that become saturated through evaporation. Evaporation is a process that frequently occurs at inland salt lakes, along sea margins, and within partly isolated marine basins, particularly in arid climates (Greensmith 1989; Ashurst and Dimes 1990; Carozzi 1993). Two important evaporites used for cultural materials include anhydrite and gypsum.

Anhydrite (calcium sulfate) and gypsum (hydrated calcium sulfate) are chemically deposited rocks formed as a result of evaporation from standing bodies of landlocked sea areas, from shallow nonmarine pools, or from supratidal flats that are periodically flooded by seawater. As a result, they often contain impurities of clay, silt, sand, or carbonates, along with a variety of accessory minerals (Williams et al. 1982).

Anhydrite is readily distinguished from gypsum in thin section through differences in birefringence and form (Kerr 1977; Kuntze 1983). Anhydrite (Fig. 4.7.1) has moderate birefringence, and three good cleavages at right angles to each other. It also generally appears in blocky crystals with rectangular cleavage. There are often aggregate crystals, sometimes elongated, and more often anhedral or subhedral rather than euhedral, with parallel extinction. Gypsum (Fig. 4.7.2) has very low birefringence, and a perfect cleavage in one direction with imperfect in two others, not at right angles. It has oblique extinction, and sometimes shows a fibrous structure. The two minerals may occur together where there has been partial hydration of anhydrite by groundwater; near the surface, gypsum rocks are common. Because evaporites tend to be water-soluble, the thin-section preparation regime used for any materials suspected of including these minerals must exclude water.

The massive form of gypsum stone, often called "alabaster," is fine grained and compact, eminently suitable for carving. The term "alabaster" is sometimes reserved for gypsum rock that is white and translucent, although gypsum can also be brown, reddish, or gray, depending on the presence of impurities such as iron oxides and carbonaceous materials. It has been used as a building and sculptural material throughout the world. For example, alabaster was used in the construction of ornamental tombs and statues throughout England; as building blocks for the palace of Knossos in Crete; for sculptural work in Italy; as a sculptural material in ancient Assyria; and for sculpture in the

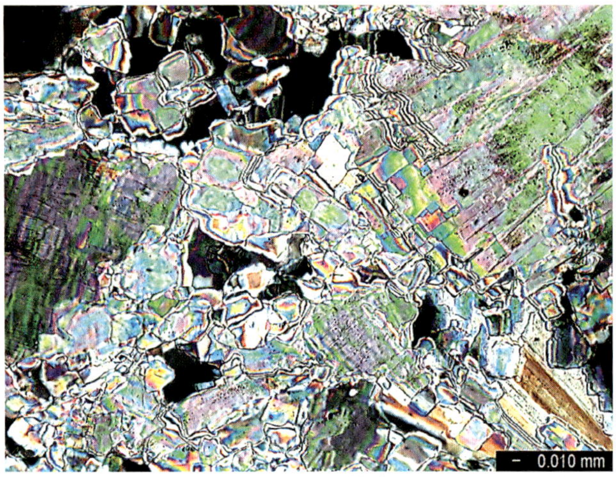

Figure 4.7.1 Anhydrite, crossed polarized light. Anhydrite has moderate birefringence and three good cleavages at right angles to each other, giving it a form with blocky crystals. The crystals are often in aggregates, are more often anhedral or subhedral rather than euhedral, and are sometimes elongated into blocky blades with rectangular cleavage.

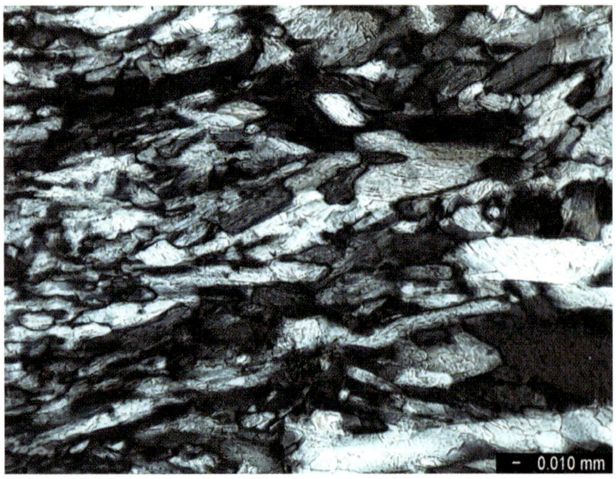

Figure 4.7.2 Gypsum, crossed polarized light. Gypsum has low birefringence, about the same as that of quartz in contrast to the moderate birefringence of anhydrite. It has perfect cleavage in one direction, but imperfect in two others, not at right angles as is seen with anhydrite. It has oblique extinction rather than the parallel extinction of anhydrite. Gypsum crystals sometimes show a fibrous structure, as in this example.

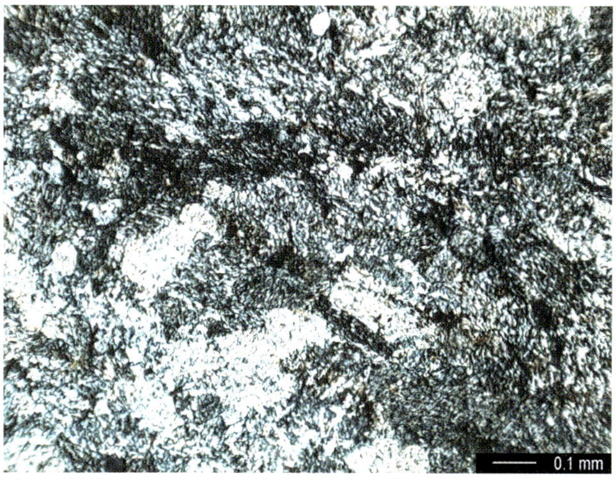

Figure 4.7.3 Gypsum, crossed polarized light. This example is fine grained with an overall fibrous texture to the gypsum, with scattered graphite that imparts a light gray color to the stone. Gypsum deposits are often associated with mud and sand deposits, so carbonaceous pigments, as well as clay impurities and scattered quartz grains, are common.

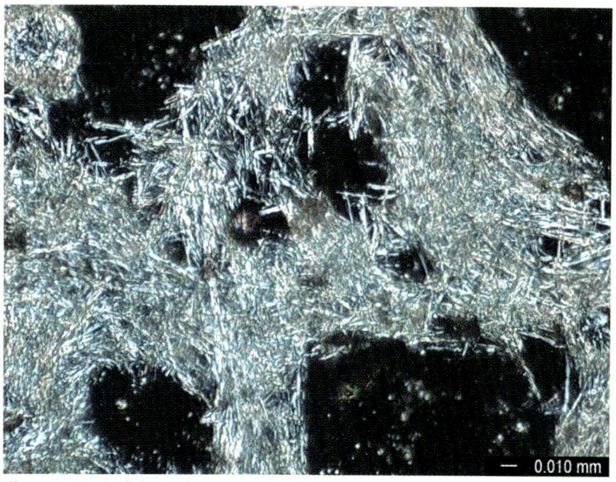

Figure 4.7.4 Calcined gypsum plaster, crossed polarized light. This example is from ancient Pylos, and shows an extremely fine-grained, fibrous form. Some recipe formulations can include bits of quartz, clay, or calcite.

Gandhara period of northwest India. However, there is some confusion in art-historical and archaeological publications regarding the term "alabaster," as it is sometimes used to refer to calcite (Penny 1993). The term is also used to refer to gypsum rocks that have an even-grained texture of interlocking equidimensional anhedra (Williams et al. 1982).

Richard Newman (1984) described a Gandhara gypsum piece, an image of the head of the Buddha dating to the first or second century, as being black with a powdery fine-grained appearance. In thin section it was seen to have many holes and voids, and was comprised of a fine-grained gypsum colored black by carbonaceous material. A few scattered quartz and calcite grains were seen, including a larger lozenge-shaped area of fine-grained calcite that may be a shell fragment.

Another Gandhara sculpture, a palette or toilet tray depicting a prince and attendants from Afghanistan, was examined in thin section and found to be made of carbonaceous gypsum (Reedy 1992b). The stone was gray and fine grained, and in thin section had an overall fibrous texture in the gypsum, with scattered graphite that imparts the light gray color (Fig. 4.7.3). Because gypsum deposits are often associated with mud and sand deposits, carbonaceous pigments are common as minor constituents of gypsum (Williams et al. 1982; Adams et al. 1997). Carbonaceous gypsum is quite common in Afghanistan, and the particular source of this sculpture has not yet been identified.

Studies of the building material taken from the prehistoric underground pit of Gortys in Crete, and probably used in the construction of buildings and/or for ornamental stone at sites on the island, identified a variety of evaporites (Dermitzakis et al. 1990). For example, near the Labyrinth area, laminated gypsum occurs, which in thin section appeared as laminae with fine alternations of micrite and argillaceous and/or organic material. Some elongated selenite crystals were also present (selenite refers to large clear crystals of relatively pure gypsum). The authors noted that this type of gypsum is often called "balatino" gypsum. In contrast, "alabaster" gypsum used as ornamental stone was finer grained, with diagenetic alterations of gypsum to anhydrite, and with impurities of clays and iron oxides. Both of these laminated gypsums also had some quartz, calcite, and dolomite present, and were sufficiently soft to be hewn but hard enough for structural use (most deposits of gypsum and anhydrite do tend to be impure). Horizontal bedding with thick laminae permitted separation of flat stones. Coarse crystalline selenites were also reported, which in thin section showed impurities that included clays and algae.

For a short period in the early twentieth century, calcined gypsum (plaster of Paris) was consolidated with a few percent wood fibers to make gypsum block for architectural use. This material was also used to fabricate precast tiles for floors, walls, and roofs; these were valued for their fire-retardant properties (Jester 1995). Plaster of Paris receives it name from the extensive beds of gypsum located north of Paris that were exploited in medieval France, however, plaster made from calcined gypsum goes back as far as the ancient world (Ashurst and Dimes 1990) (Fig. 4.7.4). Plaster of Paris is made by heating gypsum to partially drive off its chemically bound water. The mineral is then called calcium sulfate hemihydrate, or calcined gypsum. It was often ground to a fine powder, then mixed with water and poured into a mold. In thin section this material appears extremely fine grained.

5

Metamorphic Stone Materials

Metamorphic rocks are the result of processes in which pre-existing rocks undergo transformation primarily in the solid state. Transformations may include mineralogical recrystallization and/or textural alteration, often at high temperature and pressure, usually at depths below the surface zones of weathering and cementation. The pre-existing rocks must adjust their mineralogical assemblages, chemistry, and structural configurations to be closer to equilibrium in a new or changing chemical and physical environment. Metamorphic processes cause new mineral compositions to form, in new textures that often destroy earlier ones. These new textures involve new grain sizes and arrangements, although sometimes relic fabrics and composition from the original rocks may be inherited and retained in the new metamorphic rock (Best 1982; Barker 1990; Yardley et al. 1990).

The main mechanisms of change are thermal metamorphism (increased temperature), dynamic metamorphism (physical deformation due to high pressure and intense strain), chemical rearrangement through driving off of volatiles such as water or carbon dioxide (isochemical metamorphism), or through chemically active pore fluids or diffusing ions (metasomatism, also called hydrothermal metamorphism). These mechanisms of change may be found in deep-seated fault zones (dislocation metamorphism, involving dynamic metamorphism), zones of contact between magma and surrounding rock (contact metamorphism, relatively restricted in area and involving thermal metamorphism only, without enough deformation to create mineral alignments rather than new grains in random orientation), or deeply buried zones where mountain-building activity is occurring (regional metamorphism, relatively extensive in area, and which may involve elevated temperature and pressure along with physical deformation) (Dietrich and Skinner 1979; Williams et al. 1982; Yardley et al. 1990).

Certain ranges of temperature and pressure tend to produce similar assemblages of minerals; these assemblages are termed "facies." Some common ones include blueschist facies (rocks that are somewhat bluish in color due to presence of the blue minerals glaucophane and lawsonite), greenschist facies (rocks that are green due to chlorite, actinolite and/or epidote), amphibolite facies (rocks rich in hornblende plus plagioclase feldspar, usually as andesine), granulite facies (rocks rich in pyroxenes and dark red garnet with a composition between almandine and pyrope), eclogite facies (rocks rich in jadeitic pyroxene and pyrope garnet), and hornfels facies (rocks that are very fine grained and rich in pyroxenes). Many variations occur as both the original rock makeup and the temperature/pressure conditions affect the final outcome (Dietrich and Skinner 1979).

Textures found with metamorphic rocks differ from those of other types of rocks in that they are the result of crystals of various minerals competing for space as they grow within a solid medium. Some of the textures commonly found with metamorphic rocks include the presence of large crystals (porphyroblasts) of one or more minerals found in association with much smaller grains of other minerals (porphyroblastic texture), and porphyroblasts that are riddled with inclusions of other minerals, called a sieve structure, or poikiloblastic texture. Another common metamorphic texture is foliation, or the presence of a parallel structure that often creates cleavage, or planar fissilility, in the rock. Foliation can result from compositional segregation, and may undergo additional deformation in the form of microfolds or crenulations of earlier foliation. Grain alignment is called slaty cleavage when the grain size is very small, and schistosity, or schistose texture, when the grain size is coarser (Williams et al. 1982; Yardley et al 1990).

Metamorphic rocks are named partially according to the nature of the parent rock; for example, quartzite is metamorphosed quartz sand or quartz-rich sandstone, and marble is metamorphosed limestone. An important factor for naming of a metamorphic rock is the texture, especially grain size and whether or not the dominant appearance is due to minerals with preferred orientation versus random orientation. For example, a schist has relatively coarse grains that are visible by eye and are in parallel alignment, whereas a hornfels has a random fabric of interlocking grains which produce a tough rock. Specific minerals may also be added to the name of the rock when these are dominant or have geochemical significance, such as a garnet muscovite schist or a tremolite marble. In addition, there are also some special descriptive terms, often referring to conditions of metamorphism: serpentinite is composed mainly of serpentine, and granulite is a rock with equidimensional polygonal grains and minerals indicating very high temperature metamorphism (Yardley et al. 1990).

5.1 Slate, Meta-argillite, and Metasiltstone

5.1.1 Slate

Slate has aphanitic texture (so fine grained that most of the individual crystals cannot be distinguished separately with the unaided eye) (Fig. 5.1.1). It is often a strongly foliated rock, readily breaking along a plane and splitting into thin plates. The cleavage planes are due to the preferred orientation of very fine phyllosilicate grains (Fig. 5.1.2). Individual grains are too small to be seen with the unaided eye, and freshly broken surfaces have a dull appearance (Yardley et al. 1990).

Slate is produced by dynamic (compressive) regional metamorphism of even finer grained rocks such as shale, mudstone or tuff, usually formed under conditions of greenschist or amphibolite facies (low-grade regional metamorphism). They tend to be comprised mainly of micas, clay minerals, quartz, and perhaps dusts of opaque graphite (Fig. 5.1.3); accessory minerals may include tourmaline, rutile, epidote, sphene, magnetite, hematite, pyrite, or other iron oxides or sulfides. Some green slates may contain epidote, along with albite (Best 1982; Williams et al. 1982).

Commonly used as a cultural material, one of the major uses of slate is as a roofing material, as an alternative to tiles (Foster 1981; Pierpont 1987). The fine-grained appearance mimics ceramic tiles (or, more accurately, ceramic tiles mimic slate), and slate may occur in a variety of colors. Common colors include blue-gray (due to sericite and/or illite), black (rich in carbonaceous matter such as graphite), red (hematite), green (chlorite and muscovite) and brown to yellowish brown (limonite). However, most slates undergo some color change after cutting and upon exposure to weathering conditions, so current colors are not always a good assessment of original appearance (Behre 1933).

Slates are often spotted or streaked, and may contain distorted fossils. The good cleavage properties of slate, making it easily cut into thin, flat plates, as well as its strong durability and low maintenance needs (tendency to be impervious to moisture and to be chemically inert) make it highly useful for roofing, flooring, and other architectural uses (Dimes 1990b). However, the mineralogical composition of slate can vary widely from quarry to quarry, and thus there is variation in its susceptibility to weathering. For example, some slate used for roofing tiles has traces

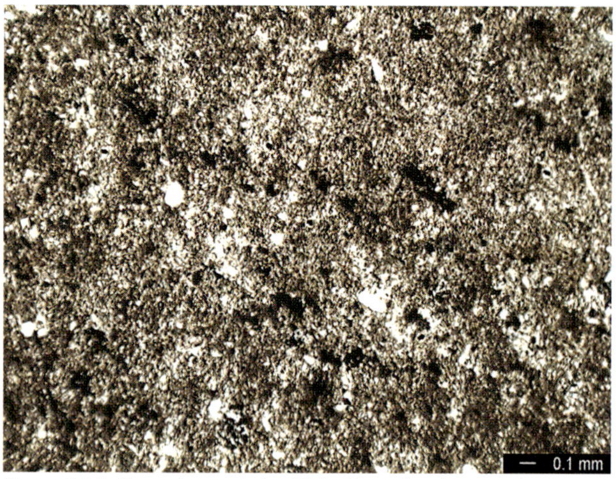

Figure 5.1.1 Slate, plane polarized light. Slate often has an aphanitic texture (so fine grained that most of the individual crystals cannot be distinguished separately with an unaided eye). Here there are a few larger grains (mainly quartz, calcium carbonates, sericite, and opaque particles).

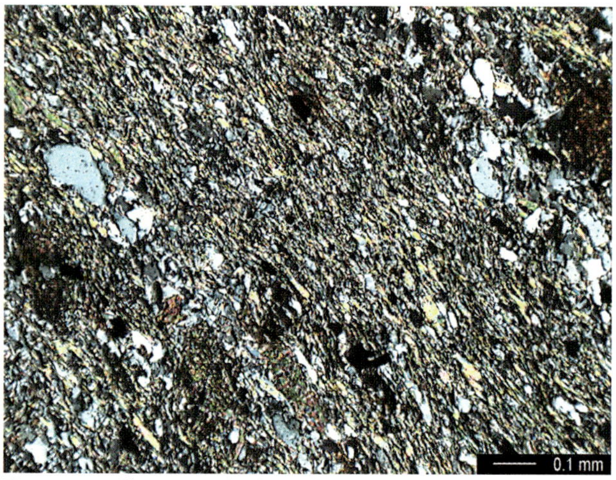

Figure 5.1.2 Slate, crossed polarized light. The cleavage planes in slate are due to preferred orientation of very fine phyllosilicate grains (muscovite mica in this case). The larger, rounded gray and white grains are quartz.

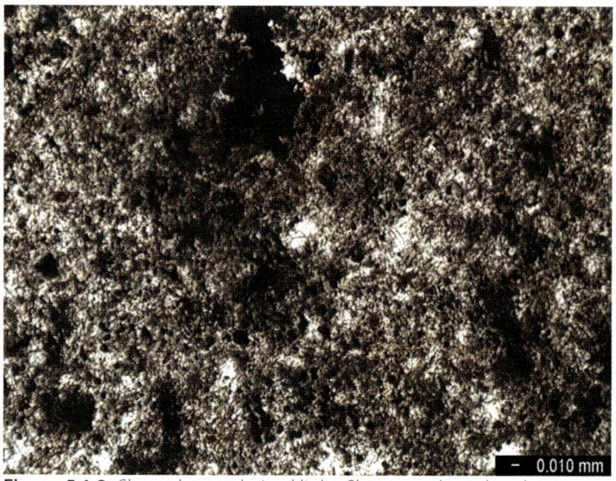

Figure 5.1.3 Slate, plane polarized light. Slates may be colored gray or black by graphite dust. The black areas are mainly fine particles and aggregates of graphite, with a few small magnetite particles. The colorless grains are mainly quartz, calcium carbonates, and sericite.

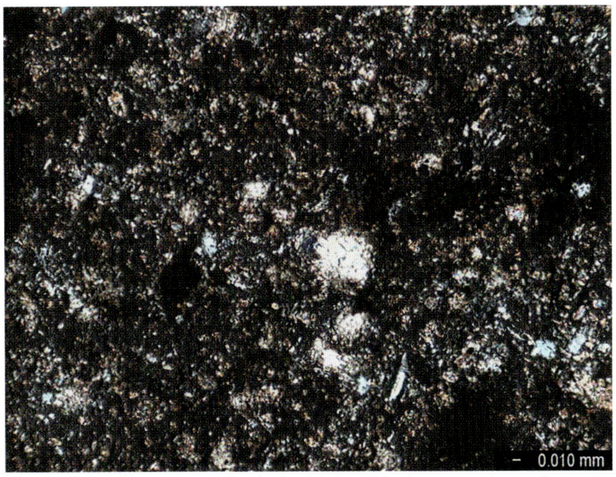

Figure 5.1.4 Slate, crossed polarized light. This slate contains carbonate minerals, which easily deteriorate. Most of the light-colored minerals in this image are carbonates, including the larger grain near the center. These highly deteriorated carbonates exhibit rounding and severe abrading of their edges.

of carbonate minerals, which easily deteriorate (Fig. 5.1.4) (Merriman 1892). As slate is often used in conjunction with metal attachments or joining materials, these metal components can also serve as a catalyst for deterioration and damage (Pierpont 1987).

Slates develop a fissile structure as the fine-grained minerals tend to become oriented with their shorter axes in the direction of pressure. Sometimes, however, areas or layers within a slate bed do not readily develop a fissile structure, and will remain a hard and dense homogeneous mass. These pieces of nonfissile material have little usefulness and are generally discarded at the quarry. Large piles of such debris can often be found at quarries—in some quarries the proportion of slate waste to useful slate material is rather large (Merrill 1891).

In addition to roofing material, slate is used for many utilitarian products such as floor tiles, walkways and steps, blackboards and school writing slates, tombstones, tabletops and shelves, sinks and cisterns, clock faces and sundials, and for mantels and fireplace surrounds. In the United States, slates have been quarried from a variety of locations from at least as far back as the eighteenth century, especially in Pennsylvania, New York, Maryland, Georgia, Maine, Michigan, Minnesota, Vermont, New Jersey, and Virginia; there are also many slate quarries located in Canada and elsewhere in the United States (Merrill 1891; Merriman 1892; Behre 1933). Sometimes slate is used as a masonry material for small buildings, as evidenced by a house in Pennsylvania that was constructed of waste blocks of slate from a nearby quarry, although weathering has greatly altered its visual appearance (Miller 1939).

Important deposits of slates are quarried for roofing material in the United Kingdom—in North Wales, Cornwall, Devon, Scotland and elsewhere They range in color from pale gray, black, green, or purple, and are very fissile, splitting easily into very thin sheets. Stone from British quarries has been exported worldwide for roofing slate. Thin-section analysis of blue slate roofing material used in medieval England was the primary research tool for studying their quarrying, transport and use (Jope and Dunning 1954; Pyddoke 1963). The analysis showed that blue slate from Devon and Cornwall was transported into southern England for use as a roofing material, where local thinly bedded limestones and sandstones had been rejected for roofing material use. An exception to the Devon and Cornwall slate use was found in one medieval village where a green slate from Belgium was used instead. Combining the thin-section analysis results with documentary evidence, the researchers were able to identify slate trade routes throughout Britain and abroad.

Slate has also been used as a sculptural material, primarily as a medium for inscriptions and shallow or deep bas-relief designs. These designs and letterings are sometimes drawn and blocked out on the surface of the stone before tooling begins; the sculptors may simply work directly on the stone with little surface preparation. The surface of carved slate is often polished and/or coated with wax after completion in order to enhance the form, designs or inscriptions. Examples of slate carvings include many pieces from ancient Egypt, such as sarcophagus lids, Koryŏ period (935–1392) Korean incised burial chests, early American tombstones, and modern slate sculptures. Cutting through what are essentially many very fine layers superimposed upon each other creates a subtle but interesting visual effect (Ubac 1956; Eliscu 1959; Boulter 1992).

5.1.2 Meta-argillite

Meta-argillite is a term that often refers to a firmly indurated (hardened) clay-rich rock that has undergone some recrystallization of original material, or a mudstone with incipient metamorphism, thus somewhere between a shale and a true slate, produced by low-grade metamorphism. Meta-argillite generally lacks the typical slaty cleavage seen with slates (Williams et al. 1982).

Wainwright and Down (1986) employed thin-section petrography, among other analytical techniques, to characterize and study deterioration of meta-argillite used by the Haida people of the Pacific Northwest coast of North America. The black, carbonaceous slate-like material from the Slatechuck Creek quarry on one of the Queen Charlotte islands has been used since the nineteenth century (or even earlier) for carved items such as pipes, bowls and plates, chests, and miniature totem poles with images of human and mythological figures; these items tend to crack and break over time. The characterization study included reference samples of quarry material to compare against artifact samples, as many carved objects lack provenance and their authenticity as Haida artifacts is suspect.

Slatechuck Creek quarry material was described by Wainwright and Down as a carbonaceous, kaolinite "slate" with an abundance of clay materials including pyrophyllite, kaolinite, and lesser amounts of montmorillonite and illite. Due to the clay content and lack of slaty cleavage, this material was identified as meta-argillite. The thin section description and illustration showed that much of the stone was comprised of veinlets of kaolinite and some montmorillonite in a very fine carbon-rich clay matrix. Some kaolinite was thought to have developed from weathered feldspar grains. Larger veinlets of kaolinite seemed to have formed as a result of alteration of the finer matrix. Natural cracks were thought to be related to the presence of the veinlets, and to postdate them. There was clear foliation, but the rock remained compact unless sharply hit. There was variation from sample to sample, especially with a variety of unusual textures and organic matter content, and some areas of the quarry contained lower quality material (with detrital minerals). Deterioration of the Slatechuck Creek material was a problem; the montmorillonite fraction was especially reactive and expanded or contracted with fluctuations in relative humidity. Cracking most often occurred at discontinuities in grain size or composition, such as at the kaolinite veinlets, especially when artifacts underwent mechanical shock.

5.1.3 Metasiltstone

Metasiltstone is rarely encountered as a cultural material. It is defined as a fine-grained, weakly metamorphosed rock that, unlike most slates, does not tend to break along distinct planes. This rock type is reported in thin-section studies of stone sculptures of India, used on some sculptures from Rajasthan (Newman 1984). Newman described one example as being very fine grained, green-gray in color, containing illite, sericite, quartz, and chlorite. He noted that the most likely source was a group of argillaceous rocks in southern Rajasthan where metamorphism increases from southeast to northwest, with rocks occurring that range from slates to amphibolite facies schists. The other example was described as being very fine grained, dark, and gray-black in color, containing illite, sericite, and quartz.

Another example of the use of metasiltstone in the production of Indian sculpture is a piece in the Ashmolean Museum (Oxford, England) that was made in West Bengal around A.D. 1050, typical of sculptural works of the Pala and Sena dynasties in Bengal and Bihar of northeastern India (Penny 1993). These sculptures range in color from a very dark gray-brown to green, and are very fine grained. Quarried in the hills of Bihar, it was also used as a building construction material.

A quartz-mica metasiltstone (called "blue phyllite") also appears as a cultural material, used mainly for medieval honestones, employed for honing metal tools. It has been identified as a material used in a variety of cities and regions of England, as well as in Germany (Moore 1978).

5.2 Phyllite

Rocks similar in composition to slates but which undergo additional metamorphism become phyllites, mineralogically similar but with a coarser grain size (yet finer grained than schist). In phyllites, micas may form large enough flakes to create a visible, silky or glossy sheen to the cleavage surface, and sometimes the coarser phyllosilicate grains are large enough to be visible to the unaided eye. Other constituents tend to be quartz and feldspars, forming visible segregation bands as they alternate with bands of muscovite, talc, chlorite, and/or graphite, accentuating the schistosity, or foliation, of phyllite (Fig. 5.2.1). This banding is due to metamorphic differentiation, in which local concentrations of minerals develop during chemical reconstitution, affected by the diffusion of fluids along a path of minimum resistance along foliation created by mechanical processes. Porphyroblasts (relatively large grains formed as a result of metamorphic recrystallization) are common in phyllites, and may include andalusite, biotite, chlorite, chloritoid, cordierite, garnet, magnetite, pyrite, staurolite, or stilpnomelane (Fig. 5.2.2). Cleavage surfaces are often less perfect than the cleavage surfaces of slates (Dietrich and Skinner 1979; Williams et al. 1982; Yardley et al. 1990).

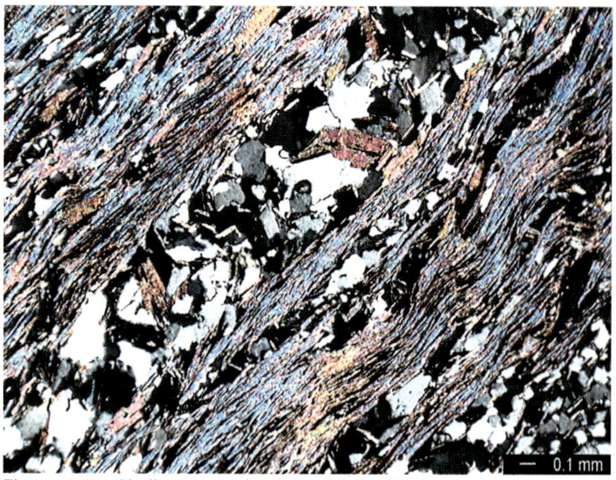

Figure 5.2.1 Phyllite, crossed polarized light. Phyllites have a coarser grain size than slate, but finer than schist. Micaceous minerals often form visible segregation bands alternating with quartz and feldspars, accentuating the schistosity, or foliation, of phyllite. In this example, wide bands of talc alternate with bands of quartz grains.

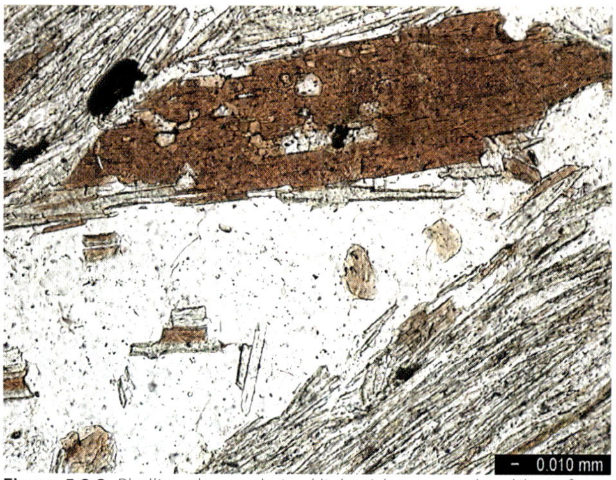

Figure 5.2.2 Phyllite, plane polarized light. A larger porphyroblast of brown biotite is visible in the upper central part of the image. Most of the rest of the image shows strings of colorless, higher relief talc with colorless and low relief (so almost invisible) quartz with scattered areas of darker chlorite, biotite, and graphite.

5.2.1 Phyllite as a cultural material

Phyllite has often been used as a building stone. For example, in Lancaster County, Pennsylvania, phyllite was quarried for that use in the nineteenth and twentieth centuries (Behre 1933). However, much earlier, phyllite was one of the building materials used in the construction of the Minoan palace of Knossos on Crete, mainly as paving slabs (Papageorgakis and Mposkos 1988).

Perhaps one of the most famous uses of phyllite as a cultural material is for the production of sculpture in ancient Afghanistan and northern Pakistan. Thin-section analysis of samples of those sculptures proved useful in provenance studies (Courtois 1962–63; Lahanier 1976; Kempe 1982; Newman 1984; Kempe 1986; Newman 1992; Reedy 1992b). Dating to the Gandhara period, the Buddhist sculptures show a clear mix of influences from both Greece and India. They are most often made of a shiny gray schistose material that can be identified in thin section as a carbonaceous phyllite (sometimes grading into schist) primarily consisting of quartz and muscovite, with chlorite sometimes present. An unusual feature is the prominent inclusions of tiny rods of chloritoid.

Comparison with samples from possible quarry sites indicates that the sources were located within the Swat and Peshawar valleys. Newman's (1992) microprobe analysis of the chloritoid inclusions enabled further grouping of sculptures that probably derived from the same source. However, much more geological work needs to be done in the Gandhara area before full provenance studies can be completed. Some impediments exist: first, although a significant amount of geological work has been carried out in northwest Pakistan and Afghanistan, many of the potential stone sources are located in remote or politically inaccessible areas: and secondly, published petrographic analyses currently exist for only a handful of Gandhara sculptures, therefore we are far from understanding the full range of materials that was used.

Courtois (1962–63) analyzed samples from ten sculptures in the collection of the Musée Guimet, and found two different types of phyllites. The first type included five sculptures from an area he described as central Afghanistan, all from documented excavations in the Kapisa Plain region (Begram, Shotorak, and Paitava). He described this rock type as a carbonate phyllite, consisting of a fine-grained carbonate-chlorite-muscovite-sericite-quartz rock with a schistose texture and having accessory rutile, apatite, garnet, epidote-zoisite, and magnetite.

The second type of phyllite was described as a carbonaceous chloritoid muscovite-sericite-quartz phyllite. This type included one sample from a sculpture from Hadda in the Jalalabad region of Afghanistan and four from sculptures attributed to the Peshawar region of northwest Pakistan (exact findspot uncertain). One of these samples also showed the presence of a strongly pleochroic yellow-brown to clear mica, and the formation of biotite. Another had accessory hematite.

Thin sections of Gandhara sculptures, mainly from the British Museum, have been studied, including eight made of phyllite (Reedy 1992b; 1997a). There were two slight variations of this sculpture material. The first type was a phyllite comprised primarily of graphite, quartz, muscovite, and chloritoid (Fig. 5.2.3). Accessory minerals included chlorite, biotite, hematite, rutile, magnetite, ilmenite, apatite, sericite, and monazite. This material was used for a variety of images such as a seated Buddha, narrative friezes, and an image of the emaciated Siddhartha. Red iron-oxide stains were common, probably caused by iron leaching during weathering of the stone. This rock type most closely matched that of the group of 12 phyllite sculptures analyzed by Newman (1984) and four of those described by Courtois (1962–63).

The other variation of this rock type had chlorite present as a major mica along with muscovite, rather than as an accessory (Fig. 5.2.4); biotite was also more common. Otherwise, the stone was the same basic chloritoid-rich phyllite with quartz and micas. The same accessories occurred, with the addition of rare stilopmelane, clinopyroxene, spinel, and

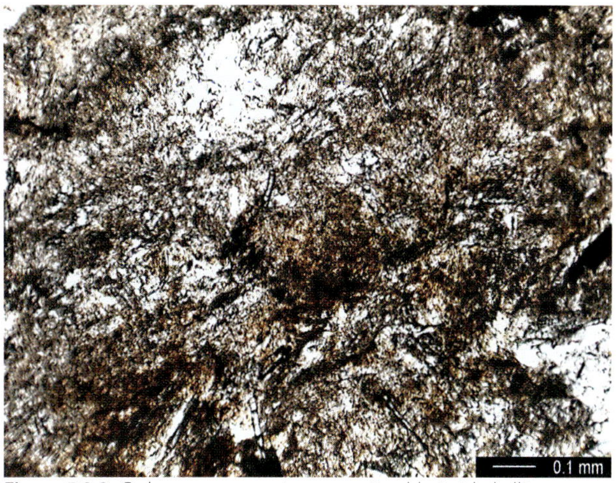

Figure 5.2.3 Carbonaceous quartz-muscovite-chloritoid phyllite, plane polarized light. This phyllite sculpture from ancient Gandhara is comprised primarily of quartz, dark strings and dusts of graphite, muscovite, and tiny rods of high-relief colorless chloritoid. Iron oxide stains are common, probably caused by iron leaching during weathering of the stone.

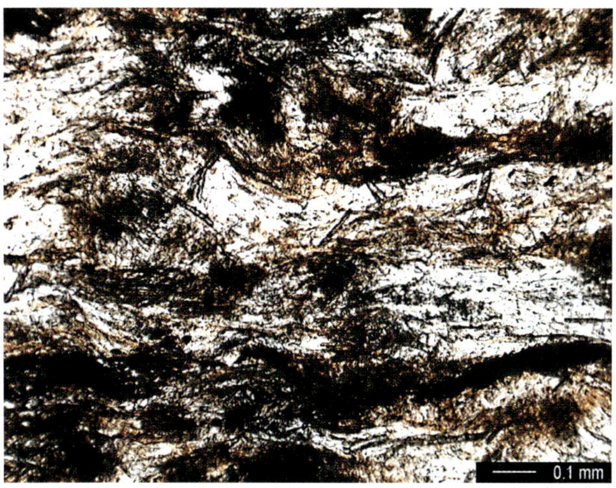

Figure 5.2.4 Carbonaceous quartz-chlorite-muscovite-chloritoid phyllite, plane polarized light. This ancient Gandhara phyllite sculpture has brown chlorite as a major mica along with colorless muscovite. There is also colorless low-relief quartz, dark strings of graphite, and small colorless, high-relief rods of chloritoid.

plagioclase feldspar (probably albite). The type was similar to one of the samples from the Musée Guimet described by Courtois (1962–32).

Similar graphite-rich, muscovite-quartz phyllites can be found in many areas of the Gandhara region, particularly in northwest Pakistan (Rehman and Zeb 1970; Newman 1984). However, chloritoid is not mentioned as being a constituent of most of those rocks. David Kempe (1982; 1986) —who pointed out that chloritoid-bearing, alumina-rich mica phyllites or schists are not common—identified one definite quarry in the Peshawar valley that could have been a source of this stone and rejected several other possible source localities. He described the quarry samples as being not of the best quality, and the quarry itself as also rather small. Thus, it was clearly one of a number of quarries that must have been exploited for Gandhara sculptures; he noted that it is feasible that the best quality outcroppings were exhausted. Kempe explored other areas in and around the Peshawar valley, but found no more sources of this particular stone, although at least one possible source was inaccessible to him.

Since large numbers of sculptures have been recovered from northwest Pakistan, the existence of many quarries is likely. Italian geologists working on architectural materials of Swat and their sources found many more outcrops and quarries of chloritoid-bearing phyllites throughout the Swat valley (Di Florio et al. 1991, 1993a, 1993b); some of these may also prove to be sources of the sculptural material. No chloritoid-bearing phyllites have been reported from Afghanistan, although most potential source areas there have been inaccessible to surveys for many years.

A typical example is shown in Figure 5.2.5, taken from a Gandhara-style Fasting Buddha image (Errington and Cribb 1992: fig. 215), attributed to the third century A.D. It can be termed a carbonaceous quartz-muscovite-chloritoid phyllite; the visual appearance of the sculpture itself is a smoky gray color with a bluish tint and prominent silvery-white mica specks. In thin section we can see that the fine-grained phyllite is composed primarily of quartz (about 60%) and muscovite (about 30%). There is a schistose fabric composed of strings of muscovite, which are often very fine grained, although larger flakes are also visible. The small anhedral quartz is scattered throughout the stone, with some veins of larger grained quartz existing in a number of places. This larger quartz is often polycrystalline, and contains inclusions of zircon and apatite.

The major accessory mineral in the stone is chloritoid. In thin section this mineral has high relief and is nearly colorless, although it may have a slight green or gray color. Birefringence is weak. The grains appear as porphyroblasts, usually subparallel to the schistosity of the stone, standing out as visible rods (Fig. 5.2.6).

Other accessories are chunks of rectangular-shaped iron-oxide minerals, mainly ilmenite with some magnetite. Strings and dust of graphite are spread throughout the sample. The surface is also covered by reddish iron-oxide stains, probably caused by iron leaching during the weathering of the stone. Other accessories include monazite and some rare biotite. A weathering layer of fine-grained clay is visible in places on the external surface, as deep as 0.32 mm with an average thickness of 0.20 mm (Reedy 2006a) (Fig. 5.2.7). The stone of this Fasting Buddha example is similar to that of most of the Gandhara sculptures examined by Newman (1984). It is also similar to stones of the Lower Swat-Buner area of northwest Pakistan described by the Pakistani geologists Rehman and Zeb (1970).

In contrast, another Gandhara-style Fasting Buddha image attributed to the third century, also showing the unique Gandhara depiction of a very emaciated Shakyamuni Buddha (Pal 1978: fig. 10), is made of a somewhat different stone. In thin section (Fig. 5.2.8) this example is best termed a kink-folded muscovite phyllite. The rock is composed primarily of muscovite (around 50%), which is generally very fine grained and defines the foliation of the rock fabric. Fine-grained anhedral quartz makes up approximately 20% of the stone.

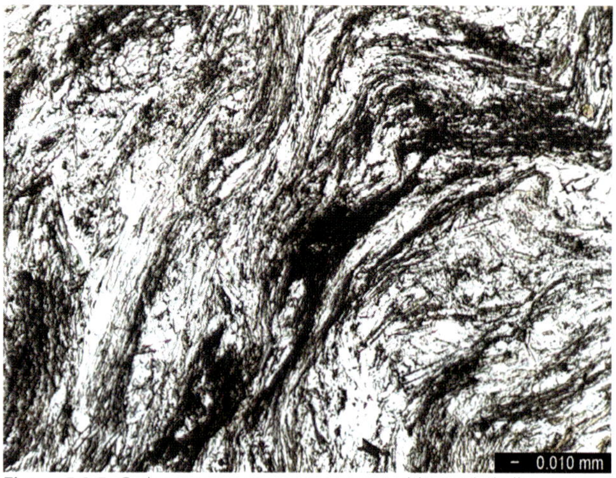

Figure 5.2.5 Carbonaceous quartz-muscovite-chloritoid phyllite, plane polarized light. This phyllite was used to make a Fasting Buddha sculpture from the ancient Gandhara period of northwest Pakistan. The fine-grained phyllite is composed primarily of colorless quartz (about 60%) and muscovite (about 30%). Strings and dusts of black graphite are spread throughout the sample.

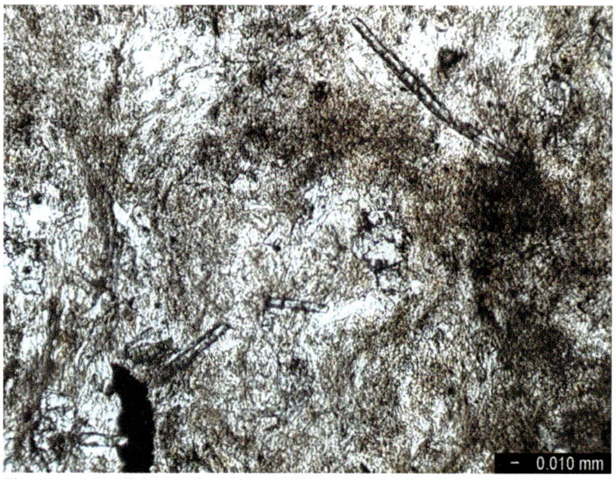

Figure 5.2.6 Chloritoid within phyllite, plane polarized light. The chloritoid grains stand out in phyllite because they have high relief, appearing as small scattered rods usually oriented subparallel to the schistosity of the stone. They are usually nearly colorless, but can also be slightly green to blue in color. The main minerals of the phyllite are muscovite and quartz, both with low relief. There are also some graphite specks and iron-oxide stains in this view.

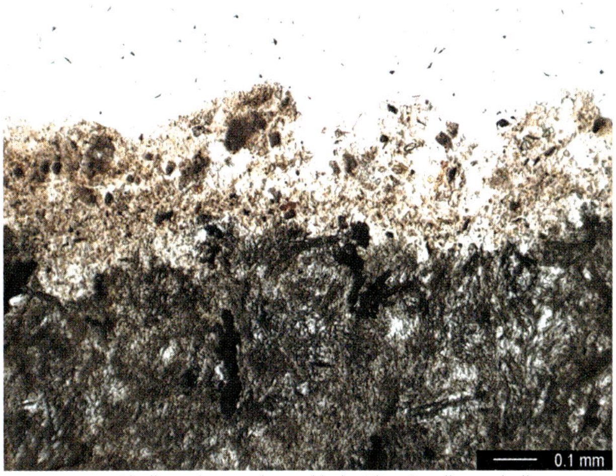

Figure 5.2.7 Phyllite, plane polarized light. This Gandhara phyllite sculpture has a weathering layer that averages 0.2 mm thickness. In plane polarized light this layer looks browner than the underlying rock, as it contains more iron oxides compared with the darker graphite and colorless quartz and muscovite in the rock below.

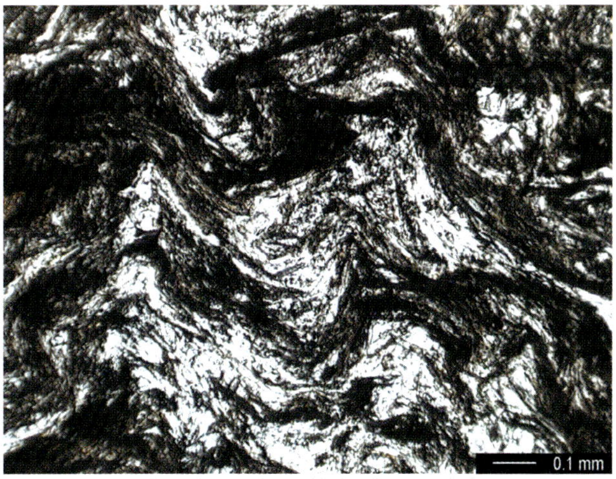

Figure 5.2.8 Phyllite, plane polarized light. This sample from a Gandhara-style Fasting Buddha phyllite sculpture shows localized kink-folding in a muscovite phyllite. The fine-grained muscovite defines the foliation of the rock fabric; there are also some fine-grained anhedral quartz, and dusty opaque graphite. Chloritoid is absent in this rock.

A dusty opaque graphite occupies about 30% of the sample. This graphite and graphite dust spread throughout the rock account for the overall gray appearance of the stone.

A predominant texture of the rock is the pronounced localized kink-folding, caused by a combination of elevated temperatures and mechanical strain during metamorphic formation of the rock. The presence of this pronounced kink-folding, as well as the absence of the chloritoid inclusions that distinguish other Gandhara phyllite sculptures, indicate a source different from the majority of Gandhara works. No weathering layer is seen on this piece, either.

Although chloritoid phyllites were used for architectural purposes (stupa umbrellas) in Swat (Di Florio et al. 1991, 1993a, 1993b), those examples differ from the chloritoid phyllites used in sculpture. The architectural stones have lenticular concentrations of rust-colored carbonates with limonite, crenulations, and coarser grain size. Di Florio et al. pointed out that chloritoid phyllites, with or without carbonates, blackish or bluish, are very common throughout the Lower Swat valley.

Interestingly, a chloritoid phyllite was also used as a sculptural material in northeastern India (West Bengal and Bangladesh) from the late fifth through the twelfth centuries (Asher 1987). The material was identified as part of Asher's work to test a dark gray or black stone used extensively as a sculptural material in northeast India. This stone has been given a wide variety of names by art historians on the basis of visual appearance (basalt, schist, sericitized slate, chlorite, etc.). Asher used thin-section analysis to identify a correct terminology and to determine if the sculptures were all carved from the same stone, or from varying materials.

Almost every sculpture from Bengal examined was carved from a black phyllite composed of quartz, muscovite, chloritoid microphorphyroblasts, and lesser amounts of chlorite, zircon, tourmaline, and opaque minerals. Chloritoid content ranged from 7% up to 30%. Earlier art historians had hypothesized that the stone used for these black sculptures came from the Rajmahal hills, an idea that was widely published. However, Asher discovered that phyllites were not found there, only basalt. Instead, the most probable source was identified as the Dalma hills in Singhbhum district, Bihar, from where it could have been transported by river into Bengal. Sculptures in Bihar, however, were carved from a wide variety of stones and not just the chloritoid-bearing phyllite.

Another cultural use of phyllite is for the production of utilitarian lithic objects such as knives and scrapers. Vaughn (1993) used thin-section analysis to characterize prehistoric lithic artifacts from northeastern Alabama. Fifty worked artifacts from a wide range of archaeological sites were compared with local rock samples; the source outcrops were identified and the information used to infer early trade and migration routes. Some of the artifacts were found to be made of phyllite.

Vaughn discovered that two major rock formations served as the sources of the lithic materials. Phyllites from the Hillabee Formation (a low-grade metamorphic rock) were comprised of mainly ferromagnesian minerals (such as iron actinolite and epidote) and feldspars, with minor amounts of chlorite and quartz. The Heflin phyllite, in contrast, consisted of layered iron actinolite, sericite, and stretched/sutured quartz. Phyllite was also chosen as a material for honestones (sharpening stones) in the Viking world. A quartz-mica phyllite termed "purple phyllite" was used and traded during Scandinavian–English interactions of the ninth–fifteenth centuries, although the source is uncertain (Crosby and Mitchell 1987).

5.3 Schist

Schists have medium grain size (often visible to the unaided eye) and are characterized by strong schistosity, or foliation, referring to a parallel alignment of grains (Fig. 5.3.1). Schists tend to have an abundance of platy minerals (muscovite, biotite, chlorite, chloritoid, antigorite or talc), and often have a laminated structure due to mineral assemblages of different composition segregating into alternating layers parallel to the schistosity, created by metamorphic differentiation of initially homogeneous rocks. Schists also tend to have a marked parallel alignment of elements such as rod-shaped mineral aggregates, prismatic crystals, etc.; this feature is termed "lineation" (Dietrich and Skinner 1979; Williams et al. 1982; Yardley et al. 1990).

Schists are often named according their mineralogical composition, such as garnet muscovite schist, graphite schist, chlorite schist, or hornblende schist (Fig. 5.3.2); overall composition, such as ferruginous schist or garnetiferous mica schist (Fig. 5.3.3); or by some aspect of their appearance (e.g. spotted schist) (Dietrich and Skinner 1979). For example, blocks of stone described as mica schist were used in the construction of masonry buildings in Pennsylvania, while other buildings were made of a schist described more specifically as sericitic quartz schist (Behre 1933). Honestones found at four major Viking Age sites in northern Europe were identified as quartz-muscovite schist (Fig. 5.3.4) (Mitchell et al. 1984), as were honestones used and traded during the period of Scandinavian–English interaction (ninth–fifteenth centuries A.D.) (Crosby and Mitchell 1987). However, there were several identifiable types of quartz-muscovite schist that varied in texture and accessory minerals.

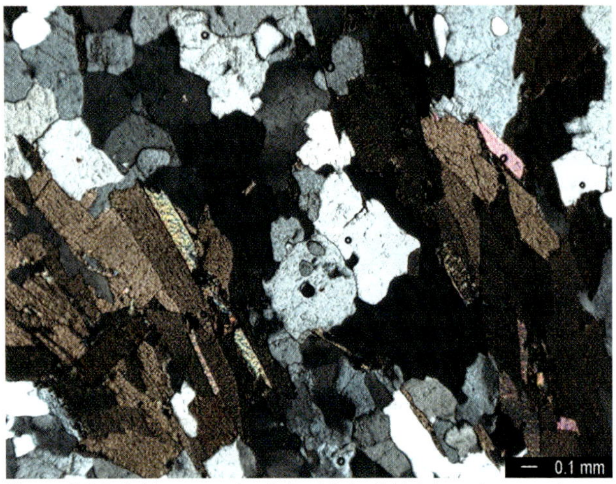

Figure 5.3.1 Schist, crossed polarized light. Schists have larger grain size than phyllites. As with phyllites, they are also characterized by the presence of a parallel alignment of grains. This mica schist has mainly biotite (brown in crossed polarized light) interspersed with quartz and some plagioclase feldspar; there are also some higher birefringent flakes of muscovite.

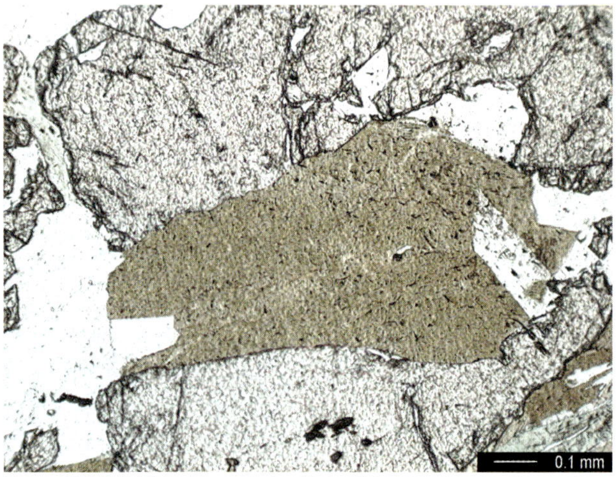

Figure 5.3.3 Schist, plane polarized light. This schist is named according to its overall composition, a garnetiferous mica schist. In plane polarized light the garnet, which is the most abundant mineral in this field of view, is pale brown and shows high relief. The micas include biotite, the yellowish-brown mineral near the center of the field of view, chlorite, a less abundant green mica in the lower right corner, and some colorless to pale green muscovite flakes. Quartz is also present, and is colorless in plane polarized light.

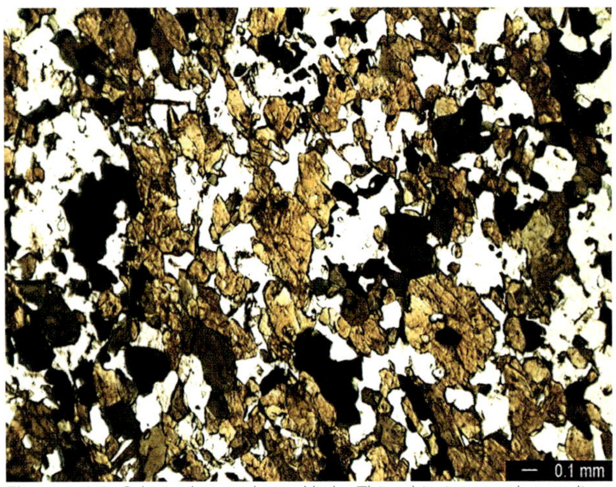

Figure 5.3.2 Schist, plane polarized light. This schist is named according to mineralogical composition, so is termed a hornblende schist. The greenish-brown and yellowish-brown flakes that predominate in the rock are hornblende, interspersed with quartz and plagioclase feldspar (colorless in plane polarized light). This schist is strongly foliated and clearly banded.

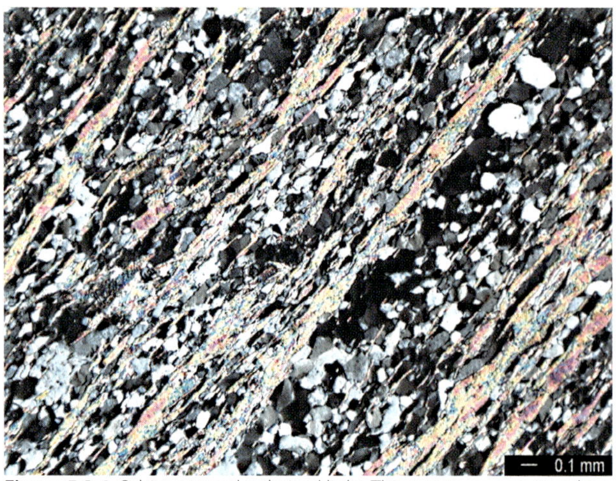

Figure 5.3.4 Schist, crossed polarized light. This quartz-muscovite schist consists of narrow, alternating layers of low birefringent, fine-grained quartz and high birefringent strings of muscovite mica.

5.3.1 Talc schist and steatite/soapstone

Schist comprised predominantly of talc is called talc schist if the talc flakes have a preferred orientation (Fig. 5.3.5). However, if the rock is instead massive (Fig. 5.3.6), it is often called "steatite" or "soapstone" since talc is very soft and feels like soap. Since soapstone is easy to cut and carve, and can take high temperatures, it has been used for a variety of objects including decorative items (ornaments and statuettes) and functional ones (cooking bowls and griddles, stoves, fire-resistant utensils, vessels, oil lamps, moulds for bronze casting, beads, or floor tiles and architectural elements) (Dietrich and Skinner 1979; Gibbins 1988; Storemyr 1997; Truncer et al. 1998; Santi et al. 2005).

So soft that it can be scratched with a fingernail, steatite, or soapstone, has a long history as a carving material (Plehwe-Leisen et al. 1992). Since it withstands high temperatures and thermal shock well, it has found use in components for smoking pipes, cooking pots, lamps, stoves, crucibles, and casting molds, as well as in lining of furnaces. Steatite can have a wide variety of colors ranging from green to various shades of gray. In powdered form it is used for a variety of products such as talcum powder (Weinstein 1958; Ritchie 1973; Kempe and Harvey 1983).

An important early use of steatite was in the manufacture of beads at Harappan sites of South Asia in the third and second millennia B.C. For example, Hegde (1983) reported a find of 34,000 small beads within one cache which were found to be a talcose steatite. During manufacture, the beads were heated to a temperature at or above 850 °C, which caused a mineralogical change leading to the formation of cristobalite (a high-temperature form of silica), enstatite, and alumina. As a result, the hardness of the steatite was greatly increased. The softness of the unaltered steatite was a great advantage in carving objects such as seals and larger beads; it could also be used to prepare a paste for forming small beads. Heat treatment would then create a hard material that would be much more durable and able to handle the constant use required of the many utilitarian objects made of steatite.

The technology used to produce steatite beads and other ornaments in the Indus valley area, 7000–200 B.C., was explored in detail by Bouquillon et al. (1995), and the local innovations and changes over time in technology were outlined. They noted that steatite was one of the main materials used at these early sites to produce beads and stamp seals, and that these artifacts were usually fired and often glazed. In mineralogy, the body of the beads was comprised mainly of enstatite and cristobalite, typical of the changes undergone by talc when fired at about 1100 °C (they proposed a higher firing temperature than Hegde). Crystals of diopside or forsterite were seen growing in the copper-rich glaze areas; copper oxide dendrites and copper silicates such as chrysocolla could also been seen. Halite and gypsum crystals were sometimes present as a result of burial conditions. Bouquillon et al. proposed a progression in mineralogy over time as the beads move from being more crystalline in the earlier Chalcolithic period up to vitreous by the mature Indus period (with the earlier beads having more diopside, chrysocolla, and some trydimite, another high-temperature form of silica, up to only forsterite and glass in the later beads).

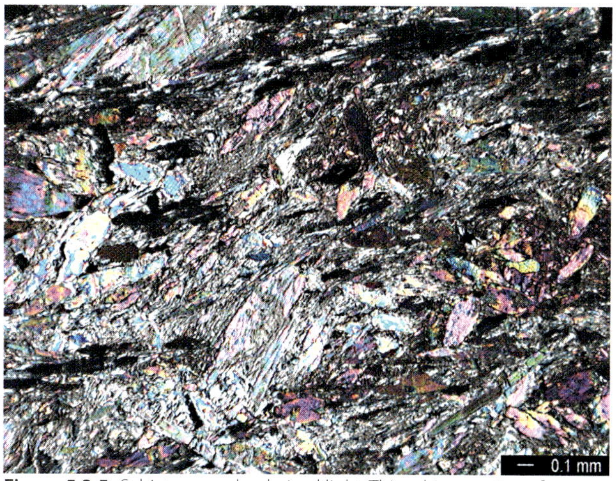

Figure 5.3.5 Schist, crossed polarized light. This schist consists of alternating layers of talc and slender, subparallel prismatic crystals of tremolite that is altering to talc. In crossed polarized light, the fibrous shreds of talc show very strong birefringence, whereas that of the tremolite is lower (moderate to strong). The preferred, rather than random, orientation of the talc here is typical of talc schists.

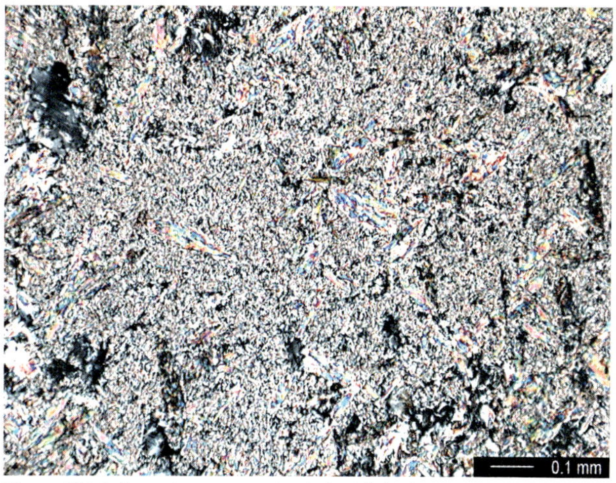

Figure 5.3.6 Soapstone, crossed polarized light. In contrast to the preferred orientation of talc flakes in talc schist, soapstone, or steatite, is massive. Thus the small talc flakes here have a random orientation.

Steatite, or soapstone, took the place of pottery in the Norse settlements in the Atlantic area at Viking period sites (Butler 1989). Butler described the steatite as having talc as the primary constituent, with carbonate minerals making up most of the remainder of the rock composition. He noted that the talc component made it easy to carve with metal or stone tools, and provided high resistance to heat, an advantage for cooking utensils. He reported that earlier in Norway, in the Neolithic, steatite-tempered pottery was used, while steatite itself was used mainly for carvings and for metal-casting molds. By the Celtic Iron Age, steatite vessels started to appear, the earlier ones following the forms of pottery of the period. By the Viking period, however, steatite vessels had replaced pottery vessels altogether. Pottery made a comeback in the Middle Ages, but most was imported rather than locally made.

In addition to cooking vessels, Butler reported that in Norse Shetland, where local steatite deposits were found, a wide variety of other utilitarian objects was carved from steatite. These objects included lamps, baking plates, loom-weights, linesinkers, spindle whorls, drill whorls, metal-working tools, toys, gaming boards, and jewelry. Butler carried out a program of thin-section petrography and elemental analysis to try to match artifacts to quarries, but was unsuccessful. He found that the basic mineralogy and elemental composition of Shetland steatites were all very similar, and accessory minerals and trace elements tended to vary as much within individual sources as between sources. He concluded that the most successful approach was comparing overall visual appearance in hand sample.

Steatite, or soapstone, was used as one of the building rocks of historic monuments in the city of Ouro Preto, Brazil (Ferreira 1992). The steatites used contained one or more of a range of minerals: talc, magnesium-chlorite, tremolite-actinolite, carbonates, sulfides, magnetite, hematite, or goethite. A wide variety of accessory minerals was also present.

Soapstone was identified by Di Florio et al. (1991, 1993a, 1993b) as a building stone at the Pānr I Sacred Area of Swat valley, northern Pakistan. It was described as a talc-schist with ankerite lenses and nodules. Ankerite is $CaFe(CO_3)_2$, a rhombohedral carbonate that is usually massive. In this case, the soapstone contained many fine cracks that have filled with ankerite, and was closely associated with a chlorite schist that was also used in construction. The latter rock had a schistose texture, consisting mainly of chlorite with some magnetite crystals.

Storemyr (1997) illustrated a thin section of soapstone from a quarry where it was obtained for use in Nidardos Cathedral (Trondheim, Norway), originally built in the Middle Ages and restored in later campaigns. The thin section of a cracked specimen showed a pyrrhotite inclusion that had oxidized to form goethite, and fissures preferentially associated with relatively stiff tremolite needles. Storemyr noted that soapstone, abundant in Norway, was an important construction material used throughout the country. For the cathedral, it was used as masonry blocks and for decorative elements. A variety of quarries used as sources for the stone, and their weathering characteristics, is discussed in Storemyr (1997).

Storemyr (2004) also discusses the processes involved in weathering of soapstone used in an outdoor context, such as Norwegian medieval stone architecture. He noted that of several hundred medieval stone buildings constructed in Norway, many have local and regional soapstones used for decorative purposes, and a few structures were even built with soapstone as a primary construction material. He found that stone that weathered extensively during the Little Ice Age in Norway (fourteenth–nineteenth centuries) has been relatively stable during the past one hundred years, probably due to fewer damaging frost events. He also found that architectural designs can impact weathering rates; air pollution in cities, until recent reductions in sulfur dioxide emissions, also had an adverse effect on weathering. Alkaline salts from Portland cement, sulfate from air pollution and some stones, and chloride from acid increased salt loads in many buildings during the nineteenth and twentieth centuries. Such salt sources have led to severe weathering and deterioration especially in instances of water leaks and run-off problems. He recommended that all of these factors be considered in developing conservation treatment options for specific buildings.

Santi et al. (2005) studied 41 archaeological artifacts from eight medieval sites in central-eastern Italy made of soapstone, focusing on the type called *pietra ollare* that indicates a group of Alpine metamorphic rocks, mainly of the greenschist facies. This material was often used for production of stoves and fire-resistant utensils for cooking and preserving food. In addition to the archaeological samples, ancient and modern quarries of *pietra ollare* in the Alps (the Valchiavenna) were also sampled. Thin-section petrography was done to characterize the modal mineralogy and textures of the rocks. This soapstone was identified as being a fine-grained, magnesite-bearing talc schist, and there was marked foliation of the talc and chlorite. The magnesite was anhedral to subhedral, and opaque minerals were also visible. Modal mineralogy and texture of the artifacts and those from the Valchiavenna were found to match each other very closely. Additional work by x-ray diffraction (XRD) analysis, scanning electron microscopy (SEM), and elemental analysis supported the thin-section petrography findings.

The term "soapstone" is sometimes applied to quartz-sillimanite-mica schist when the sericite is especially abundant and creates a "soapy" feel somewhat like that of talc. For example, such rock was quarried in Pennsylvania and called "soapstone" although it actually contains only minute amounts of talc. It was used as a steel furnace liner and for paper filler (Miller 1939).

Collins (1977) described a pale green chlorite-talc schist containing accessory calcite and albite that was used to sculpt a medieval graveyard effigy of a man in armor in Scotland. The probable quarry was local, and Collins noted that it was an ideal sculptural material, as it occurred in large flat slabs, was relatively easy to carve, and yet not too fissile or easily weathered. Quarries of this stone were near sea transport locations, so it was also a good trade material. Newman (1984) found talc-chlorite schist used in the production of sculptures from southern India; for example, an image of Krishna playing the flute from Mysore, Hoysala period (twelfth–thirteenth centuries). A sculpture of a Jain king, from the Deccan area of southern India, tenth century, was made of a talc-chlorite-calcite schist, and an image of Kubera from Mysore, Hoysala period (early twelfth century) was seen in thin section to be a carbonate-anthophyllite-talc-chlorite schist.

Like the term "greenstone," which is used in the literature to refer to a wide variety of green-colored rocks (Santallier et al. 1998; D'Amico 2005), the term "soapstone" is not very useful since it too has been used to refer to multiple rock types including talc schist. To complicate matters further, some writers differentiate between "steatite" as a rock composed mainly of talc versus "soapstone" that includes talc plus other minerals such as chlorite, amphibole, phlogopite, antigorite, magnetite, and carbonates, and which has a soapy or greasy feel to it. Some geologists refer to the exceptionally pure talc as "talc of steatite grade" (Page 1951; Kempe and Harvey 1983) whereas other researchers combine steatite with massive talc plus other minerals into one category called alternately "steatite" and "soapstone." Other geologists do not recognize the terms "steatite" or "soapstone" at all, simply calling such deposits "talc" versus "talc schists" (Williams et al. 1982).

5.3.2 Greenschist, greenstone, and epidiorite

The terms "greenschist" and "greenstone" are often applied interchangeably to green-colored metamorphic rocks, with greenstone most often referring to a massive form lacking in visible schistosity (Fig. 5.3.7).

Curtis (1959) described one of the stones used to make implements at the archaeological site of La Venta in Mexico as a greenschist. It is composed mainly of pumpellyite, a calcium-magnesium-aluminum silicate often found in metamorphic rocks, and which tends to be associated with glaucophane. It is usually found in minute fibers or narrow plates, and has a bluish-green color (Dietrich and Skinner 1979).

Some greenschists contain epidote and actinolite along with chlorite (Yardley et al. 1990). This composition gives the rock a color ranging from dark green to yellow-green. Collins (1977) identified some medieval grave slabs in Scotland as being made of this rock, also variously called greenschist or greenstone. It is described as dark green in hand sample, and varying in texture. Crystals of plagioclase feldspar and of hornblende are often clearly visible. In thin section the feldspars were seen as randomly oriented laths, suggesting relic igneous texture. Large crystals of leucoxene extended up to 1.4 mm in length. The chlorite was pleochroic, from pale green to pale yellow. Collins did not find matching rock deposits where this stone might have been quarried.

Williams and Heizer (1965a) described the stone used for two Olmec monuments: one was a chlorite-muscovite-schist with stringers and lenses of coarsely granular quartz; the other was a chlorite-actinolite-epidote-albite-sphene schist with many small porphyroblasts of magnetite. Other analyses of Olmec metamorphic rocks included a pumpellyite schist, muscovite-actinolite schist, and an actinolite schist (Fig. 5.3.8) (Curtis 1959). As Williams and Heizer noted, all these schists are characteristic of the greenschist facies of metamorphism.

Newman (1984) found a fine-grained chlorite-actinolite schist used as a sculptural material in southern India, dated to the Hoysala period (twelfth century). Other sculptures from the Chola period of southern India (ninth century) were simply described as being greenstone or greenschist.

Another greenschist example is a chlorite-magnetite schist used to produce some Gandhara sculptures and smaller sculpted trays (Newman 1984; Reedy 1992b). These rocks are the product of regional metamorphism of igneous rocks at low temperatures. Figure 5.3.9 shows a thin section of a tray from ancient Afghanistan. The stone is green with

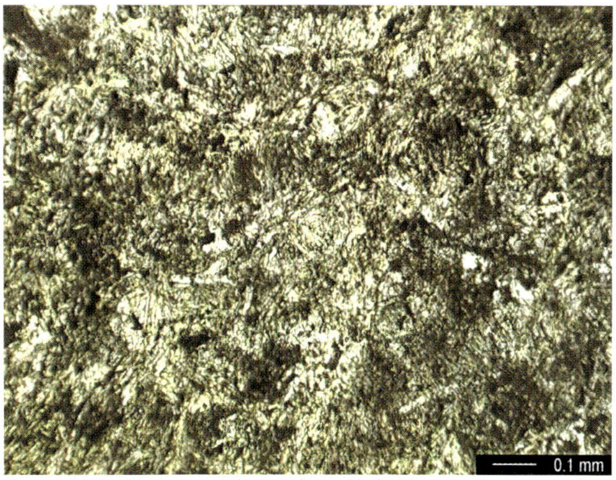

Figure 5.3.7 Greenstone, plane polarized light. The term "greenstone" may be applied to rocks of a variety of compositions, but generally refers to green-colored metamorphic rocks that have a massive form lacking in visible schistosity. In this example, much of the rock consists of tiny blades of the amphibole actinolite, colorless to green in plane polarized light, with some pleochroism apparent.

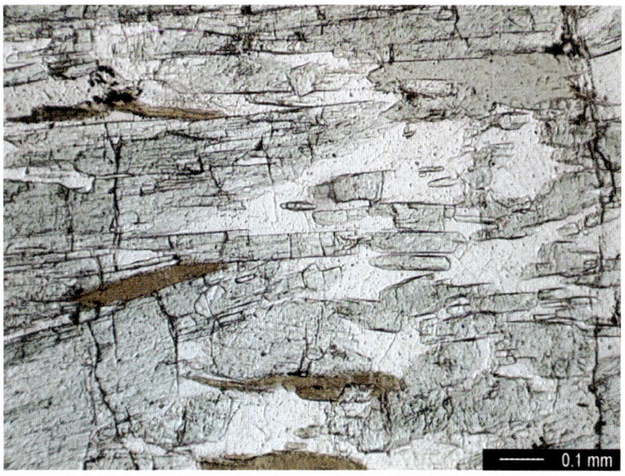

Figure 5.3.8 Actinolite schist, plane polarized light. Most of the field of view contains coarse blades of faintly green actinolite. Interspersing the layers of actinolite are colorless quartz, green chlorite, and brown biotite mica. There are also a few scattered black grains of magnetite and pyrite, and tiny high-relief grains of rutile.

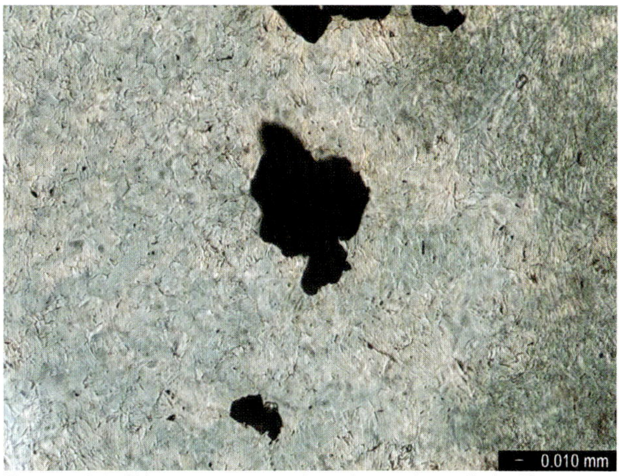

Figure 5.3.9 Chlorite-magnetite schist, plane polarized light. This sample is from a Gandhara sculpted tray from ancient Afghanistan. Most of the stone is chlorite, which ranges from light brown to green in plane polarized light, appearing as scaly, fibrous, or fan-shaped masses rather than as flakes or plates (probably the iron-rich prochlorite form). The large black inclusions are magnetite.

a blackish patina, and is relatively coarse grained. In thin section the stone is seen to be comprised primarily of chlorite. In plane polarized light the chlorite is pleochroic from a very bright green to a light brown. It appears as a scaly or fan-shaped fibrous mass rather than as flakes or plates, thus, it is probably the iron-rich prochlorite form. Magnetite, a common minor constituent of green schists, comprises about 10% of the rock.

Storemyr (1997) found greenschist used on Nidaros Cathedral in Trondheim, Norway. It is described as foliated, dense, green, and sometimes quite soft due to a high chlorite content (30–40%). It was used mainly for ashlars, moldings, and decorations. In addition to the chlorite, it contains 15–25% hornblende, 10–20% biotite, 10–15% dolomite, 5–15% quartz, 5–10% plagioclase feldspar, and lesser amounts of talc, calcite, pyrite, and titanite.

D'Amico and Starnini (2006) discussed greenschists used for prehistoric axe or adze blades in Italy. They are described as being very fine grained and containing albite, chlorite, epidote, actinolite, ore minerals, and relic jadeite and garnets. However, this material appeared relatively infrequently, with the majority of the axe and adze blades being made of other metamorphic rocks.

5.3.3 Blueschist

The term "blueschist" refers to a schist that usually is not actually blue in hand sample, but is more often gray. The color derives from an abundance of sodic amphibole, typically glaucophane or crossite (Yardley et al. 1990). Glaucophane schists (Fig. 5.3.10) were quarried in the Western Alpines for polished stone implements of the Southern French Neolithic, identified as part of a major study on Neolithic tool collections and potential source rocks in France, Italy and Switzerland (Ricq-de Bouard and Fedele 1993). In that study, more than 2,000 tools and tool fragments from a wide variety of important archaeological sites were studied in thin section. XRD analysis served as a supplementary technique to verify identifications when rocks were extremely fine grained, and thus more difficult to characterize according to optical properties.

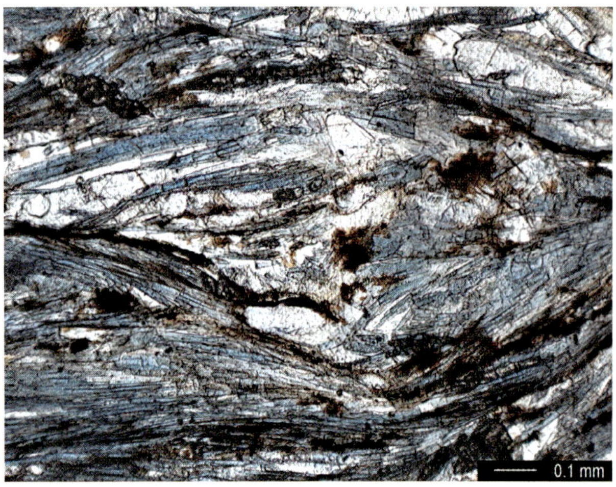

Figure 5.3.10 Glaucophane schist, plane polarized light. Glaucophane, the main component of this schist, is pleochroic, going from neutral to blue or violet in plane polarized light. It occurs as prismatic crystals or columnar aggregates that have fairly high relief.

Many of these Neolithic implements were made of glaucophane schist, characterized by the presence of the blue sodic amphibole, glaucophane. Ricq-de Board and Fedele concluded that the presence of lawsonite and pumpellyite within the glaucophane schists indicated that likely sources could be pinpointed fairly specifically. This thin-section work permitted the construction of detailed models for routes of circulation and exchange of rock material and implements in the Western Alpine and Southern French Neolithic. Patterns of marked cultural choice emerged regarding selection of implement materials, leading to some extremely interesting conclusions that were not always consistent with previous beliefs based upon observations that did not include thin-section analysis.

5.3.4 Schist as a cultural material

Architectural uses

Schist is a common building material. For example, one of the construction materials at La Venta (Mexico) was identified as muscovite-actinolite schist (Curtis 1959). This schist was used for facing block, as well as for rough stone pavements. A calc-chlorite-albite schist was used as a building material in the Western Highlands of Scotland, and a chlorite schist was used at Akrotiri, Greece for building and flooring applications (Collins 1977; Kempe and Harvey 1983). A garnet-hornblende-biotite-muscovite-quartz schist from Norway was used locally, and also exported to many parts of Europe to be used for architectural purposes. A mica-quartz schist from Mount Bracco, Italy, with a reported quartz content of 95% or more, also had wide use as a flooring or cladding material (Dimes 1990b).

Collins (1977) characterized late medieval monuments of the Western Highlands in Scotland, including crosses, gravestones, and effigies of the fifteenth–sixteenth centuries. He found ten rock types, calc-chlorite-albite schist being the most common. Other schist types include a tremolite-chlorite-schist, a biotite-muscovite-schist, and a chlorite-talc-schist. Collins was able to identify the probable sources for all of these materials.

In eastern India, a garnet-sillimanite-graphite schist was used in the construction of Konarak Temple; this schist is also termed "khondalite" (Singh et al. 1995). The term "khondalite," which originally was intended as a shortened name for a specific quartz-garnet-sillaminite-graphite schist, later referred to a more general group of high-grade metamorphic rocks found from Orissa in eastern India down to southern India (Newman 1984). However, Banerji (1982) discussed these rocks in detail, recommending that the term khondalite be discontinued as the definition has become so overextended as to be simply confusing rather than helpful. For example, the name khondalite has been adopted to refer to a series of quartzites, granulites, schists, marbles, and other stones including a garnet sillimanite rock in Sri Lanka; the Khondalite System refers to this extensive group of metamorphic rocks (Kularatnam 1979).

One of the building materials in the Gandhara sites of Swat valley, northern Pakistan, is a calc-schist (Di Florio et al. 1991, 1993a, 1993b). It has a schistose texture, and consists of carbonates, muscovite, and secondary quartz and feldspar. Sometimes the stone is black due to presence of organic matter; sometimes reddish due to iron oxides.

Sculpture

Schist is also an important sculptural material, used by sculptors in ancient Egypt (Goedicke 1964; Kempe and Harvey 1983), for example. In Sierra Leone and Guinea, chlorite schist has been used for producing carved heads (Allison 1968), and a number of sculptures from northeastern India (Bihar, West Bengal, Orissa) and Bangladesh were made of schist. Newman (1984) identified a wide variety of schist types used for sculptural purposes in those regions, including an andalusite-biotite-quartz-mica schist in the Gupta and Pala-Sena periods, and a biotite-quartz-mica schist and a chlorite-mica schist (with and without quartz) utilized for Pala-Sena sculptures.

For Gandhara sculptures, in addition to the phyllite examples previously discussed, some images and sculpted architectural panels were made of schist very similar to the phyllite examples (containing distinctive chloritoid needles). Lahanier (1976) reported that a Gandhara portrait bust of a king was a bluish-gray chloritoid-muscovite-sericite-quartz schist, with accessory carbonaceous material, biotite and hematite. Kempe (1982) discussed the use of a chloritoid-paragonite-muscovite-quartz schist, with accessory plagioclase, biotite, and carbonaceous material. Kempe (1986) later identified an ancient quarry near Sakhakot in northern Pakistan as a possible source.

Other uses

Neolithic stone tools were the subject of an extensive thin-section study conducted by Schwarz-Mackensen and Schneider (1983a, 1983b). They analyzed 2,500 stone tools from the region of Braunschweig in central Germany as part of a study to identify the main sources of raw material used in the production of Danubian stone axes and hatches in central Europe. A frequently occurring stone type was an actinolite-hornblende schist. Schwarz-Mackensen and Schneider's well-illustrated thin-section studies were supplemented with some XRD analyses. They were able to conclude through petrographic comparisons to geological sources that this schist came from the western Carpathian Mountains and the Balkan region, and were able to rule out an origin in Saxony, the previously assumed source. This work showed that the tool materials were transported along trade rotes for distances of more than 700 km.

Schists were used for similar stone tools in Italy, where they have been identified in use for axe or adze blades, other types of functional implements, and ornaments (D'Amico and Starnini 2006). Some of the types of schists utilized for cutting blades included glaucophane schists containing glauchophane or crossite as the major mineral, along with zoisite-epidote, chlorite, garnet, lawsonite, albite, ore minerals, and residual sodic-pyroxenes; and actinolite-hornblende schists. Materials used for other tools and ornaments included paragonite schists, consisting of close to 100% paragonite; and chlorite schists.

5.4 Gneiss

Gneisses are weakly foliated and imperfectly fissile metamorphic rocks that are coarser than schists, and mainly comprised of quartz and feldspars (Fig. 5.4.1). They contain more minerals of granular habit (such as quartz, feldspars, hornblende, or pyroxene) (Fig. 5.4.2) than phyllosilicates. Phyllosilicates are often biotite or muscovite mica whereas gneisses often have porphyroblasts of garnet (Fig. 5.4.3), or uneven foliation marked by streaks of hornblende or micas surrounding sections of quartz and feldspar (Best 1982; Williams et al. 1982). Gneisses may be named according to texture (e.g. augen gneiss refers to rocks with ovoid megacrysts of feldspars) (Fig. 5.4.4); overall appearances (e.g. banded gneiss); the minerals besides quartz that might be

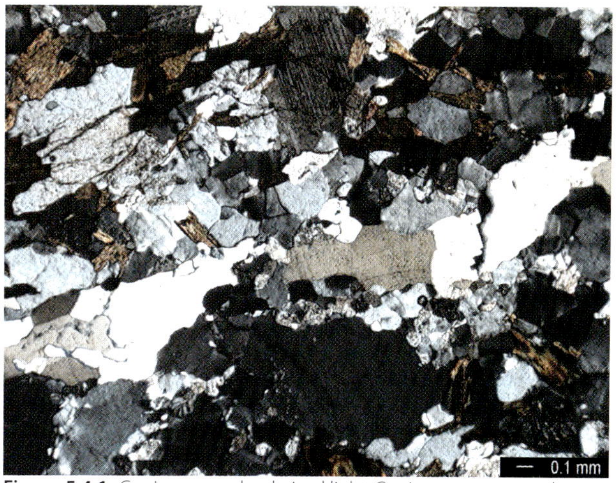

Figure 5.4.1 Gneiss, crossed polarized light. Gneisses are coarser than schists, and mainly comprised of quartz and feldspars. This example also contains some biotite.

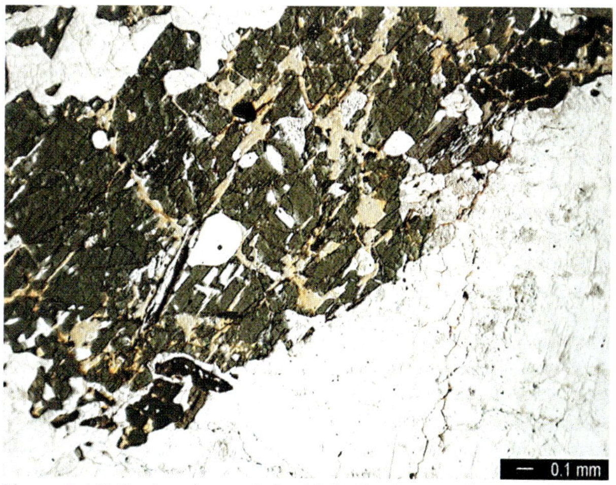

Figure 5.4.2 Gneiss, plane polarized light. Gneisses tend to contain more minerals of granular habit, such as this dark green hornblende, rather than phyllosilicates. The hornblende is surrounded by colorless quartz, with a few small grains of brown biotite.

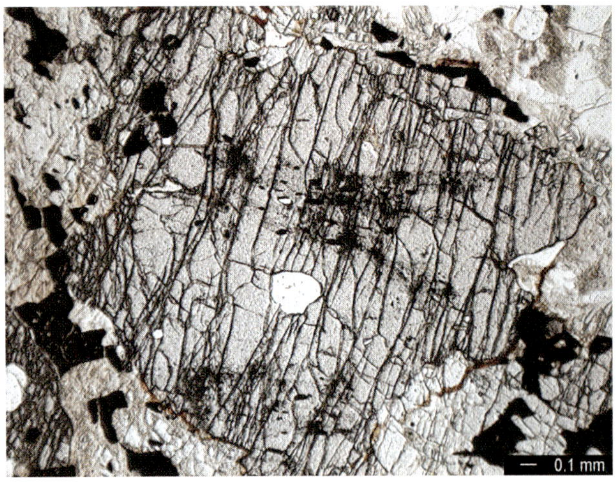

Figure 5.4.3 Garnet porphyroblast within gneiss, plane polarized light. The large garnet in the center of the field of view has a pale brown color and high relief, with typical irregular fractures and inclusions (of magnetite and quartz here) visible.

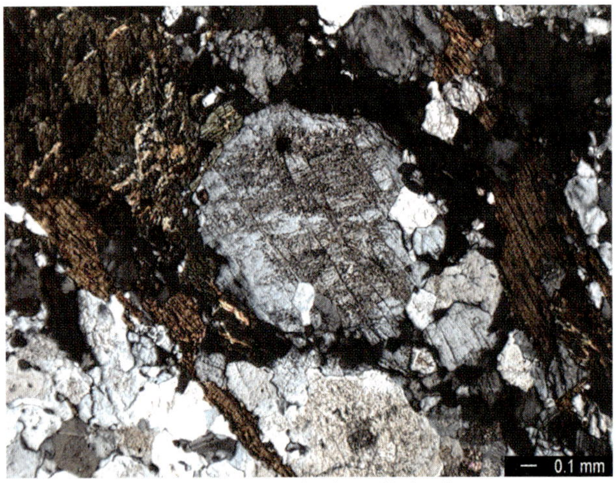

Figure 5.4.4 Augen gneiss, crossed polarized light. The term augen gneiss refers to a gneiss containing ovoid megacrysts of feldspar. The plagioclase ovoid in the center of the field of view is heavily altered and has some antiperthite (alkali feldspar intergrowths).

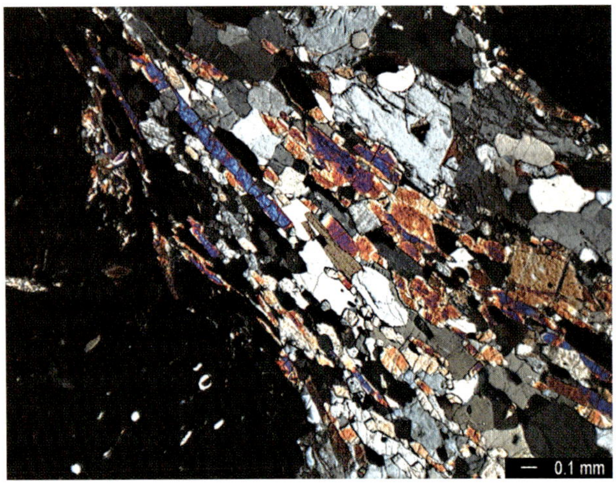

Figure 5.4.5 Sillimanite garnet gneiss, crossed polarized light. Sillimanite appears as lath-like, elongated crystals, often somewhat bent, with transverse fractures. It has moderate birefringence, in contrast to the surrounding low birefringent quartz and the dark garnets at the edges of the field of view.

present (e.g. biotite hornblende gneiss, or sillimanite garnet gneiss) (Fig. 5.4.5); or composition of the precursor rocks (e.g. granite gneiss) (Dietrich and Skinner 1979; Yardley et al. 1990).

5.4.1 Gneiss as a cultural material

Architectural uses

Gneiss is sometimes used as a building material. Confusingly, it is often called granite, due to similarities in composition and overall appearance except for schistosity (Merrill 1891; Dimes 1990b). The alternating harder and softer layers are sometimes said to render gneiss less suitable for building and sculpting than many other stones (Kempe and Harvey 1983), although the property of splitting with relatively flat surfaces is useful for construction and decorative purposes. Gneiss has been quarried for a variety of construction purposes in many parts of the United States for historic buildings (such as Pennsylvania, Maryland, and Massachusetts) (Merrill 1891).

For example, gneiss from Port Deposit, Maryland has been used extensively in Pennsylvania: for a house in Kennett Square; a school and residence in West Chester; several buildings at Swarthmore College (mica gneiss with an attractive glistening silvery surface); and many homes in and near Elizabethtown (a granite gneiss). In many other parts of Pennsylvania gneiss has been extensively quarried for use as a structural stone; it was employed often in and near Philadelphia (Stone 1932; Miller 1939).

At the Olmec site of La Venta (Mexico), gneiss was one of the materials used as a construction stone (Curtis 1959). It appears, for example, in rough stone pavements underlying a jaguar mosaic mask in the Southwest Platform. The material is an actinolite-epidote gneiss.

Gneisses were used in the building of giant Buddhist statues in Sri Lanka, as well as for the carving of ancient caves, and for construction materials such as paving, roofing, stone canopies, steps, etc. (Kularatnam 1979). The Sri Lankan gneisses include a biotite gneiss complex, described by Kularatnam as being a fine-grained, streaky banded gneiss with well-developed foliation. It is generally comprised of quartz and orthoclase with some plagioclase and biotite; sometimes hornblende and pink garnet are also found. Composition and texture vary somewhat on different parts of the island.

Di Florio et al. (1991, 1993a, 1993b) reported detailed petrographic studies of rocks used in the construction of monuments and masonry at the Gandhara-period Pānr I Sacred Area in the Swat valley, northern Pakistan. Interestingly, they found that each period in the life of the site was characterized by a specific lithotype (all available locally). A variety of gneisses was used at the site described as:

- Granitoid gneiss: light colored, medium grained and consisting of quartz, feldspars, muscovite, biotite, and garnet that is evenly distributed and may be as large as 1–2 cm in diameter. The micas form thin, somewhat elongated, lenses.

- Garnet augen gneiss: similar to granite gneiss except for the presence of "eyes" of idioblast crystals of feldspars. Muscovite and biotite may sometimes be very abundant.
- Flaser-gneiss: augen gneiss that is rich in biotite, with extremely elongated and aligned feldspar "eyes."
- Garnet-tourmaline-aplitic gneiss: very light colored, fine grained with low schistosity due to scanty mica content, mainly consisting of quartz, feldspars, garnet, and tourmaline (occurring in small, black prisms that are mostly aligned).

Sculpture

Although not a popular sculptural material due to its coarse-grained appearance, gneiss is found as a sculptural material in Africa. For example, many Yoruba stone figures, such as those in the community of Eshure, are carved from a local gneiss, described as a granite gneiss (Allison 1968).

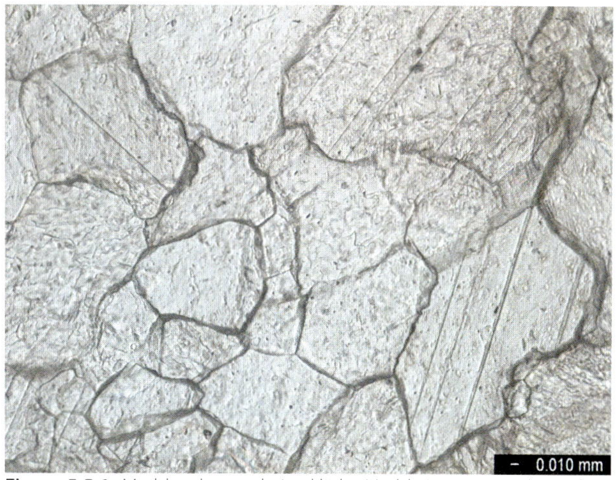

Figure 5.5.1 Marble, plane polarized light. Marble is composed mainly of metamorphosed limestone and/or dolomite, with the calcium carbonates recrystallized to produce an interlocking mosaic of grains, with original sedimentary structures and fossils no longer visible. Calcite and dolomite grains are colorless in plane polarized light, and relief varies with direction.

5.5 Marble

Marble is metamorphosed limestone and/or dolomite. It is comprised mainly of calcite and dolomite, recrystallized to produce an interlocking mosaic of grains, with the original sedimentary structures and fossils no longer visible (Fig. 5.5.1). It may often contain additional accessory minerals such as brucite, diopside, epidote, clinozoisite, feldspars, forsterite, graphite, grossular, humite, periclase, phlogopite, pyrite, quartz, scapolite, serpentine, sphene, spinel, talc, tremolite, vesuvianite and/or wollastonite (Figs. 5.5.2 and 5.5.3). Even within the calcite and dolomite content there can be variation between marble sources, in grain size and texture (such as degree of suturing of grains, shape of grains, and uniformity of grain size and shape) (Figs. 5.5.4 and 5.5.5) depending upon the temperature, pressure and length of time of formation of the stone (Dietrich and Skinner 1979; Ashurst and Dimes 1990).

Although marble is often chosen for use on the basis of color (with pure white usually favored), many other characteristics (such as texture, grain size, and inclusions) can affect its quality and durability. For example, fine-grained marbles often take a polish better than coarser grained marbles. And, marbles with more tightly interlocking crystals may be more durable than those with more loosely interlocking ones, where water can move more rapidly along the grain boundaries (McGee 1999).

In art and architectural descriptions, any carbonate rock that can take a polish may be called "marble," so descriptions by nongeologists may have to be scrutinized. Although commonly thought of as a pure white rock, marble can have a wide variety of colors, with either a uniform or splotchy appearance. The purer marbles have a hardness of about 3 on the Mohs scale, so are easy to work. Sculptures tend to be made of more pure white marble, as are many buildings and decorative applications. For example, in nineteenth-century Pennsylvania, small rectangular blocks of pure white marble were used for accenting trim on dark brick row houses

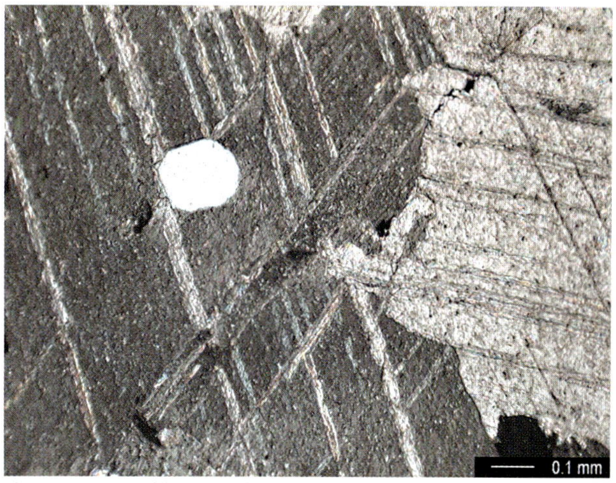

Figure 5.5.2 Marble, crossed polarized light. The rounded, low birefringent grain is a quartz accessory in this marble. The much larger carbonate grains have pearly gray and extreme pastel interference colors, and show rhombohedral cleavage and twin lamellae.

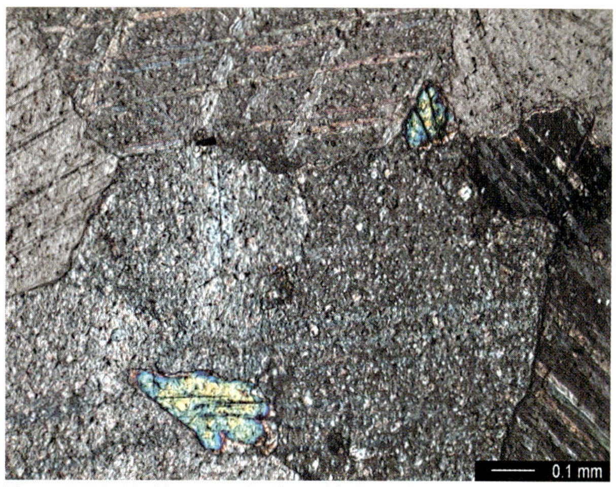

Figure 5.5.3 Marble, crossed polarized light. In this case, there is accessory diopside. In crossed polarized light the upper second-order interference colors of the diopside contrasts with the pearly gray sheen and pastels of the carbonates.

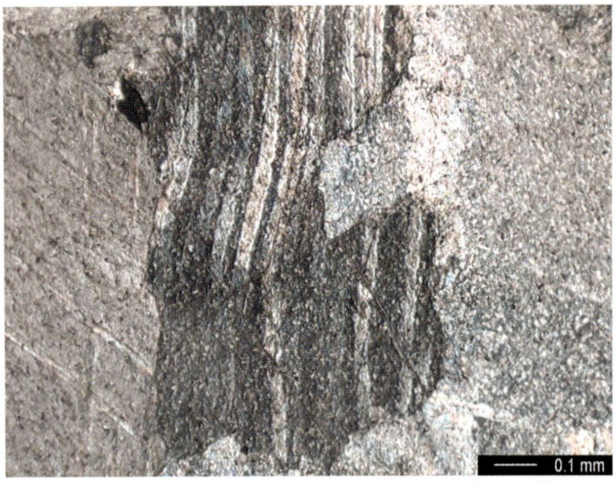

Figure 5.5.4 Marble, crossed polarized light. This coarsely crystalline marble shows sutured grain boundaries and curvature of twin lamellae (indicating intercrystalline curvature of the crystal lattice).

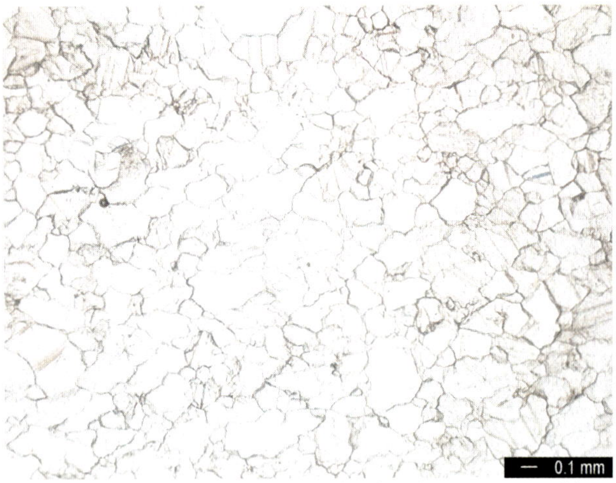

Figure 5.5.5 Marble, plane polarized light. This white Vermont marble is relatively fine grained with a granoblastic texture (having equigranular grains oriented at random) and is free of inclusions.

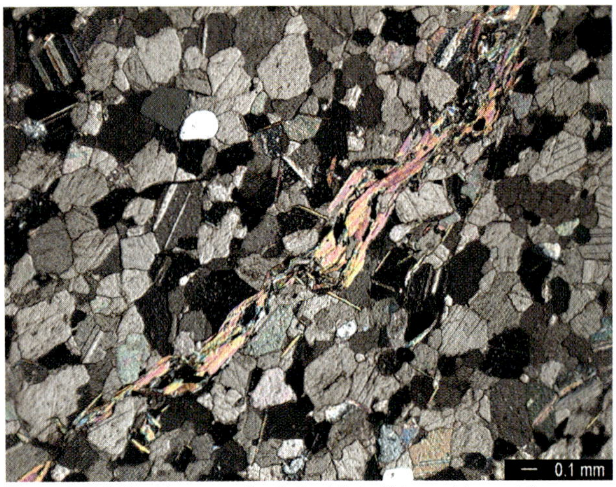

Figure 5.5.6 Marble, crossed polarized light. This marble contains some shapeless, low birefringent quartz as well as platy aggregates and shreds of high birefringent talc, characteristically bent. There are also some isolated prismatic crystals of tremolite within the calcium carbonate matrix.

(Stone 1932). However, many other marbles are found used in architecture, flooring, and other decorative uses—a wide variety of marbles was used for the construction of medieval French monuments, including white, gray and veined marbles, some probably local (from the Pyrénées) and others probably imported from Italy or Turkey (Blanc 1995).

Contact marbles are formed when limestones and dolomites undergo high-grade contact metamorphism, producing a granoblastic texture (equigranular with grains in random orientation rather than in a preferred orientation) (see Fig. 5.5.5). Many such marbles contain little or no quartz, and instead have impurities such as forsterite, humites (magnesium silicates), periclase (magnesium oxide), brucite (magnesium hydroxide), spinel (aluminates, ferrites, and chromites of magnesium and iron), corundum (aluminum oxide), and some rare silicates of calcium and magnesium. Rocks containing more quartz have a different set of impurities present, such as wollastonite, diopside, tremolite, and talc (Fig. 5.5.6). As metamorphic temperatures rise, the resulting mineral assemblage is partly due to the temperature and pressure achieved, and partly due to the beginning proportions of calcite, dolomite, and quartz (Kerr 1979; Williams et al. 1982).

Higher grade marbles differ from the lower grade ones in accessory minerals, especially for marbles derived from carbonate sediments with clay impurities and siliceous components. For example, accessories may include diopside, amphibole, phlogopite or biotite, plagioclase or microcline feldspar, quartz, sphene, apatite, graphite, and a variety of calcium-aluminum silicates (such as grossular-andradite, clinozoisite or zoisite, and scapolite). Additionally, some marble deposits can show evidence of post-formation stress or deformation—curvature of twin lamellae may be visible, indicating intracrystalline curvature of the crystal lattice (see Fig. 5.5.4) (Williams et al. 1982).

5.5.1 Marble characterization and source identification

White marbles

Marble, particularly from the Aegean region, has received much attention from thin-section petrographers. A systematic description (now outdated) of the petrography of classical marble quarries was published more than a century ago (Lepsius 1890). That work described features such as grain size, extent and orientation of metamorphic foliation, color, and presence of trace minerals. Later work expanded knowledge of the petrographic characteristics of specific quarries of the classical world (e.g. Washington 1898) and added the description of features such as preferred orientation of calcite grains, form of grain boundaries, and the degree and manner of recrystallization.

The potential practical usefulness of thin-section analyses for characterizing and sourcing marbles used for cultural materials was recognized early. Marinos (1948) used thin sections to examine structural differences among marbles from various sources in ancient Greece: he showed how variations in petrofabrics might be related to source areas, and discussed how that information could be used to identify the origin of marbles used in ancient sculpture and architectural works.

Weiss (1954) further explored the potential of petrofabric analysis in thin section. He illustrated seven specimens of marble from quarries in Greece, showing how marbles can have distinctive fabric appearances. He explored how this information might potentially be useful in characterizing fragments excavated at an archaeological site, discerning which fragments might have come from the same piece, and deducing which fragments might have been located in certain positions within a piece (thus helping restoration and reconstruction efforts).

Herz (1955) also discussed how petrofabric data might be useful in assembling archaeological marble fragments with inscriptions, and in determining provenance of buildings and statuary marble. Later, Renfrew and Peacey (1968) reviewed inaccuracies and inadequacies in the earlier work. Some of these problems were due to incomplete petrographic descriptions, the use of subjective methods of analysis that did not include blinding or randomization, and sample sizes too small to characterize adequately the variation found in marble quarries.

In thin section, the main constituent of marble is, of course, the carbonate fraction. Distinguishing between calcite and other carbonates such as dolomite can be difficult, however, stains such as alizarin red S can aid in that differentiation. Marbles also vary in the average and maximum size of the carbonate grains, in grain shape, and in other textural characteristics (Barbin et al. 1995). Examples (Lapuente 1995) are homoblastic texture (an equidimensional mosaic of calcite grains with straight and gently curved boundaries); heteroblastic texture (exhibiting nonuniform calcite grain size); polygonal, strained, elongated, or layered grains; and grain boundaries may be straight, sutured, or curved. There may also be a variety of accessory minerals present (such as quartz, feldspars, micas, epidote, graphite, and iron oxides) that can be useful in characterizing a particular source and in distinguishing it from other quarry sources.

A complete petrographic description of a marble thin section may include an overall assessment of the coarseness and evenness of size of the carbonate grains; their minimum, maximum and average size; shape of carbonate grains and their grain boundary qualities; textural features of the carbonate grains (such as presence of polysynthetic twinning); qualities of the overall rock fabric (such as consisting of a dense mosaic of equant grains); amount of calcite versus dolomite; a listing of any accessory minerals present along with their relative amounts; features of any banded areas (such as containing elongated and medium-grained carbonates with a different suite of accessory minerals or a different concentration of the same ones as found in the stone as a whole); whether or not any pore structures are visible as small voids or cracks; and the presence, average thickness, and characteristics of any weathering layer (Harvey 1977).

Newman (1984) conducted a study of stone types used in architecture and sculpture of India using thin-section petrography as the major analytical tool. His goals were to characterize the range of rock types used, provide correct terminology for museum collections, and, where possible, identify possible stone sources (using a combination of analysis of samples collected from possible sources in India and existing geological reports). He identified marble used for a minority of objects from a number of regions within India. For example, a Pallava column from southern India was made of a periclase marble, formed by contact metamorphism of a dolomitic rock. Copper and lead minerals were also found in the marble (the area is known to contain copper and lead deposits). Objects and buildings from the Rajasthan and Gujarat areas of western India included some made of marble: a variety of quarries was worked for buildings and decorative arts of the Mughal period, one of the most famous being the Taj Mahal in Agra. Newman discussed some pure white dolomitic marbles. Calcitic marbles were used in other areas of central and northern India.

A few other examples of Indian marble uses include the Jain Temple at Mount Abu, Rajasthan (western India), eleventh century and later, carved completely out of white marble, with exquisite decorative details such as highly complex carving of ceilings that incorporate sculptures and geometric patterns. From this same period in western India (Rajasthan and Gujarat), there are also many free-standing sculptures in white marble, including ritual scenes, images of Saraswati (the goddess of learning and wisdom), Jain *Tirthankara* (a term for one who has attained perfect knowledge), and secular images such as royalty, dancers, and musicians (Huntington 1985; Pal 1988).

As the number of sampled quarries has increased, it is now recognized that no one method of analysis alone is sufficient for provenance studies of classical marbles, except when the possible source locations and comparisons are very limited. Inherent overlaps found when many potential sources are compared inhibit the effectiveness of petrographic data. Therefore, thin-section analysis is most useful for marble when used in combination with other methods (such as isotopic, elemental, XRD, cathodoluminescence, or other analytical techniques), particularly if multivariate statistical analysis includes two or more types of data (e.g. Lazzarini et al. 1980; Moens 1987; Germann et al. 1988; Moens et al. 1988a, 1988b; Roos et al. 1988; Moens et al. 1989; Lazzarini 1990; Moens et al. 1990, 1992; Cabanot 1995).

Petrographic features found most useful in those studies include the shape of calcite crystals, minimum, maximum and average calcite grain size, and, to a limited extent, the presence and relative abundance of accessory minerals. The thin sections can also be stained with alizarin red S in dilute hydrochloric acid to distinguish between calcite and dolomite. Since, unlike dolomite, calcite is readily soluble in the acid, only the calcite stains (Hutchison 1974).

Lazzarini et al. (1980) combined thin-section petrography with stable isotope analysis to characterize and distinguish marbles from ancient quarries in Italy, Greece, and Anatolia. They noted that grain size is an important thin-section characteristic to measure because it depends on the rate of nucleation and growth of crystals, relating to metamorphic conditions of formation of the rock. It is also related to the amount of pre-crystallization strain and the presence of impurities, and thus to the characteristics of precursor rocks. They recorded the shape of the boundaries between calcite crystals as this is also tied to the history of formation of a marble. In addition, they noted the texture of the marble, features such as bending of polysynthetic twins, and accessory minerals present.

In another example, Moens et al. (1988b) used a multi-method approach that included thin-section petrography, elemental analysis, and stable isotope analysis to develop a comparative database of ancient marble quarry sites in the Mediterranean region, including all major exporting sites from the middle of the seventh century B.C. to the sixth century A.D. Extensive and careful sampling was done within each quarry site, although this particular paper reports only on a selection of sites where white calcitic marbles were quarried (174 samples total). Thin-section work included identifying the main and accessory minerals (with staining to distinguish calcite from dolomite), characterizing the texture, and determining the maximum grain size of calcite crystals. This latter characteristic proved to be especially helpful in separating marbles from different districts. Sometimes, but not always, this characteristic alone was enough to distinguish one marble source from another. However, combining the petrographic data with elemental and isotopic data allowed the best separation between possible sources, although some areas of overlap still existed.

Lazzarini et al. (1988) conducted an in-depth, thin-section study of white marbles used in a variety of ancient monuments in Rome. Their goal was to document the types and range of marbles used, and to identify provenance of the stones. They began with preliminary hypotheses about provenance based on archaeological and historical data, which were generally supported by the thin-section analyses. Reference samples were limited to sources known to have been used in antiquity, and monuments to those of known archaeological or political importance. The possibility that some marbles might have come from little known or exotic sources not yet represented in reference analyses was noted by the authors. They also noted that at that time the number of reference samples was too small to permit reliable statistical analyses. Nevertheless, detailed petrographic descriptions were given for each monument sampled, along with the most likely provenance based upon the data available.

Data included an overall color index (Munsell), a very useful criteria compared to subjective qualitative descriptions usually seen (such as white, cream colored, off-white, gray, etc.). For the thin sections, the presence or absence of accessory minerals was reported for quartz, potassium feldspar, other feldspars, epidotes, graphite, and opaques. Average grain size was given, along with a description of grain boundary shapes (such as curved, straight, embayed or sutured). The overall fabric was categorized as either homoblastic or heteroblastic, and additional petrographic observations regarding texture were listed (such as layered, polygonal, strained, traces of recrystallization, etc.). The presence or absence of dolomite was determined by XRD analysis; isotopic analysis provided the Ca/Sr ratio.

Germann et al. (1988) focused on differentiating between two marble-producing sites of the Cycladic islands, Paros and Naxos, important sources of marble for Greek sculpture and architecture. They were particularly interested in being able to identify Greek mainland architectural stone that had been imported from the Cyclades, as an indicator of close connections at crucial time periods. Sources at Paros and Naxos were analyzed, and compared with samples taken from buildings at Delos and from the "treasuries" at Delphi. They extensively mapped the marble deposits and ancient quarry sites on Paros and Naxos, and analyzed 104 samples. For petrographic data, they found that the average grain size as determined from 100 calcite grains (using an image analysis program) was an especially useful discriminating criteria. Naxos marbles showed a considerably larger grain size than those of Paros (a median of 2.2 mm with a range of 1.3–3.0 mm versus a median of 0.45–1.0 mm from three Paros quarries, which ranged from 0.035 to 1.1 mm). There were also differences in development of twin lamellae in the calcite. Only a few accessory minerals were found (dolomite, quartz, muscovite, and a few other silicates).

Thin-section data were augmented by elemental and stable isotope analysis. Physical properties of the stones were also determined (compressive and tensile-bend strength, etc.) to identify their usefulness and properties as construction materials. The authors found that internal variation at Paros and Naxos was so great that they overlapped almost the entire range of characteristics known for the ancient Mediterranean. Thus, marbles previously identified as of problematic source attribution might well come from these Cycladic sources, which have favorable physical properties. It was suggested that provenance studies of marbles need to focus on adequate characterization of potential sources, and comparison of objects to sources that are geologically and historically reasonable, with the use of multivariate combinations.

Naxos and Paros were also the focus of study—along with other white marbles (those with less than 10% dolomite) from Pentelikon and Thasos—by Roos et al. (1988). They analyzed a total of 80 samples from the four quarries. The goal was to develop a database of reference material for provenance studies, produced from thin-section petrography, elemental analysis, and ultimately (but not in this paper), stable isotope analysis. The larger work included analysis of samples from many other quarries as well, using carefully representative and systematic sampling procedures. The thin-section work included measuring the size and shape of calcite grains, and identifying and estimating abundance of accessory minerals. Detailed thin-section descriptions were given, including texture (homoblastic texture, an equidimensional mosaic of calcite grains with straight and gently curved boundaries; heteroblastic texture, exhibiting nonuniform calcite grain size); average diameter and maximum size of calcite crystals; evidence of deformation, strain, and recrystallization; and accessory constituents (type and relative abundances).

However, they found that only the data for maximum grain size were helpful in separating quarry districts. A series of hierarchical cluster analyses comparing various pairs of quarry districts (such as Pentelikon and Naxos) was presented, which interestingly combined data on elemental composition with petrographic data (although using only a limited subset, such as manganese and iron composition along with maximum grain size). Even then, quarry separation was not 100%, due to within-quarry variation.

Data from different analytical approaches were also combined in the work carried out by De Paepe et al. (1988) to try to match four classical marble sculptures to quarry sources, utilizing reference materials. Isotopic analyses were presented in one plot of $\delta^{13}C$ and $\delta^{18}O$; cluster analyses combined data for up to ten elements (as logarithms of concentration) and one feature from thin-section petrography, maximum grain size of calcite grains. The authors concluded that it was possible to identify the most likely source area for the sculptures, and that this source was in agreement with art-historical evidence.

Dean (1988) also highlighted the variation that can be found in one outcrop area, necessitating a multivariate approach. She discussed Carrara marble of Italy, noting that thin- section petrography showed a wide range of grain sizes and textures over ten quarries studied. Some contained only calcite; others were mainly dolomite; yet others were mainly calcite with traces of dolomite plus a variety of other accessory minerals. The calcite tended to be fine grained, but some marbles had a random orientation of calcite crystals while others had a pattern of alternating crystal size with preferred orientation and elongation of calcite crystals. In order to better characterize the Carrara marble and specific quarries within it, Dean included elemental and stable isotope analysis. Although Carrara marble could be distinguished from other similar Mediterranean marbles used in antiquity, identifying specific quarries was problematic due to the heterogeneity of the marble. Nevertheless, she concluded that the combination of stable isotope analysis and thin-section petrography was quite helpful in many cases.

Moens et al. (1988a) compared white marbles from four quarry districts in Italy (Carrara) and Turkey (Proconnesus, Dukimeion, and Upak), combining thin-section petrography with elemental analysis (16 major, minor, and trace elements). A total of 94 samples was analyzed. All of the marbles were relatively pure calcite, so the accessory mineral component, including dolomite, was always under 10%. There was much variation in accessory minerals found (including plagioclase feldspar, muscovite, opaque minerals such as iron sulfides, potassium feldspars, quartz, epidote minerals, and carbonaceous inclusions), although even within a single quarry there could be significant variation. Two quarries had homoblastic texture, one had a heteroblastic texture, and one a mortar texture (large crystals surrounded by small crystals). There was variation in maximum grain size, but some of the ranges overlapped for some of the quarries.

Both petrographic and elemental data showed significant overlaps between quarry districts. Nonetheless, Moens et al. (1988a) found that by using a multivariate cluster analysis with selected attributes they could discriminate between each pair of districts. In those analyses, maximum grain size proved to be an important contributor to discriminant functions.

In another example, Moens et al. (1989) compared white marbles from Carrara, Italy (27 samples) with those from Marmara, Turkey (24 samples) using both thin-section petrography and elemental analysis. Although the compositional ranges for each site showed a large spread, and those ranges overlapped for most elements for the two sites, separated ranges were found for three elements (K, Mn, and Hf). Likewise, maximum grain size for calcite was a feature that separated the two sites, although data were also collected on presence and relative abundance of accessory minerals, presence of carbonates other than calcite, and shape of calcite grains. Qualitatively, the authors noted that Carrara marble usually had an equigranular texture, with sizes ranging from 0.6 to 1.3 mm; in contrast, Marmara marble usually had large and small grains intermixed, with a range of 2.2–3.6 mm. It is unclear whether or not any statistical analyses were employed using the thin-section data.

Barbin et al. (1989) combined thin-section analysis with cathodoluminescence to characterize 91 samples of white marbles from five important ancient quarrying areas. By combining differences in color, intensity, and distribution patterns seen in cathodoluminescence with thin-section data on texture and grain size, they were able to identify seven marble types. The islands of Thasos and Paros each had two marble types, whereas Pentelikon, Carrara, and Usak-Kavacik each showed one main type of white marble.

Moens et al. (1990) combined thin-section analysis with elemental and isotopic data to perform provenance determinations of ancient white marble sculptures. They noted that such provenance studies have major implications in studies of economic history as well as in the history of art. For example, they cited the fact that since the nature of the stone can seriously affect carving possibilities, characterizing the stone and the available quarry deposits can help in understanding to what extent style might have been imposed by the nature of the material available to artists in a specific locale. Provenance studies can also contribute to studies of trade and artistic networks; they may also help in identifying forgeries or past restorations.

The authors discussed their long-term work carried out on ancient quarries of Italy, Greece and Turkey, including all major exporting centers from the middle of the seventh century B.C. to the sixth century A.D. These quarries probably represent most, if not all, sources of ancient marble in the Mediterranean world that one might expect to find with architecture and sculpture uses. They found that no single analytical technique could adequately characterize a source and separate it from all other possible sources. Instead, they concluded that a combination of thin-section petrography, elemental analysis, and isotopic analysis was required.

Moens et al. (1990) emphasized that the reliability of such characterization and provenance studies largely rests on the sampling procedures. Since the ancient quarries themselves are gone (the actual material used has been removed for the most part), they sampled not just from the walls outlining the extent of the quarries, but from the entire area where a quarry had been located, in order to ensure that the full variation of a marble outcrop was captured. Only unweathered areas were analyzed, so that original material was included in the characterization studies, although weathered areas were used to examine changes that the material underwent as it deteriorated.

Their thin-section analysis included staining with alizarin red S, which showed that most ancient quarries held calcitic rather than dolomitic marble. Since dolomitic marbles also tended to have trace element patterns quite different from those of calcitic marbles, they were treated separately. Many accessory minerals were found, including plagioclase and other feldspars, muscovite, quartz, epidote minerals, and opaque minerals. However, the type or amount of these accessories did not seem to vary much. Another problem noted was that, altogether, the accessory component tended to be no more than 5% of the rock volume. Thus, whether or not a particular mineral would be seen, or seen more than rarely, within a thin section depended upon random chance in the sampling process. Therefore, they concluded that the accessory component was not likely to be very helpful in characterizing white marbles or in conducting provenance studies of that material.

Textural differences were also recorded. For example, stone from some quarry sites showed a well-developed homoblastic texture, while stone from other sources was predominantly heteroblastic. But the most useful criterion for discriminating between quarry sites was found to be the maximum size of the calcite grains (MGS). Data were recorded on the lowest value found (in millimeters), the highest value, and the median. For some areas, MGS ranges were overlapping, but in other cases well-separated ranges allowed the use of MGS to discriminate between quarries. When the MGS data were included in statistical analyses with data from elemental and isotopic work in a combined multivariate analysis, the ability to discriminate between quarries and to match artifacts to a probable source increased considerably.

Thin-section analysis was combined with XRD and isotopic analysis to characterize the marble used in the construction of the twelfth-century cathedral of Bari, a Romanesque monument of Apulia in southeastern Italy (Calia et al. 1992). The cathedral itself was made of limestone, but there were many decorative elements made of white marble, probably imported. Samples were taken from the great apsidal window, a very important feature of the cathedral, for analysis. The combined analyses indicated that the marble came from various areas, including Proconnesus, Pentili, Carrara, and another as-yet-unidentified source area. The thin sections were used to identify accessory minerals other than calcite/dolomite (which included quartz, muscovite, feldspars, graphite, and other opaque minerals). Data were also recorded on maximum grain size, crystal boundary types (related to the type of metamorphic event that produced the marble, including straight, curved, sutured, or embayed), type of structure (such as homoblastic, heteroblastic, polygonal, or mortar), also related to conditions of metamorphism, and additional observations such as presence of layering, foliation, or other fabric characteristics.

Blanc et al. (1995) characterized the white marbles from the site of la Bauve à Meaux. They published petrographic descriptions of 17 samples (with some photomicrographs), including information such as grain size, shape, and textural characteristics for the carbonates; they also identified accessory minerals (micas, feldspars, and quartz). Their groupings were augmented by additional data from isotopic analysis and cathodoluminescence. All three approaches pointed toward the same result—that there were four distinct marble groups represented at the site.

Moens et al. (1995) found that maximum grain size, although by itself not sufficient to completely distinguish between marbles of different areas of the Mediterranean region, was nonetheless a parameter that was useful in some situations. For example, this criterion did allow them to differentiate between rocks from Carrara and those from the island of Marmara, and to distinguish between samples from Thasos and those from all other regions except Paros and Naxos. Maximum grain size, combined with data from other analytical approaches, was often useful in characterizing marble specimens (Kokkorou-Alevras et al. 1995). In other cases the lowest, medium, and highest grain size were used in analysis, along with other thin-section characteristics (Lapuente 1995).

Dooley and Herz (1995) used grain size data, together with other thin-section data, for a statistical analysis aimed at characterizing and differentiating early American marbles so that provenance studies could be carried out for sculptures and architectural materials made of marble. Samples were collected from 20 quarries in Vermont and from four in Massachusetts; these analyses were added to data previously collected from one Vermont quarry and one in Georgia. The goal was to see if the various quarries could be distinguished from each other and from Mediterranean marbles, especially Carrara marble that is visually similar to some Vermont marbles.

Along with stable isotope data, thin-section petrography was employed to characterize each quarry on the basis of type and abundance of accessory minerals, range and average grain size, degree of recrystallization, and abundance of each carbonate type (calcite was differentiated from dolomite with alizarin red S staining). Although the categorical data (type of accessory minerals, texture characteristics, etc.) and abundance measures of accessories were not included in any statistical analyses, average and range of grain size was used, as were isotopic ratios, for a hierarchical cluster analysis. The isotopic data were not combined with the petrographic data for statistical analyses; instead separate dendrograms were constructed.

The authors concluded that isotopic ratios of the New England marbles were clearly different from those of the Mediterranean region, however, they were more difficult to distinguish from each other. The petrographic data and the use of statistical analyses were helpful in that regard. Each quarry seemed to be unique, representing slight variations in metamorphic conditions, when variations in carbonate type, accessory minerals, color, grain size, and texture were taken into account. Quarries of higher metamorphic grade tended to be coarser in grain size and lighter in color, with a higher percentage of silicate minerals and with more areas of recrystallization compared with quarries of lower metamorphic grade.

Lazzarini et al. (1995) conducted a characterization study of Carystian marble (Cipollino Verde) from ancient quarries of southern Euboea in Greece, using thin-section petrography, accompanied by XRD, elemental, and isotopic analyses. The 30 largest quarries were chosen for sampling, with 90 samples analyzed. Mineral composition was found to be quite homogenous for the region as a whole. Calcite, the most abundant mineral, often showed stressed and deformed polysynthetic twinning and cleavage. Epidote (clinozoisite) was usually the most abundant accessory mineral. Quartz was scarce, occurring as small grains with undulous extinction, often in association with plagioclase feldspar (albite, generally twinned). Micas included muscovite with small amounts of fengite (a high-silica variety of muscovite, confirmed by XRD and microprobe analyses) occurring as needles, often in clusters or bundles and associated with chlorite; biotite was rare, occurring as tiny flakes. The abundance of chlorite ranged widely, and it occurred as both flakes and as needles of varying size. Other accessory minerals included apatite inclusions in quartz and albite, sphene, and iron oxides; more rarely found were rutile, tourmaline, and zircon. Graphite was seen in the gray-colored specimens.

Considerable variability was reported in the texture of the marble. Most common was a polygonal homoblastic fabric formed by an equidimensional mosaic of calcite crystals with occasional polysynthetic twinning. There were also microbeds of lepidoblastic areas. The rock fabrics were mainly lineated with iso-orientation of sheet silicates and elongation of the calcite crystals.

Lazzarini et al. (1995) were able to conclude that the Carystian marble could be classified as an impure marble derived from marly limestones by low-grade metamorphism. The presence of biotite and fengite, with absence of kyanite, indicated the rock formed at a temperature of 430–460 °C, and the presence of fengite along with the absence of aragonite indicated a pressure of 6–9 kbars. The authors concluded that the qualitative petrographic study was not useful in distinguishing the various quarries, so recommended future modal analysis and quantitative evaluation of textural parameters. Nonetheless, even with the qualitative approach, they were able to identify key differences among many of the quarries on the basis of both texture and mineral composition (variable abundance of many of the accessory minerals).

Lapuente and Turi (1995) characterized marbles from Portugal, focusing on the major ancient quarry sites. The characterization of about 100 samples from 19 quarry locations included thin-section petrography, XRD, and isotopic analyses. The thin sections were used to characterize the overall texture (recorded as heteroblastic, homoblastic, mosaic, polygonal, mortar, strained, elongated, and/or layered), the primary grain boundary shape (recorded as straight, sutured, or curved), and maximum grain size, along with minimum and average. Alizarin red S was used to distinguish between calcite and dolomite, and the mineralogical composition obtained from XRD analysis was confirmed and marked as main component, accessory, or trace (minerals found were calcite, dolomite, quartz, potassium feldspar, plagioclase feldspar, unidentified feldspar, epidote, diopside, and amphibole).

Brown and Harrell (1995) used thin-section petrography to characterize the marble of a Roman quarry in Egypt. Visually, the marble was described as white and fine grained. The primary minerals, in order of abundance, were calcite (occurring in blocky aggregates up to 0.8 mm in size), brucite (in fibrous aggregates also up to 0.8 mm in size), dolomite (as rims around brucite measuring up to 0.1 mm in width), and quartz (with grains up to 0.4 mm). There were veins up to 5 mm wide of pure white, fine-grained calcite free of brucite, together with other veins that were dark gray and sometimes contained graphite. Scattered orange and yellow staining was due to weathering of iron impurities. Color banding (red, orange, yellow, green, gray, and black) was seen at the margins of the deposit. Modern quarrying activities have destroyed most of the traces of ancient working.

Lazzarini et al. (1997) used thin-section petrography combined with XRD and isotopic analyses in a research program on identification of white marbles used in the most important monuments of ancient Rome, reporting on results for the Portico of Octavia (built in 147 B.C. and later enlarged, decorated, and restored). Forty-seven samples of white marble were taken from the monument and employed in thin-section analysis and XRD; 28 samples were used for isotopic analysis. For thin sections, the authors recorded data on presence/absence and relative abundance of accessory minerals (quartz, muscovite/sericite, feldspar, graphite, and ore minerals), maximum grain size (in millimeters), fabric characteristic (homoblastic, heteroblastic), fabric type (polygonal, mosaic, mortar, lineated, etc.), and shape of the borders of calcite grains (such as curved to embayed or sutured). They found that Carrara marble was used in the enlargement made under the emperor Augustus, Pentelic marble during reconstruction by the emperor Domitian, and Proconnesian marble during restorations by Septimius Severus, confirming historical sources and in line with results of analyses on other Roman monuments of these periods.

McGee (1999) thoroughly characterized the Colorado Yule marble, quarried in Marble, Colorado, and used in the construction of the Lincoln Memorial of Washington, D.C. Four grades of the marble exist, ranging from pure white to a more mottled appearance due to inclusions of accessory minerals. McGee characterized quarry samples of all four grades, as well as samples taken from the Lincoln Memorial and from an early quarry dump site. Thin-section analysis was accompanied by SEM with elemental analysis, and electron microprobe work. She found that the calcite grains, the predominant constituent, were irregularly shaped, and had a considerable range of grain sizes. She chose four to eight areas of each thin section at random to measure the size of calcite grains, and arrived at an average of 300 μm, with a range of about 50–1,000 μm. These data were similar for all four grades of the marble.

McGee found that the Yule marble is mainly calcite, with accessory quartz in the form of rounded grains (7–400 μm) in clusters that may include other accessory minerals such as muscovite or phlogopite micas, feldspar, pyrite, sphalerite, sphene, apatite, zircon, and rutile. In the grades containing the highest amount of accessory components, inclusions occurred as clusters. With the exception of quartz, all accessory minerals were finer grained than the calcite component. Today, inclusions of quartz and feldspar protrude from the surface, because they are more resistant to weathering than the surrounding calcite grains.

Some surfaces of this marble remain in very good condition, while even adjacent surfaces can be quite deteriorated, with a softening and disaggregation and "sugaring" of the calcite grains. This variation in deterioration did not appear to correlate with the presence of inclusions or variations in localized weathering. Instead, McGee noted the variation in grain size and texture, due to formation of the deposit by contact metamorphism. Areas of the marble that were closer to the intruding heat source could have been more extensively recrystallized, resulting in variations in grain boundary characteristics.

Schmid et al. (1999) used a quantitative fabric analysis and multivariate statistical analysis (discriminant analysis) to distinguish white marbles from ancient quarries, supplemented with other techniques when needed. Fabric analysis included variables such as grain size and shape. The quantitative approach had the goal of defining quantifiable parameters based on geometrical features observed in thin section. In this example they analyzed 63 thin sections from 14 quarries in Greece, Italy, France, and Turkey. Grain boundary networks of 100–450 grains were drawn from direct projections of thin sections, rotating the microscope stage and using both plane and crossed polarized light so that the maximum number of grain boundaries were seen and drawn. From the grain-boundary network tracings, they computed four variables: area, major and minor axes, and perimeter for each grain with the aid of an image analysis program (NIH Image). Following from that, they computed axial difference (major–minor axis) and perimeter/area for each grain.

Additional data calculated involved grain distribution within a thin section, rather than internally within a grain. They included, for each of the above variables, the mean, standard deviation and skewness (calculated after logarithmic transformation of the data for each thin section). Use was also made of the PARIS-factor, which the authors noted describes the ratio between convexity and concavity of a particle outline (grain boundary, determined using a computer program called '+SHAPES"). It used an input of X–Y coordinates representing particle outlines.

The most accurate discrimination between quarries was achieved with two variables, mean (logarithmic) major axis, and PARIS-factor. Where there were overlaps, cathodoluminescence contributed additional data useful for characterizing marble source. It was noted that the color and pattern of luminescence depended mainly on trace elements, whereas the fabric of the marble was a marker of the history of deformation and recrystallization of the rock, so the two types of data represented very different and independent sources of information.

Lazzarini and Cancelliere (2000) combined thin-section petrography with stable isotope analysis to characterize white marbles from two ancient Roman quarries on the islands of Fourni and Skyros in Greece. These were previously unrecognized ancient quarries, important because the database used for sourcing samples from sculptures and architecture was still incomplete regarding data for all possible ancient quarries. These two quarries had relatively pure marble, with distinctive features. The usual characteristics of texture, carbonate grain characteristics, and accessory mineral content were described in detail and illustrated in photomicrographs. The Fourni marble had a lineated fabric, a relatively large grain size, and a high amount of accessory minerals, especially mica and chlorite. The Skyros marble had a very regular equilibrium-polygonal fabric and a very fine grain size.

Tykot et al. (2002) concluded that one exception to the need for a multi-method approach to sourcing ancient marbles of the Mediterranean world exists with Cape Vathy on the island of Thasos. They identified this source as the only significant one for dolomitic marble in the Mediterranean region, and found that stable isotope analyses of all classical sculptures are consistent with a Thasian origin, thus indicating that thin-section petrography is not necessary. At the same time, Lapuente et al. (2002) found that some of the dolomitic marbles from the Malaga province of Spain were isotopically consistent with Thasian dolomitic marbles. In this case, the Spanish marbles could be discriminated from the Thasian ones by using petrographic criteria (maximum grain size and texture characteristics) in conjunction with isotopic and cathodoluminescence data.

Lazzarini and Antonelli (2003) combined petrographic and isotopic analysis to characterize 25 samples of white and gray marbles from ancient quarry sites on the island of Tinos in the Cyclades in Greece. The marbles are homogeneous and belong to the same geological unit. They described the thin section appearance as being mainly a heteroblastic, lineated fabric formed by frequently iso-oriented calcite crystals with short trains of sericite and concentrated veins of carbonaceous matter/graphite. The calcite crystals had mainly curved boundaries, with a strong kinking on the traces of polysynthetic twinning at two of the quarries. Maximum grain size varied from 0.41 to 2.50 mm. Accessory minerals found included quartz, tiny needles of sericite, and ore minerals often with associated limonitic masses, apatite, and occasionally, epidote. Dolomite was sometimes present in trace amounts. It was found that these marbles from Tinos were easily distinguished from other marbles used in antiquity, and thus it was possible to match some architectural marbles with a source on Tinos.

In their investigation of marble artifacts excavated at the Roman burial area of Faschendorf in Carinthia, Austria, Unterwurzacher et al. (2005) also combined petrographic and isotopic analyses. About 160 marble fragments were recovered, including architectural components and fragments of reliefs and of life-sized statues. Dating was relatively easy due to the presence of ceramic finds in the same strata. However, the source of the marble was in question because, while marbles were being imported at that time from the Mediterranean area, some local eastern Alpine marbles were also being used. Thin sections were cut perpendicular to the metamorphic schistosity. Maximum grain size was determined by measuring the largest cross-section dimension of calcite/dolomite grains, with 150 measurements taken per thin section. To determine the mean grain size, the widest cross-section dimension of 150 randomly selected grains was measured. Results for the analyzed artifacts were very similar, indicating a single source for the raw material. Combining these data with carbon and oxygen isotope analyses permitted assigning a probable provenance to a local source, the Gummern Marble Complex in Carinthia, Austria.

Red marbles

Red marbles, as well as white, have been characterized through thin-section petrography. Lazzarini (1990) focused on red marbles used in the ancient Mediterranean world. Red marble was especially prized during Roman Imperial times, as it was the same color as the cloth used by the nobility and the emperors; however, it was used in various contexts throughout antiquity and into the Renaissance, Baroque, and Neoclassical periods. Red marble of uniform color was available only in small blocks, and was considered a precious stone. It could vary from deep red to a red-violet, and may have had a white banding.

Lazzarini concentrated on an in-depth characterization of a commonly used red marble type quarried in the Peloponesos, often named *rosso antico*. Within this marble source, the red marble outcrops represented only a small portion of the overall outcrop. Four localities were sampled. Comparative studies were done on two other red marbles, a Carian marble and an Aphrodisian marble. Analytical techniques included thin-section petrography, reflected light microscopy (to examine opaque minerals), XRD, and elemental analysis.

He found that in thin section the four localities of the *rosso antico* marble were very similar. It was described (and illustrated) as being mainly calcite, with curved to embayed boundaries. Accessory minerals included: quartz (subangular to subrounded, sometimes polycrystalline, filled with inclusions, and/or exhibiting undulous extinction); plagioclase feldspar (often showing both albite and Carlsbad twinning, and of albite composition); mica, in the form of tiny needles; chlorite, also in tiny needles and associated with mica, or forming veins with a felt-like structure; hematite, mainly in veins but also dispersed throughout the rock as a dark red-brown powder (and accounting for the red color of the marble); and sometimes epidote or limonite.

Overall texture was described as crystalloblastic and homoblastic. Sometimes there was a local lepidoblastic schistosity in connection with the mica and chlorite veins. A layering was seen in sections parallel to the schistosity, often transitioning to an area of preferred orientation of small equidimensional calcite crystals. There were also areas of microcrystalline calcite along with areas of larger calcite. Schistosity was emphasized by hematite veins connected to quartz and plagioclase feldspar. If white banding was present, those areas usually had a granoblastic heteroblastic texture, a larger grain size, and fewer accessory minerals.

In comparison, the Carian marble was very similar both in mineralogy and in texture, but had a more marked preferred orientation and layering; its calcite grains often showed some distortion in polysynthetic twinned crystals, and the plagioclase feldspar showed albite but not Carlsbad twinning. Despite this, it was very difficult to distinguish it from the *rosso antico* marble.

The Aphrodisian marble differed more markedly from the other two. It had a granoblastic homoblastic polygonal fabric with triple points in calcite crystals. Mineralogically, it was quite different from the other two marbles: lacking chlorite and plagioclase feldspar; having calcite with straight to curved boundaries, often embedded with particles of hematite, and sometimes oriented according to cleavage and twinning planes; mica was present as isolated needles or in strings, sometimes with hematite inclusions; quartz appeared as isolated subrounded grains or as polycrystalline grains; potassium feldspar was present as small subrounded grains; and hematite was very abundant either as semi-transparent and resinous brown-red or opaque particles. XRD analysis was done to confirm mineral identifications. Elemental analysis showed that manganese may be present in hematite particles, accounting for the sometimes violet hue found with *rosso antico* and the Carian marble, but not with the Aphrodisian marble.

Gorgoni et al. (1992) combined thin-section petrography with stable isotope analysis to characterize the red *rosso antico* marble, along with some white-gray marbles also quarried from the Mani peninsula of the Peloponnesus region in Greece. Four localities with *rosso antico* outcrops were sampled: Platsa, Paganea, Prophitis Elias, and Marmaro. Several quarries were also sampled for white-gray marbles in these areas, as they have not been as well studied as the *rosso antico*.

The four *rosso antico* localities proved to have very similar mineral and textural characteristics. Accessory minerals included quartz, plagioclase feldspar, mica, and chlorite; however, there was some variation in amounts (Marmaro red marble had a higher plagioclase feldspar and lower quartz content, and the Prophitis Elias marble was quite rich in mica). For all four localities, foliation was common, sometimes emphasized with dark hematite veins. There was a crystalloblastic-homoblastic texture, with mica and chlorite strings and concentrations. Layering or areas of preferred orientation was often seen. Calcite crystals had curved or embayed boundaries, and showed an average grain size of 0.8 mm with a maximum of 1.3 mm. Areas of microcrystalline calcite were also seen, as well as areas of larger sized calcite grains.

Hematite was dispersed as a reddish-brown powder as well as being in veins. Quartz was found in subangular/subrounded crystals or in polycrystalline form, and sometimes showed undulous extinction and inclusions. The plagioclase feldspar was of albite composition, in small crystals, and was often twinned with both albite and Carlsbad twins. Micas were in tiny needles (sericite form of muscovite), and often grouped in strings; chlorite was also sometimes formed in strings. Other accessory minerals included epidote and limonite. Where white banding occurred, such areas had a grano-heteroblastic texture with a larger grain size and greater purity.

The white marbles of the Mani peninsula differed in that they were almost purely calcite in composition. They showed a heteroblastic fabric, deformation twins in calcite crystals, and sutured boundaries. Accessories included quartz, white mica, and iron ore. The gray marbles were the most homogeneous from outcrop to outcrop. They had variable grain size, generally ranging from small to medium, with heteroblastic lineated fabrics, calcite crystals with curved to embayed boundaries which were sometimes sutured. Accessory minerals included quartz, plagioclase feldspar with albite twinning, iron ores, and white micas. Color variations were due to variation in the amount of dispersed carbonaceous matter or graphite.

Despite the fact that virtually all modern studies of marble include a wide variety of analytical techniques, Cooper (1986) emphasized that thin-section petrography remains an important contributor to characterization and provenance studies of marble. He noted that much information is visible and can serve as a first step in distinguishing between possible sources. Information can be gleaned about crystal shape, twinning, size and orientation, grain boundary characteristics, layering features, overall texture of the fabric, and type and texture of accessory minerals.

Gorgoni et al. (2002) are developing a reference database for the main Mediterranean marbles used in antiquity that will hold a variety of types of information for each marble, including thin-section petrography and carbon and oxygen stable isotopic analyses. For the thin-section data, in addition to the usual MGS measurements of calcite or dolomite crystals, they include a description of the typical fabrics and crystal boundary shapes for each of the marbles. For the MGS data, instead of the traditional simple distribution bars, they propose box bars indicating not only the extreme values and the median, but also a range defining 70% clustering of all single values, on the basis that this is statistically much more significant for provenance attributions. To reduce overlap on the diagrams of isotopic values, separate diagrams are shown for fine-grained marbles (average MGS less than 2 mm) and for the medium- to coarse-grained ones (average MGS over 2 mm).

5.5.2 Marble deterioration mechanisms and conservation treatments

Microweathering and other forms of marble deterioration may be visible in thin sections. For example, weathered marble tombstones have uneven surface layers on the exposed edges that are apparent in thin sections, as a result of the gradual detachment of some grains on surfaces exposed to the environment (Fig. 5.5.7). Deterioration products may also form along the surface or in crevices created by the detachment and removal of grains (Fig. 5.5.8). Gypsum and anhydrite may begin to form in such crevices as the stone is exposed to sulfur pollutants in the atmosphere (Fig. 5.5.9).

Young and Ashmole (1968) published a benchmark study on authentication of ancient marble objects that encompassed studies of deterioration and weathering layers (patina), combining a variety of analyses that included thin-section petrography. Central to this work was the development of a collection of samples from objects of known authenticity as well as those from recent sculptures

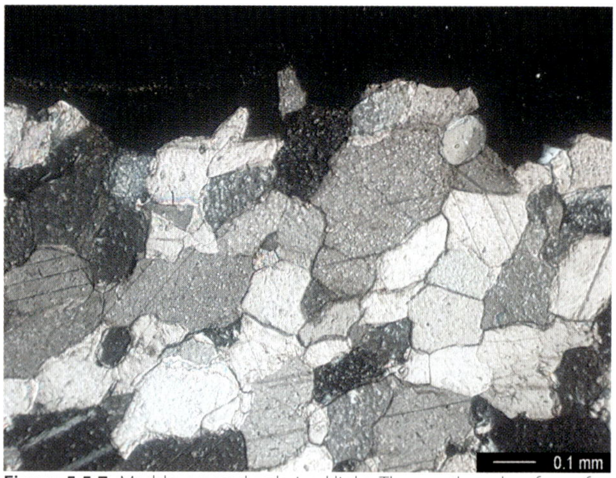

Figure 5.5.7 Marble, crossed polarized light. The weathered surface of this marble tombstone from Laurel Hill Cemetery, Philadelphia (Vermont marble) shows an uneven surface resulting from the gradual detachment of calcium carbonate grains on the exposed and deteriorating surface.

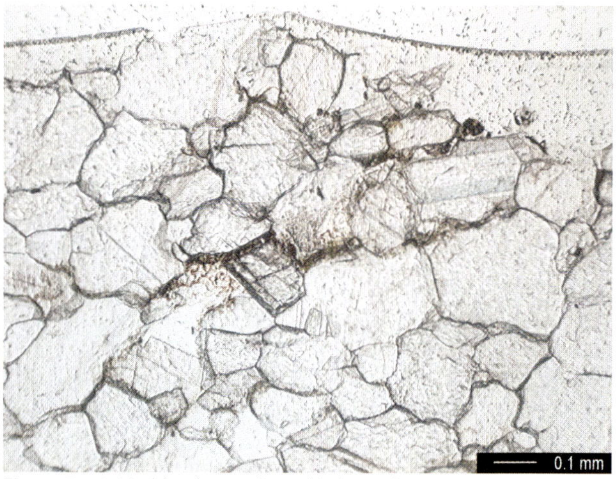

Figure 5.5.8 Marble, plane polarized light. Dark, microcrystalline deterioration products are forming on and near the surface of this marble as weathered carbonate grains become loosened or detached.

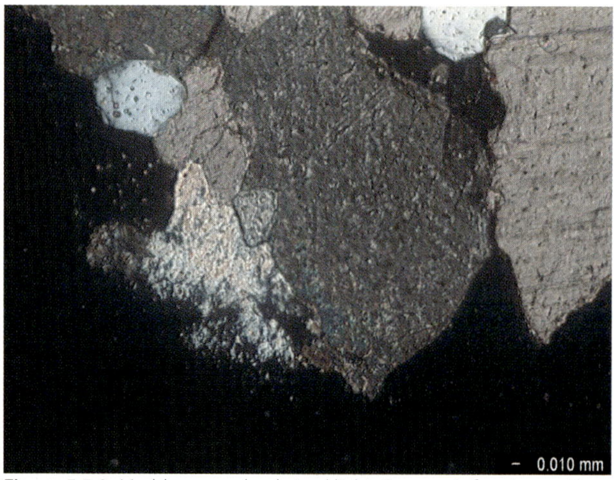

Figure 5.5.9 Marble, crossed polarized light. Gypsum is forming on the surface of this weathered marble tombstone as the calcium carbonate is exposed to sulfur pollutants in the atmosphere. The gypsum has low interference colors (gray, white, and straw yellow), and is unevenly grained with anhedral aggregates and some areas with a fine fibrous structure. There is also a gray quartz grain on the upper left, and pearly gray carbonate grains.

and known fakes. Most recent sculptures or fakes do not have a well-developed patina, unlike most objects of known authenticity. They showed how, in plane polarized light, clear differences can be seen between deteriorated and nondeteriorated marbles.

For example, a known forgery of a supposed Archaic Greek head lacked any weathering crust. In contrast, a fifth-century B.C. Greek relief showed a relatively thick, dark, micritic calcite crust on top of the calcium carbonate crystals. Another sample illustrated a thin section from the surface of a Greek marble sphinx dating to the sixth-century B.C.; it too showed a dark fine-grained micritic layer over the large, clear, calcium carbonate crystals below. In this case the layer was less even, having formed deeper in recessed areas of the surface. A thin section taken from a Roman marble sarcophagus dating to the second–third century also had the same dark, fine-grained micrite layer over the crystalline marble surface. It too was unevenly formed.

Other tools more commonly used today, such as electron microscopes, can provide additional information not visible in the traditional thin sections that Young and Ashmole studied. On the other hand, image analysis tools now readily available to use in conjunction with digital images captured at the polarizing microscope can greatly improve our ability to discern the weathering layer boundary and to obtain quantitative measurements such as minimum, maximum, and average thickness. Although a scanning electron microscope or electron microprobe can also give elemental composition information, which may be quite helpful in identifying all components present, a thin section is still useful because it can provide an image of a relatively large area of the surface and can aid in identifying minerals and petrographic textures. Young and Ashmole did note that marble patination was not found to be useful for determining an exact date of sculpting of the piece, but could be a useful tool, in conjunction with other examination methods, for authenticity studies.

Thin-section petrography has been widely used as a means of viewing deterioration layers on marble sculptures or architectural elements, and of aiding identification of the deterioration components. For example, Laurenzi (1971) illustrated thin sections of a marble bas relief, in both plane and crossed polarized light, showing a marble (with quartz and pyrite accessory minerals) that had a deep deterioration layer, mainly comprised of hydrous calcium sulfate (gypsum).

Allesandrini et al. (1976) included thin-section petrography among other techniques used in the investigation of mechanisms and extent of deterioration of marble quarried at Candoglia and used in the construction of the facade of the *duomo* in Milan. A relatively large number of samples were available from the facade, as well as comparative samples from the original marble quarries. Thin-section studies included calculation of mineralogical composition, expressed as volume, determined by an approximate modal analysis. The marble from Candoglia varied in color—depending upon color, calcite comprised 80–95% of the stone. Accessory minerals included quartz (3.5–10%) and phlogopite mica (up to 1.5%), as well as lesser amounts of pyrite, pyrrothine, pyroxene (fassaite, an augite with low iron content), epidote, chlorite, and plagioclase feldspar. Alteration layers included calcite, quartz, albite feldspar, secondary gypsum presumed to be the product of interaction with pollutants, and components derived from pigeon droppings (such as phosphates and nitrates).

Structural as well as mineralogical features were also observed in the thin sections. For example, the pink quarry marble had pyroxene crystals that were deeply fractured, with calcite refilling the empty spaces, and plagioclase showed an extremely altered form, sometimes almost completely altered. These structural features were seen in the deteriorated samples as well. However, these samples did differ in the fact that they showed less cohesion of grains below the altered surface (down to a depth of 2 mm below the crust). White alteration products were due to the decay of calcite crystals (because of exposure to atmospheric agents such as water and carbon dioxide), corroding along preferential faces of the rhombohedron forming masses of lamellar crystals; in contrast, pigeon droppings and black deposits could be characterized as true "crusts." The pigeon droppings resulted in an irregular and porous crust of 200–300 μm in thickness adhering to the surface but not penetrating the underlying calcite crystals; the gypsum crystals (formed due to SO_2 from air pollution) did not adhere as well to the surface and had a maximum thickness of 1 mm. One sample had a much thicker gypsum crust (up to 4 mm) —in that case, gypsum crystals formed in several zones and penetrated between the underlying calcite crystals. That phenomenon was probably due to a complex cycle of deterioration involving formation of gypsum, freezing and thawing cycles, condensation and evaporation, and loss of cohesion of grains.

Del Monte et al. (1981) used thin sections to examine surface alteration layers on marble in several urban areas of northern Italy. The marbles had deteriorated as a result of exposure to airborne carbon particulates. The authors illustrated layers 200–300 μm in average thickness, comprised of authigenic gypsum crystals (with acicular structure and a fibrous radial texture), residual calcite, minor traces of quartz and feldspars (probably from soil dust), and a significant concentration of spherical, opaque carbonaceous particles that had become incorporated within the stone surface area. This layer rested atop unaltered marble and appeared to be causing further deterioration inward. A correlation was found between the thickness of the altered layer and the amount of carbonaceous particles embedded within the layer. Chemical analysis of the particulate fraction showed a high sulfur content. The sulfur in the airborne-pollutant-derived deposited particles reacted with the marble calcium carbonate surface when water was present, to transform it into gypsum. The optical microscopy work showed that the carbon particles become embedded within the gypsum, and thus continue to induce deterioration as long as they have sulfur to release, as the structure of gypsum crystals is favorable to the retention of water.

Margolis and Showers (1988) studied surface weathering crusts and patinas on ancient Greek and Roman artifacts, with the goal of identifying mechanisms useful for authenticity studies. Polished thin sections were examined from ancient sculptures, a variety of quarry samples, and modern weathered outdoor marble garden sculptures. Most of the analyses focused on SEM, microprobe, XRD, and stable isotope analysis; however, the thin-section examinations confirmed those results. The authors noted that during the working of a marble surface by a sculptor, the outermost layer of crystals underwent fracturing, creating a conduit for water penetration and thus for increased deterioration as the suspended clays and dissolved ions in the water promoted recrystallization of calcite (into micrite, the microcrystalline form of calcite, with some relics of the original crystal forms). Other components of the weathering crust included clays, zeolites, iron and manganese oxides, calcium oxalate, and gypsum. Over time, the weathering layer progressed further and further into the interior of the stone, with a gradual transition. Iron-oxide staining from iron-containing minerals in the marble was sometimes seen often well below the surface layer of micrite. Acid cleaning of marble sculptures or surfaces can remove much of the evidence needed for authenticity determinations.

It was noted that signs of accelerated weathering, indicating a forgery, might include sudden transition from a surface of micrite to an unaltered fresh interior accompanied by lack of any evidence of tool marks, and iron oxides just on the surface and not in the interior (from an applied solution and not the result of alteration of minerals within the marble). Margolis and Showers discussed how dolomitic marbles are more resistant to weathering, but will sometimes have a layer of calcite that is evidence of the natural process of dedolomitization. They found this evidence on outcrops at ancient marble quarries as well as on Archaic Greek statues; Margolis (1989) later used this evidence of apparent dedolomitization to authenticate a controversial ancient sculpture. Later, others (e.g. Pruesser 1993) found that in fact the sculpture did not show evidence of dedolomitization, as the apparent calcite crust had been misidentified and was instead calcium oxalate.

Wang and Zhu (1988) employed thin-section petrography, among other techniques, to study the mechanisms of deterioration of black marbles used as a sculptural and building material in China, and to test new treatments designed to prevent such deterioration. The black marbles were comprised mainly of calcite, with some quartz, pyrite, dolomite, and fluorite, with carbon and clay creating the black color; fossil relics remained in some of the marbles. Sulfuric acid rain created deterioration in the form of gypsum formation. Treatments tested included a pre-treatment of the surface to change gypsum into calcium carbonate, leaving a surface that was more porous than those of original undeteriorated surfaces. That surface proved good for coating tests. A three-component condensed lacquer was tested, intended to serve as a barrier to further exposure to acid rain pollutants.

A review was published of how weathering layers develop on marble surfaces (both during burial and during aboveground exposure), and of how such layers have been used in authenticity studies (Newman 1990). Newman contrasted the mechanisms that occur with calcitic marbles with those that occur with dolomitic marbles. In addition to what can be viewed in thin sections, he discussed the use of other techniques such as SEM and microprobe analyses. He noted that even long-term weathering layers on marble are not very thick, generally only a fraction of a millimeter. Samples for thin sections should be cut as cross-sections that include both the surface layer and some of the inner unaltered portion of the stone. In thin section, the weathering layers are usually visible, although identifying specific minerals within the fine-grained layers is difficult. Preparing the samples without cover slips allows them to be analyzed later in a scanning electron microscope or electron beam microprobe, where elemental composition of the layers can be ascertained, and some structural details may be more visible. Many of his illustrations featured thin sections viewed with back-scattered electrons, with shades of gray that are a function of atomic number.

Newman found that weathering layers of calcitic marbles from burial environments were mainly recrystallized calcium carbonate (forming through reprecipitation of calcium freed by dissolution of the marble surface combining with carbonate from the stone or from groundwater), often incorporating clay and iron oxides from the burial soil. Grains of calcite from the rock or other minerals from the soil may also be trapped within the weathering layer. Weathering layers formed in modern polluted environments featured calcium sulfates, as the calcite reacted with sulfur dioxide or sulfuric acid in air. The boundary between interior stone and weathering layer may be smooth or highly eroded, indicating differences dependent upon aggressiveness of the deterioration processes. Weathering layers on dolomitic marbles were often comprised of fine-grained and compacted calcium carbonate (dedolomitization), and there was frequently a sharp boundary between this calcitic layer and the underlying dolomite layer.

Newman determined that whether or not the study of such layers can aid authenticity studies depends upon the presence of sufficient experimental data to demonstrate that such layers should always be present on ancient marble surfaces, in the absence of any significant cleaning, and that layers of similar appearance cannot be applied or artificially induced so convincingly that a faked "weathering layer" cannot be detected. Furthermore, he noted that it is highly unlikely that measuring the thickness of weathering layers will ever lead to more exact dating methods, because the rate of formation can vary so much depending upon burial conditions (even localized variations within a site) and upon subsequent storage or display conditions.

Margolis and Showers (1990) also illustrated thin sections of marble weathering layers. They claimed that long-term naturally weathered marble should show the greatest amount of iron oxide-clay staining below the micrite alteration layer, and should extend well into the matrix of the marble rather than being concentrated on the marble surface as would happen if a stain were artificially introduced. They also proposed that naturally weathered layers may show remnants of original tool marks, whereas artificially precipitated layers probably would not. Further, they showed, in plane polarized light, a thin section taken from an ancient Cycladic sculpture that seemed to reveal relicts of the form of the original crystal outlines within the weathering layer, more evidence for slow *in-situ* alteration versus a rapid artificial process. Reflected light helped them to identify additional features of thin sections, such as lichens, iron staining, manganese oxide, and gypsum crusts.

Thin-section petrography was included by Agarwal (1991) as one of his analytical tools in studying the deterioration of lithic materials, including marble, about 70 years after they had been used in the construction of the Victoria Memorial Building in Calcutta, India. Marble was used as an outer cladding on this brick structure. It was a white marble, quarried in Rajasthan, western India, with some veins or areas of coloration of gray or blue which are not overly noticeable, but instead tone the white color; some bonding stones have darker blue color areas. Historic literature indicated that the marble was specifically selected to be free from iron pyrites, and from any visible flaws or defects. Yet, it is now visibly weathered, turning to light brown or dark brown in color on external surfaces exposed to wind (especially at corners and the undersides of ledges protected from rainwater), reducing the brilliant white appearance. Dark brown or black deposits are also visible under balustrades and cornices.

Agarwal used thin-section petrography to identify three classes of marble found on the building: a coarse, pure dolomitic marble; a coarse, impure dolomitic marble with twinned xenoblastic crystals of dolomite, and containing prismatic tremolite and quartz as accessory minerals; and a medium-grained, pure dolomitic marble with twinned dolomite and very rare grains of quartz. In all three types, dolomite and calcite were the main constituents, with some recrystallization and polygonization seen. Where present, the tremolite needles and quartz grains showed preferred orientation. Weathering characteristics Agarwal expected to see in thin section were the removal of carbonate material from grain margins and significant deposits of chemical compounds brought in through circulating or permeating solutions. These characteristics were not, however, observed. Instead, there was only a very thin, superficial deposit of gypsum on the stone surface, too thin to be detected in thin section but which could be identified through XRD analysis. Lead and carbon were also detected in the deposit, along with a small amount of hydrous calcium nitrate. The data were all consistent with the deposit forming through reaction with sulfurous material in the atmosphere, probably from automobile exhausts. The result was a slight discoloration on the surface, but no major mineralogical or structural deterioration had occurred. Gentle cleaning techniques proved sufficient to remove the very superficial deposition.

Helmi (1992) studied the weathering of Egyptian marble, a material used for Pharaonic period statues, and for a building material in the Coptic period and in Islamic architecture. Thin sections of various specimens were illustrated. He noted that marble formed through dynamic metamorphism would be likely to have the grains in close contact, while those formed by regional metamorphism would be more likely to have grains separated by intergranular heterogeneous phases. The first type would be stronger and more durable than the second. This difference may account for differing durabilities of two of the marble specimens studied. They also differed in the presence of dolomite as well as of accessory mineral types; differential weathering of the two specimens was also connected to these differences.

Marble weathering of a Roman column on the Via Appia at Brindisi in Italy was studied by Zezza et al. (1992). After 2,000 years of exposure to the elements, the column exhibits extensive pitting. A variety of analytical techniques was assembled to examine the deterioration, including optical microscopy of thin sections. The marble had a banded texture, alternating between small crystals (0.3–0.8 mm) and medium to large ones (1.5–2.5 mm). Pitting tended to start in the smaller crystals. This finding supported ideas about marble corrosion phenomenon being associated with rock texture. Lichens appeared in some places as secondary invaders of the pits, taking advantage of water, airborne particles, and organic material deposited there, rather than being the originators of the pits.

Verges-Belmin (1994) examined thin sections taken from a severely deteriorated statue made of Carrara marble. Thin sections were made using oil as the lubricant instead of water, in order to retain details of water-sensitive materials such as clays or soluble salts. The core of the marble itself was comprised of calcite crystals that showed intergranular fractures increasing in length, width, and interconnectiveness as they approached the outer surface. Near the surface, the fractures could be as wide as 5–30 μm, and were partially filled with recrystallized calcite and gypsum rims. The surface of the sculpture was completely covered with a black crust.

In plane polarized light, Verges-Belmin was able to see that the main sulfation zone was made up of two distinct layers. The outer, irregular layer, which ranged from 0.1 up to 3 mm in thickness, contained many allogenic particles that impart a dark appearance; it also incorporated quartz grains (probably from soil deposition). The inner layer ranged from 150 to 200 μm in thickness, and appeared clear because it had none of the allogenic particles. This inner layer was comprised of clear gypsum, and seemed to be the result of a pseudomorphism process (transformation of calcite into gypsum without any change in shape or volume of the calcite crystals, with even cleavages and grain boundaries being preserved). It was concluded that the typical transformation of calcite into gypsum as a result of sulfation processes from particulate pollution occurred only in the outer black layer. Transformation of calcite into gypsum within the inner layer apparently resulted from a four-step process that involved the following: thermal microfracturing of the marble, which penetrated as far as 4 cm, but did not penetrate the black layer which seemed to have formed later; localized inner formation of gypsum rims lining the boundaries and cleavages of calcite crystals; dissolution of calcite grains by sulfuric acid and the formation of hollow gypsum boxwork; and precipitation of gypsum within all the voids left (gypsum boxwork and thermal cracks). Verges-Belmin highlighted the implications of this inner pseudomorphism for stone conservation, which was that it helped retain surface details of the original sculpture just beneath the visible black crust, and therefore every effort should be made to retain any inner clear layer. In this case, it was conserved with consolidants and protective agents, while the black crust was removed from the outer surface.

Abd El-Hady (1994) studied the deterioration of decorative marble tiles used on Manluk and Turkish buildings (thirteenth–fifteenth century) in Cairo. In addition to characterizing the marbles found (by mineral composition, structure, texture, size of calcite grains, crystal forms and grain contact types) he also illustrated the type of microcrack development, internal cleavage and formation of micropores, and, ultimately, detachment and disintegration of crystals and crust formation, as seen in thin section. The severe deterioration was attributed to cyclic variation in temperature and relative humidity, groundwater, and air pollution.

Patterns of pore spaces that indicated developing deterioration were seen in thin sections illustrated by Fassina et al. (1994). This work was one component of their research on factors contributing to damage of marble structures and sculptures in the Mediterranean coastal environment. They showed thin sections in crossed-polarized light, accompanied by digitally enhanced line drawings outlining the pattern and extent of pore spaces.

Heller and Herz (1995) used thin sections, along with other methods, to study the role of oxalates in the weathering of dolomitic marbles, a factor potentially important in authenticity studies of apparently ancient marble sculptures. Weathered surfaces of dolomitic marble samples from Thasos, Greece (exposed for more than 1,500 years) and from Lee, Massachusetts (exposed for no more than 50 years) were compared. Experiments were also carried out using samples of dolomitic and calcitic marble, submerging them in solutions containing potato mold, to identify any reactions with oxalic acid. In addition to examination of thin sections, analyses included elemental and isotopic analyses, XRD, and SEM. The primary goals of the study were to test the validity of authenticity studies based on various structural, chemical, and biochemical processes that may affect marble surfaces, and to identify weathering mechanisms that might be useful in authenticity studies involving dolomitic marbles.

The fresh interior was compared to the surface, with a focus on looking for differences in structure, mineralogy, chemical and isotopic composition, and effects of biological attack from molds, lichens, and fungi. By studying how dolomite reacted to various weathering agents, they hoped to see if it was possible to identify the approximate length of time that a stone had undergone weathering, thus allowing for meaningful authenticity studies to be performed. Specifically, they focused on effects produced by the formation of a surface layer of calcite, indicating a long-term process of dedolomitization has occurred, and on chemical and physical deterioration caused by organisms such as lichens as they penetrate into microfractures. These organisms secrete acids that can etch the rock surface and produce soluble metal complexes below the organic encrustations. Oxalic acids secreted from the organisms are also thought to react with carbonates to produce new calcium oxalate minerals such as whewellite or weddellite.

Thin sections showed that the marble from Thasos was medium to coarse grained, almost pure dolomite with small amounts of quartz and calcite. The grains were unequal in size, anhedral, with sutured grains and mortar between grains. Weathered crusts could range from 0.2 mm up to 3.0 mm in thickness. The surface was penetrated by a variety of organisms, clay, and airborne materials. In contrast, the Massachusetts marble was fine to medium grained, the dolomite was equigranular with distinct angular boundaries, and there were minor occurrences of white micas. The surface was noticeably less altered or weathered than the Thasos marble, with few organisms visible.

Although full conclusions relied on a combination of a variety of analytical techniques, thin-section petrography facilitated observation of the weathering layers. The Thasos stones, which underwent a long period of natural exposure, showed pitting, the presence of loose and floating grains, granulation, microfracturing of the surface, and deep penetration of weathered material into microfractures and between grains and cleavage faces. In contrast, the Massachusetts marble, which underwent a relatively short period of exposure, had few microfractures, little surface pitting or irregularities, few or no floating grains, and little granulation; however, the uppermost grains were more exposed due to partial recession around them. Minimal foreign material was seen penetrating the surface layers compared with the more extensive surface penetration seen on the Thasos marble.

Ultimately, Heller and Herz did not find any evidence of dedolomitization. Instead, organisms such as lichens appeared to play a major role in the deterioration of the marble from Thasos, whereas climate conditions in Massachusetts had promoted mechanical and chemical weathering. Calcium-rich weathering products found on the surface and in veinlets of Thasos marble were not carbonates, but calcium oxalate in the form of whewellite, resulting from the actions of lichens, molds, and fungi. Experiments with dolomitic and calcitic marbles in oxalic acid baths showed relatively fast reactions with whewellite forming on dolomite and extensive surface damage on calcite. The Thasos marbles, with surfaces that have been exposed far longer, were more deteriorated than those in Massachusetts, but even 50 years of exposure had caused some samples to have clay particles or debris that were penetrating into interior areas. The authors concluded that weathering of dolomitic marble can vary greatly depending on source and exposure conditions, so using surface weathering evidence for authenticity determinations may be problematic.

Vendrall-Saz et al. (1996) studied marble patinas formed on monuments in the Mediterranean area, investigating to what extent they were related to microbial and biogeochemical pathways of formation. They noted that most patinas were multilayered: orange to gray layers contained mainly calcium carbonates and oxalates with some phosphates, while dark gray to black layers were mainly gypsum and sometimes entrapped airborne particles. The thin sections illustrated included one showing a two-layer patina on marble from the cathedral of Tarragona, with a sparitic calcite layer overlaying a micritic calcite one; both were atop the undeteriorated marble itself.

Weiss et al. (1999) reported that different marble types used for building the eighteenth-century Marmorpalais in Potsdam, Germany, showed variation in degree of deterioration. The authors studied grain fabric and texture of the marble types to determine whether there was a connection between those features and weathering susceptibility. Understanding such connections is important because restoration of deteriorated areas involves replacement with fresh marble from the same sources, but rapid degradation can occur with some types of materials. In addition to the weathered building stones, they examined fresh marble specimens from Poland, the Czech Republic, and Italy (sources of marble for the building's construction, and where restoration materials are obtained).

Differences were found in average grain size, grain boundary geometry, and texture. The research incorporated digital image analysis to quantify the data, to facilitate a comparison among marble types, and to look for any correlations with other parameters such as mechanical strength. For the image analysis, thin sections in three mutually perpendicular orientations were used. Representative cross-sections of the thin sections were drawn by hand, and then digitized. Grain boundary lines were analyzed for a variety of parameters including area, grain size, elongation, shape, preferred orientation, and Paris factor (which defines the irregularity of grain boundaries, with smooth boundaries giving a value close to 1 and higher numbers for more curved boundaries). The thin-section analyses were supplemented with cathodoluminescence work to highlight additional features such as open grain boundaries or refilled microcracks, and with neutron diffraction to characterize texture (lattice preferred orientation of the calcite crystals, an important feature to quantify in order to predict spatial distribution of thermal expansion). The authors concluded that different grain fabrics and textures, and their combinations, led to different types of weathering mechanisms (such as powdering versus breakouts along microcracks). Knowledge of these relationships can help in the selection of appropriate (more stable) conservation and restoration materials.

Alberti et al. (2000) were concerned about the heavy deterioration occurring on marble building stones of one of the castles built by Emperor Frederick II in the thirteenth century, Castelmaniace in Syracuse, Sicily. Most of the building was made of limestone, but a monumental portal and great window were decorated with white and colored marble brought to the site from diverse sources throughout Greece, Italy, and Asia Minor. Deterioration included scaling, exfoliation, powdering, fissuring, discoloration, salt efflorescences, biological growths, and black incrustations. Laboratory studies were conducted to identify the causes and mechanisms of deterioration prior to restoration attempts. The two main causes of deterioration were found to be crystallization of soluble salts and thermal changes due to the high humidity in the warm marine environment. Different marbles reacted differently due to differences in fabric and stone properties such as porosity. Thin sections were helpful in revealing the extent and type of deterioration, while also serving as a means for characterizing marble fabrics and properties.

For example, one marble type had microcracks that followed the natural lineation of the material. Thin sections of another marble showed an oxalate patina covered by a gypsum layer; gypsum and particulate matter, found at the bottom of a flake, were responsible for its detachment from the marble surface. Yet another marble showed strong decohesion (with visibly enhanced marking of the boundaries of calcite crystals), microfissuring, and sulfation. The migration of calcium oxalate from the surface of the stone to the central fissure could also be observed in thin sections. After fully cataloging and characterizing the deterioration evidence for each type of marble, Alberti et al. identified sodium chloride (from the marine environment) and gypsum (from air pollution) as the most problematic salt species. Thermal changes due to exposure to direct sunlight exacerbated deterioration.

Galanos and Doganis (2003) employed thin sections to study the orange-brown surface layer (epidermis) found on the Parthenon sculptures. The epidermis preserved surface evidence for craft techniques (such as tooling, polishing and finishing marks, and incised lines used for measurement in the initial building stage of the monument), and the incised decoration was highlighted by the epidermis, which also contained medieval graffiti. The authors noted that the presence of an undisturbed epidermis indicated the best preserved surfaces of a sculpture, and thus can help reveal original artist intent. They also compared undamaged epidermis on sculptures still at the Parthenon with surface layers on Parthenon sculptures that were cleaned at the British Museum in the 1930s, in order to assess the degree of damage caused by that cleaning intervention.

The epidermis was identified as an artificially applied layer, not simply a natural weathering characteristic of the marble. The thin sections showed the layer "bleeding" into the uppermost layer of polished marble crystals. It had a glassy, even surface of uniform thickness of about 150 μm, and was better preserved in sheltered areas. The authors hypothesized that this layer was a result of an ancient treatment that was probably repeated at various times during antiquity. It may have been an early method of preventive conservation, part of the finishing process, or a way of toning the glare of freshly quarried marble. The removal of the epidermis from sculptures that were cleaned has resulted in removal of craftsmanship documentation such as traces of polychromy, and tool and finishing marks, it has changed the aesthetic effects of the sculptures in subtle but important ways, and exposed the marble surface to deterioration.

5.6 Amphibolite

Amphibolites, formed by regional metamorphism of moderate-to-high grade, are dark colored (green, gray, or black) due to their primary visual constituents of hornblende and plagioclase feldspar (Fig. 5.6.1). There may also be lesser amounts of other minerals such as epidote, chlorite, biotite, garnet (almandine), quartz, kyanite, sphene, rutile, apatite, opaque oxides or sulfides, and other minerals. Sometimes

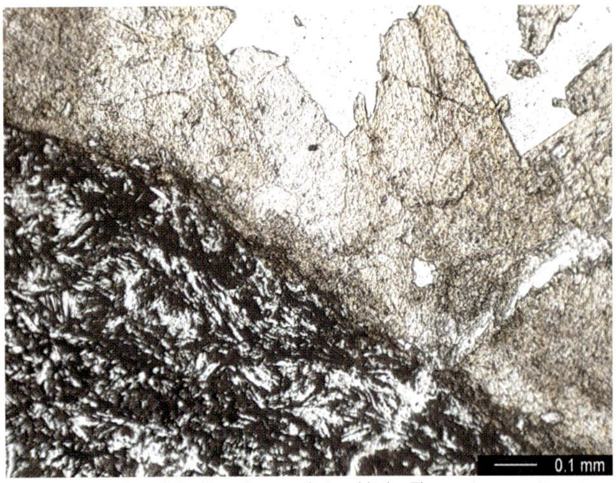

Figure 5.6.1 Amphibolite, plane polarized light. The main constituents of this rock are a pale brown hornblende with typical amphibole cleavage, high relief, and prismatic but poorly terminated crystals, and tiny blades of plagioclase feldspar, colorless in plane polarized light. Other constituents include fine-grained black opaques, colorless quartz, and some granular, yellowish epidote.

amphiboles other than hornblende are also present. Amphibolites tend to be medium to somewhat coarse grained, although they can also be rather fine grained. They may be clearly foliated, with an almost schistose foliation visible to the unaided eye, or sometimes they are completely lacking in visible foliation with an irregular fracture on broken surfaces. However, in thin section amphibolites usually show preferred orientation of hornblende crystals, the amphibole from which this rock's name is derived (Dietrich and Skinner 1979; Williams et al. 1982).

5.6.1 Amphibolite as a cultural material

Curtis (1959) used thin-section analysis to identify the materials used for artifacts and architectural stone at La Venta, Mexico. One sample, of a rock used for rough stone pavements underlying a jaguar mosaic mask, was described as an actinolite amphibolite. In this case, the amphibole was not hornblende, but actinolite, which comprised most of the rock.

McPherron and Milovanović (1974) reported thin-section analysis of amphibolite stone tools used in the Neolithic in Serbia. The amphibolite was described as consisting mainly of hornblende, plagioclase feldspar (labradorite-bytownite), and iron ores. The primary texture was granoblastic, and the rock was very fine grained. It was noted that metamorphic processes have replaced the primary hornblende, and some plagioclase, with a needle-like amphibole that is very pale green, almost colorless. This stone was probably selected for hardness, advantageous for tool use.

Amphibolite was identified (though thin-section petrography) as one of the materials used for late medieval monuments in the West Highlands of Scotland (Collins 1977). The rock was black and coarse grained, with hornblende crystals up to 5 mm in length. In thin section the hornblende, the primary constituent, was green; other major constituents included biotite and plagioclase feldspar (probably labradorite). Accessory minerals were sphene, apatite, and zircon.

Schwarz-Mackensen and Schneider (1986) found that amphibolite was one of a wide variety of metamorphic rocks used in the Late Neolithic period for production of stone axes in the Harz Mountains. The thin sections showed that the rock was comprised mainly of hornblende, plagioclase feldspar, and quartz (with undulous extinction). Accessory minerals included apatite, biotite, epidote, muscovite, chlorite, and titanium minerals. Asher (1987) found a fine-grained amphibolite used (although not abundantly) as a sculptural material in eastern India, identified through thin-section petrography. However, the stone was not described in further detail.

Faccenna et al. (1993) studied building materials used in the Swat valley of northern Pakistan, focusing on materials of the Sacred Area of Pānr I, a site near Mingora dating to the Gandhara period. Their goals were to classify the rock types used, identify nearby rock outcrops through geological surveys, locate the areas and quarries where the architectural stone derived, and identify any differences in materials usage over three building periods at the site. One of the rock types found was a dark green schistose amphibolite. Major minerals present were hornblende, with a dark green color and prismatic habit, quartz, and feldspars. Small round crystals of garnet were also sometimes found, however, this material seems to have been used only occasionally—it was found only with a few slabs, blocks used to construct plinths, wall blocks, and step risers.

5.7 Hornfels

Hornfelses form by contact (thermal) metamorphism resulting in solid-state crystal growth in rocks surrounding an igneous intrusion (as a result of magmatic heating). They can be formed by metamorphism of a wide variety of rocks, and so may incorporate an extensive range of mineralogical composition. The minerals usually grow in a random pattern of interlocking grains, because of the absence of strain during growth. Thus, they usually have no foliation, and if fine grained may even break with conchoidal fracture; sometimes there is a slight foliation with some parallel alignment of mica flakes. Visible foliation may be due to relic bedding (from the parent rock) comprised of differences in mineral composition, but without fissility. Most often hornfelses are nearly homogeneous, microcrystalline to fine grained (Fig. 5.7.1), and may have scattered, relatively large and irregularly oriented porphyroblasts of a wide variety of minerals (Fig. 5.7.2). These larger grains often contain many minute inclusions, giving some hornfelses a spotty appearance. The random fabric of interlocking grains and absence of deformation give a rock that is dense, hard, and tough. Hornfelses are named according to mineral content (e.g. diopside hornfels, or hornblende-plagioclase hornfels) or according to chemical characteristics (e.g. calc-silicate hornfels) (Dietrich and Skinner 1979; Best 1982; Williams et al. 1982; Yardley et al. 1990).

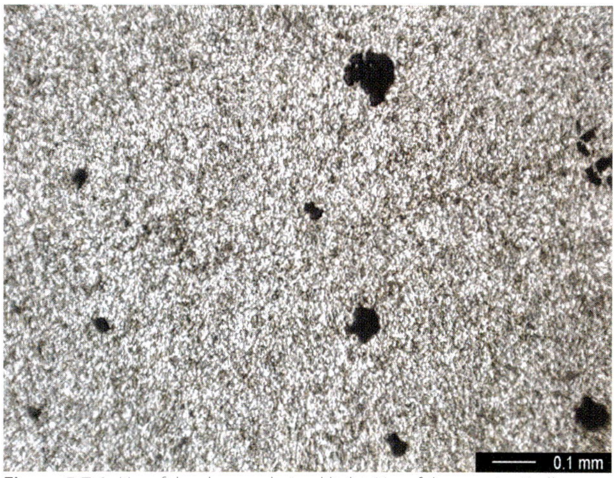

Figure 5.7.1 Hornfels, plane polarized light. Hornfelses are typically microcrystalline to fine grained. Here the tiny grains are minutely intergrown brown biotite and colorless quartz and muscovite. Larger grains of black magnetite and pyrite are scattered throughout this fine-grained matrix.

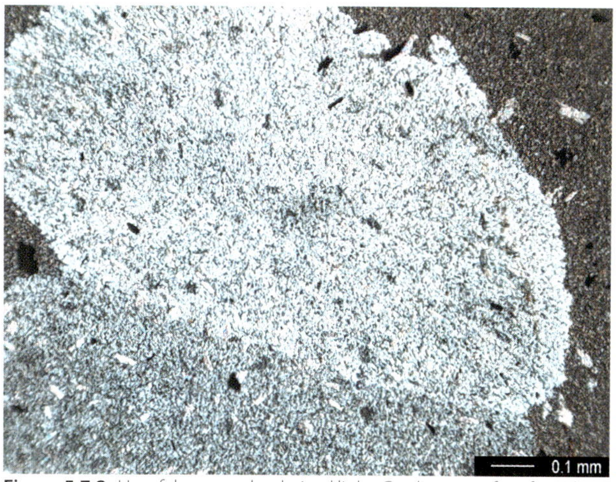

Figure 5.7.2 Hornfels, crossed polarized light. Cordierite is often found in hornfelses as large porphyroblasts such as this one. It shows weak birefringence, similar to that of quartz, and this form of twinning is common for cordierite. Typical for hornfelses, it also shows a sieve-like structure (poikiloblastic texture), and is altering to sericite. The fine-grained matrix in the background, seen on the upper right, is comprised mainly of biotite, quartz, and muscovite, with the latter including some slightly larger grains. Magnetite and pyrite are scattered throughout the rock.

5.7.1 Hornfels as a cultural material

Hornfels has been identified as an important material used for ancient Mesopotamian cylinder seals. Berry (1969) noted that hornfels is a good material for this use because it is soft enough to shape and carve, yet hard enough to resist wear during the lifetime of use of a seal and sufficiently tough to resist breaking when dropped. In addition, the fine-grained and homogeneous character of most hornfels is advantageous, creating a smooth surface that wears evenly and tends not to chip. However, due to the small size, presence of design patterns, and great value of cylinder seals, taking even small samples for thin-section analysis is generally only possible for fragmented or damaged seals.

Hornfelses were used in the production of stone axes, probably due to their hard, dense fabric with lack of fissility. For example, an actinolite-cummingtonite hornfels was used as axe material in Australia, cordierite hornfels (with or without biotite, chiastolite, or garnet) was used in New South Wales, and a calc-silicate hornfels was the primary axe material in Killin, Britain (Kempe and Harvey 1983).

Asher (1987) found hornfels used as a sculptural material in fifth–twelfth-century eastern India (Bihar). Sculpture specimens were characterized by thin-section petrography, and described as being medium (0.2–0.5 mm) to coarse (0.5–1.0 mm) grained, with granoblastic texture. Major minerals found were muscovite, quartz, and chlorite—both with and without minor amounts of cordierite, opaques, and plagioclase—and accessory amounts of biotite, apatite, epidote, sphene, and zircon.

5.8 Quartzite

Quartzite is metamorphosed siliceous sandstone (such as quartz arenite) or chert. It is generally nonfoliated and consists mainly of quartz. The quartz is recrystallized under metamorphic conditions to produce an interlocking mosaic

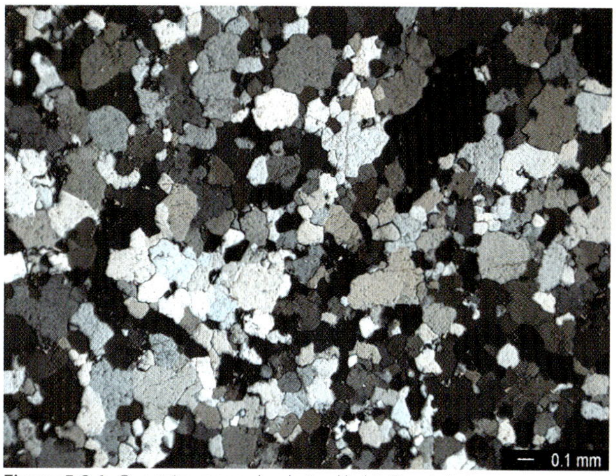

Figure 5.8.1 Quartzite, crossed polarized light. Quartzite is metamorphosed siliceous sandstone or chert. It is usually nonfoliated and consists mainly of quartz. The quartz recrystallizes under metamorphic conditions to produce an interlocking mosaic of grains; thus this is a very hard and durable material.

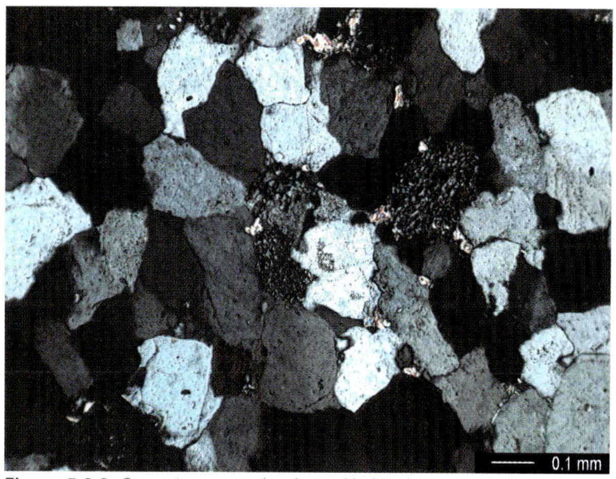

Figure 5.8.2 Quartzite, crossed polarized light. This example includes a few grains of metamorphosed chert (microcrystalline quartz) and some scattered grains of higher birefringent interstitial sericite.

of grains, making a very hard and durable material (Fig. 5.8.1). Upon breaking, fractures go through rather than around the tightly interlocked grains. Accessory minerals typically include feldspars and opaques, although a variety of metamorphic minerals may be seen as well (Figs. 5.8.2, 5.8.3, and 5.8.4) (Best 1982; Williams et al. 1982; Ashhurst and Dimes 1990; Yardley et al. 1990).

5.8.1 Quartzite as a cultural material

Architectural uses

Quartzite, although often a very hard material, does have a long history of use as a cultural material. Generally it is used because its hardness is a sought-after property, or because of its reddish appearance that can take a polish. For example, early thin-section studies aimed at identifying sources of the stones used to build the famous site of Stonehenge in England showed quartzite as one of the materials used to form monoliths (Storey-Maskelyne 1878; Thomas 1923), and it was used in ancient Mexico in the construction of La Venta, Tabasco, where a black quartzite is found forming rough stone pavements underlying a jaguar mosaic mask at the Southwest Platform (Drucker et al. 1955). There, thin-section analysis showed angular and poorly sorted grains of quartz.

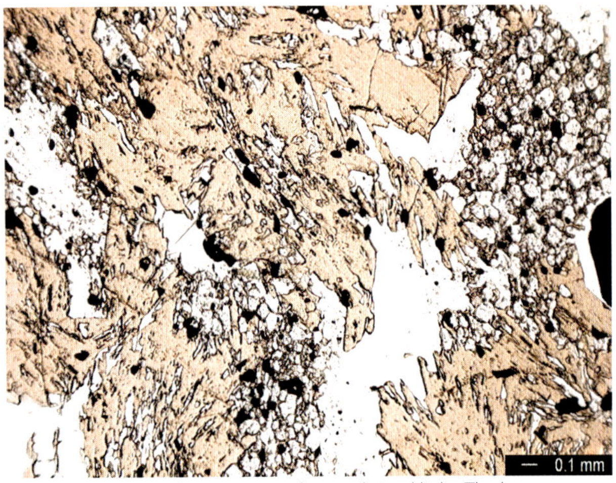

Figure 5.8.3 Staurolite quartzite, plane polarized light. The large anhedral staurolite porphyroblasts in this quartzite have a poikiloblastic (sieve-like) texture with irregularly arranged inclusions of quartz, typical of this mineral. It is a pale yellow color, with high relief. The colorless quartz is highly variable in grain size in this example, with some small granular crystals and some large anhedral ones. There are also many small black grains of magnetite present.

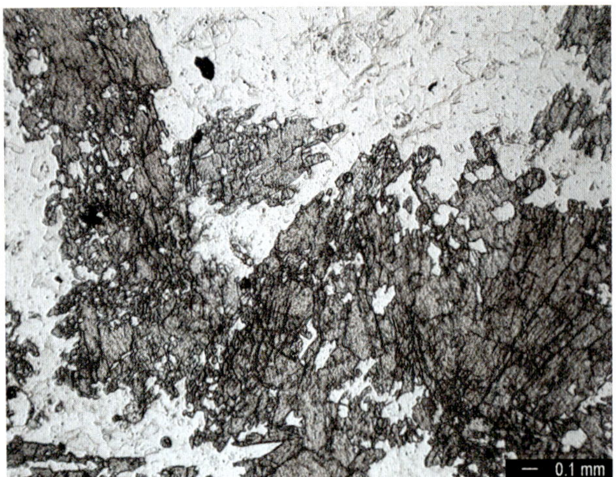

Figure 5.8.4 Kyanite quartzite, plane polarized light. The large kyanite porphyroblasts (some of which have a poikiloblastic texture) are in the form of broad elongate tabular crystals with a pale somewhat blue color and high relief. They characteristically have one perfect cleavage, one imperfect one, and a cross parting at angles of 85 degrees to the length of the crystals. The quartz component is colorless, and there are a few scattered black anhedral magnetite grains.

Quartzite rubble, set into mortar, was a common construction material for nineteenth-century buildings in central Pennsylvania; alternatively, regular blocks of quartzite were quarried from local deposits and used for masonry construction. A variety of quartzite compositions has been used here, for example, one quarry near Avondale, Pennsylvania (Chester County) produced a micaceous quartzite. The abundant mica gives a glint to the stone, and iron staining is said to impart a pleasing color variation. Quartzite quarried in Lancaster County (Greentree quarry) has some areas that glisten with white mica, others that are banded, some light colored, and some stained brown with iron oxides. The Chickies quartzite in Montgomery County, widely used as a building stone and for furnace linings in steel mills, is described as only in part a true quartzite and in part actually a quartz-sericite schist (Stone 1932).

De Geyter (1985) utilized thin-section analysis to characterize siliceous building stones in northern Belgium. A quartzite from Tienen was the most important building stone used for many Romanesque and early Gothic monuments. De Geyter found that this stone could be distinguished by its hardness and resistance to weathering, equigranular texture, absence of glauconite and chlorite minerals, and frequent presence of silicified plant roots and fragments of silicified wood. The quartz grains were closely packed, generally 100–200 μm in width, with secondary overgrowths. There were a few minute euhedral quartz grains firmly locked between larger quartz grains. The cell walls of the silicified roots and wood were clearly visible, with cells filled with microcrystalline quartz.

Ferreira (1992) described quartzites used as building stones for historic monuments in Ouro Preto, Brazil. One type of quartzite was also found in a location near the town, where partially worked blocks still remain as evidence that the site served as a quarry in the past. Thin-section analysis showed that the main mineral constituents were quartz and sericite; accessories were limonite, hematite, and magnetite. The other type, also quarried near the town, was mainly quartz with the sericite component having parallel orientation. The only accessory mineral was limonite.

Fitzner et al. (1993) studied deterioration of quartzite masonry of a Baroque monument in Minas Gerais, Brazil. Samples were taken from the monument stone and from quarry sites nearby. Characteristics relevant for weathering were documented using thin-section petrography combined with several other analytical techniques. The main mineral component was found, of course, to be quartz, although grain size varied significantly. Other minerals present were micas (20%) and iron minerals. Relic sedimentary structures could be recognized, with the original sedimentary rock most likely being a clay-rich sandstone. The micas showed preferred orientation giving an almost schistose texture. The schistosity appeared to be responsible for flaking, scaling, and exfoliation detachments now occurring on the monument. Furthermore, the quartz grains were only weakly bonded, resulting in granular disintegration. As weathering proceeded, total porosity increased, and gypsum began to form in pores below the surface, contributing to scaling and detachment of material from the stone.

Sculpture

The stone used in the construction of the Colossi of Memnon, two colossal statues located on the plain of Thebes in Upper Egypt, was identified as a ferruginous quartzite (Heizer et al. 1973). It was noted that in addition to being extremely hard, this stone can take a high polish, and has an aesthetically beautiful appearance. Six quartzite quarries were located in Egypt from which the stone might have been derived, although some may not have been able to provide the large high-quality blocks used in the statue's construction. Some areas of restoration could be traced back to the Roman emperor Septimius Severus (third century). A combination of chemical analysis (using neutron activation analysis) and thin-section petrography was done in order to try to pinpoint the source(s) of the stone. However, discussion of the statues by these and some subsequent authors is somewhat confusing because they oscillate between calling the stone "quartzite" and "quartzose sandstone." As a result, it is unclear whether or not the stone is actually the metamorphic rock, quartzite, or a sandstone in which lithics that include chunks of quartzite have been cemented in a sedimentary process.

Asher (1987) found quartzite sometimes used as a sculpture material in eastern India (fifth–twelfth century). Visually, this quartzite appeared glassy and purple to dark blue in color. In thin section it was seen to be comprised almost entirely of coarse (0.5–1.0 mm) intergrown quartz grains. There were frequent inclusions of a very fine-grained (≤0.01 mm) tourmaline as the main accessory mineral.

In the discussion by Stross et al. (1988) of Egyptian quartzite from the quarry near Cairo (Gebel el Ahmar) and that from the Aswan quarry, they noted that both quartzites consisted mainly of quartz grains coated by ferric oxide and cemented by extensive quartz overgrowths around the detrital grains. The two quartzites differed in texture and trace mineral components. The detrital grains of Gebel el Ahmar were more rounded than the Aswan quartzite, and most samples from the Aswan quarry contained trace amounts of anatase (titanium dioxide) formed in the sediment after deposition. The authors reported that in thin section the two quartzites could easily be distinguished from each other on the basis of grain shape, supported by the presence or absence of anatase.

Tools

Quartzite was one of the grinding-stone materials at a Predynastic site in Upper Egypt (Mahmoud and Bard 1993), and, even earlier, was used in prehistoric flaked-stone technology to produce tools more than two million years ago in eastern Africa (Toth 1987). These tools were produced by starting with quartzite blocks that were struck with a cobble hammer, resulting in a core and flakes or fragments. Some flakes might be further retouched. Early stone tool use in the Perth basin of Western Australia also included exploitation of hard quartzite as a material (Glover 1984). Quartzite is a common stone in this area, and was used for both small and large tools—in fact, quartzite has been used as one of the most common materials for early stone tools worldwide (Kempe and Harvey 1983).

Shipley and Graham (1987) employed thin-section analysis to characterize rocks utilized for ground stone tools (*manos*, or grinding stones) at the Maya sites of Seibal and Uaxactun in Guatemala. They then matched these rock descriptions to outcrops, and had some success matching specimens to outcrops in Belize. Among the variety of rock types identified as a material used for the grinding stones was one quartzite example. According to the thin-section description, it consisted mainly of equigranular recrystallized polycrystalline quartz. Also present were lesser amounts of calcite, and abundant minute granules of calcite and hematite. The thin section exhibited faint banding due to variations in grain size and concentration of inclusions.

5.9 Serpentinite

Serpentinite is a product of low-grade contact metamorphism, often involving hydrothermal alteration of peridotes (magnesium-rich rocks consisting predominantly of olivine and pyroxene), or sometimes of dunites (a peridotite with 90–100% olivine) or pyroxenites. They can be green, gray, black, or reddish in color. Several morphologic varieties of serpentine may be seen in these rocks (chrysotile, antig-

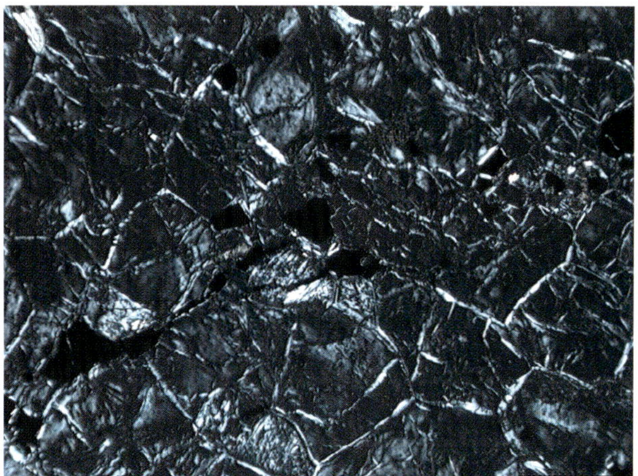

Figure 5.9.1 Serpentinite, crossed polarized light. Serpentinite is comprised mainly of the serpentine group of minerals (hydrous magnesian silicates) which have low birefringence and are usually gray in crossed polarized light. Serpentinite sometimes shows the mesh texture seen here, in which fine-grained serpentine minerals of very low birefringence are divided into small blocks by veinlets of serpentine showing slightly higher birefringence.

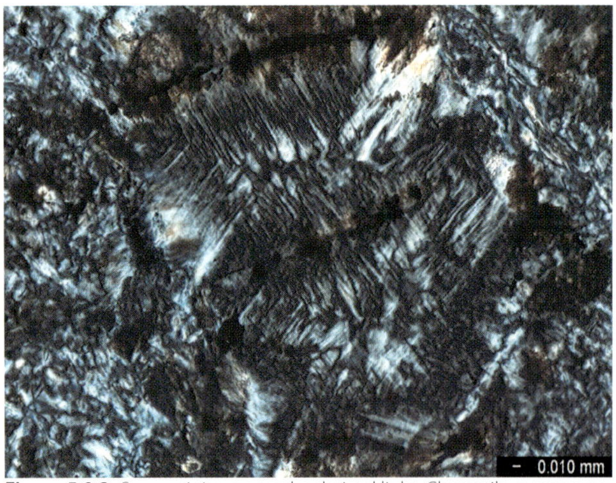

Figure 5.9.2 Serpentinite, crossed polarized light. Chrysotile, a common serpentine form, can be distinguished from other serpentine minerals because it is fibrous, and often occurs in cross-fiber veinlets.

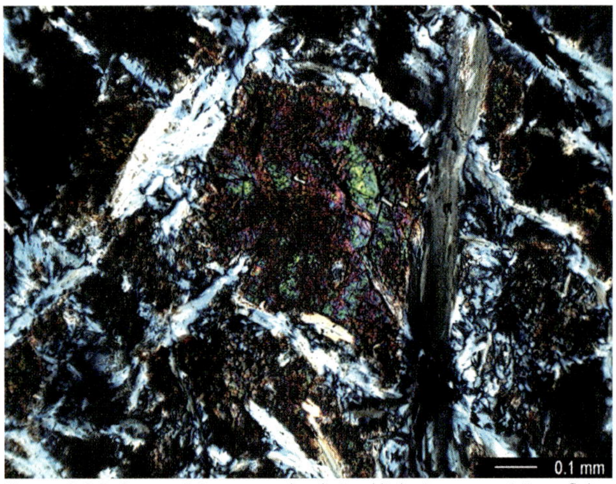

Figure 5.9.3 Serpentinite, crossed polarized light. In serpentinite, a flaky micaceous-like antigorite may replace olivine and/or pyroxene from an original peridotite rock. Olivine and pyroxene (in this case clinopyroxene) remnants are often preserved while much of the remainder of the rock alters to antigorite. The low birefringence of antigorite contrasts with the high birefringence of the clinopyroxene remnants.

orite, and lizardite) although they often cannot be reliably distinguished in thin section. This serpentine comprises most of the fine-grained rock. Serpentinite sometimes has a distinctive mesh texture (Fig. 5.9.1), in which a low birefringent, fine-grained serpentine is divided into small blocks by thin veinlets of serpentine of slightly higher birefringence (MacKenzie and Adams 1994).

Chrysotile, a common serpentine form, is fibrous (Fig. 5.9.2) so can be distinguished from antigorite and lizardite in thin section. Iron often precipitates as dusty opaque iron oxide outlining the grain boundaries of the original peridotite rock. Pyroxene crystals from the original rock may be preserved as pseudormorphs filled by antigorite, a form of serpentine that occurs as an aggregate structure (bastite). This aggregate, comprised of a flaky micaceous-like antigorite, may replace either or both olivine and pyroxene. Olivine and pyroxene remnants are often preserved even when much of the previous rock has been altered to serpentine (Fig. 5.9.3). Chromite from the original rock may remain unaltered or may form secondary streaks and veinlets; brucite may also be present as small flakes, along with some unaltered olivine. Other secondary minerals that may be present include talc, actinolite, tremolite, dolomite, magnesite, or a wide variety of other minerals (Kerr 1977; Dietrich and Skinner 1979; Best 1982; Williams et al. 1982; Yardley et al. 1990).

5.9.1 Serpentinite as a cultural material

Architectural uses

Serpentinite has found wide use as a building or decorative stone. For example, in nineteenth-century Pennsylvania it was used for facing buildings in and around Philadelphia. This stone has a pleasing and interesting green appearance but the quality of the stone is highly variable; deposits can be either massive or banded (for example, with calcite veins) and the stone can be soft or hard, colored a relatively clear light or dark green, or quite mottled (spotted green and white varieties are called "ophiolites," "ophicalcites," or "ophites" by various writers).

In Pennsylvania, serpentinite is found serving as masonry stone or building rubble, or as flooring. Serpentinite is also called "greenstone" here, although elsewhere "greenstone" refers to a wide variety of other rock types (D'Amico 2005). The stone did not have widespread popularity because it deteriorated rapidly in the polluted city atmosphere. Problems included major color change and severely spalling surfaces. Nineteenth-century row houses built of serpentine masonry and white marble trim demonstrate the much faster deterioration of serpentine; it shows deep pitting and spalling while the marble trim appears fresh and unaffected (Stone 1932).

Serpentinite was used as one of the building materials for the Minoan palace of Knossos on Crete (Papageorgakis and Mposkos 1988). Here it ranged widely in color, depending on mineral composition and degree of weathering. However, in all cases the main mineral present was serpentine, sometimes with patches of calcium carbonates and/or chlorite. Compact and hard, it has good mechanical strength, but is not very resistant to weathering.

Blanco Varela et al. (1992) studied surface decay on serpentine used in the construction of the Descalzas Reales Convent in Madrid, Spain. Part of the structure was recently remodeled. The early material was a dark green serpentine, while the new material used was a green serpentine with marble streaks. Shortly after the restoration work was completed, deterioration became evident: the old serpentine showed the formation of flakes, while the new material showed the formation of blisters. In thin section, both materials have a fibrous structure and similar texture; they differ mainly in percentage and composition of secondary minerals, with the new serpentine having veins filled with large calcite crystals, as well as chlorite phenocrysts. The blisters seen on the new areas were identified as gypsum, probably originating from mortar used in construction. An ionic exchange process caused crystallization on the surfaces of both serpentines.

Storemyr (1997) found serpentinite used as one of the construction materials in Nidaros Cathedral. It was comprised of 40–55% serpentine, along with 20–30% chlorite, 5–20% talc, 5–10% each dolomite and magnesite, and 5% biotite. There was also a wide variety of trace minerals present. Derived from the Bubakk quarry, it was described as bluish in color after exposure (although remaining relatively unaffected by salts), was massive, and had many small brown carbonate aggregates with a few thin veins.

Sculpture

In thin-section analyses of ancient Gandhara artifacts (Reedy 1992b), two objects were found to be carved from serpentinite. One object was a triangular relief panel, depicting a triton (ichthyocentaur), probably from the Peshawar valley in northern Pakistan. The stone had a medium gray coloring with a greenish cast to it. It was relatively fine grained, so the surface of the sculpture had a smooth and polished appearance. The primary constituent was the antigorite form of serpentine. Large chunks of magnetite comprised about 5% of the rock. Some of the smaller opaques may be chromite. Irregular, subparallel streaks of chromite marked a mild foliation of the rock.

Another panel carving depicting a group of marine figures was also made of serpentinite. The visual appearance of the stone was of a dark, very smooth surface. Black flecks were visible on the arms of some of the figures, and especially on the background panel between the figures. Along the bottom edge a thin top layer has flaked off in large areas, marking the foliation of the rock. Few tool marks showed on the surface, although there were some long, continuous parallel lines visible along the sides of ridged areas. In thin section this panel appeared to be very similar in material to the triton panel. Most of the stone was the antigorite form of serpentine. Chunks of black magnetite comprised approximately 10% of the stone.

While the source of Gandhara serpentinites is not known, some possibilities can be proposed. Although comparative quarry samples are not yet available, Rehman and Zeb (1970) reported that serpentinite rocks were found near Mingora in Swat. Magnetite and other iron oxides were found in association with the serpentine, sometimes as distinct aggregates and sometimes in small aligned granules. Serpentinites were also found in Lower Swat Kohistan, west of Alpurai village (Jan and Tahirkheli 1969), containing primarily the antigorite form of serpentine, with accessory iron oxides and carbonate. Serpentinite was also available in Afghanistan, in both Kandahar and Zabul provinces, where most of the sources are currently being mined by the local people (Shareq et al. 1977).

Other uses

Serpentinite has also been used for carving vases and bowls. At a Late Bronze Age temple at Amman, a number of stone vessels were found and their materials identified (Hankey 1976). No sources of raw materials for these vessels have been found in the area around Amman and it is thought that they were import materials brought in from Egypt and Crete. Included among these materials were serpentinite examples, in two varieties. One type was very fine grained, green to gray and mottled, able to take a smooth or glossy polish. It is of a type used widely in Egypt, and found from Predynastic times onwards as a vessel material. The other type was black with small white specks, and also polished to a high gloss. In thin section this appeared as a medium-grained serpentinite, consisting of pale brown, finely veined serpentine crystals up to 3 mm in size, forming as pseudormorphs after olivine, and filled with scattered granules of magnetite. Also seen were fragments of altered olivine in a network of colorless fibrous serpentine (identified with the aid of XRD analysis as lizardite) accompanied by scattered magnetite.

Serpentinite was also an important carving stone used in the Canadian Arctic by Inuit, or Eskimo, carvers (Gibbins 1988). The stone most often exploited here has a yellowish-green color. It consists mainly of serpentine, talc, and chlorite, often with relics of olivine and pyroxene.

In the Venezuelan Andes, serpentinite was carved into pendants. Thin-section analyses showed that the main mineral present was antigorite, with minor amounts of talc, magnetite, and calcium carbonates. Serpentinites were also important carving materials used by native peoples in the American Southwest region, and in Mesoamerica. For example, the Olmecs carved ceremonial axes out of green serpentinite, and the Aztec, Toltec, and Maya peoples carved a variety of small objects out of serpentinite. A talc-rich serpentinite was used to carve winged sculptures and pendants in the Venezuelan Andes (Wagner and Schubert 1972; Kempe and Harvey 1983).

Garza-Valdes (1991) performed thin-section petrography (among other analytical techniques) on pre-Columbian objects from Mesoamerica that had been identified as jade (a rock made of predominantly of the mineral jadeite, a sodium-aluminum silicate in the pyroxene group, or nephrite, a variety of tremolite-actinolite). Some of the objects studied were found to actually be serpentinite, most often comprised primarily of antigorite. Accessory minerals included olivine and magnesium pyroxenes. For example, a dark green pectoral was found to be serpentinite. Since this stone tends to weather easily, the object showed severe etching and white deposits, and was deemed authentic. Another seated figurine made of serpentinite and carved in Olmec style had no secondary mineral formations on the surface, just remnants of polishing compound; the object was identified as a modern carving.

Serpentinite was one of the stones quarried by the Romans in Egypt (Brown and Harrell 1995). It is described as a mottled yellowish green and dark green, with a fine-grained texture that looks medium to coarse grained because of the presence of large crystal aggregates. The primary minerals present are, in order of abundance: antigorite, clinochrysotile and lizardite, and dolomite. Accessory minerals are magnetite, talc, and tremolite. The rock is highly altered, with serpentine and other minerals replacing original olivine and pyroxene.

Serpentinite was used to make decorative axes in Slovakia during the Neolithic (Hovorka and Illášová 1995). Most axes of the region were made of other metamorphic rocks (amphibole schist or nephrite); those made of serpentinite differ in size and shape, and so are thought to be decorative or ceremonial rather than functional. Since this rock has a relatively low hardness compared with other rocks available, this hypothesis of ceremonial functions for the serpentinite axes is further supported. Thin-section analyses of samples taken from some of the axes showed the rock to be fine grained (less than 0.5 mm grain size), massive, and composed mainly of antigorite in bladed form with random orientations of grains of uniform size. Accessory minerals included talc, magnesium chlorite, carbonates and magnetite, and there were a few relic grains of clinopyroxene. Comparisons with possible sources indicated a local provenance.

Serpentinites were also used in Italy for the production of prehistoric axes or adze blades, as well as ornaments and a variety of tools (D'Amico and Starnini 2006). Sources were located in northwestern Italy, the central and eastern Alps, and possibly the Apennines. These serpentinites are usually characterized by the presence of antigorite or lizardite, magnetite, chlorites, and/or residues of calcium-pyroxenes, amphiboles, and rare olivine.

6

Pottery Product Types and Inclusions

Traditional ceramics are formed primarily of clays, usually with varying amounts of impurities or additives such as sand, calcium carbonates (shell or crushed limestone or marble), other crushed lithics, crushed ceramics, organic materials, or volcanic ash. When mixed with water, the clay can be shaped into various forms, then dried to retain that shape. Firing creates a harder and more permanent object. Pottery is by far the most common ceramic cultural material, and has been studied extensively, so three chapters (6–8) are devoted to it in this book. Pottery is usually defined as encompassing fired wares such as storage containers, cooking vessels, vases, pots, bowls, cups, or plates. Other ceramic or unfired clay-based materials such as sculpture and molded objects, architectural bricks and tiles, and casting cores for bronzes will be discussed in the final chapter (9).

Clay minerals, when mixed with the appropriate additives, have the property of being plastic when wet so they can be formed into shapes without cracking or breaking, and will retain the shape after the forming process. They harden as they dry, without losing shape and ideally without cracking, and should be strong enough in the dry state to be handled without deforming. They then harden permanently when fired, creating utilitarian products (Ryan 1978). Ceramic products have sometimes been termed "artificial rocks" because of the unique processes and alternations the raw materials undergo (Echallier 1983).

Clay minerals are sheet (layer) silicates, as are micas, but their crystals are microscopic and can usually only be observed visually with the aid of an electron microscope. Therefore, mineral identifications in thin-section petrography of pottery focus mainly on larger grained natural contaminants to clay or on additives, or on evidence relating to technology of production. When clays are heated to high temperatures, that energy causes loss of chemically combined water within the crystal structure, forming new high-temperature minerals, and the material becomes permanently hardened. Some of these new minerals may be identifiable in thin sections.

Clay minerals form as a result of weathering and decomposition of silicate rocks as they interact with air and water on the earth's surface. Different parent rocks and different decomposition processes produce different types of clays. Primary (residual) clays are formed by natural chemical action on rocks such as granite, basalt, limestone, or shale. These clays tend to be relatively pure and are found in or near their place of origin where the parent rock broke down. They tend to be low in plasticity, coarse, and full of small fragments (such as quartz, feldspars, and micas) from the parent rock. Clays may also be secondary (also called sedimentary clays). These clays are transported by water, wind, or glacial action to a site away from their original place of formation. As they travel, they pick up new impurities which affect their properties. These clays tend to be more plastic than primary clays due to the greater fineness of particles caused by friction during transport, and to the addition of organic matter picked up along the way. The original parent rock impurities are usually lost (Goffer 1980; Henderson 2000).

Kaolinites are one of the most widely occurring and extensively used clays. They can be found as primary clays, formed by an advanced stage of weathering of granitic rocks high in feldspar and quartz, or of micaceous schist. The clay mineral kaolin results, a mineral that forms flat, hexagonal plates, and is high in alumina (Al_2O_3) as well as silica (SiO_2). They are often found in warm regions with high rainfall and good soil drainage, leading to acid leaching that removes most elements, especially bases such as calcium, magnesium, iron, sodium, and potassium, thus leaving the clay enriched in silica and alumina. Primary kaolinite clays tend to be relatively coarse, therefore they are not as plastic as many other clays. They are often free of iron impurities, so are white when fired, and have a high firing temperature (i.e. they are refractory). A famous use of kaolin is in Chinese porcelain: this clay was named after the village of Gaolin in Jianxi province of China where deposits were first exploited for the production of porcelain (Qingzheng 2002). Kaolinite clays can also occur as secondary clays, in which case they become more plastic and pick up new impurities as the original parent rock fragments are lost.

Smectite clays, such as montmorillonite, are not as highly weathered as kaolinite. Additional alteration, especially leaching from high rainfall, high temperatures, and good drainage, will transform smectites into kaolinites. They are less refractory (i.e. they will melt at lower temperatures) than kaolinite clays. Their small particle size and a tendency to adsorb water between layers results in clays with high shrinkage that crack as they dry. But, their expanding structure with capacity to adsorb ions is good for slips or paints.

In illite clays, about one-sixth of the silicon is replaced by aluminum, leading to a charge deficiency balanced by potassium ions (K^{1+}) and sometimes by Ca^{2+}, Mg^{2+}, and H^{1+}. The charge deficiency is in the outer silica layers close to the surface rather than in the interior alumina layers as in smectites. Thus these clays are nonexpanding and have poorly defined flakes of small size. These clays are often found in marine deposits; in an alkaline marine environment, other clays may form illites.

Clay minerals, as sheet silicates, have a plate-like structure combined with a very small size. Because of their structure, they have a large surface area and plasticity results from the interaction of clay and water on the surfaces. The finer the particle size of clay the greater the plasticity, since the fine size means a larger number of particles, and thus surface area, is present. As the wet clay is dried, the water in the pores between particles is removed through capillary action. However, the fine particles, while good for plasticity, will have a finer capillary network, causing water to evaporate from the surface faster than it can be supplied from the interior. The result is uneven shrinkage, a strong moisture gradient, warping, and cracking. Cracks may be invisible until firing. One solution is to ensure the presence of a fraction of coarse-sized particles (either as a natural accessory to clay or as an additive, called "temper") to increase the pore size and promote water movement from interior to exterior. Facilitating the drying process also prevents trapped pore water from suddenly bursting out during firing, damaging the vessels (Goffer 1980; Rice 1987). Gibson and Woods (1997) pointed out that the need for a coarse component is especially important when open firing instead of kiln firing. Open firing leads to a less controlled, rapid temperature rise and closer contact between ceramic and fuel, so a structure that allows steam to escape easily as firing begins prevents damage from occurring.

Most raw clays contain over 50% impurities (such as sand, organic material, talc, micas, and iron oxides), thus need to be processed (purified) to increase plasticity. Clays should contain, or have added as temper, inclusions that are somewhat angular and in a range of sizes, in order to correct problems of stickiness (of too pure, fine clays), to improve working properties (of drying, shrinkage reduction, and firing), and improve functional qualities (such as heat resistance). For example, some clays absorb water rapidly, making them sticky and unmanageable for the potter. It is difficult to achieve the best consistency for plasticity with such clay—it may not hold a shape well or have a long enough working life for fabrication steps to be completed. Other clays have a tendency to dry out too quickly, and crack during working. The addition of a temper material helps to counteract both of these tendencies and create a material that best meets the working needs of potters (Birks 1998).

The best additives also depend on the natural components found in the clay and the intended firing regime. For example, clay with quartz-rich sand requires a relatively high firing temperature to fuse and form a hard ceramic material. The production of low-temperature pottery (700–900 °C) requires alkali fluxes to help lower the required melting temperature. Feldspars, calcium carbonates such as calcite, and calcium oxide (lime) are typical fluxes (Velde and Druc 1999).

In addition to the natural inclusions found with clay deposits, and the changes to that material which result if temper is added, there are many steps in processing a clay and in fabrication of an object that can change the appearance and properties of the material. For example, after clay is quarried, it may need to be purified by removing stones and other contaminants; this may involve drying, crushing, and sieving the clay. Levigation is a process that uses water-filled basins or a series of traps to allow large particles to settle out while the finer ones remain in suspension.

A variety of physical techniques of clay preparation then help eliminate air bubbles and excess water, to prevent damage in later drying and firing stages. These techniques may include wedging (repeatedly cutting the clay with a wire and pressing it together), kneading, foot-treading, folding, and pressing. If a clay is too stiff, plasticity may be increased by adding substances such as yogurt, beer, starch, or acids like vinegar; alternatively, acidic or bacteria-containing ingredients may be added and the clay allowed to age until the right consistency is reached (Henderson 2000).

Many fabrication methods are possible. Hand-building methods include thumb-pinching clay to form vessels, slab-building using a roller on a flat surface and then joining the slabs with a clay slip, or coil-building and then joining the coils. Molded vessels are made by pressing wet or leather-hard clay into a mold, using a release material such as ash, sand, or powdered clay so that the molded vessel can be extracted for drying and firing. Wheel-built vessels are shaped on a turntable, with one hand turning the support while the other shapes the vessel. Wheel-thrown pots are shaped using a wheel that can be turned mechanically, such as by the feet, thus freeing both hands to shape the pot (Henderson 2000).

Surface decoration can be achieved by smoothing, burnishing, or roughening the surface with various tools; scraping and trimming with a knife or other tool to remove imperfections and irregularities; creating surface decorations with various tools such as ropes, combs, stamps, roulettes, bone or shell pieces, or many other possible tools; adding clay to form appliqué decoration; adding slips before firing (liquid suspensions of fine clay particles often of a different color than the vessel itself) or washes (a thin pigment or lime-based solution usually applied to an already fired pot). Glazes are made with powdered glass prepared using the same colorants and firing conditions employed to color glass vessels. A binder (such as animal glue) and wetting agent (such as water) are used to prepare the powdered glass in liquid form—it can either be painted on the vessel, or the vessel can be dipped in the liquid glaze (often after an initial, or bisque, firing to harden the clay, make it stronger, create a better surface for the glaze, and eliminate shrinkage for the glaze firing). The vessel is then fired at a high enough temperature to form a glass over the surface, which tends to fuse into the clay body, providing a long-lasting decorated surface. Paints, enamels, and inks may be added under, in, or over the glaze, or sandwiched between two glaze layers.

Thin-section petrography of pottery and other ceramic materials draws its epistemological basis from sedimentary petrology, and when applied to cultural materials can be used to explore features of microstructure that relate both to the original geological deposit and to the processing steps (Krishnan 2002; Krishnan and Shah 2005). In addition to identification of minerals and lithics, ceramic petrographers often try to discern the presence of natural versus added components. A variety of features is present in ceramic materials solely because of processing steps, and their interpretation differs from interpretation of natural sedimentary materials.

Fabric characterization utilizes a range of descriptive criteria to arrive at a complete description of the microstructure of the material, including features from the original raw materials and any modifications made during processing and pottery fabrication steps. Often this analysis involves visual characterization of thin sections into different fabric groups. Some of the characteristics that may be recorded in a fabric characterization include appearance, color, and degree of birefringence of the clay matrix; features of nonplastic inclusions such as mineralogical composition, shape, sphericity, size range, sorting, homogeneity, and orientation; and number and shape of voids or pores (Krishnan and Shah 2005). Sometimes characterization, description, and comparison is the primary or only goal, however, more often the characteristics recorded are used to make inferences about sources of raw materials, processing and vessel fabrication steps, aspects of technological style, or deterioration mechanisms and preservation implications.

There is a wide variety of research questions pertaining to ceramics that can be addressed through thin-section petrography (L. D. Jones 1986; Freestone 1995; Whitbread 1995; Velde and Druc 1999; Stoltman 2001). Some of these are specific to pottery vessels rather than other forms of ceramic materials, so architectural materials, figurines and sculpture, and other nonvessel ceramics are discussed separately in Chapter 9. Riley (1982) provides a good discussion of the importance of having both the archaeological and the petrological assumptions clear and explicit so that petrographic analysis is clearly related to the questions of interest. Some of the research questions that ceramic petrography may tackle include:

- Identifying and describing the type of ceramic product.
- Characterizing and conducting provenance studies on inclusions.
- Identifying clay selection and processing choices.
- Reconstructing forming methods.
- Inferring firing conditions.
- Deducing intended function and identifying evidence of use.
- Interpreting voids and porosity data.
- Studying decoration and surface/clay body interfaces.
- Exploring evidence of technological choice, variation, and change.
- Studying deterioration products.

Thin-section petrography as it relates to these questions in pottery studies will be explored in this and in the following two chapters.

6.1 Characterizing Ceramic Types

On occasion, simply describing the type of ceramic product is useful for addressing a variety of questions about the production and function of pottery. In addition, sometimes the type of ceramic body or product delimits the types of features that can be observed or the range of research questions that may be appropriate. Major ceramic types discussed here include earthenware, stoneware, porcelain (hard and soft paste), and stonepaste/fritware.

6.1.1 Earthenware

Earthenware is porous and low fired (usually around 800–1100 °C), and water-permeable. There is often a high proportion of nonplastic constituents, either natural or deliberately added, which creates a high degree of porosity (Fig. 6.1.1). The addition of fluxes such as calcium oxide can facilitate melting, and help create a more durable material. There can be a range of colors (generally red to cream), and the clay is more or less unvitrified, although earthenwares with a more intermediate firing temperature may be denser and partially vitrified (Fig. 6.1.2). Earthenware may be slipped or glazed,

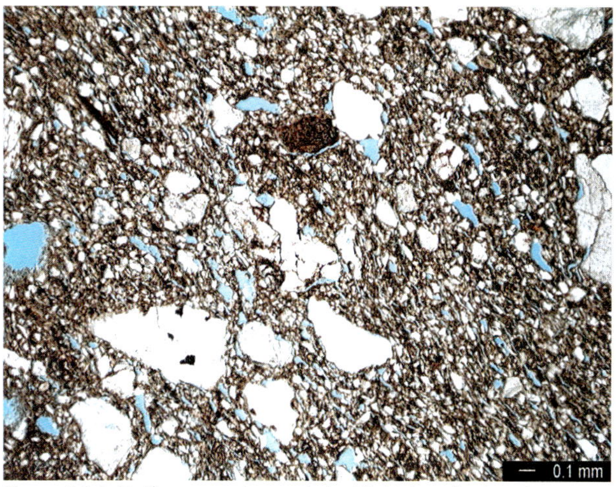

Figure 6.1.1 Earthenware, plane polarized light. Low-fired earthenwares tend to have a high proportion of nonplastic constituents, either natural or deliberately added, which creates a high degree of porosity. This example of a low-fired ware from China shows high porosity (made visible by the blue mounting resin) and many nonplastics within the clay matrix.

however, the bond between a glaze and body is often poorly developed, looking like distinct layers when sectioned.

Sometimes the term "terracotta" is used to refer to unglazed low-fired bodies made from clay that fires to red or cream (usually at or below 900 °C). It tends to be coarse and porous (Fig. 6.1.3). Much of the earliest fired pottery in all areas of the world fall into this category. Terracotta may be covered with a slip, but is usually not glazed as glazes do not adhere well to this mainly unvitrified ceramic material. However, the terms "terracotta" and "earthenware" overlap, and distinguishing between the two is often impossible. Terracotta is sometimes used to refer to earthenware nonpottery ceramics, such as architectural materials (tiles) or sculpture.

The thin section in Figure 6.1.4 is American, from Virginia, dating to 1720–45. The clay is very red, with some large chunks of hematite. It exhibits two populations of quartz size: large ones from the sand temper additive and small ones that are impurities of the clay. The long elongated and tapered pores are shrinkage cracks appearing from release of gases; many of these were induced by the burning off of organic material, which also left reduced areas around some of the pores. The somewhat parallel orientation of the elongated pores and the dark stringy areas, imparting a laminated appearance to the clay, is an indication of pressure applied in fabrication, probably from wheel throwing.

Another earthenware sherd (southern Delaware, 300–800), seen in plane polarized light in Figure 6.1.5, has a large-grained sand component mixed into a very iron-rich clay with reddish color. Oxidized firing conditions that burned out most of the abundant organic matter enhanced the

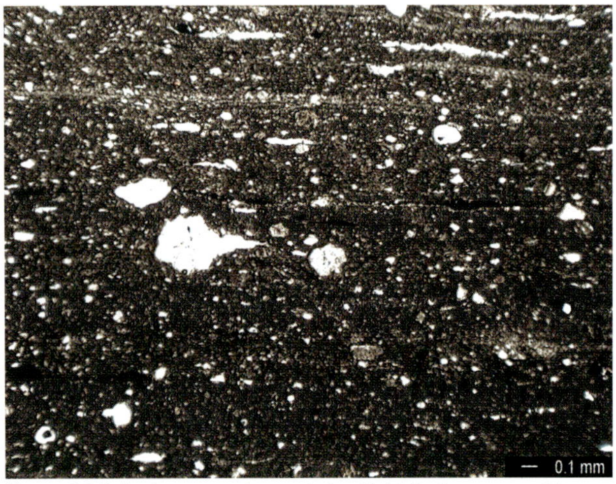

Figure 6.1.2 Earthenware, plane polarized light. Earthenwares with a more intermediate firing temperature may be denser and partially vitrified. This English earthenware (from 1675–1725) has less porosity and fewer nonplastics, with a higher proportion of clay in the matrix, than the example in Figure 6.1.1.

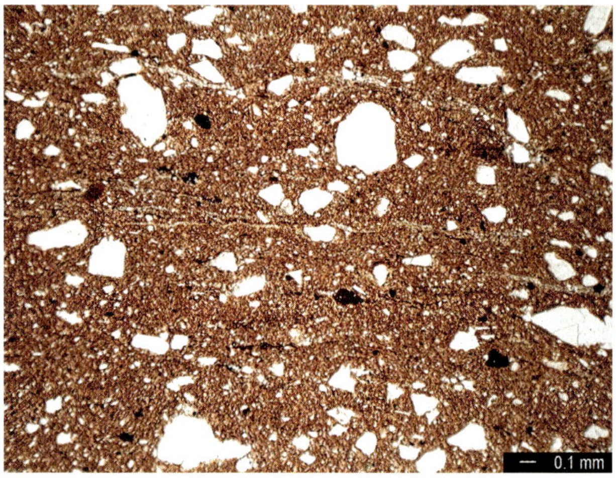

Figure 6.1.4 Earthenware, plane polarized light. This earthenware example is from Virginia, 1720–45. The clay is very red, with some large chunks of hematite. There are two populations of quartz (some large ones mixed in with the small ones that are original to the clay), and long narrow elongated pores (probably shrinkage cracks from release of gases). The somewhat parallel orientation of pores and dark stringy hematite-rich areas of the clay give a laminated appearance, probably from pressure applied during fabrication.

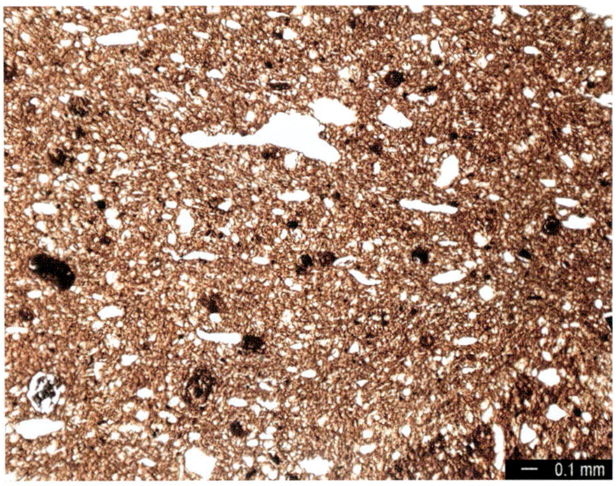

Figure 6.1.3 Terracotta, plane polarized light. The term "terracotta" is sometimes used to refer to unglazed low-fired earthenwares made from clay that fires to a reddish color, as here. It tends to be rather coarse and porous. Many of the pores are somewhat elongated, and deep reddish to black chunks of hematite are visible.

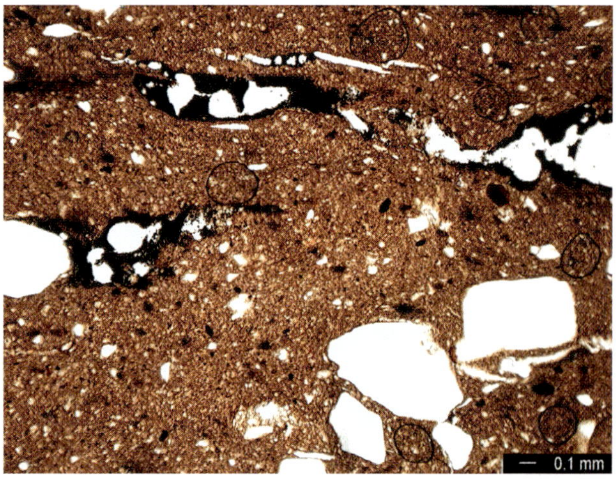

Figure 6.1.5 Earthenware, plane polarized light. This earthenware example (southern Delaware, 300–800) has a large-grained sand component, and an iron-rich clay matrix. Oxidizing firing conditions that burned out most of the abundant organic matter enhanced the orange-red color of the iron oxides. Some organic matter remains in place as a charred powder.

orange-red color of the iron oxides. Some of the organic material remains as charred powder.

Figure 6.1.6 shows a thin section of an English earthenware mug, dating to the sixteenth century. There is also much quartz present, which is of a relatively large size. The color of the matrix is more of a brown than a red, and the piece represents the higher end of earthenware firing temperatures. The material is less porous than some of the other examples, and has been able to take a glassy glaze over the surface. However, the glaze has spalled off in some areas, leaving only interrupted segments of it in place over a very thin clay slip.

Although many examples of earthenware are dark brown or reddish, iron-rich clays, white unglazed earthenwares were made in China as far back as the thirteenth to eleventh centuries B.C. (Harrison-Hall 1997b). The clay used was secondary kaolinite, derived from primary kaolinite deposits. These deposits lie underneath a layer of loess, and are relatively free of mineral impurities such as iron. Sometimes slips were used to create a pristine white surface appearance. In the European example illustrated (Fig. 6.1.7), the earthenware is fine grained, not too porous, and light brown in plane polarized light (while being a somewhat "dirty" white in hand sample). It has scarce iron-oxide minerals and carbon, and the quartz component is fine grained and relatively uniform in size. Because of the presence of a relatively high percentage of nonplastic inclusions in earthenware, many examples of this material are discussed in section 6.2 below, which focuses on characterizing and interpreting these inclusions.

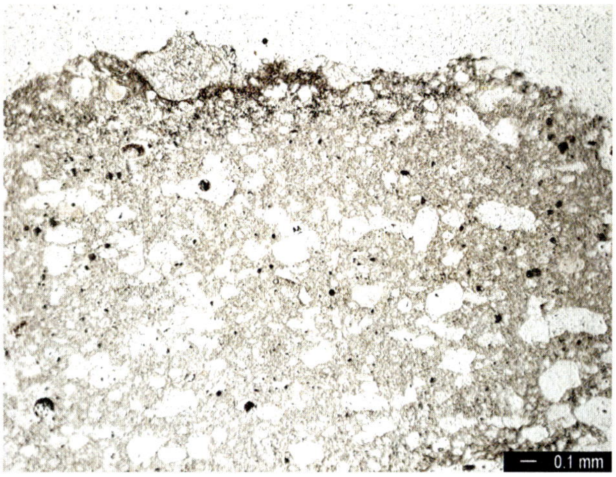

Figure 6.1.6 Earthenware, plane polarized light. This English earthenware mug, dating to the sixteenth century, has an abundant quartz content. However, the clay matrix is not very porous, and has a brown rather than red color. A glassy glaze was applied to the surface, but it has spalled off in many places, leaving interrupted segments of glaze on top of a thin clay slip.

6.1.2 Stoneware

Stoneware is typically fired at temperatures between 1100 and 1300 °C. It tends to be off-white, cream colored, gray, tan, or light brown. Although earthenwares grade into stonewares, stoneware is usually considered to be denser and stronger, more vitrified, and more impermeable to water than earthenware (Fig. 6.1.8). When glazed, the glaze and body tend to be fused (Fig. 6.1.9). The clays used for stoneware often contain, naturally or as additives, some fine quartz and minerals such as alkali feldspars that can serve as a flux. The raw materials generally contain fewer iron oxide minerals than are usually seen in earthenwares, consisting mainly of the clay, micas, feldspars, and quartz, so the finished product generally has a light color. Vitrification produces clay bodies with a high density and low porosity. Stonewares are often known for their strength (Henderson 2000).

Deposits of clay ideal for making stoneware were prized, and in most places are probably more rare than deposits suitable for making lower fired wares. For example, some nineteenth-century New Jersey potteries thrived due to their

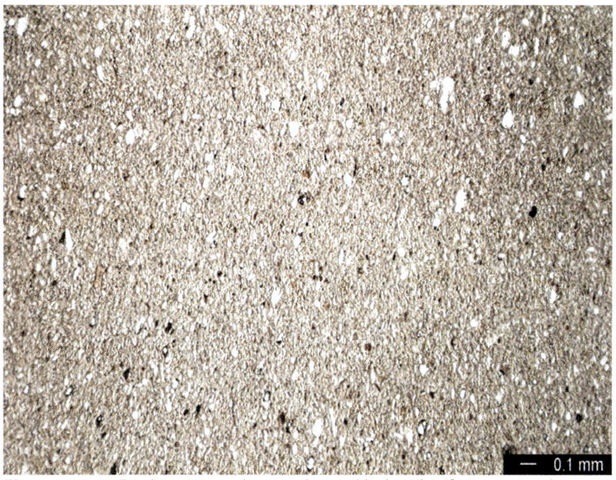

Figure 6.1.7 Earthenware, plane polarized light. This finer grained earthenware has a whitish color in hand sample, and in plane polarized light the clay is light brown and not very porous. There is a relatively fine-grained quartz component of fairly uniform size. Only a small amount of iron oxides and carbonized organics are visible. This earthenware is on the high end of firing temperatures, grading into stoneware.

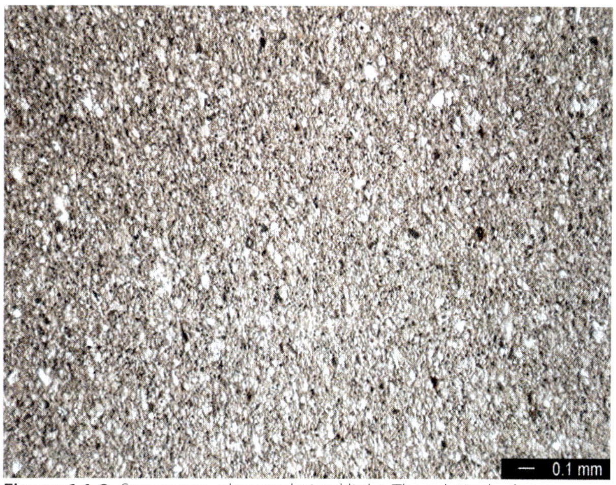

Figure 6.1.8 Stoneware, plane polarized light. The relatively dense material of this stoneware has a clay matrix that is partially vitrified. In hand sample, the material is off-white, as it contains fewer of the iron-oxide minerals that are often typical of earthenware. There is actually a gradation between the higher fired, cleaner earthenwares and the lower fired stonewares, rather than an abrupt line of difference.

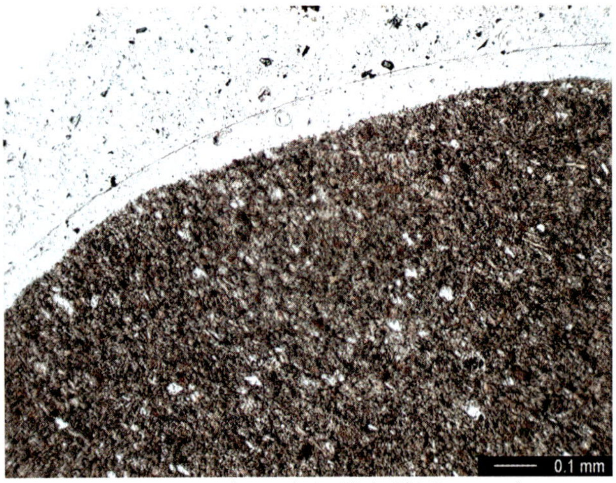

Figure 6.1.9 Stoneware, plane polarized light. This "ironstone" stoneware plate has a dense clay matrix that is undergoing vitrification. Fine-grained quartz is scattered throughout. Along the edges of the plate a thin, clear glaze layer is well bonded to the clay surface. A very thin dark line marks the outer edge of the glassy glaze.

proximity to the Morgan clay bank. Potters from Maine to Georgia imported and used "Morgan clay" because it was thought to make more durable and attractive stonewares. The gray clay bodies were usually salt glazed and decorated with surface designs (incised, molded, stamped, slip trailed, coggled, or brushed on with colorants of cobalt, manganese, or iron oxides) (Racine 1997).

Schneider (1989a) examined 53 thin sections of northern Mesopotamian stoneware dating to about 2700 B.C. He had encountered two very different stoneware products—one group made with noncalcareous raw materials, and a smaller group made with calcareous clays—so was comparing these two products. He noted that minerals visible in thin sections of the ceramics may be original to the raw material used, or may have formed during firing. Under lower firing temperatures, micas and feldspars could still be seen (Fig. 6.1.10). The calcareous stonewares had rhomohedric pores visible, resulting from the decomposition of carbonates during firing, and the noncalcareous stonewares seemed to have been made from very fine clay that naturally contained mineral and rock inclusions of about 30–40% by volume. He identified the presence of abundant quartz and plagioclase feldspars; the latter were partly melted in the more vitrified samples. Small amounts of micas and mullite were also found, along with microcline feldspar in one of the two groups. The other group had a few sherds with cristobalite (a high-temperature form of silica, Fig. 6.1.11) and hercynite (an iron aluminum oxide member of the spinel group). It was noted that modern stonewares also contain mullite, which increases strength of the ceramic, and cristobalite.

The three main crystalline forms of silica are quartz, tridymite, and cristobalite. Although at temperatures above 573 °C quartz inverts from the alpha (low-temperature) form to the beta (high-temperature) form (Herz and Garrison 1998), upon cooling the beta quartz inverts back to the alpha form, so that all such quartz seen in thin sections is alpha quartz (Kerr 1977). Quartz is the stable form of silica up to about 870 °C. Above that, high-temperature forms of silica (cristobalite and tridymite) will begin to form, if the firing temperature is held constant for a long enough period of time. Both of these minerals are also seen in volcanic rocks. In thin section, cristobalite has extremely weak birefringence, and may show a mosaic structure. It occurs as minute square crystals or aggregates with moderate relief. Tridymite usually occurs as minute crystals that can be six-sided, thin, tabular, with moderate relief, and often twinned; it may also occur as a porous crystalline aggregate, and wedge-shaped, twinned crystals are characteristic. Birefringence is very weak (Kerr 1979; MacKenzie and Guilford 1981). If high temperatures are sustained for a sufficiently long period of time, these silica phases can melt, along with the feldspars, and form glassy (amorphous, noncrystalline) areas. However, the changes from one form to another are sluggish, and may require the presence of various impurities and the maintenance of higher temperatures for some time;

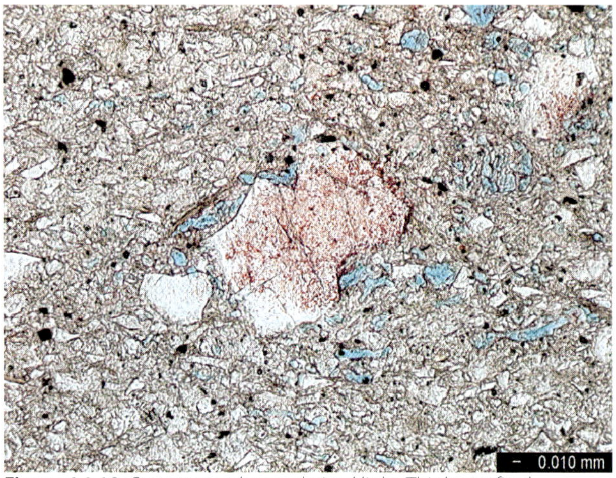

Figure 6.1.10 Stoneware, plane polarized light. This lower fired stoneware contains some large grains of feldspar; the red stain here helps to identify plagioclase feldspar. The blue areas are voids, which show the blue-dyed epoxy below, indicating areas of porosity. Some fine-grained tabular brown micas (chlorite) are also visible within the clay matrix, along with scattered black grains of magnetite.

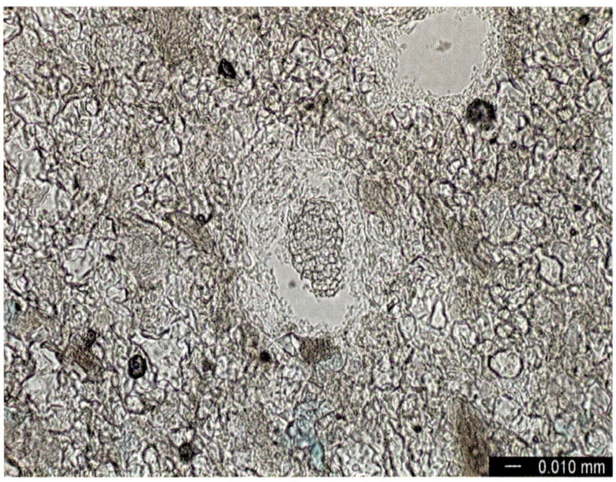

Figure 6.1.11 Stoneware (Guan-glazed celadon), plane polarized light. In the center of the image, a small area of silica glass (where quartz has melted) has an aggregate of minute squares (cristobalite) in its center. Another round silica glass area is seen to its upper right. Blue areas are pores which reveal the blue epoxy below. The matrix is comprised mainly of partially vitrified clay, pockets of tiny mullite needles, small grains of quartz, and a few scattered tabular grains of sericite.

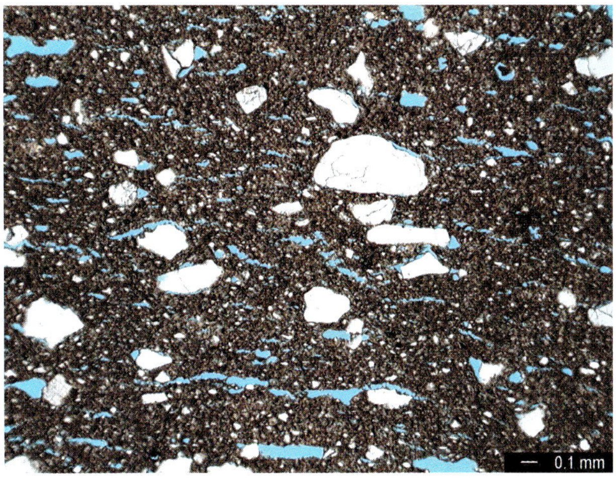

Figure 6.1.12 Stoneware (German, seventeenth century), plane polarized light. Historical evidence indicates that in the Rhineland, raw clays were refined then sand was added as a temper. As a result, the vitrified clay contains a uniformly fine-grained sand, primarily quartz. The large-sized sand additive is mainly quartz (including some that is polycrystalline) with a lesser amount of plagioclase feldspar. This example is quite porous (blue areas).

reverse changes on cooling occur very slowly as well. Rates of conversion from one form of silica to another are also affected by the fineness of the material (Ryan 1978).

German stoneware was an important product during the thirteenth to nineteenth centuries (Gaimster 1997a, 1997b). Large amounts of it have been excavated and found as trade items in Europe, Britain, and America. New kiln designs capable of reaching temperatures of at least 1150–1250 °C, along with ample wood supplies, enabled the industry to develop. In the Rhineland, raw clays were refined, and sand was added as a temper (Fig. 6.1.12). Objects were made on a fast wheel, in one action, with handles, spouts, and other separate pieces added at the leather-hard stage. Stoneware clays tend to be highly plastic, therefore a good material for use in molding and applying relief ornaments, which became a standard method of decorating German stoneware products. German stoneware was used for a variety of functional vessels such as those for storage, serving, and drinking, as well as for water pipes and architectural tiles. In the early sixteenth century, salt glazing was developed; prior to that ash glazing was common.

Doherty and Maske (1998) studied stonewares made in the Takatori kilns of Japan (Chikuzen province, Kyushu) during the seventeenth to nineteenth centuries. They found that the clay bodies had undergone significant vitrification, resulting in the formation of secondary minerals and "bloating" pores (due to the strong fluxing effect of the iron-rich materials). Relict primary minerals included quartz, feldspars (highly corroded), and magnetite-ilmenite. Smaller quartz grains were sometimes converted to cristobalite along the edges, but many larger quartz grains were unaltered, with euhedral shapes (typical of acid or intermediate igneous sources outcropping in Kyushu). Ferruginous clay pellets were also seen within the clay matrix.

Celadon, made in China and Korea, is a renowned stoneware product. Celadons are characterized by the use of a green glaze over a gray or whitish stoneware body, giving a jade-like appearance. Some celadons were made with a porcelain body, or with stoneware that grades into porcelain. This is especially true for celadon-glazed wares that are thinner, denser, and whiter than some other celadon stonewares, so some of the mineralogy encountered is similar to that for porcelains. However, celadons are often more porous than most porcelains. Because the visual effect achieved by the celadon glaze and body combination can occur with materials that may be either stoneware, porcelain, or something in between, celadon is sometimes called "porcellaneous stoneware." The ceramic bodies may contain, in addition to some glass, varying proportions of quartz, mullite, and cristobalite, and rare muscovite (or sericite), biotite, or feldspar (Guozhen and Lingxiang 1986; Koh Choo et al. 1999). Celadons are sometimes classified as porcelain rather than stoneware if mullite crystals are present (Kerr and Wood 2004). Although much information about glaze composition is obtained through elemental analysis and/or high-magnification scanning electron microscopy (SEM), some aspects of glaze technology can be viewed in thin sections (for example, number and distribution of bubbles, presence of unreacted quartz, concentration of anorthite crystals, and the thickness and evenness of the glaze layer).

Kingery and Vandiver (1986) discussed a Song dynasty Longquan celadon jar in detail. A modeled dragon on the jar had randomly oriented porosity, indicating that very little extension of the clay took place during modeling, contrasting with the groups of pores in the ceramic body itself, aligned in an upward spiral direction as a result of the force required to raise the form. At the base, where there was little extension, no pore alignment was seen. The surface pattern was incised by cutting away clay, with the incisions visible in cross-sections. The glaze was thicker in the more deeply incised areas. Many tiny mullite crystals were found in the matrix, some larger mullite crystals forming as sericite and feldspars melted (Fig. 6.1.13), and some partially dissolved

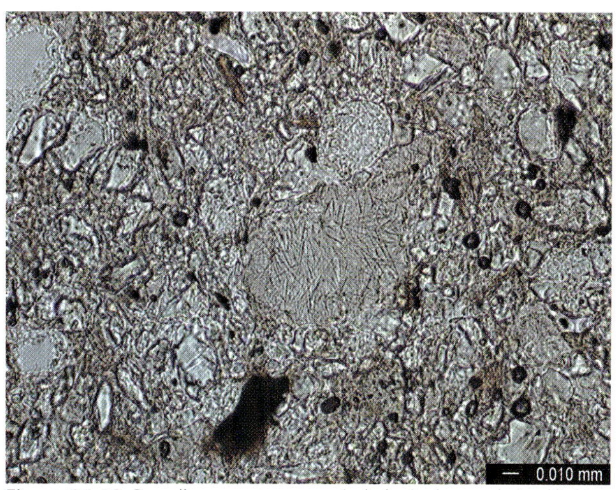

Figure 6.1.13 Porcellaneous stoneware (celadon), plane polarized light. The large feldspar grain in the center of the field of view in this celadon stoneware has melted to form glass plus needle-shaped mullite grains crisscrossing over the glass area. Above this area is another round bubble of glass. The matrix is heavily populated with glass and tiny mullite grains. There are also remnants of unmelted quartz, and some tiny tabular micas (sericite and chlorite). One larger dark and smudgy spinel is located below the central grain, with a few smaller ones scattered throughout the matrix.

quartz remnants surrounded by glass (Fig. 6.1.14). In the glaze, fine needle-like anorthite and rounded or chunky wollastonite crystals formed, and some residual quartz particles remained.

Guozhen and Lingxiang (1986) illustrated thin sections of Chinese celadon glazes, showing needle-like anorthite crystals developed in the glassy glaze, ranging up to about 40 μm in length (Fig. 6.1.15); these crystals are also seen in most celadons at the body–glaze interface. The authors also illustrated a thin section of a Longquan plum-green glaze, with many bubbles in the glass which varied widely in size from about 20 up to 130 μm. Their Guan ware example had the acicular anorthite crystals, residual quartz particles, and bubbles (Fig. 6.1.16). In contrast, a Yinqing glaze had no crystals or bubbles developed within it. As a result, the glaze was transparent with a high surface gloss, clearly showing the carved low relief designs on the ceramic body. Finally, an imitated Guan ware (from the Qing dynasty) showed, in thin section, an interlayer between the body and glaze where anorthite and mullite crystals had developed and interlocked the two layers.

Gas bubbles in the celadon glazes (Fig. 6.1.17) can be used to deduce the presence of a firing atmosphere at a temperature where the glaze became too viscous for the bubbles to escape (Henderson 2000). By the time the high-temperature glaze was heated to the maximum in a particular firing regime, gases released by components of the ceramic body and glaze would have ceased, and the main contribution to gas released would then be from the kiln atmosphere itself. Zhongtang (1992) demonstrated that the size of the gas bubbles increased by the time of the Song dynasty while their density decreased.

Yue celadons, made in southern China, contained iron oxide which turned the ceramic body gray upon reduced

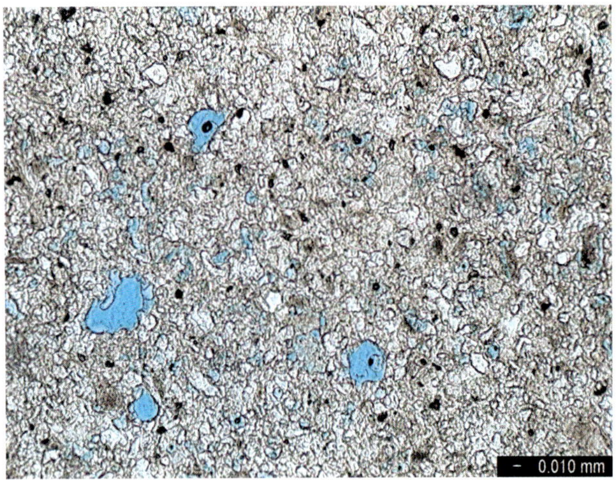

Figure 6.1.14 Porcellaneous stoneware (Longquan celadon), plane polarized light. The clay matrix in this Longquan celadon has broken down into a network of fine mullite grains, with some glass. Many of the quartz grains are partially dissolved. The blue areas highlight the areas of porosity.

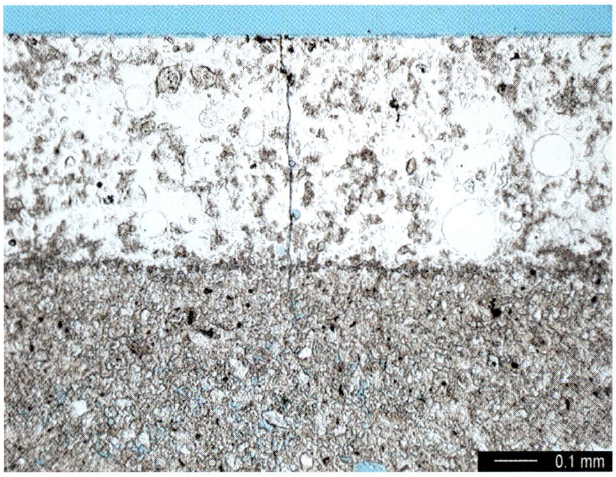

Figure 6.1.16 Porcellaneous stoneware (Chinese celadon), plane polarized light. This celadon Guan glaze shows extensive crystal development. The fine brown areas are clusters of acicular anorthite. There are also many clear round bubbles of varying sizes, and small clear residual quartz particles. The crack in the glaze extends from the surface all the way into the clay body.

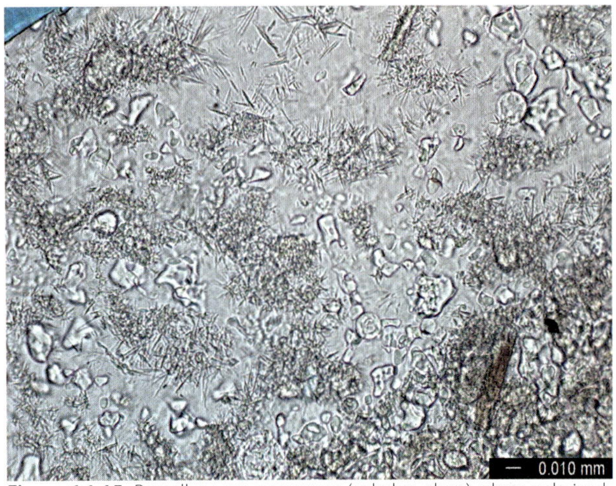

Figure 6.1.15 Porcellaneous stoneware (celadon glaze), plane polarized light. Many clusters of acicular anorthite appear throughout this celadon glaze. There are also a few small round glass bubbles, some low-relief anhedral quartz remnants, and some small, higher relief round or chunky precipitated wollastonite grains (actually pseudowollastonite).

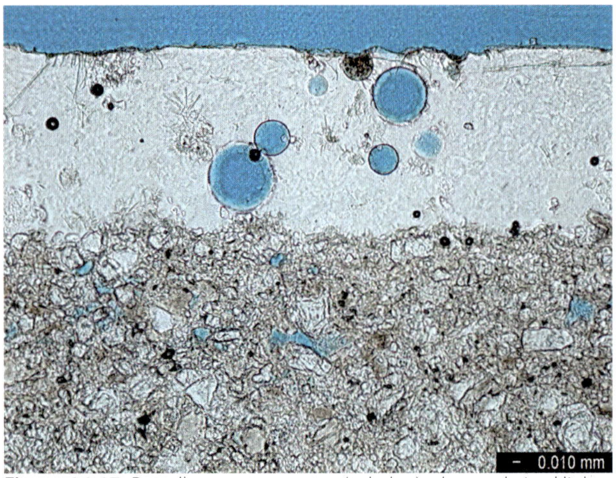

Figure 6.1.17 Porcellaneous stoneware (celadon), plane polarized light. Gas bubbles in a celadon glaze are a result of the firing regime; the glaze was too viscous for the bubbles to escape. A few bubbles incorporating acicular anorthite are also visible in the glaze, and the somewhat porous ceramic body has many clear quartz grains.

firing. Firing temperatures were relatively low in comparison with northern Chinese celadon wares. Celadons of the Koryo dynasty in Korea, while inspired by Yue celadons, used very different materials and technology to achieve the same appearance (Vandiver et al. 1989a). Different proportions of silica, natural fluxes, and alumina are also found in various Chinese celadon wares from different dynasties (for example Yaozhou wares versus Longquan wares).

Considerable variations are found within Korean celadon wares, as well. For example, some kiln sites retained good control in choice of raw materials, while others had more experimental operations with varying raw materials (Koh Choo et al. 1999). Visible variations included differences in uniformity and fineness of the ceramic body: those with uniform texture have scarce pores that are very small and are distributed uniformly, whereas other wares have more abundant and larger pores. Sometimes the pores are long and thin in shape, while in other cases they are rounder. Some have larger inclusions of remnant quartz grains or black iron-rich minerals, whereas others are more refined and are free of such inclusions. Wares also differ in the thickness of the glaze layer, ranging from thin (less than 200 μm), medium (200–400 μm), or thick (more than 400 μm).

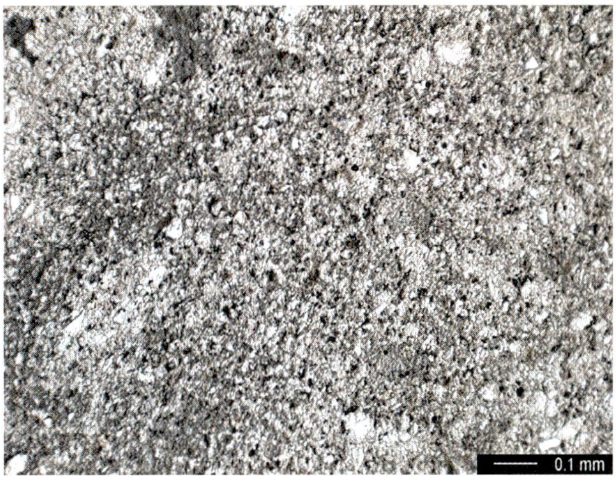

Figure 6.1.18 Porcelain (hard paste), plane polarized light. Porcelain fires to a hard, dense, relatively nonporous material that is white in color. This example of hard-paste porcelain (from a modern Korean porcelain piece) was mounted on blue epoxy, but is so dense and nonporous that no blue is visible.

6.1.3 Porcelain

Hard-paste porcelain

First developed in China (possibly during the Han period 206 B.C.–A.D. 220), hard-paste porcelain is also called "true porcelain." One definition of hard-paste porcelain is that it is made from naturally occurring ingredients as opposed to soft-paste porcelain that includes some synthetic ingredients (such as glass or a glassy prepared frit) (Owen 2001). Porcelain fires to a hard, dense, nonporous, translucent white ceramic body that is usually glazed (Fig. 6.1.18).

Chinese porcelain is often a mixture of white-firing kaolin clay plus a material known as "porcelain stone" or "China stone." Porcelain stone, often kaolinized itself, is a mixture of quartz and very fine white mica (sericite or hydromica) with some feldspar, although the exact composition and proportions vary (Fig. 6.1.19). The sericite improves plasticity of the kaolinite, and serves as a flux during firing; this porcelain stone material may be used alone or, for improved firing properties, mixed with additional kaolin (Qingzhen 2002; Kerr and Wood 2004). Similar materials were also exploited in Japan (Fig. 6.1.20) and elsewhere (Impey 1996). The material is fired in reducing conditions at higher temperatures than most stonewares, although at lower temperatures than most soft-paste porcelains.

At high temperatures and extended firing periods, many small needle-like mullite grains form from recrystallization

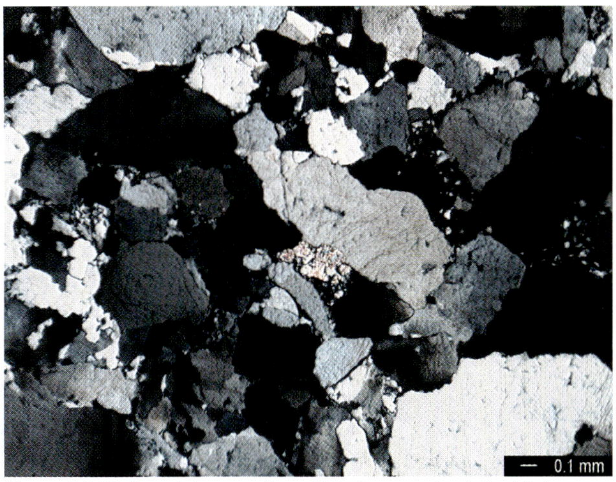

Figure 6.1.19 Porcelain stone (from China), crossed polarized light. Porcelain stone deposits are found in many areas of China. It contains mainly quartz, sericite, feldspars, and kaolinite. In this example, the small high birefringent flakes in the center of the field of view are sericite. The large low birefringent grains that make up most of the slide are quartz (some are polycrystalline or microcrystalline) and feldspars. Fine-grained clay occurs as weathering products on the feldspars and between grains.

Figure 6.1.20 Porcelain stone (from Japan), crossed polarized light. This Japanese porcelain stone (used for a Kutani kiln during 1860–1871) is finer grained than the previous example from China. The many tabular crystals of sericite are very minute; the fine-grained quartz and feldspars are the main components visible in crossed polarized light.

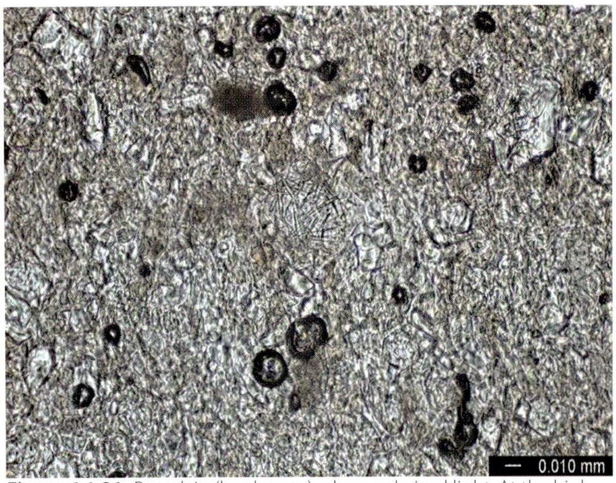

Figure 6.1.21 Porcelain (hard paste), plane polarized light. At the high temperatures and extended firing temperatures achieved by many hard-paste porcelains (such as this tea bowl from Japan), an extensive network of interlocking grains of tiny mullite needles can form in random directions within the matrix. The clear areas are glass with remnant quartz grains; there is also some smudgy dark spinel.

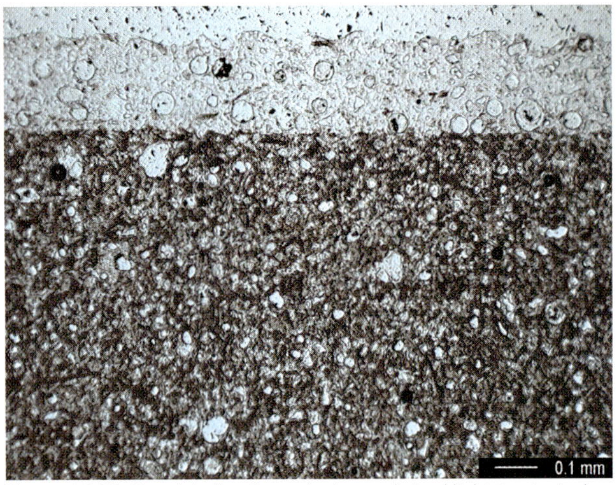

Figure 6.1.22 Porcelain (hard paste), plane polarized light. In a typical Chinese porcelain (this one is dated to 1785), in plane polarized light the ceramic body tends to have a relatively homogeneous appearance. The glaze layer appears somewhat thick and distinct.

of the aluminum-rich kaolinite clay. The mullite needles form in random directions, and create a network of interlocked grains that contribute to the hardness and strength of porcelain (Fig. 6.1.21). Mullite is relatively rare in natural geological materials, because it forms at conditions of high temperatures but low pressures, but is extremely important in high-temperature ceramics. When kaolinite is heated, it transforms into mullite: at about 550 °C kaolinite dehydroxylates to form metakaolin; metakaolin decomposes at about 950 °C to form spinel, amorphous silica, and a minor amount of mullite; above 1000–1200 °C the phases are abundant mullite and amorphous silica. Mullite grains in sintered bodies like porcelain can be acicular, prismatic, or needle-like in shape, although thin sections of modern refractory materials fired at very high temperatures (1700–2000 °C) show tabular mullite crystals, and kaolinite fired at 1400–1600 °C forms very tiny mullite "whiskers" (Schneider et al. 1994).

Other changes occur above 1100 °C, as clay minerals begin to break down and new, high-temperature minerals begin to form. Which minerals will form depends upon the raw materials, the firing conditions, and length of firing. Common minerals, in addition to the mullite, are high-temperature forms of silica, calcium silicates such as wollastonite, calcium ferrosilicates such as hedenbergite, and spinel ($MgAl_2O_4$) (Herz and Garrison 1998).

The glaze firing is done at lower temperatures, but for long firing times. That regime, combined with a relatively porous body, means that glazes tend to be partially crystalline and to become partially absorbed by the clay body. The glaze and body fuse together, increasing the strength of the material and preventing spalling of the glaze.

In a typical Chinese porcelain (Fig. 6.1.22), in plane polarized light, the ceramic body tends to have a relatively homogenous appearance. It appears relatively coarse in thin section, and is crisscrossed by tiny needles of mullite, randomly oriented. In plane polarized light, the glaze layer appears somewhat thick and distinct. However, in crossed polarized light (Fig. 6.1.23) the quartz-rich body and the glaze appear to be almost one continuous material from

Figure 6.1.23 Porcelain (hard paste), crossed polarized light. In crossed polarized light, the quartz-rich body and glaze of this Chinese porcelain appear to be almost one continuous material from glaze to ceramic body.

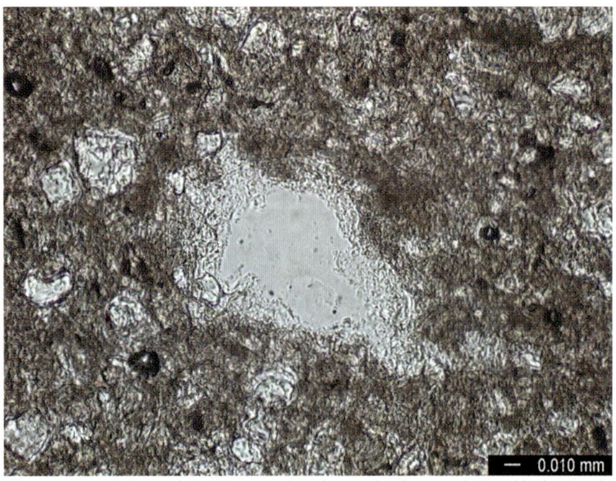

Figure 6.1.24 Porcelain (Chinese hard paste), plane polarized light. Pools of amorphous silica glass form as quartz dissolves.

glaze to ceramic body. The glaze is seen to be filled with quartz crystals.

Kingery and Vandiver (1986) illustrated, with scanning electron micrographs, an example in which the original quartz in the clay body, which was fired for several hours at about 1250–1300 °C, partially dissolved in the surrounding liquid, leaving residual rounded quartz particles surrounded by pools of amorphous material where the viscous liquid solidified as glass. There were some pools of silica-rich glass where the quartz had completely dissolved. These amorphous pools of silica glass are also visible in thin sections (Fig. 6.1.24).

Variation in Chinese porcelains existed between regions and between wares, as a result of variation in raw materials, recipes, firing temperatures, kiln atmosphere, and other production steps. For example, Guozhen and Xiqiu (1985) reported that northern porcelain bodies during the Tang and Song periods were made mainly from clay minerals rather than with the use of China stone. The resulting high-alumina and low-silica ceramic bodies consisted largely of mullite and a small amount of cristobalite along with glass and "worm form" clay polymorphs. In contrast, in the south, a mix of kaolin and China stone was used, resulting in a larger amount of quartz crystals and a smaller amount of mullite, along with feldspars. The authors noted that Ding whitewares had mullite crystals that were well developed, homogeneous, and interlocked with each other in the body. In contrast, southern kilns such as Jizhou, Jingdezhen, and Dehua had wares with large quantities of quartz, mullite, and sericite. Some of the glaze differences they reported between northern and southern wares could be seen in thin section: more bubbles, quartz relicts, and anorthite crystals in the interaction layer between glaze and clay body for the northern whitewares with fewer of these present in southern whitewares.

After a great deal of experimentation, including development of a similar soft-paste porcelain (see the next section below), Europeans were finally able to approximate the appearance and functional properties of Chinese porcelain, with the first European hard-paste porcelain made in 1708 (by Böttger in Meissen, Germany). Böttger partially fluxed a white clay with lime; a firing temperature of 1350 °C was required (later lowered by fluxing with feldspar instead). The technology of correct formulations and firing conditions spread quickly to the rest of Europe, although with much variation in exact recipe ingredients as each manufacturer attempted to keep their successful recipes and procedures proprietary (Atterbury 1982; Kingery 1986).

The appearance of European hard-paste porcelain in thin section tends to be quite different from that of Chinese porcelain. Figure 6.1.25 shows an example of a thin section of mid-eighteenth century Meissen hard-paste porcelain in plane polarized light. Few quartz grains, as well as few pores, are visible as clear areas within the brownish-gray material with extensive mullite development. Meissen porcelain glazes were almost bubble free (unlike many of the Chinese glazes), as well as free of crystal development, with no clay/glaze interaction layer. This was due to the clay paste recipe that was rich in kaolin clay and highly vitrified; this paste was nonreactive with the glaze (D'Albis 1985). Long needle-like grains of mullite can be seen crossing the clay in random orientations, a hallmark of true porcelains. A very large number of those mullite grains make up the bulk of the ceramic matrix. In crossed polarized light (Fig. 6.1.26), rod-like mullite grains have gray to pale yellow interference colors. There are fewer quartz grains than in typical Chinese porcelains. Furthermore, the Meissen glaze is completely isotropic and lacks the crystalline component seen in most Chinese porcelain glazes.

The European hard-paste porcelain dinnerware in Figure 6.1.27, as seen in plane polarized light, has more of a quartz (some converted to cristobalite and tridymite) and feldspar component than the Meissen porcelain. There is also reddish-brown biotite and spinels, in addition to abundant mullite. A dark enamel is sandwiched between two clear glaze layers (Fig. 6.1.28), and the glaze is very glassy, with

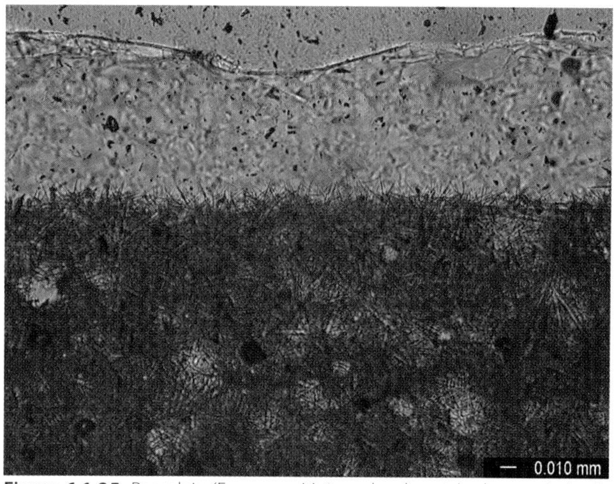

Figure 6.1.25 Porcelain (European Meissen hard paste), plane polarized light. This Meissen porcelain has a ceramic body that has many long, needle-like grains of mullite crossing in random orientations. The glassy matrix has few quartz crystals. The glaze is very glassy, with no crystals and very few bubbles.

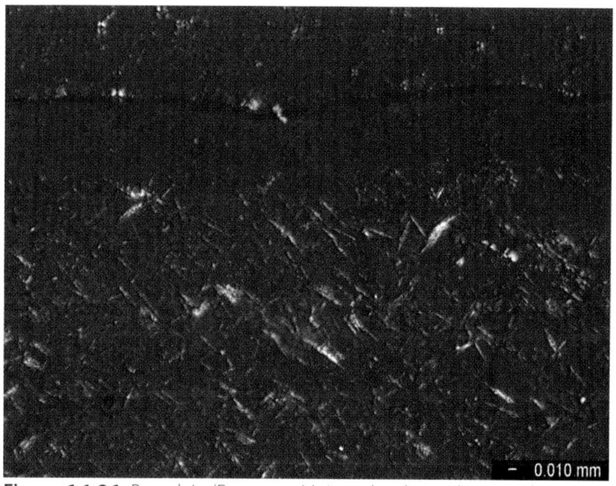

Figure 6.1.26 Porcelain (European Meissen hard paste), crossed polarized light. In crossed polarized light, it is clear that there are very few quartz crystals in this European Meissen porcelain, unlike most of the examples from China, Japan, and Korea. The mullite rods have gray to pale yellow interference colors. The glassy glaze is isotropic with few crystals apparent.

very little crystalline component. The firing temperature and the composition of the glaze and clay are such that there is only a thin interaction layer with the ceramic.

Soft-paste porcelain

This material was developed in Europe in an early effort to try to imitate Chinese porcelain, which was being imported into Europe as early as the fourteenth and fifteenth centuries, and had become hugely popular by the sixteenth century (Kingery and Smith 1984). Soft-paste versions were produced in Florence in 1575–87 ("Medici" porcelain), and French soft-paste porcelain had attained the translucence and whiteness of Chinese products by 1695 at St. Cloud (the factory transferred to Sèvres in 1756). Kingery and Smith pointed out that this material is derived from earlier fritware (see the next section below).

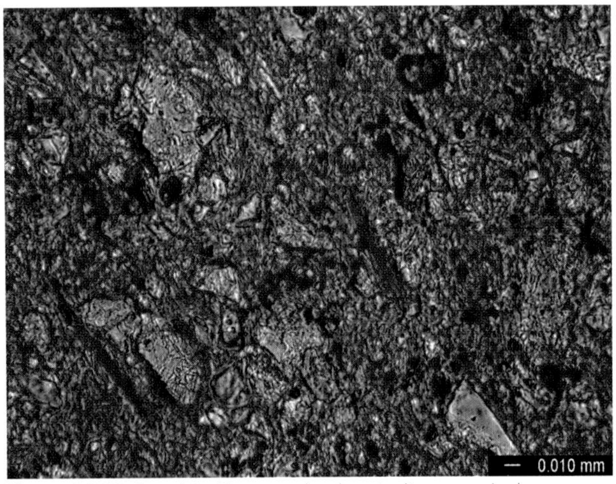

Figure 6.1.27 Porcelain (European hard-paste dinnerware), plane polarized light. This example of European hard-paste porcelain has a glassy matrix with some quartz (partially converting to cristobalite and tridymite), feldspars, spinel, and reddish-brown biotite, along with abundant mullite.

The Medici soft-paste porcelain had a white body consisting of fine sand, white clay, and a glaze sinter made of sand mixed with salt and wine lees (sediments that form in the bottom of wine caskets during fermentation, rich in tartaric acid and tartrates) (Kingery and Smith 1984). After biscuit firing, an underglaze of manganese-oxide and cobalt-oxide paints was added, followed by a lead-silicate glaze. The material thus consisted of angular quartz in a glassy matrix that included some cristobalite/tridymite and fine pyroxenes. Kingery and Vandiver (1986) illustrated an example of French soft-paste porcelain where the original quartz in the ware, which was kept for as long as 90 hours in a fluid alkali-silicate liquid at temperatures of about 1000 °C, changed to multiple grains of another silica form (cristobalite), which dotted the matrix and in some cases surrounded some of the remaining quartz cores. Fine chunky or blocky crystals of a calcium silicate, wollastonite, were also distributed within the glass of the matrix.

French eighteenth and early nineteenth century soft-paste porcelains, made with a sintered alkali-silicate material containing quartz, largely transformed into cristobalite and/or tridymite, fine wollastonite crystals, and glass (Fig. 6.1.29). Later nineteenth-century soft-paste porcelains made at the Manufacture Nationale de Sèvres sometimes used this earlier material, or alternatively a siliceous glass mixed with silica and a lime-rich clay. This latter method does not result in the tridymite-cristobalite development, but instead has a larger fraction of quartz (Kingery and Smith 1984). Another factory studied by Kingery and Smith (run by M. Naudot at Raincy in the nineteenth century) used a material similar to that of the eighteenth century, but with the alkali reacted with the silica by sintering; that process resulted in extensive development of wollastonite crystals.

Many other recipes and processes were tried, and successful ones kept secret for as long as possible. For example, one recipe (D'Albis 1984) from eighteenth-century Vincennes and Sèvres called for 60% (wt. %) Fontainebleau

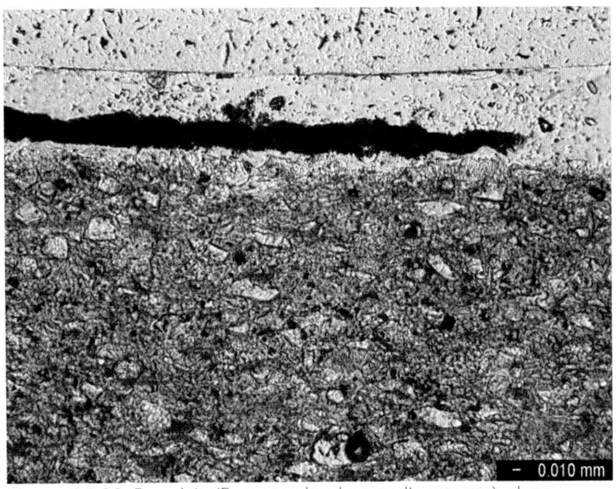

Figure 6.1.28 Porcelain (European hard-paste dinnerware), plane polarized light. This is the same example of European hard-paste porcelain as in Figure 6.1.27, shown at a lower magnification. In this view, a dark enamel can be seen sandwiched between two clear glaze layers. The lower glaze layer is very thin, with only a very thin and partial interaction layer that has developed between the ceramic body and the glaze. Both glaze layers are very glassy, with few crystals.

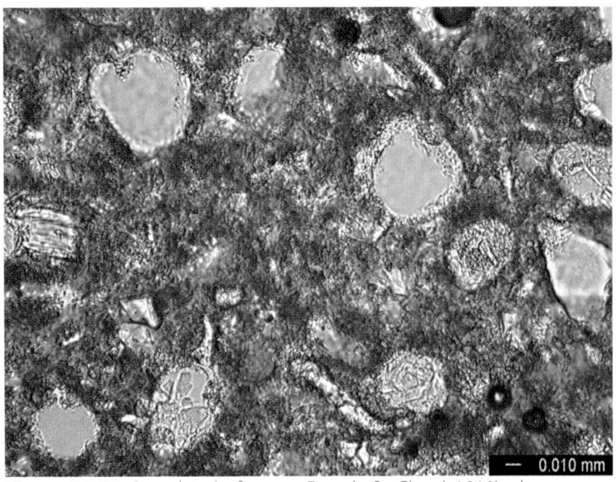

Figure 6.1.29 Porcelain (soft paste, French, St. Cloud, 1810), plane polarized light. This French soft-paste porcelain has clear pools of silica glass, many with edges and "encrustations" of minute cristobalite (squares) and/or tridymite (wedges). The matrix is very glassy, with a few quartz remnants and small grains of clear wollastonite in elongated rods or columnar, blocky aggregates.

sand, 21.7% "mineral crystal," 7.2% Gabelle (sea) salt, and 3.7% each of rock alum, calcined gypsum from the Montmartre quarries, and soda from England or preferably Alicante. The raw ingredients were crushed into a fine powder and then fritted in the bisque furnace for 30–35 hours. The resulting opaque white, hard material was then cooled and crushed, and used in the paste recipe, which consisted of 75% by weight of the frit, 16.7% of Spanish chalk, and 8.3% of calcareous soil from Argenteuil.

D'Albis noted that the "mineral crystal" (probably potassium nitrate or saltpetre), sea salt, and soda of Alicante (sodium carbonate) were water-soluble alkaline fluxes. Calcined gypsum added another flux, calcium oxide, and rock alum added aluminum; these last two ingredients gave whiteness to the product. An organic binder was included to make a good modeling paste for molding figurines or other pieces. Soft-paste porcelain in general had especially good properties for molding. Figure 6.1.30 illustrates a thin section taken from one soft-paste porcelain example: it is rich in sand so would not have been very plastic—clearly this material was more suited to use in molds rather than wheel-forming fabrication processes.

Owen (2001) discussed early British soft-paste porcelains, and showed that they fall into three categories: bone ash, soapstone, and frit porcelains. He analyzed many examples from a variety of factory sites. For example, at the Worcester porcelain manufactory, early (prior to the 1750s) porcelains were siliceous and contained substantial amounts of bone ash; later ones were made from a paste containing soapstone (steatite, a talc-rich metamorphic rock) and flint glass (Owen 1997, 1998). The soapstone recipe is said to have produced a harder, whiter and more durable ware that did not crack when boiling water was poured into it. Owen et al. (2000) also identified porcelain from the Vauxhall china works (1751–64) in Britain as being made with Cornish soapstone as an additive. High-temperature forms of silica were found, with tiny diopside crystals, and coarser grained quartz.

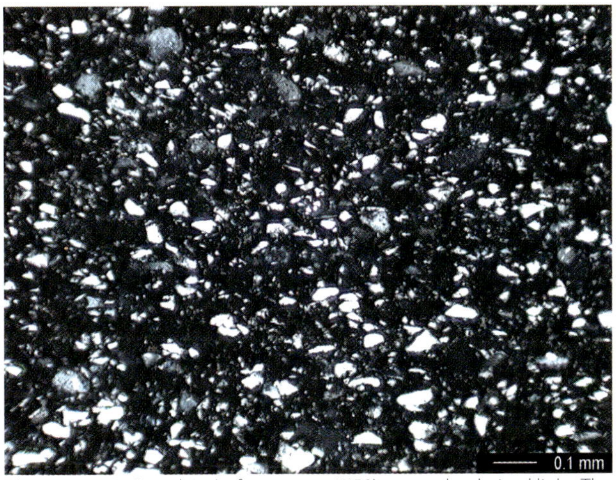

Figure 6.1.30 Porcelain (soft paste, ca. 1750), crossed polarized light. The high sand content (mainly quartz, white and gray interference colors) of this soft-paste porcelain from the 1750s is visible in crossed polarized light. The high sand content affects plasticity, making the material better suited to use in molds than to wheel-forming fabrication methods.

Two porcelain works in Wales also showed differing materials: primarily bone ash porcelain at the Nantgarw works and a modified bone ash recipe at Swansea that also included ground flint in place of quartz sand and a higher proportion of clay (kaolinite) (Owen et al. 1998). Minerals reported to be present in the glassy matrix for these products include quartz, tridymite, cristobalite, the calcium-rich plagioclase feldspar anorthite, a high-temperature alkali feldspar (sanidine), and pyroxenes (such as enstatite) resulting from dehydration of talc. Some sherds contain whitlockite, a tricalcium phosphate derived from dehydration of bone ash, which appears as small (less than 100 μm in diameter) ovoid grains.

Bone ash porcelain was a popular European ware that is actually intermediate between stoneware and porcelain. It is comprised of a low-refractory clay plus additional alkali minerals to create a flux and improve melting (firing is at 1100–1200 °C), producing a relatively strong and dense body (with a porosity as low as 0.5–1.5 vol. %) that is semi-vitreous to vitreous, and off-white in color. Sometimes called phosphatic, or "bone china," it may have up to 50% bone ash added to the clay to serve as the flux. The animal bone ash, rich in calcium, imparts a pure white color and produces a translucent body that was highly prized. Originally, only bones of oxen were used, but eventually bones of other animals were also calcined for use. The calcination process is important: over-calcined bone has little or no organic material remaining and cannot contribute much to the plasticity of the clay body, while under-calcined bone may contain so much organic material that frothing may occur in the body slip, leading to defects in the ceramic product (Ryan 1978).

Kingery and Smith (1984) hypothesized that the idea of adding bone ash came from its use in milk glass, so it was originally added to a more traditional soft-paste porcelain frit as an attempt to produce a whiter material. Although the porcelain was of fine quality, the paste produced by phosphatic bone ash recipes is difficult to fire (it has a tendency to distort or sag during initial high-temperature firing in the biscuit kiln), producing much kiln wastage and creating often unprofitable products (Owen and Sandon 1998; Owen et al. 1998; Owen and Morrison 1999; Owen et al. 2000).

While the calcium served as a flux, lowering the temperature needed to melt the silica, phosphorous in the bone would also serve as a glass former, promoting the development of amorphous glassy areas within the ceramic body. Other components typically include silica polymorphs and plagioclase feldspar (anorthite). Figure 6.1.31 is a thin section of a fine-grained English bone ash porcelain from the 1780s, in plane polarized light. The small round black particles within the clay body are the remnants of the very fine bone ash that did not get combined with silica to form glass. Owen et al. (1998) showed that these particles are whitlockite, a tricalcium phosphate derived from dehydration of the bone ash. Owen et al. (2000) further noted that the original bone ash component (hydroxy apatite) progressively dehydrates to tricalcium phosphate over the temperature range of 1000–1400 °C while releasing calcium oxide. This CaO reacts with clay and silica minerals ultimately to

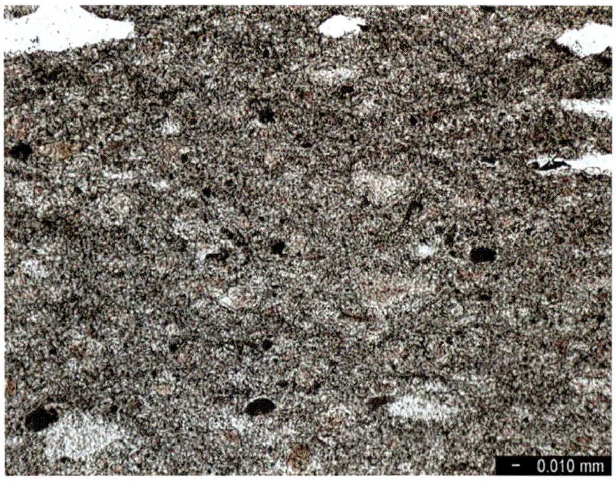

Figure 6.1.31 Porcelain (soft paste, English bone china, 1780s), plane polarized light. The small rounded black particles in this English bone china of the 1780s are the remnants of very fine bone ash that did not combine with the silica to form glass. These particles are whitlockite, a tricalcium phosphate derived from dehydration of the bone ash. The larger, very clear areas are pores; small rounded, chunky, and elongated clear areas are tiny quartz and plagioclase feldspar grains.

form small anorthite crystals. However, if the animal bones were calcined at high temperatures, ground and mixed into the clay paste, then an additional form of calcium, such as calcite, must be present in order for anorthite to form.

The mineralogy observed in archaeological porcelains can be used to formulate hypotheses about the raw ingredients used in the paste and about firing conditions (although thin-section data may need to be supplemented with other analytical methods) (Owen et al. 2000). For example, anorthite-bearing phosphatic wares were probably made using calcined bone ash, ground quartz or flint, small amounts of potassic flux (such as potash), and enough clay to lead to the formation of plagioclase feldspar; some siliceous pastes in Britain were made using a Cornish "china stone" (hydrothermally altered granite) which has a highly variable composition, mixed with kaolinite; the presence of enstatite (magnesium silicate) and diopside (calcium-magnesium silicate) in a steatite (soapstone) porcelain can provide information about the composition of the paste and firing temperature.

Soft-paste porcelain recipes, however, were difficult to fire properly, leading to considerable kiln wastage. This material also had poor mechanical properties compared with hard-paste porcelain (especially strength and thermal shock resistance). Thus, by the end of the early part of the nineteenth century, production of this material decreased, often replaced by hard-paste porcelain products (Kingery and Smith 1984).

6.1.4 Stonepaste/fritware

Stonepaste is a ware consisting of crushed quartz combined with a small amount of very plastic clay (such as white montmorillonite) and an alkaline crushed glass frit. It was extensively used in the Islamic world (especially in Persia, Egypt, Syria, and Turkey) (Mason and Golombek 2003; Watson 2004). The small amount of clay served as a flux, helping the crushed glass (or prepared frit) to melt during the firing process and create a fused glassy network that would fill out the areas around the quartz crystals. This stonepaste material is also known as fritware. In the microstructure (Fig. 6.1.32), the crushed quartz particles (often with some accompanying feldspars) are held together by a glass formed from the clay-frit mixture. The ceramic body tends to be very porous (Fig. 6.1.33). Unlike faience, a small amount of clay, and perhaps an organic binder or salt, is added so the object can either be thrown on a wheel or formed in a mold (Kingery and Smith 1984; Canby 1997). Both coarse-grained and fine-grained wares may be seen, varying in color (white, yellow, pale brown) (Keblow Bernsted 2003).

Kingery and Vandiver (1986) illustrated the microstructure of a Persian lusterware object. Grains of crushed quartz were visible, often rounded from having partially dissolved into solution during firing. These quartz grains were held together by silicate glass that formed from the frit and clay

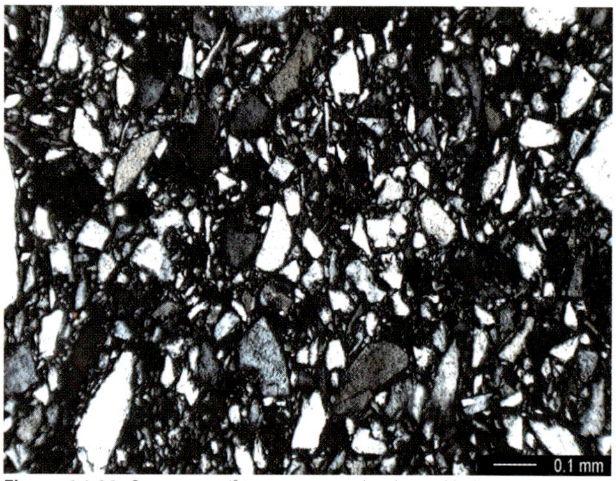

Figure 6.1.32 Stonepaste/fritware, crossed polarized light. In stonepaste, or fritware, crushed quartz particles (often with some accompanying feldspars) are held together by a glass formed from the clay-frit mixture. This Persian example shows how little interstitial component is present, and how isotropic and glassy it is; most of the material consists of the quartz component.

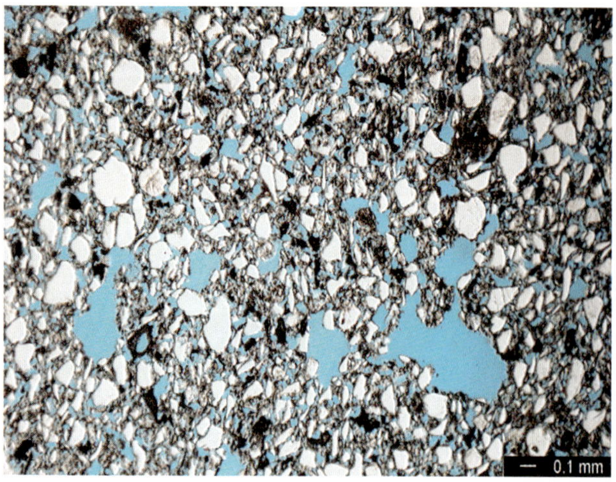

Figure 6.1.33 Stonepaste/fritware, plane polarized light. The ceramic body of a stonepaste/fritware object tends to be very porous. In this example the blue epoxy areas indicate the extensive porosity.

components. The small amount of iron present dissolved in the glass, so the resulting object is white. The alkaline-lead glaze that was fired simultaneously with the ceramic body is seen to have penetrated into/bonded well with the body itself. Small particles (of tin oxide) in suspension in the white glaze served as an opacifier. Overglaze lusters were formed by painting on a metal-containing compound and refiring in a mildly reducing atmosphere; tiny copper, silver, or gold particles formed in the glaze.

Although hard, stonepaste pottery can be very porous, and so is relatively weak. A clean white slip provides the background for translucent glazes (Fig. 6.1.34). It is thought that the development of stonepaste was a response to the popularity of Chinese porcelain, and thus an attempt at creating a hard white ceramic body that would make a good ground for overpaint decoration. Visually, this attempt was successful, although neither the hardness nor the translucency of Chinese porcelain was achieved (Henderson 2000).

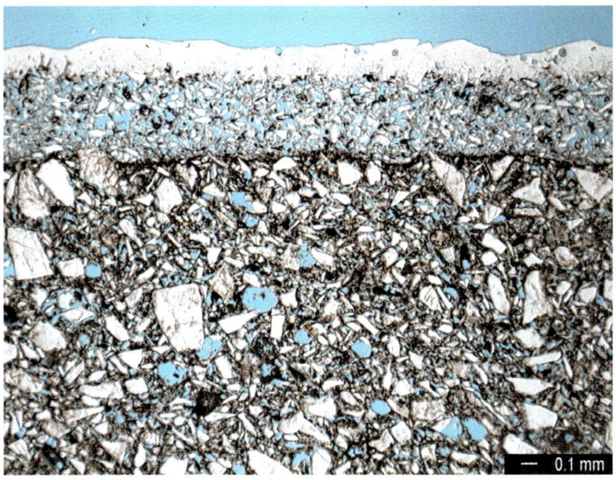

Figure 6.1.34 Stonepaste/fritware, plane polarized light. As is typical, this stonepaste ware is very porous (pores are marked by the blue epoxy mounting resin). On top of this porous, quartz-rich ceramic body sits a thin clay layer. Between this clay and the outer transparent, glassy glaze is a layer of very fine-grained quartz and clay.

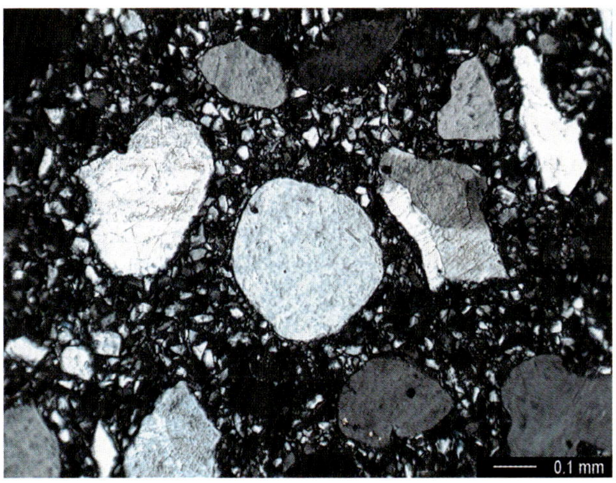

Figure 6.1.35 Stonepaste/fritware, crossed polarized light. This porous stonepaste/fritware has two populations of grain size. Many of the larger quartz grains are rounded, while the finer ones are mainly angular.

An example is Turkish Iznik ware, used in Ottoman Turkey between about 1480 and 1650 (Henderson 1989; Tite 1989; Henderson 2000; Paynter et al. 2004). Most analyses of Iznik wares have been by SEM. These show the pale ceramic body with a network of quartz grains, recrystallized silica polymorphs (tridymite or cristobalite) within a matrix of glass, frit, voids, and some interstitial glass around smaller particles and forming a bond between adjacent quartz grains. There is usually a clearly delineated thin slip layer on top of that (containing smaller silica crystals than the body and less frit), a slip-glaze interaction layer (silica crystals within the glaze) followed by a thin translucent glaze layer (which sometimes has some relict silica crystals suspended in it, along with a few small tin crystals). Some of these components can be seen in thin section as well. The thickness of the interaction layer varies, and chronological changes can be seen. For example, a mid-seventeenth-century vessel, made when the production of Iznik ceramics was in decline, had coarser crushed quartz with maximum size of about 400 μm instead of the 100 μm used in sixteeenth-century Iznik ceramics; bone ash was used as the calcium source in the later object (and phosphorous as a glass former) (Henderson 2000).

Mason and Golombek (1991) employed thin-section analysis of sherds (including excavated materials and wasters) to compare fifteenth–sixteenth-century Islamic blue and white stonepaste pottery from Persia (Iran), Syria, and Egypt that was a stylistic imitation of Chinese blue and white porcelains. Medieval recipes and ethnographic studies showed that stonepaste was made with about ten parts powdered quartz, one part glass frit, and one part fine white clay. However, the authors were able to discern regional and production center differences in raw materials.

The Persian products contained a considerable amount of quartz with characteristics other than regular, clear quartz–instead the quartz was polycrystalline, cloudy with fluid or inclusions, microcrystalline (chert), or cryptocrystalline (chalcedony). A variety of nonquartz minerals was also present. The angular quartz material was large in size and appeared to be made of crushed cobbles or pebbles. Variations were also found at sites within Iran indicating differences in raw materials at different production centers. For example, the site of Siraf on the Gulf Coast had stonepaste with 50–60% quartz, but most of the quartz was chalcedony. At Nishapur, stonepastes had about 50% quartz, along with 3–5% felsic volcanic lithics with large crystals of potassic and plagioclase feldspars, consistent with volcanic inclusions found in other types of ceramics from this area. A group of objects from northeastern Iran had about 50% quartz that was a mixture of clear, cloudy, and polycrystalline grains, with about 5% chert. And, finally, samples from Kerman were similar to the northeast Iranian ones except that grain size was consistently less than 0.2 mm, and the chert component was not present.

In contrast, stonepastes from Syria were made with a clean, mature quartz sand. Some of the grains were rounded, and ranged from 0.2 up to 0.5 mm; other grains were finer and angular, and were perhaps the result of crushing of the sand (Fig. 6.1.35). All the sand grains were clear, there was no cloudy or polycrystalline fraction. The Egyptian stonepastes were also made with sand, but in this case the raw material was an impure, silty detrital sand deposit, consisting of subrounded grains with a wide range of grain sizes.

European fritware included tin-glazed majolica ware. Calcareous clays were used that fired to pale colors and were easily masked with a white, tin-opacified glaze. During production, the alkali was made insoluble by sintering it with silica beneath the firing chambers of the kiln; the reacted alkali silicate was then milled with a tin oxide-lead oxide mixture to form the white glaze ground over a yellow lime-rich clay. The resulting microstructure is more like Chinese porcelain than the earlier Turkish and Persian fritwares; more glass was also obtained (Kingery and Smith 1984; Thornton 1997).

Keblow Bernsted (2003) illustrated thin sections of Persian fritwares (Kashan type). Many grains of quartz and feldspar made up the fine sand fraction. Fracturing and browning (calcining) of the quartz and feldspar grains also appeared, which she proposed suggested a high firing temperature. This fracturing made it difficult to ascertain the shape and degree of rounding of the finest fraction, but the largest grains appeared to be angular and crushed (Fig. 6.1.36). There were clearly two quartz/feldspar size fractions: grains varied in size from 20 up to 250 µm, with most being in the silt range (about 25 µm). Diopside had formed in cavities, as sheaf-like or radial aggregates of needle-like crystals. The matrix was heavily sintered. There was a relatively low porosity, with oval to spherical macropores (around 40 µm) and very small micropores (around 5 µm) with a rim of glass. She concluded that firing temperatures may have been 1050–1100 °C.

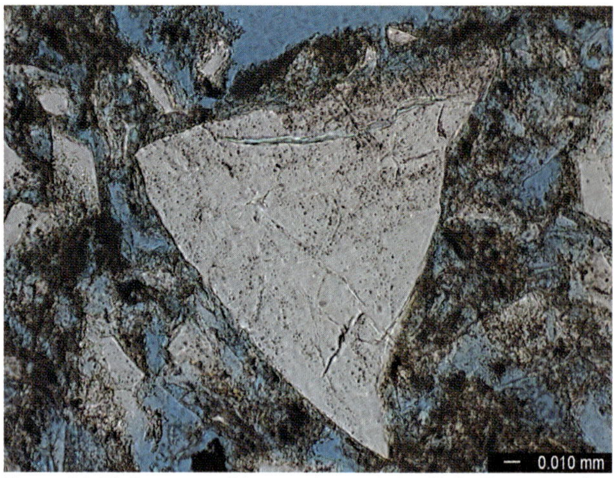

Figure 6.1.36 Stonepaste/fritware, plane polarized light. Fracturing of quartz grains in a stonepaste/fritware can indicate relatively high firing temperatures. This very angular quartz grain exhibits several fracture lines.

6.2 Characterizing Inclusions in Clay

Since the early twentieth century, a large number of successful characterization and provenance studies have been carried out with ceramic thin sections, an approach that has become routine (Peacock 1970; D.F. Williams 1983). Other methods of analysis are often used in conjunction with thin-section petrography, but thin sections still provide crucial basic information not available through other analytical methods. The majority of thin-section studies involve analyzing the nonplastic inclusions to characterize a material and, when possible, comparing the inclusions with potential geological sources (for exact pinpointing of raw material sources), or at least comparing inclusions with nonplastic materials from other ceramic specimens (for identifying and studying the relationships between objects that might have a common source). These studies have involved ceramic traditions from a wide range of regions and time periods. Often the research on nonplastic inclusions is complemented by examination of other features observable in thin section that may be helpful in characterizing, grouping, or identifying sources of vessels or sherds (Whitbread 1995). A small subset of successful studies is reviewed here and in the following two chapters.

There are a restricted number of clay minerals (often called "plastics" because they are plastic when wet), and most cannot be definitively characterized through thin-section analysis alone because of their small size. Although they often comprise 50% or more by volume of a ceramic artifact, they are usually only about 1 µm thick, so cannot be specifically identified in thin sections. Instead, the clay fraction appears as an amorphous groundmass or matrix. Some methods for characterizing the clay paste component of a ceramic thin section—and interpreting what that component may indicate regarding materials selection and processing steps—do exist. Some were incorporated from soils science, while others have been developed within the field of archaeological ceramic petrography. Those methods are discussed in Chapter 8.

However, the most important focus of thin-section petrography of pottery is on identifying, quantifying, and interpreting the nonclay inclusions. These inclusions improve working, drying, and firing properties of the clay, and can also affect the use properties of the finished ceramic material. Beyond identifying the inclusions present, useful interpretations may require identifying a variety of characteristics such as abundance (frequency of inclusions), size (average or modal, or most common, as well as the grain-size distribution), internal distribution of particles, morphology, and alteration mechanisms. For example, a rounder shape for quartz suggests a longer history of weathering and transport. In the various examples given here and in the following chapters, some approaches are qualitative, identifying what minerals or other components are present; some are semi-quantitative, estimating some parameters such as abundance and shape using visual estimation charts, often ones developed specifically for working with ceramic materials; and some are quantitative, using a variety of different methods for point counting or for digital image analysis (Rice 1987; Stoltman 1989; Matthew et al. 1991; Orton et al. 1993; Rice 1996; Gibson and Woods 1997; Reedy 2006a).

6.2.1 Methods for characterizing inclusions

In addition to identifying the nonplastic constituents that are present in the clay matrix, it is often useful to calculate the area percentage of inclusions and the relative amount of each constituent, as well as other parameters such as average grain shape and the minimum, maximum, and average grain size. Traditional methods of obtaining this information include measuring features with a micrometer or eyepiece graticule (Shepard 1968), doing visual estimations with the aid of published estimation charts (Williams et al. 1982; Matthew et al. 1991), and using variations on point counting, where the minerals or features are selected for counting and/or measurement by recording what appears at the intersections along a superimposed grid (Stoltman 1989; Goins and Reedy 2000). More recently, comprehensive computer packages for digital image analysis have emerged as promising tools for improved characterization and interpretation of nonplastics in ceramic materials (Reedy 2006a) (Fig. 6.2.1).

Visual estimation of features such as volume percent, size, or shape has the advantage of being relatively fast, so that it is feasible to include more thin sections, and thus improve the statistics regarding variation within and between thin sections of a single object and between objects of the same category. Visual estimation charts specific to archaeological ceramics have been developed, which improve the usefulness of this approach (Matthew et al. 1991). Some charts have been specially constructed for certain particle shapes and sizes to make the comparisons easier and more accurate (Flügel 1982). However, a great weakness is that not all researchers are adept at visual estimates, and thus accuracy and reproducibility may not be very high.

In point counting, a grid, or procedure for systematic sampling, is established. At fixed intervals, any grain that falls at that point is counted, and attributes such as particle type, size, and shape may be recorded. Differences in approaches include the counting intervals selected, and researchers may differ on whether or not a grain should be counted more than

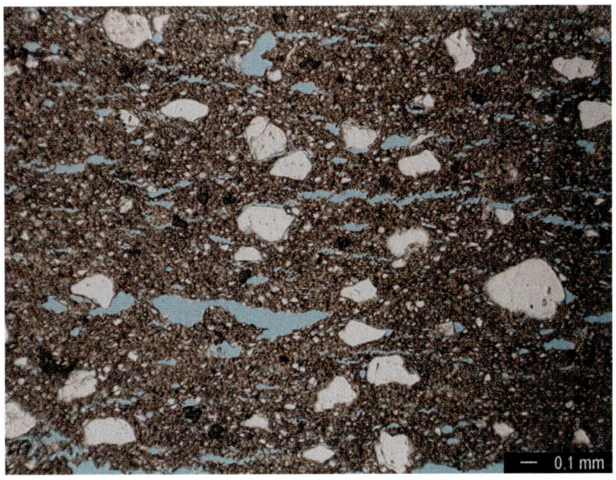

Figure 6.2.1 German stoneware, seventeenth century, plane polarized light. This image shows another field of view of the same object that is illustrated in Figure 6.1.12. Digital image analysis (averaging the two fields of view, and using the Image-Pro Plus software package) shows that the area sand amount is 13%, pores are 4%, and clay matrix is 83%. Sand size ranges from 0.006 mm up to 0.669 mm. There is clearly a bimodal distribution supporting the hypothesis of use of sandy clay with an additional sand additive (probably those grains measuring 0.1 mm and above).

once if it is large enough to intersect more than one point on the grid. Other methods include counting the grain nearest a grid point if no grains fall exactly on the point (Streeten 1982). Variations include area, ribbon, or line counting: area counting incorporates all of the grains in a specified area of the thin section; in ribbon counting, the thin section is moved along a line or band, and all countable grains that fall along that line are identified and classified by type and size; and in line counting, the thin section is advanced along equally spaced parallel lines, with all grains that fall under the cross-hair recorded until a preset total number of grains (usually 200–400) has been reached. Counts for each category of interest are then totaled to compute the percentage of each component present. The main purpose of all of these counting methods is to obtain an unbiased estimate of the constituents of the sample, expressed as an area or volume percentage, or as the percentage of individual grains.

Middleton et al. (1985) concluded that area counting and ribbon counting gave very similar results, whereas the mean diameters found through line counting and point counting were biased towards the larger grains. They noted that point counting was originally adopted in thin-section petrography as a means of calculating the relative areas of each of the mineral phases present (proportional to the numbers of points intersected by that phase). Thus, the multiple-intercept approach (allowing a grain to be counted more than once if it extends to more than one point on the grid, while counting nothing if no grains fall at a particular grid point) is an indicator of the relative area comprised by each mineral phase (area %), and is not a good indicator of the number of grains with that diameter (a number frequency). They further noted that the modifications to point counting of "single-intercept" or "nearest grain" approaches still do not give good frequency data, and adversely affect the accuracy of area percent measurements. Line counting also was found to be problematic–a drawback of line counting is that the number frequencies of grains cannot be precisely correlated with area, volume, weight, or even percentage (Stoltman 2001).

The methods recommended by Middleton et al. as being rigorous, reproducible, and having physical meaning are: area or ribbon methods for frequency distributions, with the areas or ribbons widely spread along the thin section to give as representative a count as possible, and ribbons having a width of at least twice the diameter of the largest grain; and point counting, with one measurement taken whenever a grain falls on a grid point, for determining the area percentage of particles of different diameters, using a relatively coarse grid to avoid repeat intersections as much as possible. Point counting has advantages for traditional petrographic analysis using an eyepiece graticule, and is readily understandable in terms of volume, clay mixing, and physical properties of prepared clay, but area counting is better suited to digital image analysis. The authors recommended dropping line counting and variants on point counting. They further noted that as few as 50 grains can give adequate information on mean diameter, but more grains may be needed for analyses of the size distribution patterns of those particles, or if measures other than mean diameter are sought (such as skewness and kurtosis) (Leese 1983).

Stoltman (2001) described point counting, sometimes referred to as the Glagolev-Chayes method, as effectively superimposing a grid over the thin section and recording all observations made under the cross-hair at every grid intersection point. These observations may be grains, clay matrix, or voids. Rather than actually superimposing the grid, the microscope stage is advanced back and forth at specific increments. He noted that while geologists may count 1,000 or more points per thin section, the nature of ceramic thin sections versus those of rock materials means that some changes in procedure are needed. Most ceramic thin sections have relatively coarse mineral grains within an amorphous or very finely crystalline matrix, so count totals are usually lower for ceramic materials.

As long as certain counting procedures are followed, the basic theory of point counting is that the area counts made in thin section observations will reliably estimate the actual volumetric proportions of each constituent. This relationship has been established through experimental work by geologists (Chayes 1954a). Accepted procedures include selecting an appropriate sampling or counting interval—the interval chosen will vary depending upon the material under study. It is generally held that the point distance should be larger than the largest grain fraction to be included in the analysis, while at the same time, the smaller the interval distance, the greater the number of observations and thus the greater the precision of the analysis. For ceramic thin sections, the problem is that there are often coarse mineral grains present, so selecting counting distances larger than those would leave too few total points counted to ensure a reliable analysis. Stoltman (1989, 2001) recommended a compromise measure of using a 1 mm sampling interval. He found that this distance tended to work well for ceramic thin sections, usually giving a total count of 100–300 points, and that this choice was likely to give results that were reliable within a range of ±3.5%.

Stoltman further noted that for volumetric calculations, volume is proportional to area, so there is no problem presented by the fact that large grains might end up being counted more than once. However, for recording properties of individual grains such as mineral type, size, and shape in order to calculate grain class, it is preferable to include each grain only once. With 1 mm counting intervals, these separate records are relatively easy to maintain. With very small counting intervals, tracking individual grain data is more difficult because of the large number of repeat counts.

With ceramic materials, there are times when the presence or absence alone of specific inclusions is not enough to characterize a ware adequately or to distinguish it from wares made in other workshops or locations. For example, for ceramics originating in an area of predominantly sedimentary geology, where quartz content is high while other mineral suites are restricted, products from an entire region may be very similar in nonplastic composition. In such cases, textural analysis can be a useful tool for characterizing the fabric of ceramic materials with quantitative data. Fabrics of similar mineralogy may be distinguished, for example, by differing grain-size distribution. Drawing on techniques used by sedimentary petrographers, textural analysis can include quantifying the percentage of various inclusions, roundness and sphericity (usually of quartz grains), and mean grain size. These data can then be incorporated into multivariate statistical analyses. Other measurement parameters derived from sedimentary petrography that have been found useful in ceramic petrography include the measure of the spread of grains over different size classes, skewness as a measure of the degree of symmetry of the distribution (relative spread of the data on one side or other of the mean), and kurtosis as a measure of the "peakedness" of the distribution curve (Darvill and Timby 1982).

Flügel (1982) reviewed some of the standard grain-size classification schemes that have been used, and highlighted the advantages of using the phi scale. Phi units correspond to the negative logarithm to the base 2 of the grain size measured in millimeters, so a grain size of 1 mm equals 0 on the phi scale; smaller grain sizes have positive phi values while larger ones have negative values, and calculation of statistical parameters are simplified. Flügel reviewed problems with the practice of converting thin-section data to sieve data, and did not recommend doing so (although others, such as Fieller and Nicholson (1991) did convert grain measurements into size categories based on the mesh sizes used in geological sieve stacks). Grains may also be put into size categories on the standard Udden-Wentworth scale, where silt size is less than 0.0625 mm (62.5 μm), and sand size is greater than 62.5 μm (Chandler 2001).

In the past, these measurement parameters, as with percentage calculations, also involved either time-consuming individual measurements or faster but less reliable visual estimation methods. For example, shape (such as rounded, subrounded, subangular, or angular) is often estimated for quartz grains using visual guides for estimating degree of roundness (Pettijohn 1975; Scholle 1979; Pettijohn et al. 1987). Selection of grains to measure or estimate can be done rigorously by point counting, or faster but less accurately by trying to identify grains typical for the thin sections.

Flügel (1982) noted some of the sources of errors that can arise, including the problem of measuring smaller grain diameters than are really present simply because of the random way in which the sample may be cut, the fact that in many point-counting methods larger grains will have a higher probability of being counted and thus the resulting distribution curves may be slanted towards coarse grain sizes, and the overall effects of grain shape and orientation on whether or not higher or lower values may be obtained than are actually represented by the true grain sizes. He described some approaches to try to get around these problems. However, since he also noted that for accurate results 300–500 grains need to be counted per thin section, depending upon the type of material being analyzed, clearly modern image analysis approaches may be quite helpful.

Image analysis systems incorporate many of the same measurements that have been used by petrographers for decades using manual methods, however, a considerable saving in time is achieved by using analysis algorithms in conjunction with a computer. The earliest systems often required in-house programming, were instrument-specific, or were limited by computer memory and processing speeds. Today, a variety of comprehensive image analysis packages is available that make use of advances in microcomputers and software library developments. User interfaces simplify the application of complex operations (Hetzner 1998; Goins and Reedy 2000).

In computerized analysis of digital images taken through the microscope, one can highlight or mark for analysis certain minerals or features of a specific size range, shape, color, or contrast (Fig. 6.2.2). A wide range of data on those highlighted minerals or features can then be collected rapidly and simultaneously (such as number of grains, range and average size, degree of roundness or sphericity, range and average length of axes, length of perimeter, etc.). Identification of specific minerals and textures present must still be done mainly through petrographic training and experience. However, even automatic identification of some textural characteristics has been successfully demonstrated (Wang 1995).

In digital image analysis, protocols for differentiating (or segmenting) the components and features of interest are needed. For example, a mineral might vary in size, color, and shape within a thin section or it may be very close in size, color, and shape to another mineral in the thin section. A good image analysis software package easily allows one to manipulate image contrast and brightness and apply a variety of different filters and algorithms to the image in order to enhance features of interest. Once these features can be separated reliably by the software, decisions then need to be made regarding what data should be collected in order to answer specific research questions. Some of the work on image analysis performed by sedimentary geologists is relevant to ceramic studies, and can provide some useful ideas (Schäfer and Teyssen 1987; Protz and VandenBygaart 1998; Francus 1998; Lagrou et al. 2004).

Whitbread (1991) wrote an image analysis program called Grain-Size Analysis (GSA) to help in distinguishing particles to be measured. He found that in crossed polarized light, ceramic fabrics tend to be rather dark, limiting the ability of automatic image analysis to identify boundaries of some components. For example, composite inclusions such as lithic fragments can present problems in differentiating between boundaries of the lithic fragment and those of individual grains within the fragment. Thus Whitbread used a combination of conditions including plane polarized light, crossed polarized light, and crossed polarized light with a gypsum plate, in order to clarify the boundaries of some features. Modern comprehensive image analysis packages help to overcome some of these limitations by providing a variety of approaches to aid in separating the components of interest (Reedy 2006a). Recent innovations in petrographic microscope equipment (Fueten 2006) also promise to help maximize the usefulness and success of image analysis work.

Once an image analysis package has been selected, a variety of other factors need to be considered as part of good image analysis practice. Some of these factors include experimental design and statistical validity issues, calibration of images, image quality (which can have a major impact on analysis results), and methods for image enhancement or preprocessing (Reedy 2006a). The time-consuming and tedious methods of quantitative petrography for characterizing ceramic materials lend themselves well to digital image analysis. For example, Forte (1994) incorporated image analysis of thin sections in characterization studies of Etruscan bucchero ceramics and found it helpful when addressing a variety of questions including identification of production areas.

Velde and Druc (1999) presented results of digital image analyses of common ware sherds from central France. The goals were to count grains of specific types, estimate the amount of nonclay material present, and estimate grain sizes and relative amounts of different-sized grains. In this case, digital images were not taken directly at the microscope, but were obtained by scanning photomicrographs. Quartz and feldspar grains were enhanced through a binary conversion process that converts those grains to black and the clay matrix background to white. An image analysis program (not specified) was then used to compute the surface area of the particles, perimeters, and length over width; frequency histograms were produced with the data by the use of a separate spreadsheet. One potential problem with the printing and scanning of photomicrographs is that calibration issues have to be fully considered, especially if size measurements are important (Reedy 2006a).

Using a digital microscope camera and an image analysis package (such as Image-Pro Plus), when studying a sandy ceramic thin section it is fast and easy to focus on the nonclay sand fraction and look at its size distribution. Once calibration and preprocessing steps have been completed, one can use a procedure in which the bright objects (the sand) within the darker clay matrix are highlighted and measured. Histograms can then be viewed for visual evaluation of size distributions. Figure 6.2.3 and the corresponding histogram, Figure 6.2.4, indicate a continuous distribution of sand grain sizes. Figure 6.2.5 and its corresponding histogram in Figure

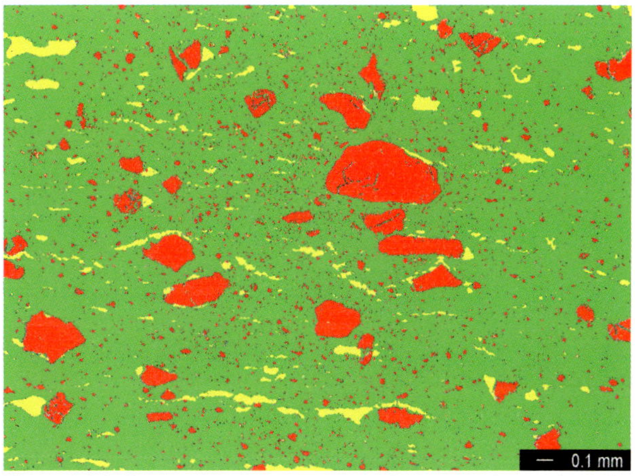

Figure 6.2.2 German stoneware, seventeenth century, plane polarized light image from Figure 6.1.12 with the sand content highlighted in red and the pores in yellow (the clay matrix is marked with green), for digital image analysis using the Image-Pro Plus software package. The results from this analysis, averaged with those from a second field of view, are given in the caption for Figure 6.2.1.

6.2.6, in contrast, point to the presence of two distinct populations: one population has a similar spread to that of Figure 6.2.3, although with a somewhat larger starting size, while the second population consists of a concentration of larger grains. The first population of grains may be a natural constituent of the clay, with the second representing a separate temper additive. The data, along with other parameters such as total area percentage of the sand component, the minimum, maximum, and mean length of the sand particles, and minimum, maximum, and mean quartz roundness (perimeter divided by area), or other measurements, can be exported to a statistical package for multivariate statistical analyses.

A similar approach, although working with scanned photomicrographs, was taken by Bouchain and Velde (2001) in a study of grain-size distribution by image analysis of thin sections from Gaulo-Roman common wares excavated in France. All specimens studied had a similar mineralogy of quartz, potassium feldspar, muscovite, and biotite with occasional accessory minerals. A two-mica granite was suggested as a precursor, with a common origin for both the clay and grit resources, probably coming from sandy micaceous deposits found along a river bed near the site. An in-house program was used for a binary conversion process, so the brown clay was converted to white background while the translucent grit was converted to black. Estimation of grain size for the grit particles was done by computing the percentage of the area of the photograph occupied by various size categories. Grain-size distribution curves were then plotted.

Bouchain and Velde concluded that three types of grain distribution were present. In one, grain sizes were evenly and symmetrically distributed on plots that compared log size to

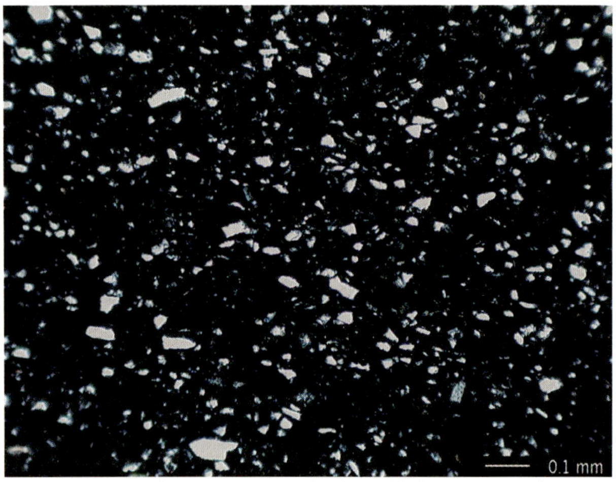

Figure 6.2.3 Sandy ceramic, crossed polarized light. A ceramic thin section analyzed using "automatic bright objects" procedure in the Image-Pro Plus software package, to select the bright sand grains within a dark clay matrix.

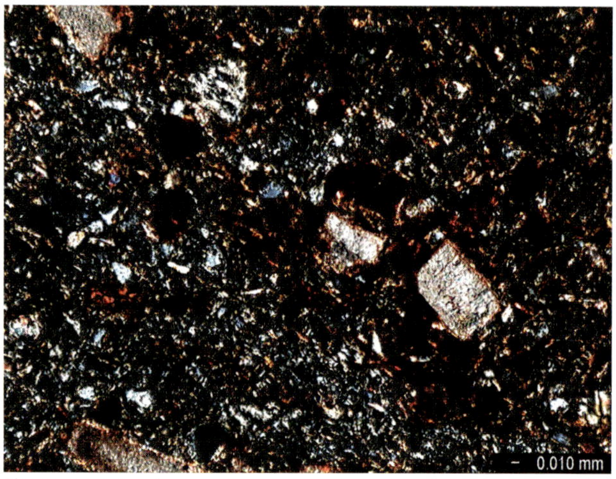

Figure 6.2.5 Sandy ceramic, crossed polarized light. A second ceramic thin section analyzed using the Image-Pro Plus "automatic bright objects" procedure to select the bright grains within the darker clay matrix.

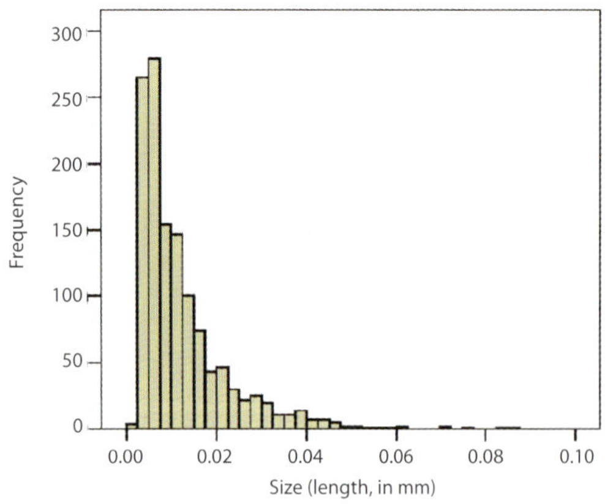

Figure 6.2.4 Histogram resulting from analysis of the thin section in Figure 6.2.3, showing a continuous distribution of sand grain sizes, ranging from 0.002 to 0.09 mm, with the largest number of grains falling between 0.003 and 0.02 mm. Such a continuous distribution of grain sizes indicates that the sand was probably a natural constituent of the clay, rather than being a deliberate addition of temper material.

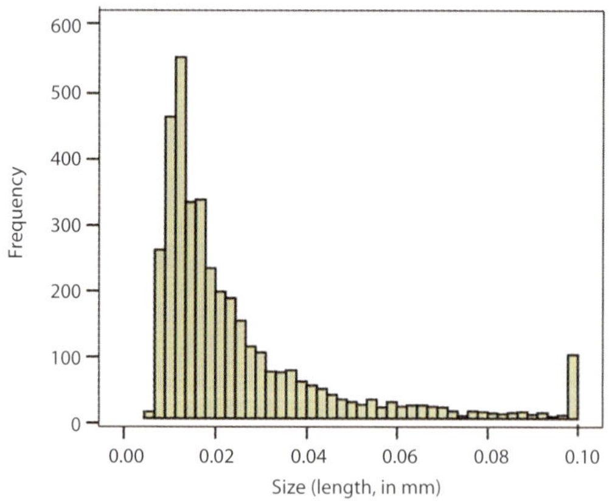

Figure 6.2.6 The histogram resulting from the analysis of the thin section shown in Figure 6.2.5, showing two distinct populations of grain size rather than one continuous distribution. Most of the grains are small, a few are larger. One background population shows the same continuous distribution of grain sizes as seen in Figure 6.2.4, although with a starting point of larger grains. This group ranges from 0.006 to just above 0.09 mm, with most grains falling between 0.008 and 0.05 mm. A second group is concentrated at larger sizes of 0.10 mm or larger. The first population may be a natural constituent of the clay, with the second representing a separate temper additive.

area percent, indicating a single origin for sediments used to make the paste. In the second, the same regular distribution appeared, but there were also some very large grains with a clear size cut-off, as might result from sieving of the river sediment to produce the grit. The third was similar to the second, but the large grains included some that were much larger than the others, which Bouchain and Velde interpreted as indicating the addition of a temper of varying grain size. Thus different methods of preparing paste and temper apparently were used for a single ceramic ware.

Another approach compared the total surface area of the larger grains with that of the smaller grains, keeping in mind that there will be a range of diameters for both added and naturally occurring grains because of two-dimensional cutting of the material represented by a thin section (not all grains get cut at the center). The two statistical approaches combined can be used to characterize the amount of temper additive. Even when the interpretation of grain-size distributions is unclear, these distributions provide one way of characterizing the ceramics of a specific site or region and comparing them to those of other sites or regions (Velde and Druc 1999).

The degree of rounding, particularly for quartz grains, is one indication of how far the material has been transported, and thus can help to pinpoint geological origins of inclusions or group materials according to similarity or differences. Roundness may be estimated using visual comparison charts prepared by sedimentary geologists, which generally categorize grains as angular, subangular, subrounded, or rounded. Image analysis packages permit a more quantitative measure of roundness (Figs. 6.2.7 and 6.2.8, for perimeter over area), and rapidly provide information on minimum, maximum, and average roundness for a selection of phases and size fractions. The degree of roundness is relatively meaningless as a measurement applied to all inclusions or temper additives, because different minerals and lithic types differ in hardness. Thus the characteristic is primarily useful when applied solely to the quartz component.

Rice (1977) emphasized that conclusions drawn from a study of ceramic sherds may be virtually useless if not accompanied by comparable analyses from potential raw materials, as well as information on the geology of the region of interest. She noted that such data provide a context or a standard of reference against which sherds can be compared. The comparative data permit interpretations to be made that cannot otherwise be made. One example of this point is the ethnographic study of potters in Quinua, a modern village in the southern Andean Highlands located near the ancient site of Huari carried out by Arnold (1972). He analyzed raw materials that the potters used for their clay and temper sources, as well as samples collected from the clay and temper mines, and found that the Quinua potters did not use clearly distinguishable clay materials versus "pure" temper additives like crushed rock or sand. Instead, both materials contained both plastics and nonplastics, differing only in the proportion of each. The material they added as temper contained more nonplastics than clay, whereas the material they used as clay contained more clay than nonplastics. Clay to which they chose not to add any temper contained a high proportion of naturally occurring mica, so would have good working properties without any need to add nonplastics. In contrast, the clay to which added temper was added contained very little except for clay minerals, so would need an nonplastic additive. Arnold recommended avoiding any implications of deliberate addition when nonplastics are encountered in sherds, because it is difficult to discern what was added and what was the result of judicious materials selection.

It is clear that some nonplastics may occur naturally within a clay deposit (which may be selected by a potter specifically for the qualities imparted both by the clay and by the nonplastic components), while others may be deliberately added by the potter in specific proportions to produce the desired functional qualities. Some researchers reserve the term "temper" (or, less often, "filler" or "opening materials") for deliberately added nonplastics (Freestone and Hughes 1989; Gibson and Woods 1997), while others define temper

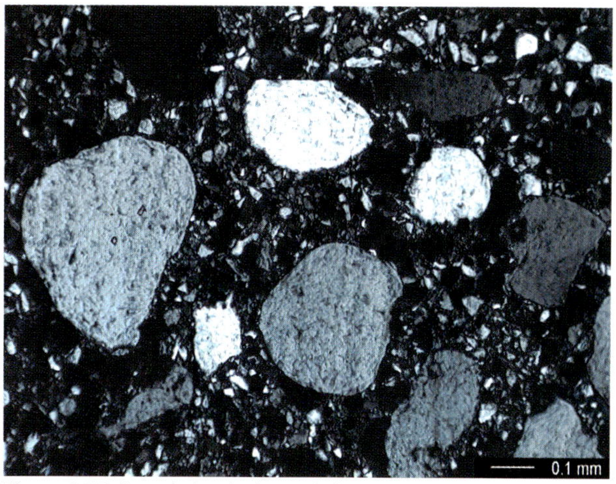

Figure 6.2.7 Rounded and subrounded quartz grains in Near Eastern ceramic (Nippur, Islamic kiln finds), crossed polarized light. Mean roundness for grains larger than 0.02 mm is 3.2 (measured by Image-Pro Plus). A perfectly circular object would have a roundness measure of 1.0; as objects become more and more irregular, they will have larger values.

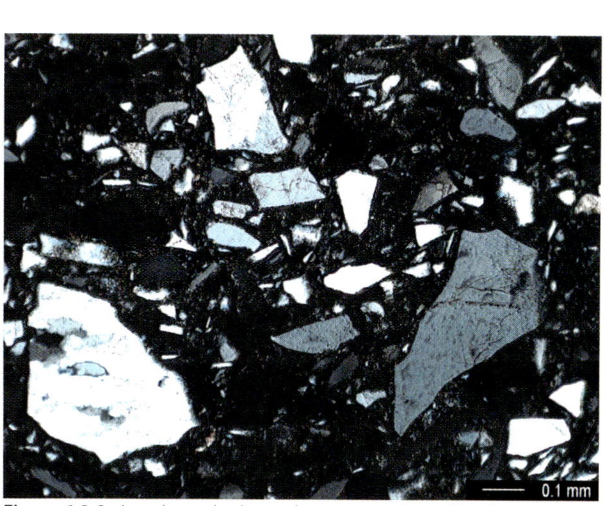

Figure 6.2.8 Angular and subangular quartz grains in Near Eastern ceramic (Raqqa, Syria, twelfth century), crossed polarized light. Mean roundness for grains larger than 0.02 mm is 7.5 (measured by Image-Pro Plus). A perfectly circular object would have a roundness measure of 1.0; as objects become more and more irregular, they will have larger values.

as encompassing all components of diameter greater than 0.015 mm (Maggetti 1982). The term "opening material" derives from the function of temper during firing, especially when the increased porosity of a tempered clay body facilitates the release of gases and thus prevents damage to the ceramic. An added temper can sometimes be recognized by angularity, from crushing, versus the rounding of transported sand, and by polymineralic composition of minerals from rock fragments, as opposed to single, separate crystals in clay (Gibson and Woods 1997). However, sand naturally occurring in clay can vary in composition, sorting, and level of transport, and sands are often added as the temper. Rye (1981) pointed out that a small portion of the nonplastic fraction may actually be parting agent (such as sand or ash), used by traditional potters to prevent sticky clay from adhering to the surface where kneading, tempering, and other clay preparation steps take place. It may appear as a minor component, incorporated into the clay if used before kneading, and only on the surface if used later during vessel fabrication.

Since it is not always easy to determine whether or not inclusions were added to the clay, or what proportion was added versus what was naturally present, the definition of temper that encompasses all of the larger nonclay components is often the most practical. Whether an inclusion was naturally present or was added, the fact remains that the type and amount of this component affects the working, drying, firing, and use properties of the clay, and the presence, type, and amount of these inclusions within the clay matrix is a reflection of the technological style of a workshop.

In the study by Giot and Querré (1987) of medieval ceramic wares from Brittany, they found that most Breton pottery was made with clay from a source that showed little transport or redeposition. Thus natural temper constituted almost half of the volume of the clay. Even when other nearby clay sources were used, sands and arenite reflecting the neighboring natural temper-rich source were deliberately added as temper.

Even if difficult, it can be very helpful to try to deduce whether or not there is an added temper if one is interested in characterizing technological styles as closely as possible, or in identifying possible raw material sources. Distinguishing an added temper from natural inclusions in the clay can lead to inferences about production and exchange that are otherwise not possible. To better understand the range of practices for clay processing and nonplastic additions in pre-Spanish pottery from Peru, Ixer and Lunt (1991) compared thin sections made from archaeological sherds to local clay deposits as well as to modern local pottery made with local materials by known production methods. The archaeological sherds came from three groups of pottery (Inca Fine Ware, Inca Utilitarian Wares, and Killke Ware) from the Cusichaca valley, Peru, dating to the Late Horizon period (ca. A.D. 1450–1532). The lower limit for recording data on nonplastics was set at 15 μm. For particles that size and larger, the particles were identified and a qualitative assessment made of their relative abundances, internal grain size of polycrystalline rocks (coarse, medium, or fine), roundness/angularity, degree of weathering and nature of bounding surfaces (smooth and planar or irregular and dented). Maximum length for the nonplastic fraction was measured and a grain-size analysis was performed, counting the first 150–180 grains per thin section. Data were plotted on the phi scale using a class width of ½ phi, and then used in a statistical analysis.

Results showed that the Inca Fine Wares, like modern fine wares, were made with a very clean (sedimentary clast-free) clay. The clays must have been either naturally clean, or if alluvial clays were used, these were sieved and/or levigated to clean up the clays. A crushed basalt-andesite temper (of nonlocal origin) was then added. The Inca Utilitarian Wares had a unimodal distribution and a statistical means that varied for vessel type (phi of 3.53 for the cooking pots and 3.38 for other utilitarian wares). They were similar to the modern coarse wares, but had much variation and did not form a cohesive group. They were made by adding heterogeneous sands or crushed sands/gravels to clays with variation in nonplastic content, and could be either local or nonlocal. The Killke Wares differed from modern coarse and fine wares and from most Inca Fine Wares. They had crushed feldspathic rocks (nonlocal) added to clean to dirty clays with variable nonplastic content.

Peña (1992) analyzed four clays used by potters in Vasanello, Italy, to make cookware vessels. One source showed abundant nonplastics in the silt to fine sand range, mainly subangular to rounded grains of quartz, with scattered iron oxides and detrital fragments of mica, plagioclase, well-rounded sandstone, and some orthoclase and olivine. Samples of the clays used by the potters were fired as briquettes, along with samples of ceramics made by the potters, so that firing changes in the original nonplastic composition could be observed. A second source also had abundant nonplastics in the silt to medium sand range, but these consisted mainly of subangular to rounded grains of quartz, some mica and iron oxides, and rare subangular to rounded grains of alkaline volcanic material (sanidine, augite, and fragments of volcanic rocks). Peña suggested that the presence of rounded quartz grains indicated a sedimentary or secondary clay rather than a primary clay formed in place by the weathering of the volcanic parent material. The third source was found to be a sedimentary deposit containing a mixture of materials of both alkaline volcanic (from weathering of nearby trachitic-phonolitic tuff) and pre-volcanic (polycrystalline quartz of metamorphic origin). The fourth source had abundant nonplastics in the silt to very coarse sand range. These included mostly angular grains of sanidine, along with some globules of iron oxide, angular grains of plagioclase and augite, altered mica (probably biotite), and rounded fragments of volcanic rock. This suite of clays provided a good range of materials that did not require sifting, levigating, or adding temper, thus minimizing raw material processing needs.

The fourth source, rich in sanidine crystals, was especially well suited for use in cooking vessels due to its coefficient of thermal expansion. Peña noted that this clay had some fine-grained quartz and the volcanic component (sanidine) appeared in a range of sizes up to relatively coarse, illustrating the difficulty of determining whether a deliberate temper additive is present versus an unmodified sedimentary clay.

Nonplastics can accompany clay deposits either as one of the weathering products from initial formation of the clay, or as material picked up during transport processes. The potter may process the clay deposit through sieving and levigation procedures to increase plasticity or other desired qualities, removing much of the nonplastic component. This same component may be re-added by the potter in various proportions depending upon the type of ware to be made, or temper additives may be partially or completely derived separately from the clay deposit.

Sometimes it is possible to distinguish between temper material deriving from an entirely different source than the clay, by observing that major components of the larger temper additives are not found as small grains within the clay itself. Tempers added from a different source often differ from the clay matrix not just in composition, but frequently also in size or shape. For example, the larger particles may be more angular from crushing of a lithic temper material as part of processing, while the natural sand particles in the clay may be more rounded, or the clay may have angular particles while the temper may consist of worn and rounded quartz sand (Hodges 1963). A good understanding of nonplastics that naturally occur with clay versus additives is facilitated by sampling and analyzing local clay and potential temper sources. Analysis of a large selection of sherds from a site or region may also help to provide information not available from the analysis of a few sherds in isolation. For example, the proportion of temper to clay may vary considerably within specimens from a single ware, whereas the proportion of naturally occurring nonplastics to clay may remain relatively constant (Hodges 1963).

In some cases, it may be possible to detect deliberate addition of temper as opposed to naturally occurring nonplastics. Some examples are cited by Maggetti (1982). First, where calcite was present, he noted that rhombohedral calcite with sharp angular edges would only be possible if crushed calcite (as with limestone) were deliberately added to the clay; naturally occurring calcite would have rounded edges from water transport and/or from decomposition during burial. He pointed out that for quartz particles, although angularity might also indicate addition of crushed materials, angular quartz is often found naturally in clay deposits so this interpretation cannot be made reliably. Secondly, he discussed the presence of fired brick as an added temper material; one could also put grog, or crushed sherds, into this category. These materials can be distinguished from the clay matrix by differing color, mineralogy, and/or texture. Thirdly, although some organic matter can be a natural component of clay, the presence of large amounts of coarse-sized organic material such as straw indicates a deliberate addition.

Woods (1982) provided some cautionary examples that illustrate the difficulty of using size and shape characteristics to identify the presence of an added temper in the absence of geologic source information. She discussed the usual criteria of looking for size and shape characteristics that differ from other inclusions in the clay matrix. Since natural weathering of rock fragments tends to produce rounded particles, angular inclusions of lithic fragments are taken as an indication of deliberate crushing and adding by the potter. She then highlighted a case where quartz, feldspar, and hornblende inclusions in pottery from southern Britain were identified as sand temper. The fragments were large and angular as well as monomineralic. The supposition was that they derived from a naturally occurring sand that had not been transported far from the parent rock so had retained size and angularity; a sand temper was then posited. However, a geologic study of source materials later identified clays with this sand component as a natural constituent, and so in reality no temper materials had been deliberately added.

Howard (1982) pointed out that the most reliable way to relate the mineral and lithic inclusions in a ceramic material to natural clay formations, or to determine what was naturally present in the clay versus what was deliberately added by the potter, is to systematically sample clay and temper sources in the region(s) of interest. She recommended preparing and firing clay briquettes in a consistent manner to be used for thin sections that can serve as comparative known standards for thin sections made from archaeological sherds.

In Kilka's (1992) examination of thin sections of Bronze Age ceramics from Fiave, Italy, grain-size measurements were used to try to verify whether or not a temper might have been added deliberately. He collected clays on the site to study the grain-size distribution of natural mineral inclusions. The hypothesis was that a unimodal distribution, as illustrated in histograms, is an indication of a natural material with no temper added by the potter. A typical histogram for a clay rich in quartz and feldspars, one of the two main types of materials used by Bronze Age potters, had a very high percentage of very fine grains (0.01–0.05 mm) and a very small percentage of coarse grains. A calcareous clay, the other type of material selected by Bronze Age potters, also had a unimodal curve, plus a few very large grains present. Some of the sherds did not clearly show added temper but many others did have a bimodal distribution, with a typical curve for the natural temper, then a break and a second peak at a large size (such as 0.13 mm) indicating the added temper. Kilka concluded that sherds with a mainly calcareous temper present always had an added temper, those with a quartz and feldspar temper sometimes had only naturally occurring nonplastics, and on occasion it was unclear whether the temper was natural or deliberately added.

Stoltman (2001) pointed out that while it may be easier to identify deliberately added temper if certain materials were chosen (such as grog, shell, limestone, or volcanic ash), if the temper was a quartz-rich rock (often called "grit"), such as granite or quartzite, the task is more complex. Small fragments of the temper may be difficult to distinguish from the ubiquitous quartz that is a natural constituent of sand and sediments. He recommended identifying all the minerals visible in the temper, then, if possible, obtaining samples of local soils and sediments to identify which minerals occur naturally at that site or region. Even when local sand has been used as a temper material, distinguishing a natural impurity of the clay versus a separate deliberate addition is difficult.

The basic assumption, largely supported by ethnographic studies, underlying the use of an analysis of nonplastic components of ceramics in characterizing local wares and in differentiating them from imported ceramics, or for more specific provenance studies, is that potters tend to use locally available clay and temper sources (Maggetti 1982). It follows, then, that the mineralogy and attributes of a temper material should reflect the local geology of a site. In contrast, imported ceramics should be identifiable as containing temper materials not locally available. To carry an analysis this far requires not only a thorough characterization of temper materials in the ceramics, but also a knowledge of the geology of the site or of potential sources.

In some cases, identifying the minerals or lithics present may be enough to characterize a production center or identify a possible geologic source. More often, though, the mineralogy for a wide range of production centers will be similar, and simply collecting presence/absence data for various components will not be sufficient. Quantitative data, and a wide range of textural data, will most likely be needed for detailed studies of production centers or provenance.

Darvill and Timby (1982) used grain-size analysis to characterize the products found at a Roman pottery kiln complex at Purton in Wiltshire (England), in the hope that if the output from individual kilns could be characterized and distinguished from the output of other kilns, then ceramic materials from sites excavated in the areas could be matched to kiln sources and therefore used in economic studies. As a first step in the study, initial experiments were conducted to assess the sample size needed for accurate characterization of a vessel. Two thin sections were prepared from each vessel, and 100 quartz grains from each thin section were measured. A multivariate cluster analysis was used to compare samples according to the relative proportions of grains in each size class. The resulting dendrogram showed that where two samples from the same pot were taken, the density values for each were very similar and in most cases were first partner linkages. Darvill and Timby concluded that a single sample would be sufficiently representative of the vessel. Based on the grain-size analysis, two groups were clearly identified from the kiln complex. One fabric was used to make thin-walled vessels, and the other to make thick-walled vessels. It was noted that the two types of wares might have required different processing of raw materials (in clay mixing or refinement) and thus affected variability in the thin sections. Based on this work, vessels or sherds from archaeological sites in the region could be classified by form and general fabric, then checked with petrographic comparison with material of the appropriate ware from the known kiln complex. This work also highlighted the fact that unless an assessment of the full range of output from production sites is available, the premise that one cluster is equivalent to one production site is too simplistic to be valid.

Whitbread (1989, 1995) noted that in addition to techniques of sedimentary petrography, ceramic thin section analysis can benefit from the application of techniques developed by those who study soil micromorphology (an approach later seconded by Josephs (2005)). He illustrated descriptive and statistical techniques for conducting fabric micromorphology (defined as the arrangement, size, shape, frequency, and composition of components of the ceramic material). With more detailed descriptions of a ceramic fabric, fabrics can be classified into groups as part of a provenance study, with the emphasis on distinguishing between fabrics of different provenance and on matching similarities of mineral and rock inclusions to possible raw material sources. In studies focused on deducing ceramic technology, different data will be collected, with the emphasis on the relationship of a fabric and its components to possible mixing, levigating, forming, and firing actions of the potters. Whitbread pointed out that soil micromorphologists study the relationships between fabric components that include large amounts of clay matrix, in contrast to sedimentary petrographers who mainly concentrate on larger sand-sized particles.

Whitbread illustrated the use of a thin section description protocol that incorporated a detailed description of microstructure, groundmass, and micromass features. He also employed a measurement of coarse versus fine particle distribution following a method that combines a grain-size histogram with particle composition data. For example, he showed results of analysis of a Corinthian B amphora in which there was clearly a bimodal distribution with the fine grains being primarily quartz, polycrystalline quartz, and some rock fragments, while the coarse grains were mainly volcanic rock fragments, clinopyroxene, and feldspar. He concluded that there was an added temper that was well sorted and consisted mainly of volcanic material; the temper was added to a clay containing natural inclusions of fine-grained quartz and rock fragments. However, he noted that a bimodal grain-size distribution seen on a composition histogram in which there was significant overlap in composition of the fine and coarse particles would have given the misleading impression that the coarse inclusions were naturally present in the clay rather than being an added temper.

Freestone (1991) reviewed some of the advantages and problems with examination and interpretation of nonplastic inclusions in ceramic thin sections. For example, the visual image holds much more information than can generally be recorded or reported, and each analyst tends to focus on the specific research questions at hand. Freestone recommended that, in order for others to be able to use more data from published thin-section studies, six primary pieces of information should be recorded at a minimum: (1) a list of all nonplastic inclusions present above trace level, with relative abundance; (2) total concentration of nonplastic inclusions even if only in subjective terms (such as "abundant" and "rare"); (3) degree of sorting of the nonplastic inclusions, and any indication of the presence of a bimodal distribution; (4) the "typical" grain size and grain size range; (5) an estimate of roundness; and (6) color of the clay matrix and whether or not it is birefringent.

Freestone's ideas were carried to fruition in part by at least one major project, the assembly and publication of *The national Roman fabric reference collection* (Tomber and Dore 2002). This handbook includes complete descriptions of hand samples and thin sections, and photographs of cross-sections, for a variety of Roman wares (Gallo-Belgic, imported Samian, Pompeian red wares, imported fine wares, imported coarse wares, amphorae, and Romano-British wares) with many types within those wares. The thin-section descriptions included all of the information recommended by Freestone except for regular discussion of the degree of birefringence of the clay matrix (that information is given for some, but not all, entries). The availability of a resource such as this can be very helpful to archaeological petrographers, as it is possible to find a match quickly for archaeological sherds under analysis, facilitates standard petrographic descriptions, and ensures that each researcher does not have to keep starting from scratch or searching for similar materials through large amounts of scattered archaeological and archaeometric literature.

As a cautionary note, it is important to be aware of mineralogical changes that may have occurred during use or burial that may have affected nonplastic inclusions. Calcite grains may dissolve, and fine-grained secondary calcite or gypsum may crystallize, especially around the edges of pores. These changes can usually be discerned in thin section, but the analyst will need to ensure that they are taken into account in any quantitative studies of temper materials. In addition, in high-fired wares, some of the original components will have been converted into new, higher temperature minerals.

A wide variety of materials appears as natural inclusions or added temper in pottery. These include lithic inclusions that are frequently in the form of crushed rock, sand, calcium carbonates, organic material or chaff which is often carbonized after firing, grog (crushed sherds), and others. Each of these is discussed in more detail in the sections below.

6.2.2 Lithics

Lithics are small pieces of rock, or multigrain/multimineral fragments, often crushed. They can enter into the ceramic paste as one component of a temper additive, as the primary nonplastic additive, or as a natural accessory of the clay (Fig. 6.2.9). Sands often have a minor lithic component, so in order to identify the presence of a prepared lithic temper, rock fragments of the same type should be a major component of the temper material. Lithic tempers may also have separate mineral grains that have flaked off from the lithic fragments, and they often have many relatively coarse fragments, which tend to be angular because of crushing during preparation procedures (except in the case of very soft stones) (Hodges 1963; Velde and Druc 1999).

An early, famous, use of thin-section petrography to identify lithic tempers in ceramics was that of Anna Shepard (1942, 1965). In a study of the ceramics of Pecos pueblo (including over 500 thin sections and tracking of the geological resources available), she identified lithic tempers as deriving from rocks that were not available locally. She concluded that during Glaze I times, decorated wares were being imported from the Galisteo and Santa Fe regions. This finding was contrary to the previously accepted idea that the pueblos were independent economic units, and indicated a degree of specialization not previously considered. She found a number of very distinctive lithic tempers in sherds that could be traced to specific, restricted sources. For example, a devitrified tuff from the northern Pajarito contained tridymite, so served as a good marker for the pottery of that area.

De Paepe (1979) analyzed 400 thin sections of coarse wares from Thorikos in Greece. He noted that the sherd materials were impregnated under vacuum with a polymerizing plastic and dried for up to six weeks before thin sections were prepared (to prevent loss of any material), and that prior to mounting a cover glass, each thin section was stained to highlight feldspars (as otherwise, differentiating them from quartz grains can be time consuming). In order to analyze a large number of thin sections, percentages were derived through estimation rather than through point counting.

Four ceramic groups were identified based on the occurrence and relative abundances of volcanic, sedimentary, and metamorphic mineral and lithic inclusions. Lithics included fragments of pumice, lava, sandstone, limestone, dolomite, phyllite, schist, marble, quartzite, and others. All loose minerals in the clay pastes were also found as constituents in the lithics. By comparing the lithics to possible sources, De Paepe concluded that bowls and water containers were probably made locally near Thorikos, but cooking wares were probably imported trade goods. He noted that quality of cooking vessels would be crucial, because they would need to be resistant to repeated heating and sudden quenching. Most of the sherds from cooking ware at Thorikos were tempered with glassy volcanic lithic fragments, not available locally. Previously, archaeologists had argued that coarse wares should be presumed to be of local manufacture, while fine wares might be imports. However, this study showed that such an assumption is not viable.

Lithic materials may also be present as natural constituents of the clay, rather than being a deliberately added temper. For example, Morris (1982) analyzed decorated Iron Age pottery from several archaeological sites in

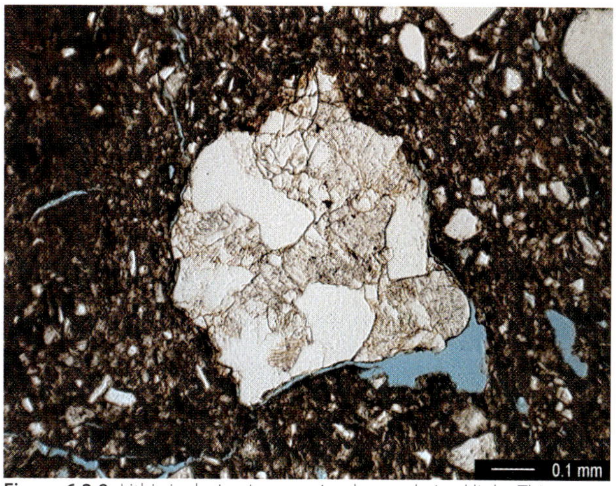

Figure 6.2.9 Lithic inclusion in ceramic, plane polarized light. This plutonic lithic inclusion (granodiorite) consists mainly of quartz (the clear grains) and feldspars (the grains exhibiting a brown weathering product). The inclusion is found within the clay matrix of an early Chinese earthenware ceramic; the blue areas mark pores.

western Britain. In thin section she found that the vesicular pottery had angular, sandy, micaceous mudstone fragments comprising 5–35% of the material; often only voids were left where the mudstone had disintegrated (since mudstone is a type of sedimentary rock that can disintegrate in water and be leached out during burial of the sherds). Using local geological data, she concluded that the mudstone was most likely a natural inclusion in the clay. Pottery made with this material was found to have a range of forms and decorations common to the region in which it was found, but it did have a distinctive distribution.

Vince (1982) discussed thin-section analyses of imported Spanish lusterwares (pottery decorated with gold luster) found in London. Andalusian lusterware (made at Malaga on the southern coast of Spain from the middle of the thirteenth century to the late fifteenth century) was tempered with schistose metamorphic lithics. Every vessel analyzed contained tabular fragments of reddish-brown schist. In contrast, Valencian lusterwares (produced in Valencia in the late fourteenth century and during the fifteenth century) did not contain the metamorphic lithic inclusions. Instead, they were fine textured with sparse inclusions of quartz, iron-bearing minerals, chert, and limestone with a matrix of finely divided calcium carbonate and anisotropic clay minerals.

Two of the temper types identified in Early Bronze Age ceramics from Bab edh-Dhra' in Jordan (Beynon et al. 1986) were crushed lithics (limestone and basalt). The limestone temper consisted of coarse, angular, tabular grains (1.5–4.0 mm in size) of fossiliferous and micritic or fine-grained limestone, with some of the fragments containing chert. The extreme angularity and wholly limestone composition helped to identify the material as crushed limestone cobbles and boulders collected from the alluvial fan surface adjacent to the site. Cobbles collected from that area—crushed and made into comparative thin sections—were found to be very similar to the limestone grains in the ceramic materials. This temper type was found mainly in cooking pots or large storage vessels. The second type of lithic temper consisted of angular, blocky fragments (1.0–4.0 mm in size) of basalt, containing relatively large grains of plagioclase feldspar, amphibole, and pyroxene. As with the limestone, the angular form of the basalt fragments and an exclusively basalt composition suggested it was an intentionally crushed and added temper material. Comparative material collected at the alluvial fan surface adjacent to the site supported the idea that this was the source of the basalt temper. This temper type was used only for the manufacture of bowls, which were usually slipped and/or burnished.

Whitbread (1986b) identified argillaceous rock fragments (mudstone) as a temper material in ancient Corinth, Greece. He ruled out similar-looking grog temper on the basis of the presence of polygonal fractures, frequent inclusions of radiolaria which matched local mudstone deposits, the composite nature of many fragments, and the lack of vessels made with a similar fabric that could have supplied a grog temper. Furthermore, the mudstone, when fired in an oxidizing atmosphere, tended to bloat more than the surrounding matrix, and turn glassy and isotropic, unlike grog. Two types of clay were used for the amphorae: a red clay derived from levigated *terra rosa* soils and a yellow clay derived from calcareous sediments. The fact that both types had similar mudstone inclusions suggested that the mudstone was deliberately added by the potters.

A lithic temper was found in an Inca-style vessel recovered during archaeological salvage work on La Puná Island in Ecuador (Isaacson and Aleto 1989). The presence of this tempering agent suggested that the vessel was not of local manufacture, and was instead of highland origin. Local ceramics had nonplastics that included mainly subangular to rounded plagioclase feldspar grains and subangular to rounded finely crystalline, weakly birefringent rock fragments lacking volcanic textures. Occasional rounded particles of volcanic rock with plagioclase microlite or lath inclusions were also seen. In contrast, the Inca-style object had a distinctly different temper. The main nonplastic constituents were volcanic lithics, identified as glassy lava, with angular shapes. The volcanic glass matrix of the lava included plagioclase feldspar laths, biotite, and hornblendes; these minerals were typical of the dacite and andesite rocks of the Ecuadorian highlands. In contrast, the temper of the local sherds derived from water-tumbled, well-rounded sand of volcanic origin that was probably collected at beach deposits, then added to the clay. The composition and angular shape of the Inca-style ceramic temper suggested it was lava that was crushed to the desired particle size and then added to the clay paste. Lava was not found on the island, so this vessel probably originated near one of the lava sources in the highlands.

Steward et al. (1990) examined thin sections from 90 sherds excavated at Robinson Ruin, New Mexico. Since most of the Robinson ceramics were grit tempered, the authors analyzed regional rock samples for comparison, collected at drainages such as those that would have been available to the potters. The ceramic thin sections could be distinguished and classified on the basis of temper: most were either a monzonite granite with abundant quartz within lithic fragments, or a syenite with little or no quartz present. This geological work suggested that the sherds tempered with quartz-rich igneous rocks most likely came from the vicinity of the nearby Capitan Mountains, while the syenite and other quartz-poor grit-tempered pottery could have come from the Robinson site or elsewhere (but excluding the Capitan Mountains). Some examples were found of materials that were clearly external to the study region (such as sherds tempered with a quartz mica schist). However, the picture was complicated by petrological gradation that hindered attempts at further subgroupings of possible production areas.

Betancourt and Myer (1994) found that five clay fabrics from eastern Crete had phyllite as the most abundant nonplastic constituent. Exposed phyllite rock deposits occur at many locations in eastern Crete, and during the Minoan Bronze Age, ceramic fabrics containing angular to subangular phyllite fragments were very common. The addition of a phyllite temper provided a number of advantages, especially to cooking vessels and large vessels used for storage. The phyllite shrinks very little during the firing process, so reduces the shrinkage of the paste as a whole. The presence of phyllite inclusions also increases the number of voids in the paste, allowing water vapor to escape during drying and firing, preventing breakage. The lentoid shape and irregular to subparallel orientation of the phyllite fragments in the clay body (Fig. 6.2.10) helps to disperse microcracks that can form from thermal shock or percussion, reducing breakage while the vessel is in use.

Thin-section analysis of phyllite-tempered wares showed that there were five different fabrics. These five fabrics were used only during specific time periods; four in the Minoan period and the fifth in the Byzantine period. In Late Minoan I, two phyllite fabrics differed from each other only in the coarseness and abundance of the phyllite inclusions. The finer fabric was used only for conical cups and other small items, while the coarse one was used for larger vessels. In Late Minoan III another two phyllite fabrics occurred. One had carbonate grains (micritic limestone) in addition to the phyllite, and the other had quartzite and associated quartz grains derived from it, along with the phyllite. The Early Byzantine phyllite fabric had temper from two different rocks, a phyllite that was different in mineral composition from that used in Minoan fabrics, and a metamorphic rock with abundant epidote. As a result of these findings, vessels made of these fabrics could be dated and assigned to the eastern part of Crete on the basis of the temper material.

Hegmon (1995) studied igneous rock-tempered ceramics from southwestern Colorado, white wares associated with the ninth-century occupation of the region. Thin-section petrography was undertaken for a better understanding of the ceramic production and the source of design variation. Slight differences in tempering materials were found at different sites, suggesting that the ceramics were produced on a small scale at many different locations. Comparison of ceramic materials with raw material sources indicated that potters in different locations were using the same criteria to select their temper. She concluded that localized white ware production was part of a widespread technological tradition in ninth-century southwest Colorado.

Hegmon focused on four attributes in her analysis: the texture of the rock used as temper; the characteristics of the feldspars (twinning, zoning, and clouding); presence and abundance of ferromagnesian minerals (hornblende, biotite, and pyroxene); and abundance of silt-sized particles in the ceramic paste. The most common rock texture was that of groundmass with phenocrysts (porphorytic). Coarser diorite-textured temper, and temper consisting of individual mineral grains, were also commonly seen. A cryptocrystalline texture was less common, and feldspars were present in all the samples. Twinning and zoning were considered to be basic characteristics reflecting formation of the minerals, not alterable by firing. Clouding of feldspars was indicative of weathering. Although a variety of ferromagnesian minerals was seen, Hegmon focused on the three most abundant ones. The silt-sized particles that were classified as abundant or not were primarily quartz. This fraction was considered to be a component of the clay and not of the added temper.

Samples of igneous rocks that might have been used as ceramic temper material were collected from some locations in southwestern Colorado. However, comparisons indicated that the primary source or sources of temper used by the prehistoric potters had not been found and sampled. Nevertheless, some interesting conclusions were reached. For example, the ceramic temper had appeared to be highly varied, but compared with the available rocks and their greater variability, the temper looked rather homogeneous. Apparently, the potters had selected only certain kinds of igneous rocks to use as temper, and the same general selection criteria seemed to be in use at different locations. Hegmon proposed that heavily altered rocks often contained calcium carbonate, undesirable because it can absorb moisture and cause the ceramics to crack and spall, so potters might have selected rocks with a certain macroscopic appearance (avoiding certain colors and white chunks).

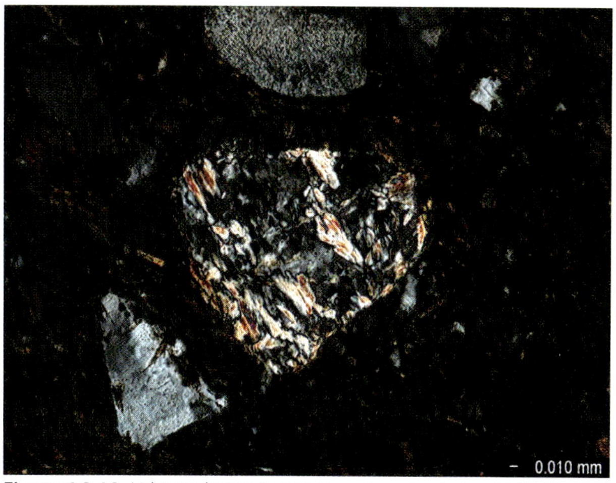

Figure 6.2.10 Lithic inclusion in ceramic, crossed polarized light. The phyllite fragment in the center of the field of view consists of fine, alternating layers of biotite mica (higher birefringence) and quartz (lower birefringence). The lithic fragment is surrounded by the clay matrix plus some quartz grains.

Mallory-Greenough et al. (1998) studied thin sections of a basalt-tempered bowl from Karnak (Luxor) in Egypt. The marl clay bowl contained some sand-sized (0.1–1.2 mm) mafic rock fragments with volcanic textures. These grains were angular, unweathered fragments apparently added as a temper. The authors noted that detrital sedimentary rocks often include lithic fragments. However, in this case they suggested that a volcanic lithic temper was added because of the high proportion of rock fragments in the sand-sized fraction, the lack of weathering on the feldspars within those fragments, and the angular shape of the rock fragments. Additional chemical analyses of pyroxene grains in the volcanic lithics, by electron microprobe, pinpointed the most likely source of the lithics as being in the Cairo area. The authors concluded that finding a vessel from Cairo in Karnak was consistent with the known movement of people and goods along the Nile.

Tomkins and Day (2001) summarized thin-section petrographic work on pottery from Knossos. A variety of lithic inclusions supported other evidence that vessels were produced at a number of different locations within Crete. The largest group of vessels contained crushed limestone temper compatible with a local provenance near Knossos. Other groups suggested a source in north-central Crete or farther away (such as a group with altered volcanic rock temper and one with phyllite temper). Most of the fabric groups occurred infrequently and most often their provenance was uncertain, although some may have originated from outside the island of Crete. One group included mostly fragments derived from granodiorite, with lesser amounts of phyllite and basaltic rock inclusions (called the "Mirabello" fabric), and had an origin about 70 km from Knossos. The authors concluded that the range of ceramic fabrics found at Early Neolithic Knossos indicated that ceramic vessels were important exchange items.

Thin-section petrography of Neolithic ceramics from the volcanic island of Yiali in the southeastern Aegean by Katsarou et al. (2002) found a variety of temper additives including obsidian. A total of 37 ceramics was sampled, along with two local sediments. Among the 37 ceramic materials studied, seven different fabrics were found. For six of the fabrics, the fine fraction was similar, consisting of a mixture of minerals and lithics. The dominant fabric had a coarser fraction of fragments of porphyritic dacite-andesite, poorly sorted with considerable variation in size. The dacite-andesite did not seem to have been selectively collected for temper material, but was part of a soil used for the temper additive. A second fabric was described as a "mixed volcanic fabric" with fragments of pumice outnumbering the fragments of dacite-andesite, again with considerable variation in size. The third fabric was described as "microcrystalline" with lithic and mineral fragments absent or very few in number. A "sorted andesite fabric" contained andesite grains that were well rounded and unimodal in size, suggesting they were sieved from soil and then added as a discrete temper material. In contrast to the above fabrics with locally available material, another fabric had sandstone and flint fragments within a clay matrix of argillaceous materials, which the authors proposed might represent imported ceramic products.

Finally, two of the fabrics included abundant obsidian. One was called "fine-grained volcanic glass fabric," and had very few coarse particles. However, the presence of obsidian in the matrix was high, along with perlitic glass, feldspars, and quartz. The authors suggested that it resulted from use of a well-sieved clay mixed with a local, well-sorted soil temper. Some organic material was also added (plant fibers). In contrast, the one termed "obsidian fabric" contained a high frequency of coarse obsidian grains, apparently deliberately added as temper material, with pumice and perlite glass. The coarse obsidian here had a consistent size and even distribution, indicating it was probably crushed and sorted to consistency. However, the specific functional advantage of an obsidian temper (Fig. 6.2.11) has not yet been determined.

Gherdan et al. (2002) used thin-section petrography to characterize Bronze and Iron Age ceramics from several sites in western Hungary. They found that most of the material contained metamorphic rock fragments, locally available. The distribution of temper fragments indicated that in most cases the metamorphic lithics were added deliberately to the raw material as a temper. A number of different groups could be identified based on the specific metamorphic material added. These materials could be localized to propose regional provenances for the various temper materials. Materials included one group with mainly talcose schist and serpentinite, one with tremolite schist, one with epidosite and phyllite, one with gneiss, and one group with nonmetamorphic materials (fine-grained quartz sand probably from alluvial deposits).

Knight et al. (2003) examined prehistoric pottery from sites in the East Midlands of Britain, characterized by coarse fabrics with angular granodiorite temper (Fig. 6.2.12). They critically reviewed the evidence for the production and distribution of prehistoric pottery tempered with granodiorite, looking at potential sources for the temper material as well as for the clay with which it was mixed. They also discussed potential advantages of granodiorite as a temper material.

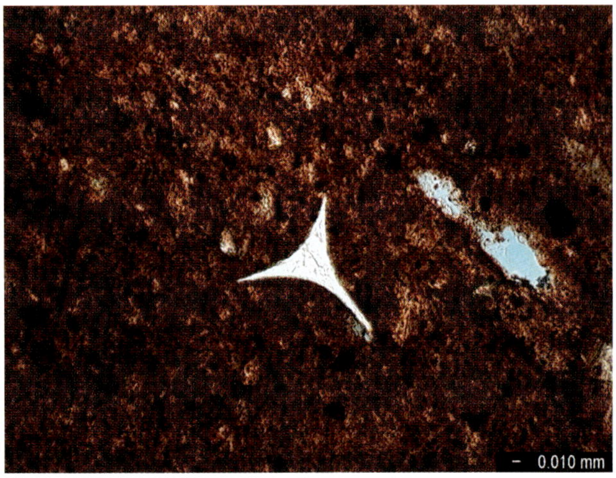

Figure 6.2.11 Lithic inclusion in ceramic, plane polarized light. The very angular, colorless fragment in the center of the field of view is volcanic glass (obsidian). The clay matrix is iron rich, with chunks of hematite and a reddish clay. The blue area is a void (pore).

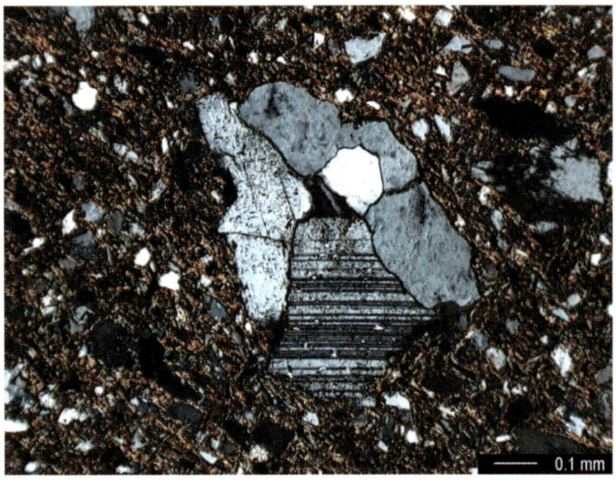

Figure 6.2.12 Lithic inclusion in ceramic, crossed polarized light. The coarse-grained, angular plutonic lithic inclusion in this ancient Chinese earthenware ceramic is granodiorite. It is composed of plagioclase feldspar (oligoclase), alkali feldspar with perthitic texture, and clear quartz grains. Much smaller quartz and feldspar grains are visible within the clay matrix.

The presumed source of the granodiorite was the Mountsorrel Complex, a large outcrop of igneous rock, principally granodiorite, near Charnwood Forest. Granodiorite is a coarse-grained igneous rock containing quartz, plagioclase feldspar, and alkali feldspar, with plagioclase being equal to or greater than two-thirds of the total feldspar content. Lesser amounts of biotite and hornblende are also found, along with accessory minerals such as sphene, zircon, and magnetite. Granodiorites from the Mountsorrel Complex have large concentrically zoned crystals of plagioclase set in granular aggregates of zoned plagioclase, quartz, and untwinned perthitic alkali feldspars. Feldspars tend to have secondary alteration resulting in minute, oxide-rich inclusions (Fig. 6.2.13). Mafic minerals include biotite with red to pale greenish-brown pleochroism and hornblende with pale yellow to pale green pleochroism. Biotite tends to be partially altered to chlorite and oxides. However, Knight et al. noted that differentiating and matching specific sources of granodiorite can be difficult when working with small inclusions in pottery.

The granodiorite inclusions that appeared to be consistent with a Mountsorrel source were generally angular or subangular in shape, suggesting that they had been deliberately crushed for use as a temper material. However, given the hardness of the unweathered parent rock, the authors proposed that a source of weathered and more easily crushed raw material would have been sought. Acceptable deposits of such material could include loose post-glacial deposits or gravel deposits derived from streams near outcrops of Mountsorrel granodiorite, as well as weathered knolls. The clays in this area included some with silt-sized quartz grains, some with argillaceous inclusions (mudstone), and some micaceous clays.

The authors also reviewed some of the reasons why potters may have chosen granodiorite as a temper material. As with other igneous rocks, it may have been added deliberately in order to reduce the plasticity of the clay and therefore make it more usable for forming vessels–it would also have given the clay added strength when wet, and acted as an aid in the release of moisture during drying and firing, reducing cracking or breaking risks. Another big advantage of granodiorite as a temper material in this area may be the ease of collection of weathered fragments from exposed deposits and the ease with which the fragments could be crushed to the appropriate size using manual pressure or gentle pounding. Preliminary heat treatment would have made the crushing step even easier, and some evidence for this treatment has been found.

Additional discussions of lithic tempers and their use in provenance studies are found in Chapter 7.

6.2.3 Sand

Sand grains in ceramics, as natural inclusions or added temper, are generally between 0.05 and 2 mm in size; the fraction between 0.002 and 0.05 mm in length is often termed "silt" and the clay particles themselves are less than 0.002 mm (Fig. 6.2.14) (Velde and Druc 1999). Experimental work (Rye 1976; Worthing 1982) suggests that one purpose of a sand temper is to improve the workability of clays (in gaining the right amount of plasticity, and having good drying and firing qualities). It also improves the thermal shock resistance of the ceramic products, even though quartz expands significantly more than the clay upon heating. Voids or pores form around quartz grains during drying and the initial firing, as the still-damp clay shrinks away from the grains (Fig. 6.2.15). Those pore spaces then provide room for the ceramic fabric to accommodate the changing size of the quartz grains during heating and cooling cycles of cooking (Middleton 1991). A random spread of quartz grains within the clay matrix increases the toughness of the otherwise brittle material by providing substantial flaw tolerance, and frictional energy is dissipated as a result of crack deflection and bifurcation. However, the optimal volume fraction of

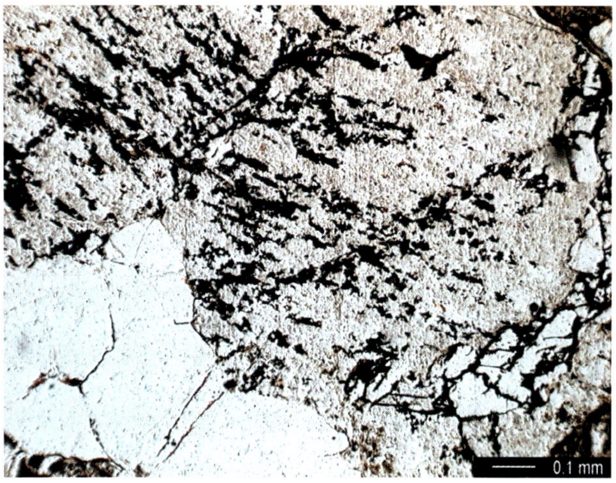

Figure 6.2.13 Lithic inclusion in ceramic, plane polarized light. This granodiorite fragment is a lithic inclusion in a sample of Owasco Corded Ware from the Upper Susquehanna valley of Pennsylvania, 900–1200. The feldspar grains have secondary alteration resulting in minute, oxide-rich inclusions. In contrast, the quartz grains in the lower left area of the field of view are relatively clear.

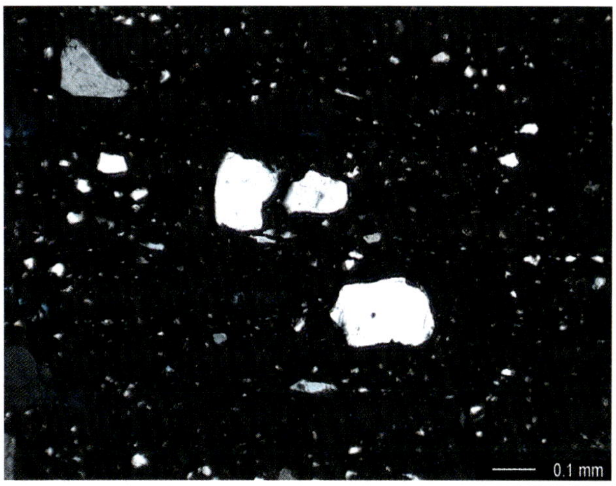

Figure 6.2.14 Sand inclusions in ceramic, crossed polarized light. Using Image-Pro Plus to measure and count the sand and silt grains in this seventeenth-century German stoneware, 498 grains were counted in this field of view. No grains below 0.002 mm were included. The silt-sized grains (0.002–0.05 mm) have a mean size of 0.01 mm, and comprise 3.5% of the area of the thin section (clay and sand only, with pores subtracted). The larger sand-sized grains (0.05 mm and above) comprise 10.5% of the area, and range up to 0.24 mm, with a mean size of 0.11 mm.

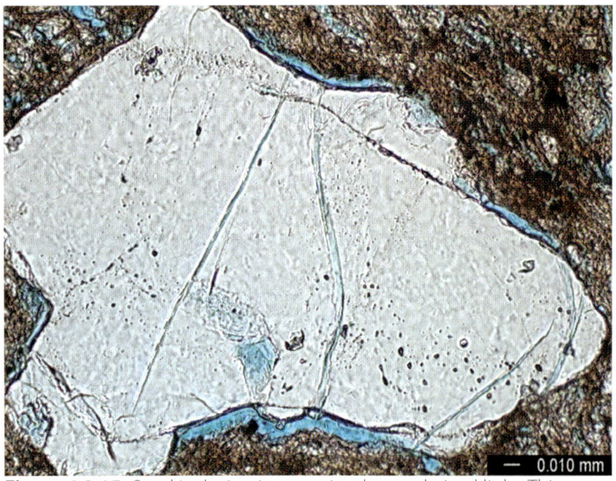

Figure 6.2.15 Sand inclusion in ceramic, plane polarized light. This quartz grain (from the same seventeenth-century German stoneware as is seen in Fig. 6.2.14) shows the formation of void or pore areas around the grain, caused by clay shrinking away from the grain during drying and initial firing stages. The pore space (blue areas) provides room for the ceramic fabric to accommodate the changing size of quartz during heating and cooling. Some cracking of the quartz is also evident here.

inclusions (independent of inclusion size) has been found to be about 20%: above that no further increase in toughness of the material is found (Kilikoglou et al. 1995).

Since quartz is very resistant to weathering, it tends to be by far the most prominent mineral in any sand component, whether occurring naturally in clay or as part of a separately added sand temper. The shape of the quartz grains can vary from very rounded to very angular, depending upon the nature of the original source material. The suite of minerals present, the percentage of each of the major components, and grain size and shape characteristics, can all be recorded for sand components in order to characterize a thin section or a group of thin sections.

If these data can be compared to thin sections of local naturally sandy sediments, it may be possible to deduce whether or not the sand in a sherd is a typical natural contaminant as opposed to originating from a completely separate sand source. For example, in medieval potteries of Poland, local sand was used to temper the clay, sometimes being sieved to produce a uniform size grade (Peacock 1985). However, if the potters mixed clays, and/or undertook levigation (artificial removal of coarse particles), then the subsequent enrichment or depletion of natural sand components can make matching of sherd sands with sediment samples very difficult (Stoltman 2001). Many thin-section studies of sand-tempered ceramic materials follow methods and theories of sedimentary geologists who study sands. For example, relevant geological research has focused on the influence of transport processes and basin geometry on sand composition (Lucchi 1985), the influence of climate and relief on compositions of sands released at source areas (Basu 1985a), methods for reading provenance of sands using studies of detrital quartz (Basu 1985b), and how to use an analysis of heavy mineral compositions in sands for provenance studies (Morton 1985).

As an example, Basu (1985b) noted that studies of thin sections made from sand (in the medium sand size range) of known sources have shown that a higher proportion of moderately to strongly undulose monocrystalline quartz grains (undulosity of greater than five degrees of stage rotation) and a higher proportion of polycrystalline quartz grains with four or more subgrains characterize sand deriving from a metamorphic source, especially lower grade metamorphic sources (Fig. 6.2.16). In contrast, sands deriving from plutonic rocks tend to have monocrystalline quartz that is nonundulose or weakly undulose (less than five degrees of stage rotation) and polycrystalline quartz grains with only two or three subgrains (also called "polygonized" quartz) (Fig. 6.2.17). These results suggest that it might be useful for ceramic petrographers to measure these quartz properties as well.

Streeten (1982) discussed some of the difficulties of distinguishing sand-tempered coarse wares in regions where the sedimentary geology leads to similar pottery fabrics over a widespread area with an absence of diagnostic inclusions. He highlighted protocols developed specifically for charac-

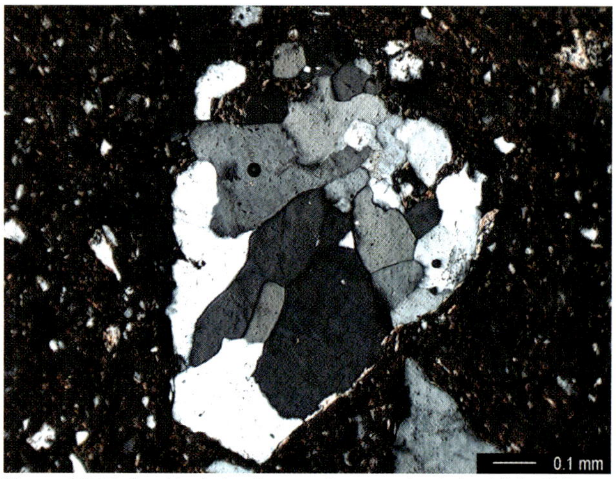

Figure 6.2.16 Sand inclusion in ceramic, crossed polarized light. This large grain of polycrystalline quartz is in an earthenware ceramic from Lancaster, Pennsylvania (Blue Rock Incised, Shenks Ferry, 1000–1300).

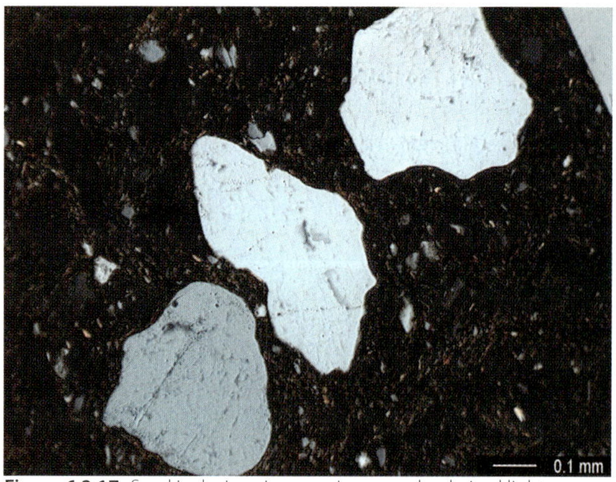

Figure 6.2.17 Sand inclusions in ceramic, crossed polarized light. The larger quartz grains in this sand temper are all monocrystalline and nonundulose. A few of the smaller quartz grains in the clay matrix are polygonized (having 2–3 subcrystals). The ceramic example is of Townsend Ware, from southern Delaware, 1000–1600.

terizing such sand-tempered ceramics. Working with materials from southeastern England, he examined close to 1,000 thin sections, including both kiln wasters and marketed vessels found in Kent, Surrey, and Sussex. The main goal was to compare pottery fabrics to each other (and fabrics from marketed vessels to those of kiln wasters). Grain-size analysis was chosen as the primary analytical method.

The original methods used were based upon those developed by sedimentary geologists to study depositional sequences of fluvial deposits for characterization of sandstones (Krumbein 1938; Folk and Ward 1957; Folk 1974; Pettijohn 1975; Pettijohn et al. 1987) and first used by Peacock (1971). The number of grains at various size intervals was quantified, with the intervals determined from the phi scale (a logarithmic transformation of the Wentworth size grades based on the negative logarithm to the base 2 of the particle size). As an example, a phi of −1 is equivalent to a metric size range of 2 mm, or very coarse sand; and towards the other end of the scale, a phi of 4 is equivalent to a metric size range of 62.5 μm, or very fine sand. Parameters of mean size, sorting (distribution of grain sizes), skewness (a measure of the asymmetry of a distribution), and kurtosis (which describes the extent to which a frequency distribution is bunched with a narrow peak in the center versus being spread towards the ends in a broad peak) were calculated.

Streeten noted that use of these size categories may hinder differentiation of fine-textured wares with a narrow range of quartz grain sizes. Another problem was that the procedure was found to be very time consuming; modifications were thus made to speed up the process and to more precisely characterize the full range of variation found at each kiln, so that comparisons to marketed vessels could be more successful. Instead of selecting random grains for measurement using point counting and a stage micrometer, images were projected at ×100 magnification onto a gridded screen with 200 mm intersections, and grains measured with a millimeter scale (consistently measuring grains along their long axis and selecting only quartz grains). Grains falling on the predetermined points (or the nearest grain in the northeastern quadrant if no grain was exactly on the point) were counted, with the thin section moved as needed until the required sample of grains was included (such as 150 grains).

The data were plotted using absolute grain-size frequency rather than statistical measures of mean size, skewness, and kurtosis. Through experimentation, Streeten found that a minimum sample of 150 grains was required to ensure that the sparse large grains in coarser fabrics were adequately represented. Within-kiln uniformity was found to be very good. He noted that characterizations were not to raw material sources, because the kiln products represented a combination of raw material characteristics and technology. His technique was useful, for example, in distinguishing products from two different kiln sites whose products were visually similar, enabling individual vessels to be linked to their probable source of manufacture.

Worthing (1982) discussed sand tempers found in ceramics from the Port Moresby area of Papua New Guinea. Sand compositions, characterized in thin section, were compared with clay and sands collected in the Port Moresby area. The raw clays were made into tablets that were fired in incremental steps up to 700 °C, and then sampled for thin section analysis. The sands were fixed in resin in small cups, and thin sections were then prepared from the hardened blocks. Point counting (approximately 1200 points per slide) was conducted, and the totals of each mineral or rock fragment present were expressed as a percentage of the total number of points (modal analysis). The clays and beach sands from known pottery-making localities in the Port Moresby area were found to each have distinctive mineralogies, reflecting the influence of local bedrock. The relative amount of chert and quartz/feldspar (C/QF ratio) was an especially useful characteristic for distinguishing between sources. Worthing noted that sourcing of sherds from the Port Moresby area can be done successfully either by using all mineralogical characteristics, or by using the C/QF ratio, or both. The advantage of the C/QF ratio is that it can be done by point counting only chert and quartz/feldspar, so more advanced petrographic skills necessary for identifying other minerals are not needed.

Beynon et al. (1986) studied tempering types and sources for Early Bronze Age ceramics from two sites in Jordan, Bab edh-Dhra' and Numeira, and found that two of them were sand. The most common temper consisted of rounded to subangular sand-sized grains (0.4– 0.6 mm) of quartz, limestone, and chert. The limestone grains were the largest, and often showed weathering or a solution rim around grain edges. The composition and shape of the grain assemblage suggested a source in the Wadis Kerak and Numeira, made of eroding sandstone, limestone, and cherty limestone. Natural transport down the wadi tends to disaggregate the sandstone into individual quartz grains, and the distance is long enough to round limestone grains yet short enough to still retain fragments of limestone. Samples of these sands were collected and studied in thin section, and were found to be identical to the ceramic temper material. In the earliest tomb wares, this wadi sand was the only temper material used in the ceramics.

A second sand temper had only quartz (0.25–0.4 mm), subrounded to angular in shape. The shape, the predominant quartz composition, and use only at the Numeira site suggested that the source was Nubian sandstone, which outcrops near Numeira. This sandstone is loosely cemented so weathers rapidly, and loose sand is available at the foot of the outcrop. Thin sections of material collected there were comparable to the ceramic temper. Over time, Bab edh-Dhra' developed from a permanent village to a more intensively occupied and heavily fortified town; Numeira also emerged as a walled town. New tempering agents were added to the ceramic industry during this period, including crushed limestone and basalt lithics, and the weathered Nubian sand. The fine-grained Nubian sand is found only in fine, thin-walled cream ware.

Lombard (1987) used point counting to study the composition of sand tempers from regionally distinct Hohokam pottery sherds in Arizona. Temper compositions from different sites were compared with maps of sand composition zones within the geographic range of the pottery, to identify a probable provenance for each sherd. The results suggested the presence of some instances of intraregional pottery exchange. At each site, undecorated pottery exhibited a larger number of sand sources than decorated pottery, indicating that the undecorated vessels were made in a wider variety of locations than decorated ones; the two types of vessels may also have been traded differently. Lombard concluded that petrographic analysis of sand temper can be useful for studying exchange of homogenous pottery in geologically diverse areas. This method also facilitates the investigation of prehistoric Hohokam interaction on a scale not previously possible.

Lombard noted that within the study area (the Tucson and Red Rock basins of Arizona), only ephemeral, and not trunk stream, sands were included in the comparative analyses, because they are the most abundant source of sands with similar size, sorting, and abrasion characteristics to the temper sands found in the ceramic samples. Samples were collected across ephemeral stream channels 5–50 m wide on *bajada* (wide alluvial) slopes. Sands were compared against a predictive model for sediment composition zones, based on bedrock geology and the characteristics of sediment transport and deposition. The sherd thin sections were cut parallel to the surface of the vessel (rather than as a cross-section of the vessel wall) to maximize the number of sectioned sand grains and usable surface area of the slide. Sections were stained for easier identification of potassium feldspars.

Point counting was restricted to particles of 0.0625 mm and larger in sherds (400 points per slide), and to the sand fraction (0.0625–2.0 mm) in the sand samples. Monocrystalline particles were counted according to mineral type, and included potassium feldspar (untwinned or simple twinned, sometimes perthitic), plagioclase feldspar, and quartz. These grains were often found within microphanerite rock fragments, usually plutonic. Lithic grains were classified as sedimentary, volcanic, or metamorphic. The point-count data were studied by converting to a variety of new variables useful for distinguishing compositional differences between sand sources. For example, ternary diagrams were used to examine the interrelationships among three variables, which were summed to 100% for each sample. Two diagrams together showed the major important characteristics of sand composition: one of monocrystalline quartz grains, total feldspar grains, and total aphanitic lithic grains; and one of total metamorphic lithic grains, total volcanic lithic grains, and total sedimentary lithic grains.

Lombard found that some sherds contained a deliberate addition of crushed gneiss. The other sherds were tempered only by sand, which was highly variable (including sedimentary, metamorphic, volcanic, and mixed sands). He pointed out the exploratory nature of this study, and noted that more detailed sourcing would probably be possible with more intensive sand sampling in the region of interest. Nonetheless, the geologic data and the sherd tempers were sufficiently comparable to allow detailed temper sourcing in his study. Zones of distinct sand composition could be predicted from knowledge of the bedrock composition and sediment transport and deposition characteristics. The maps of petrofacies were constructed after sampling and analyzing sand compositions using a point-counting approach. Sherd tempers, also quantified through point counting, were compared with the information on sand sources, and provenance assigned. It was possible to distinguish between locally made and nonlocal ceramics, and thus intraregional and extraregional trade could be inferred. For example, diversity of sources in the plain ware and red ware sherds was greater than in decorated sherds, and may indicate more localized manufacture of the undecorated wares.

Gilead and Goren (1989) studied fourth-millennium B.C. pottery from the northern Negev, Israel. Most of the Chalcolithic pottery was made with local clays with local wadi sands added, in different proportions, as the nonplastic component. The potters sieved the sands into two size groups: small for the fine ware and larger for the coarse ware. Since the sands reflect the catchment basins of the local wadis, thin-section analysis can identify where ceramics were made. Most ceramics were found to have been made in the immediate vicinity of the sites in which they were found, suggesting that many regional workshops existed, rather than a few major centers for mass production and export, and that most of the vessels were used locally.

Donahue et al. (1990) conducted point counting to obtain quantitative data on thin sections from 44 sand-tempered sherds representing eight prehistoric sites on four islands in the northern Lesser Antilles (West Indies). Three major temper components were found: sand-sized volcanic minerals and rock fragments, sand-sized fragments of carbonate skeletal material, and grog. Variation was also found in the percentage of temper, paste, and voids. At least 300 points were counted for each thin section to quantify percentages of these components, as well as of temper type. Temper amounts ranged from 12 to 55% within the samples.

Six types of particles were considered to be derived from volcanic rocks, and occurred in all of the 44 sherds. Volcanic rock fragments ranged from 8 to 90% of total temper, and were generally large (0.5–3 mm in diameter) with an angular outline. Plagioclase feldspar ranged from 4 to 86% of total temper, with grains that were intermediate in size (0.25–1 mm in diameter), having angular to subrounded outlines. Opaque grains ranged from 0 to 21%, and were small to intermediate in size (0.1–0.4 mm). Pyroxenes ranged from 0 to 19%. Olivine and quartz were also found, but in very low percentages.

When carbonate minerals were present, they were usually rounded, sand-sized grains derived from marine skeletal materials (mainly algae, forminifera, coral, and molluscan fragments). In contrast to the volcanic materials, these were found in a smaller subset of the sherds. Grog (broken fragments of pre-existing pottery) tended to have large and angular particles, with a darker color than the surrounding clay paste (due to the double firing these particles underwent). This additive was found in only a few sherds.

Voids in sherds with only volcanic tempers tended to be small and equidimensional. When carbonate grains and grog were added to the volcanic temper, voids became larger and highly elongated, with adjacent voids parallel to each other. Donahue et al. hypothesized that this phenomena indicated that sherds containing carbonate grains and grog were made of a material that was not as compact and strong as materials containing only volcanic sand or lithics as temper. The authors also concluded that three temper materials were used: volcanic sand, a mix of volcanic and carbonate sand, and a mix of volcanic and carbonate sand plus grog. The fresh, unweathered, angular nature of the volcanic grains suggested they were not naturally present in a weathered clay source, but were added to the clay as a sand temper. When small amounts of carbonate grains are present along with volcanic grains, it might be a natural component of the volcanic beach sand, but where there is a high percentage of carbonate grains, it is more likely a separate temper.

In discussing sand tempers from prehistoric Pacific ceramics, Dickinson and Shutler (2000) noted that the principle types of sand used as temper by Oceanian potters can be distinguished in thin section. Beach sands tend to have better size sorting than stream sands, due to repeated reworking of sediment on beach faces by the surf. They sometimes also have skeletal bioclastic detritus present from offshore reef tracts. The presence of well-sorted beach sands (which may have calcareous reef detritus) in a ceramic can indicate a deliberately added, rather than naturally occurring, temper. When stream sand is present, only a distinct difference in size between the sand in the clay body and the temper can imply a possible deliberate addition.

Day and Kilikoglou (2001) looked at thin sections from 56 samples collected at the Kommos kiln from Minoan Crete. They identified two groups of ceramics that included a sand fraction. The frequent, well-rounded sand grains were the dominant nonplastics within a fine clay matrix, and could be interpreted as sand temper deliberately added by the potter. They noted that this tempering practice conforms to a long tradition of paste formation in south-central Crete during the Bronze Age, with roots going back to the Final Neolithic.

Heidke et al. (2002) studied sand-tempered Hohokam pottery from central and southern Arizona. They pointed out that the diverse geology of the mountain ranges in this region has resulted in clear heterogeneity in sand compositions within well-defined drainage basins. As a result, provenance studies of pottery have often been quite successful. They chose a petrofacies approach (with each petrofacies representing a distinct sand composition zone) to map out and characterize the sands available as temper to the Hohokam potters. Existing geological maps were used, and preliminary petrofacies maps tested by collecting sand samples in the field for thin-section analysis. Quantitative data were collected by point counting using the Gazzi-Dickinson method (any mineral sand sized or larger is counted as a single mineral even if found within a lithic fragment). Statistical analyses of the point-count data helped to refine petrofacies boundaries, assign sand samples to petrofacies, and develop a mathematical function that best discriminates between petrofacies. Correspondence analysis was used as an exploratory technique, followed by

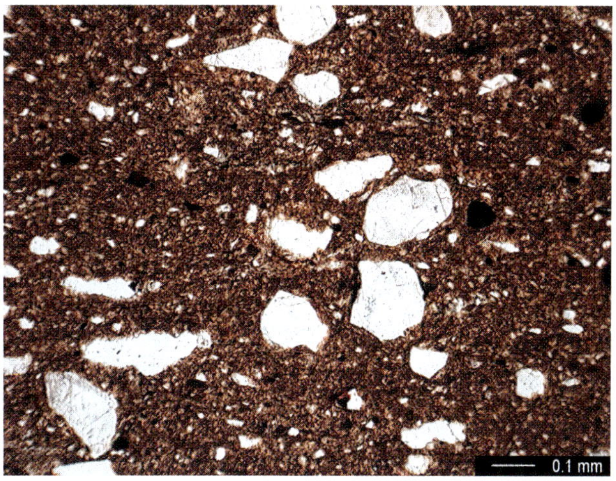

Figure 6.2.18 Sand inclusions in ceramic, plane polarized light. This earthenware from Virginia (1720–1745) has a sand temper in an iron-rich, orange-red clay. The larger grains are mainly quartz and feldspars, and there are also many small silty-sand grains within the clay matrix, along with small chunks of hematite.

discriminant analysis (after log-ratio transformation of the data). Once the iterative process of petrofacies modeling was completed, pottery tempers were then characterized and the most likely provenance identified by using the discriminant function that had been constructed with the raw materials data.

In addition to the most commonly occurring constituent of quartz, sands may also include feldspars, iron oxides, carbonates, micas, other minerals, and small amounts of organic matter. Sand-tempered wares can vary greatly in composition, amount, and relative proportions of components. Often coarse mineral grains are clearly separated from each other (Figure 6.2.18), although a small proportion of polymineralic lithics may also be present in the sand.

Sand additives can be very useful in characterizing pottery, in grouping pottery according to possible production centers, and in identifying potential geological sources, especially if comparative raw materials from the regions of interest are available. Other discussions of sand tempers and their usefulness in provenance studies or in addressing additional specific research questions are found in the following two chapters.

6.2.4 Calcium carbonates

Calcium carbonates can serve the dual function of increasing porosity and thus improving the drying and firing properties of the clay, and serving as a flux to lower the temperature required to fuse the clay body. Calcite in the form of shell is a common temper material (Fig. 6.2.19); crushed calcite or dolomite from carbonate rocks are also common additives (Fig. 6.2.20). Calcium carbonates are found especially with low-fired cooking wares, as this temper provides good thermal shock resistance (Rye 1976). Calcium carbonates can be natural constituents of a clay body (a paste is considered calcareous if it contains 6% or more calcium oxide) (Maniatis and Tite 1981).

For example, the use of calcareous clays in pottery manufacture is typical of the ancient Near East and Mediterranean

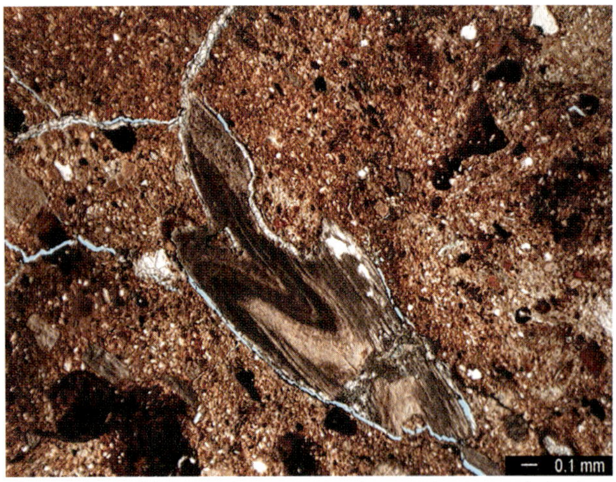

Figure 6.2.19 Calcium carbonate temper in ceramic, plane polarized light. The large chunk of shell temper here has developed some porosity around the edges (where blue mounting medium shows). This example is from China, a Jianxi shell-tempered ware.

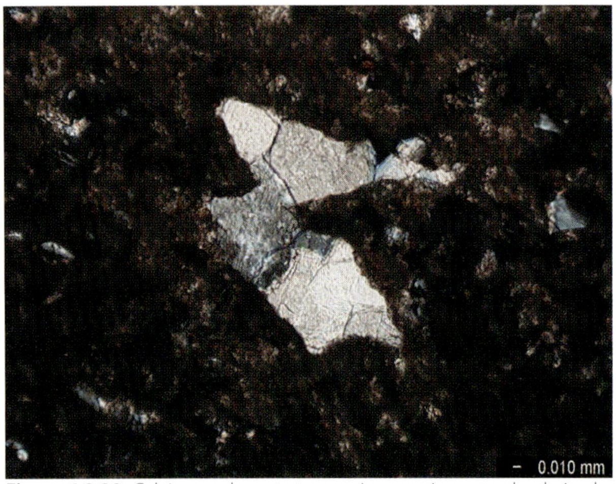

Figure 6.2.20 Calcium carbonate temper in ceramic, crossed polarized light. This angular chunk of calcium carbonate is from crushed calcite that was added as a temper to this ceramic from Iznik, Turkey. In crossed polarized light the carbonate has a pale pastel color and a gray, "pearly sheen" characteristic of calcium carbonate.

regions. In such vessels that underwent low-temperature firing regimes, calcite tends to be abundant. Those of higher temperature firing regimes may not have much visible calcite left, but will still retain high lime contents seen in elemental analyses (Freestone and Hughes 1989).

The presence of a significant amount of calcium carbonate can have a major effect on the firing properties of clay. Calcium carbonate will decompose at temperatures above 600 °C, as carbon dioxide gas is released and calcium oxide is left behind (the exact temperature depends on firing atmosphere, rate of heating, and clay composition). The calcium oxide is a good flux, but on exposure to moisture will react with water to form calcium hydroxide. This mineral has a larger volume than the original calcite crystal, creating porosity and unevenness in the structure of the ceramic body, and eventually cracking and spalling of the vessel (Shepard 1968; Hoard et al. 1995). Sometimes salt (NaCl) is added to surround the calcium carbonates and form a glass barrier that prevents too much release of gas (Velde and Druc 1999). In that case, in thin section, reaction rims are visible surrounding the shell fragments or other calcium carbonate grains; these reaction rims form a good barrier to protect the grain or fragment from "lime blowing" as illustrated by Gibson and Woods (1997).

Riley (1982) studied a large number of fine wares from the two most important sites of the Mycenaean and Minoan civilizations, Mycenae and Knossos. He found that many of those fine wares from Knossos contained abundant foraminifera, while those from Mycenae did not. Foraminifera are small marine organisms which usually secrete shells of calcium carbonate, although they sometimes build their shells up with cemented sand grains. Fossil foraminifera are usually less than 1 mm in length, but can sometimes be larger. They may accumulate in large deposits that form chalk or limestone; in the materials studied by Riley, small foraminifera were scattered within clay deposits and thus became incorporated in fine ceramic wares.

This line of research was later continued to much greater depth by Quinn and Day (2007), who noted that calcareous microfossils such as foraminifera and ostracods are sensitive paleoenvironment indicators, so potentially can provide important information about ceramic raw material sources. Drawing on methods from micropaleontology, they developed a procedure that combines thin-section analysis with a smear-slide technique (using a small scraping from a sherd), utilizing an oil-immersion microscope lens with ×1000 magnification. Their procedure permits identification of both microfossils and nannofossils. The information gleaned by this innovative approach is now being used to answer a wide variety of archaeological questions in the Aegean region regarding ceramic provenance, identification of geological sources of raw materials, clay-processing procedures, and firing temperatures.

Cooper (1982) noted that shell material used as temper could come from a fossil source (such as a shelly limestone or shelly clay bed) or from a contemporary source (such as food remains or beach material). Thin-section studies may help to identify the shell material (such as bivalves, or clams, versus gastropods, or snails) and may even facilitate identification of the place of origin of the shell. Pre-Columbian pottery from San Salvador Island in the Bahamas is composed largely of angular fragments of invertebrate shells bound by clay and quartz (Mann 1986). Crushed gastropod shells served as the temper for marly clay in ancient pottery from Oman (Echallier and Mery 1992).

L. D. Jones (1986, 1991) discussed the use of carbonate temper materials in Maya pottery. She noted that calcite temper is found at every Lowland Maya site, although the degree to which it was used varied from site to site. For example, at Tikal, calcite was the most common tempering material used during the middle Preclassic, but is not the only one (grog and volcanic ash were also used); whereas during the early Classic phase it was used much less often, volcanic ash being the dominant tempering material. In contrast, in northern Belize, calcareous tempers predominated in all periods. Calcite inclusions ranged from very fine (0.05 mm) to coarse (1 mm or larger).

Thin sections of Neolithic pottery from Franchti Cave in Greece (R. E. Jones 1986) showed that many had carbonate inclusions, mainly as agglomerations rather than as separate calcite crystals. The carbonate fragments ranged from microcrystalline to coarsely crystalline, and sometimes included recrystallized fossil fragments. Some scattered calcite crystals were also present. It was concluded that these are likely to be products of local clays since the presence of carbonate rock fragments is consistent with the limestone geology near the cave.

Kolb (1989) incorporated thin-section petrography into his comprehensive study of Kushano-Sasanian storage jars from Central Asia. Ceramic artifacts were excavated at prehistoric sites in northern Afghanistan. The main constituent of added temper was mechanically crushed dolomitic limestone, with particle size directly correlated to jar size. Massive jars had very coarse particles, in the pebble-size range (1.0– 4.0+ mm); large jars had coarse and very coarse particles (0.5–2.0 mm); medium jars had coarse particles (0.5–1.0 mm). A wide variety of minor and accessory components was also seen, consistent with the geological contexts of the Turkestan Plain of northern Afghanistan, northeastern Iran, and the southern areas of the Soviet Central Asian Republics.

Gilead and Goren (1989) found that cream ware vessels of the Chalcolithic period from the northern Negev in Israel were made with a crushed chalk temper. In thin section the ceramics were seen to be carbonate rich, with numerous microfossils (from the upper formations of the Eocene chalk of southern Israel). At one site discussed, the nonplastics were heterogeneous in size, and included various carbonates, flint, crushed bone, shells, and carbonatic sandstone. These materials were found at the site, so were assumed to have been crushed, sieved, and added to the chalky matrix. At other sites, the raw materials, especially Eocene chalk exposures, were not available, so the materials or ceramics were imported. The authors suggested that such vessels could have been connected to a custom of gift exchange involving special types of pottery, typical of many formative societies, including the Bedouin of the Negev.

Vince (1989) used thin-section petrography to study Saxon wares in the Thames valley in England. He found that for one ware the potters used a crushed limestone gravel which he could match quite well to the limestone gravels in the vicinity of Oxford. Another coarseware, misidentified as containing chalk, in fact contains a calcareous blue-green algae, found with quartz and chert in a micaceous silty clay. Vince noted that the presence of algae suggests a recent clay deposit.

Shell-tempered pottery is found widely with Native American wares from throughout eastern North America (Budak 1991) (Fig. 6.2.21). Budak pointed out that the shell additive made clay more difficult to work with, but also made the ceramic product more durable, less permeable to water, and improved properties of heat transfer from the exterior to the interior of a vessel. He discussed how the shell was burned in a fire before use, to make the material friable so it could be pulverized easily into small particles (flakes and dust-sized pieces). Once the shell was added, the clay became stiffer and less plastic, made even worse if the clay already contained a significant amount of silt or sand. Firing must be done at a temperature below that which will cause alteration of the calcium carbonate.

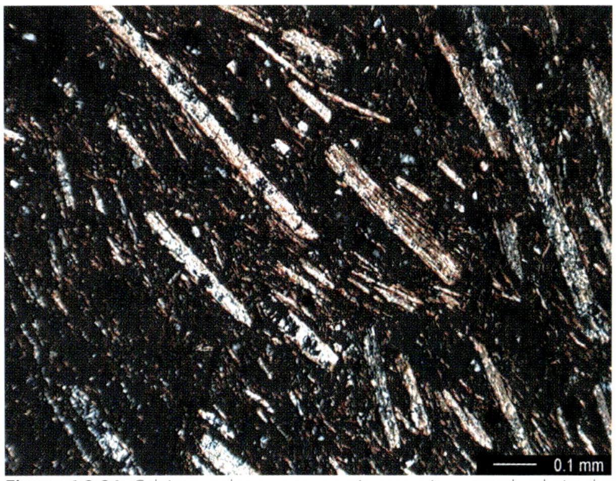

Figure 6.2.21 Calcium carbonate temper in ceramic, crossed polarized light. Shell additives in Native American wares of eastern North America were often processed by burning the shell to make it friable so that it could be pulverized into small flakes.

It was suggested (following the ideas of Schiffer 1990 and Skibo 1992) that this extra work involved was worthwhile because of the waterproofing properties of the shell temper addition. When liquids are placed in a shell-tempered pot, they penetrate the vessel walls very slowly or not at all, depending on the amount of shell additive. Grit-tempered vessels do not have this waterproofing property, unless a coating is applied to the interior, which tends to wear off with use. More heat is required to bring the contents of a permeable pot to boil than is needed for an impermeable one, thus requiring more time and fuel to cook a meal. Budak conducted a series of experiments that supported this hypothesis. He found that grit-tempered pottery has to be sealed on the interior surface in order to be effective for cooking, but shell-tempered pottery is watertight and conducts heat well. It has also been proposed (Hoard et al. 1995) that shell creates a strong ceramic composite, especially when large shell fragments are used, that is very resistant to cracking, due to the fibrous nature of the calcite in plates of burned shell.

Hoard et al. (1995) conducted experiments on limestone-tempered ceramic test specimens, comparing them to grit- and grog-tempered ones. They noted that limestone temper was common in Late Woodland pottery from the central United States, and they were interested in investigating performance characteristics that might have made that temper choice advantageous, especially given the potential for cracking and spalling after firing. As with other types of temper, the calcite temper makes sticky clay more workable. The release of calcium ions into wet clay increases stiffness. In addition, limestone-tempered ceramics are significantly more resistant to mechanical stress than grit- or grog-tempered ceramics at a firing temperature of 600 °C. The authors suggested that greater resistance to thermal shock may also be related to the thinner vessels allowed by calcium carbonate temper, and because the expansion rate of the calcite is very similar to that of clay (in contrast to quartz tempers). Since effort was required to overcome the spalling problem (such as adjusting firing conditions and length, adjusting temper particle size, and/or using a NaCl-rich clay), they proposed that the use of limestone tempers represented a technological shift in pottery manufacturing techniques.

Other discussions of the possible functions or effects of calcite tempers are found in Chapter 8.

6.2.5 Organic material

Organic materials added to clay to serve the function of temper can include plant fibers (chaff), seeds or husks, phytoliths, textile fragments, animal dung, animal hair, feathers, and other materials. Agricultural waste, such as straw or stems from crop plants (Fig. 6.2.22), may be used where pottery-making is seasonal and occurs at the end of harvesting (Rye 1981). As with other types of tempers, an important purpose is to modify the working properties of the clay. Organic material functions to improve properties of plasticity of the clay, facilitate fabrication by making a sticky, too plastic, clay more workable as the larger particles help the clay retain shape, even for large vessels, control shrinkage during drying by extending the network of pores for evaporation, and aid in the release of gases during firing (Rye 1981; Freestone and Hughes 1989). If the organic matter is allowed to burn out during firing (Fig. 6.2.23), the large pores that remain interrupt cracks that form due to thermal stress during use of cooking pots. Skibo et al. (1989) conducted a variety of behaviorally relevant tests to compare performance characteristics of untempered, mineral-tempered, and organic-tempered ceramics, using both briquettes and vessels. They concluded that organic-tempered ceramics have superior performance characteristics during manufacture of an object.

In addition, it was found that organic-tempered vessels have other properties that would make them advantageous choices: not only do the pores left after burnout of organic matter halt cracks created by thermal shock, as do mineral inclusions, but organic-tempered vessels are significantly lighter in weight than mineral-tempered vessels of the same volume, and so might be preferred by mobile groups. And, organic-tempered vessels were found to have "expedience of manufacture," meaning that a vessel can be made relatively quickly. The temper could be added to naturally wet, unprocessed, clay, and the organic material would absorb excess moisture and make the clay workable.

As the organic matter chars during firing, porosity of the ceramic increases. Under reducing conditions, the charcoal will remain at least partially in place within the pores; under oxidizing conditions organic material will completely burn out, leaving only pores behind. Thus, identifying the exact type of organic matter used is often difficult. Rye (1981) pointed out that plant fragments naturally occurring in clays vary considerably in size and tend to have irregular shapes. In contrast, organic material added to the clay by potters tends to be chopped or otherwise have a more standardized, regular size and shape.

London (1981) summarized some of the advantages of organic temper materials. She noted that the porosity of the vessel walls created as the organic materials burn out upon firing helps a vessel keep its liquid contents cool by slow evaporation through the porous walls. Thus organic-tempered vessels are advantageous for water storage in a hot climate. There are also advantages during drying. For example, organic matter may be added to the clay used for adding segments that are joined to a vessel after the basic form has been completed, to help the accessory pieces dry and fire evenly with the rest of the pot by facilitating evaporation of water. Thus, added segments such as handles may contain organic temper even if the remainder of the vessel does not.

London also discussed the relationship between the presence of organic temper and that of decorative slips and paints. A porous wall (created through the burning out of organic material) is better able to absorb evenly wet clay in the form of a slip than wares that contain mineral tempers that are not porous. Mineral inclusions will tend

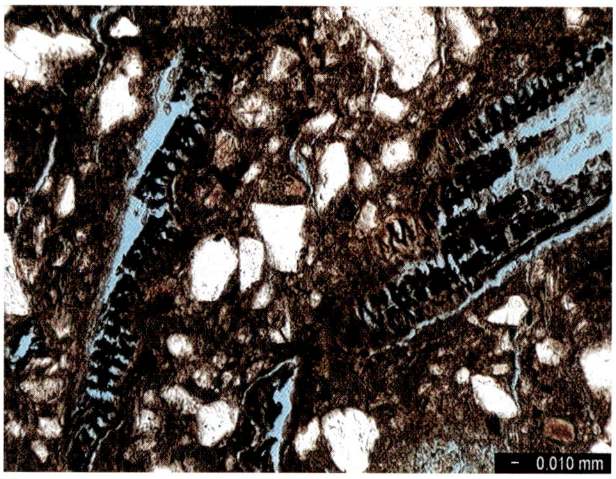

Figure 6.2.22 Organic temper in ceramic, plane polarized light. The main organic material present in this ceramic from Ban Chiang, Thailand (300 B.C.) is rice chaff, which carbonized during firing.

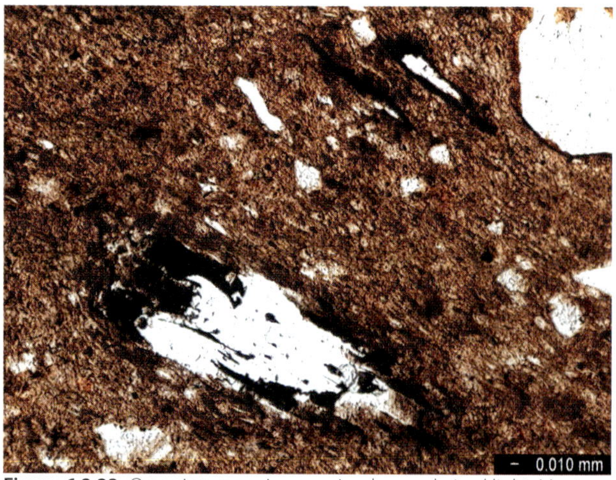

Figure 6.2.23 Organic temper in ceramic, plane polarized light. Most of the organic material in this example (Mockley Ware from southern Delaware, 300–800) burned out during firing, leaving large pores (colorless here against clear mounting resin). The clay matrix is reddish orange in color, due to the oxidizing firing conditions.

to remain visible on the surface, marring the appearance of a slip or paint. Unslipped organic-tempered pottery tends to appear coarse and rough due to the porosity remaining after burning out of organic materials, so these wares have a better appearance if slipped or painted.

Animal dung as a temper has some specific advantages. It is usually readily available year round, and as dried dung is a common source of fuel in many parts of the world, it is often on hand in many households. It is also inexpensive and requires little preparation. As an added bonus, dung has been reported to contain gel-forming hydrated organic polymers that increase the plasticity of clay, particularly if the dung-tempered clay is allowed to sit as a wet paste for a period of time before use. Skibo et al. (1989) found that the most noticeable change when dung-tempered clay is allowed to sit, or "sour," for a period of time was that the clay gained a greater wet strength; it was more rigid and better retained a shape. The presence of very fine-grained voids in the shape of seeds may indicate that dung was added to clay—even small amounts would have had positive effects on the clay (London 1981).

Simpkins and Allard (1986) discussed the use of Spanish moss fibers and scales for ceramic temper materials, which they identified in Stallings and Orange Series ceramics from the southeastern United States. They suggested that vegetable fibers would be a natural choice of temper for early potters, who would probably already be familiar with the use of fibers for basketry and net construction. Fiber would thus seem a logical choice of material to help bind clays during fabrication and firing processes. Smith and Trinkley (2006), who also studied Stallings ceramics from the Crescent site of South Carolina, found that although most of the organic material had burned out, leaving only voids, based on morphology and textures of stem fragments seen in some thin sections, Spanish moss could be identified as the most likely fiber used in vegetal-tempered sherds.

Vincent (1988) discussed the prevalence of rice husks and chaff in Southeast Asian ceramics (Fig. 6.2.24). He noted that these may be deliberately added temper, but it is possible that they were accidentally incorporated into clays during quarrying or fabrication processes. For example, he noticed that many modern potting clays in the Sakon Nakhon basin of Thailand were quarried from beneath rice fields. The rice plant material found in prehistoric ceramics from the region sometimes consisted of both rice straw and husks, and in other cases had only husks. Some wares contained abundant rice phytoliths (Fig. 6.2.25).

In the study already mentioned by Gilead and Goren (1989) of fourth-millennium B.C. pottery from the northern Negev, Israel, it was found that one group of pottery had straw as the major nonplastic inclusion, resulting in a black or gray core. The clay had a high proportion of heavy minerals (large hornblende and biotite grains, 600–700 μm), as well as augite, zircon, epidote, and tourmaline. The mineral assemblage suggested a coastal origin.

Allen (1991) studied thin sections of Bronze Age pottery from the East Midlands of England. Although most of the nonplastic materials were quartz or grog, she also found some fossilized shell fragments and a variety of organic materials. She was able to identify the source of organic material, although it was most often partially carbonized. There was a variety of material represented, including plant stems, grass stems, and wood of a broad-leafed tree such as oak, ash, or elm (appears to be wood shavings).

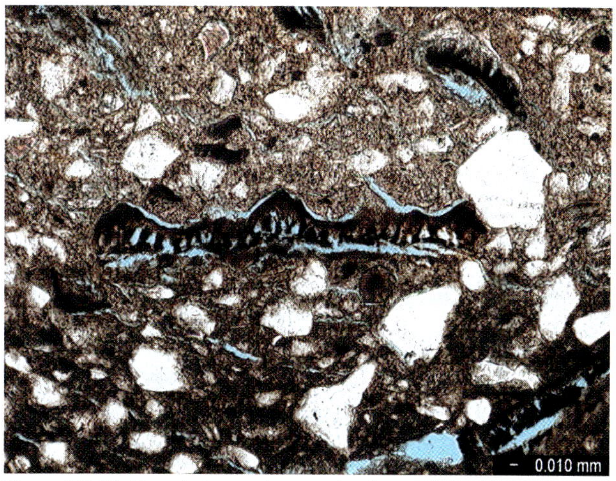

Figure 6.2.24 Organic temper in ceramic, plane polarized light. Rice husks and chaff are often seen in Southeast Asian ceramics, as in this example from the site of Ban Chiang in Thailand.

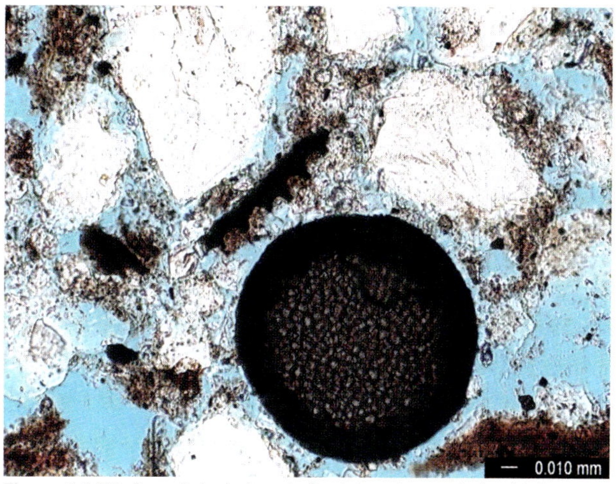

Figure 6.2.25 Organic inclusions in fired clay, plane polarized light. This sandy fired clay, collected by a potter from beneath a rice field in Cambodia, contains rice husks and chaff as well as phytoliths (such as the large round grain), and abundant quartz-rich sand.

Dajnowski et al. (1992) employed thin-section petrography, along with a variety of other analytical tools, to investigate pottery from the Neolithic site of Liangzhu, China. The pottery is characterized by a thin black finish and a layered body structure consisting of a black core that changes sharply to a gray or red band that extends to the black surface finish. The authors found that clay containing abundant charcoal was thrown and then fired under reducing conditions so that the charcoal was retained and iron was reduced. At the end of firing, or during cooling, air was allowed to enter the kiln, and a limited thickness of the outer core was oxidized. For decoration, the surface was then heavily smoked, and/or a thin iron-containing slip was reduced in smoke and burnished. Thin sections revealed that the area of color change was a diffused transition zone consistent with oxidation-reduction. Carbon particles were seen in all layers, although more were found in the core. Straw, reeds, or other organic tempering materials, known to be used in China to temper clay, were suggested as possible sources of the carbon.

Philpotts and Wilson (1994) discussed some additional advantages of an organic temper in their study of a Late Woodland vessel from South Windsor, Connecticut, that was tempered with shell fragments, quartz from stone knapping debitage, and grass. Firing temperatures were estimated to be about 950 °C, too low for vitrification to occur. The clay had a high iron content, and magnetite and wüstite were identified in the black body of the vessel, with hematite present on the exterior surface layer. The authors proposed that during firing, the exterior of the vessel was in contact with heated air in an open fire, forming hematite. Within the body of the vessel, burning of the grass temper created reducing conditions throughout the ceramic body that caused the iron oxide to be converted to wüstite, although a small amount of magnetite was also present (the temperature would have to exceed 576 °C for wüstite to form as below that there would be magnetite plus metallic iron). They suggested that the differences in oxidation state of the iron between the exterior and interior of the vessel, and the resulting mineralogical differences, had a pronounced effect on the wetting properties of the ceramic material. A drop of water placed on the red surface of the vessel soaked in and disappeared quickly, whereas one placed on the black inner surface remained as a bead of water. Thus the addition of the grass temper, in conjunction with the firing regime, created a waterproof vessel without the need for any special surface treatment of the interior.

Sestier (2005) reviewed various uses of organic tempers (such as various types of fibers, grasses, grains, seeds, moss, dung, sponges, and phytoliths) in ceramics dating from the Neolithic up to the present. Experimental work was carried out that demonstrated the positive effects that vegetal temper has on the workability of clay and on the drying properties of wet clay. The author noted that vegetal tempering of clay is a good way to cope with the natural variation of raw materials with properties that can be problematic for ceramic production.

6.2.6 Grog

The term "grog" refers to fragments of previously fired ceramic pieces used as a temper additive. These may derive from crushed sherds, or, in some cases, small balls of clay that have been dried, fired, and used as a manufactured temper material. Grog is a good temper material because it so closely matches the physical properties (such as coefficient of expansion) of the clay matrix, and is similar in composition. It is a good choice for a clay with few natural mineral inclusions but needing some temper to improve workability for forming. Grog often derives from sherds or clay of the same composition as the ceramic body in which it is used (Fig. 6.2.26), but not necessarily (Fig. 6.2.27). Upon firing, grog will have close to the same thermal characteristics as the clay matrix and thus will provide good thermal shock resistance (Hodges 1963; Porter 1964; Rye 1976; Vincent 1991). In thin

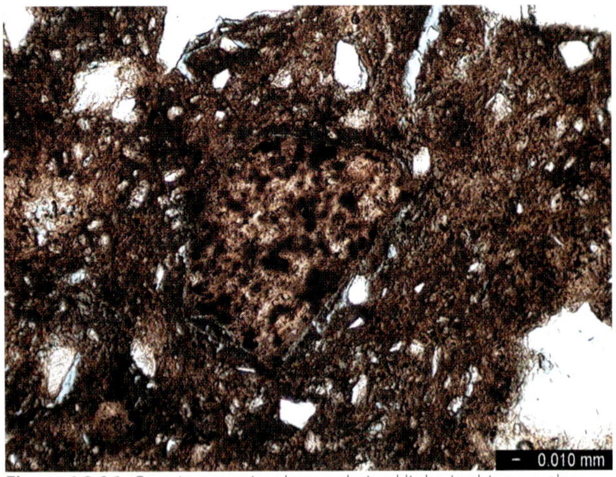

Figure 6.2.26 Grog in ceramic, plane polarized light. In this case, the angular chunk of grog in the center of the field of view has clay that is very similar to that of the matrix. The sherd is from an example of Japanese Shino red.

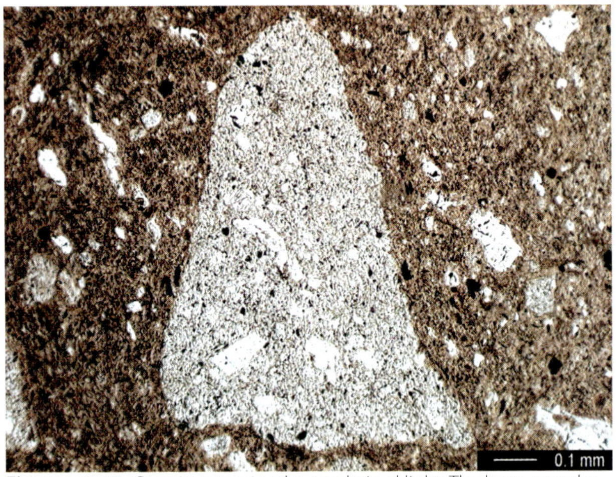

Figure 6.2.27 Grog in ceramic, plane polarized light. The large, angular piece of grog has clay that is much less iron rich and noticeably lighter in color than the clay of the surrounding matrix. This example is in a sherd from a Hopi vessel, eighth–eleventh century.

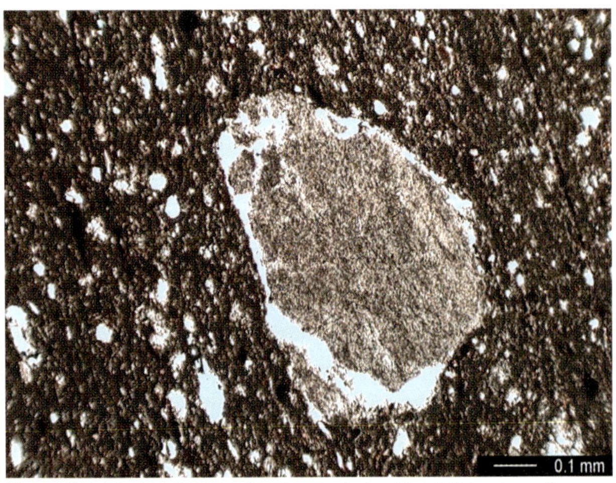

Figure 6.2.28 Grog in ceramic, plane polarized light. This piece of grog shows a typical ring of separation between it and the surrounding matrix, where the wet clay shrinks away from the grog as the vessel dries. This example is from English earthenware, 1675–1725.

section, grog is usually angular, and is often enclosed by a ring of separation between it and the surrounding matrix where the wet clay shrinks away from the grog as the vessel dries (Fig. 6.2.28) (Gibson and Woods 1997).

Grog can sometimes be difficult to distinguish from argillaceous rock fragments derived from sedimentary material such as lithified clay, mud, or silt. Whitbread (1986b) discussed the features of such natural inclusions, and noted that they could have been added deliberately by potters as temper, or could be natural inclusions in the clay material. Grog can often share many of the same characteristics as argillaceous rock fragments, although if slip is present, can be securely identified as grog.

L. D. Jones (1986, 1991), in discussing the use of grog tempers in Lowland Maya pottery, pointed out that it is best distinguished in plane polarized light, where color and textural differences between the grog and the surrounding clay matrix tend to be more apparent. Jones found that when grog was used in ceramics from Tikal, it was rarely the only tempering agent used: volcanic ash and/or calcite were usually added along with the grog.

Kelly (1979) discussed thin-section studies of grog-tempered ceramics from prehistoric New Mexico (Chupadero Black-on-White). One variant included large pieces of grog (up to 2 mm). Other variants had more finely ground grog, and were more sparsely tempered in general. One variant used ground "clinker" (pottery that is irregularly glassy, bloated, and full of holes like Swiss cheese, resulting either from high firing temperatures or from the use of clays that fuse at relatively low temperatures). In thin section, the clinker clay of the grog particles appeared partially fused rather than crystalline. In this variant, the clay matrix itself was partly fused and bubbly, suggesting a common clay source for both the grog and the clay body of the ceramic.

Vincent (1988, 1991) noted that petrographic analysis of fabrics was crucial to understanding Southeast Asian pottery. For example, many errors in classifying wares were made without petrographic data to identify what was made locally and what was exotic to the area. He further noted that clays that can be used directly without the addition of temper are rare, and that where available, natural deposits such as sand tend to be exploited. In northeastern Thailand where he conducted research, such natural deposits were not always available locally. In response, the potters chose to manufacture a temper material in the form of grog. Vincent found two variants of this manufactured grog: (1) fired clay that was crushed into particles 2–3 mm in diameter, which probably were sometimes passed through a sieve; and (2) a more complex product made by mixing hand-sized balls of clay with rice husk and chaff. The balls were then fired on an open fire and crushed. Ethnographic studies in the region were helpful in understanding the production procedures used for the archaeological materials. This latter manufactured temper eventually was adopted in other areas of the region even where sand, rice, and traditional grog tempers were available and previously exploited. This technological change occurred along with other changes that resulted from expanded exchange networks and the rise of stratified societies in the region.

Vincent also noted that tempers should not expand excessively during firing, or they will cause the pottery to crack and shatter. Sedimentary rocks tend to have a wide range of expansion rates, so where clay quality is marginal, such rock fragments would be problematic as a temper material. In his region of interest, northeastern Thailand (Khorat plateau), most of the area is located within a sedimentary plateau. Grog was therefore a suitable choice for temper additive even though it required some investment in time and fuel. In prehistoric Thailand, different kinds of grog were made: crushed potsherds, pre-fired and subsequently crushed clay balls which did not contain temper, and what he called "bleb grog," or crushed pre-fired clay that was purposely tempered with rice plant remains, especially rice husks, prior to the initial firing. The rice husk content gave the clay balls a particular bleb shape. The author highlighted the fact that artificial grog manufacture shows sophisticated knowledge of clay properties—the pre-fired grog is less susceptible to shrinkage than raw clays.

Cuomo Di Caprio and Vaughan (1993) conducted experimental work aimed at establishing criteria for differentiating between grog and argillaceous rock fragments (such as shale, mudstone, calcareous clasts, or hardened clay lumps) in thin section, as these are the inclusions most likely to be confused with grog. The features described in various publications as being useful for distinguishing grog in thin section were listed, then systematically checked using prepared clay briquettes made with known temper materials (grog or argillaceous inclusions). They found that three characteristics tended to indicate the presence of grog as opposed to argillaceous inclusions. These features included an aligned internal microstructure (Fig. 6.2.29), which includes parallel orientation of constituents and voids (parallel to the surface of the parent vessel); internal fractures parallel to the greatest length of grog fragments; and even distribution of grog throughout the host fabric. They noted that, otherwise, the only unambiguous criterion for identifying the presence of grog temper is the observation

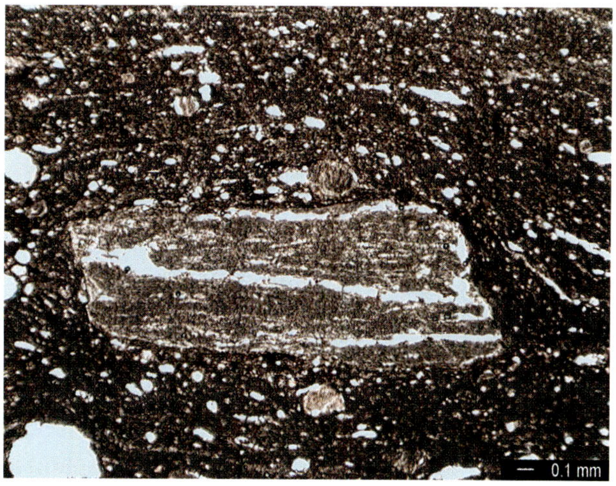

Figure 6.2.29 Grog in ceramic, plane polarized light. Grog tends to be differentiated from argillaceous inclusions by the presence of an aligned internal microstructure, which includes parallel orientation of inclusions or voids, and internal fractures parallel to the greatest length of the grog fragment. This example is of English earthenware (1675–1725).

of slip or paint on the grog particles, a relatively rare occurrence (Figs. 6.2.30 and 6.2.31).

An example of a feature previously thought to be useful but not actually helpful in distinguishing grog was the presence of shrink rims. Earlier publications had indicated this feature to be diagnostic (R. E. Jones 1986). However, Cuomo Di Caprio and Vaughan found that the presence of shrink rims can be related to a variety of variables such as degree of compaction or density, composition, and size of inclusions. They observed shrink rims around examples of both argillaceous fragments and grog. Another significant finding was that identifying the presence of grog is much more difficult when the grog particles have been ground by the potter to less than 0.25 mm in diameter.

Michelaki et al. (2002) discussed grog-tempered Bronze Age ceramics from Hungary. They reported that thin-section petrography revealed interesting differences between ceramics from two sites, Klárafalva and Kiszombor. Overall, the sherd composition from the two sites was about the same: 80–89% clay and silt matrix, 1–5% natural sand inclusions, and 5–19% grog temper. However, in the Kiszombor ceramics, grog size was distributed bimodally, which was not the case in the ceramics from Klárafalva. The authors proposed that the bimodal distribution suggested that either the potters separated their prepared crushed sherds into two sizes, or they consciously crushed them in such a way as to produce two size classes. The smaller size fraction was then used to temper thinner vessels, and the larger size was used for thicker vessels. In contrast, in Klárafalva, although the range of sizes was the same as in Kiszombor, no sorting was done. Instead, the total amount of grog was bimodally distributed.

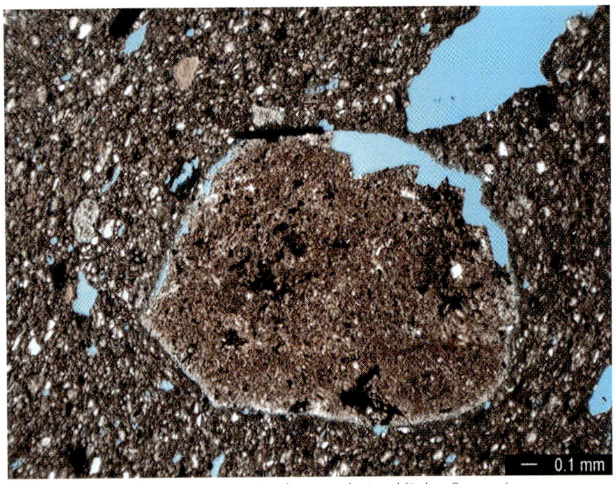

Figure 6.2.30 Grog in ceramic, plane polarized light. Sometimes remnants of a slip or paint layer on a particle is an indication that it is grog rather than an argillaceous inclusion. In this case, the grog particle has a partial paint layer on its upper edge, black in plane polarized light. This example is from Guantai Cizhou Ware, China.

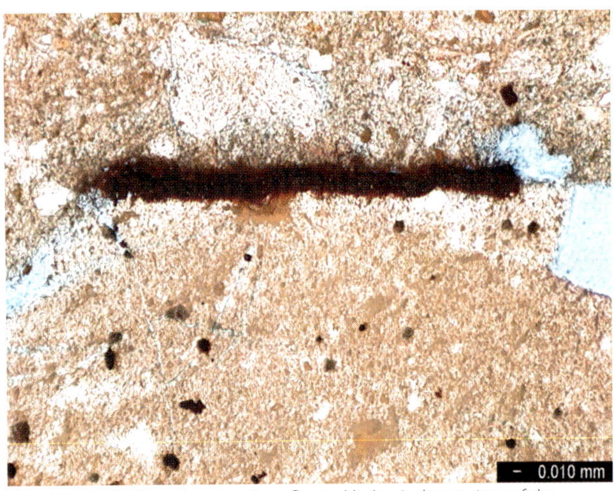

Figure 6.2.31 Grog in ceramic, reflected light. A closer view of the paint layer on the grog particle shown in Figure 6.2.30 is shown here in reflected light. As is typical of hematite, the paint layer is a steel-gray black with metallic luster, with a tendency to marginal red.

Fitzpatrick et al. (2003) utilized thin-section petrography to study prehistoric ceramics in Palau, Western Caroline islands, Micronesia (an archipelago comprising over 300 islands, both volcanic and limestone). The spectrum of temper constituents are all found locally in Palau, and suggest that the pottery was locally made. Wares were manufactured mainly using grog (atypical for Oceanian ceramic traditions, which are mainly sand tempered) or composite (mixed grog and terrigenous volcanic sand) tempers. Calcareous sand temper is absent, which is also unusual for Oceania and Micronesia, although coral beach sands are abundant in Palau. The authors concluded that the choice of grog or a composite grog/volcanic sand temper was a local cultural preference that reduced the threat of thermal breakage caused in higher-temperature firings when calcareous sand temper is used.

6.2.7 Other inclusions

Micas are often used as temper materials, whether natural or added (Fig. 6.2.32). Shepard (1953) studied prehistoric micaceous wares of the American Southwest, and noted that the mica may be a constituent of stream sand or lithics used as temper materials, or may occur in a primary clay as a remnant of rock weathering. Often, highly micaceous pastes are untempered primary clays, so the use of the term "mica tempered" often refers to what are actually natural nonplastic inclusions within clay.

Salt (primarily sodium chloride) is sometimes added by potters to their clay, in the form of ground salt or as saltwater. This addition is mainly found when calcium carbonate tempers were used for cooking pots; it helps to extend the successful firing temperature range and to prevent post-firing damage by spalling due to hydration of lime in the pottery fabric. Rye (1976) noted that evidence of salt use can be seen in pottery thin sections, and described this evidence as the presence in the fabric of cubic voids, which are psuedomorphs of the original salt crystals that formed when the vessel dried; these voids often have yellowish rims. He also stated that sometimes a thin layer of crystals will be apparent on the surface, a result of the reaction with the clay matrix of salt that was carried to the surface in solution and then deposited there as the vessel dried. This layer is sometimes termed a "self-slip" but originates from functional rather than decorative purposes.

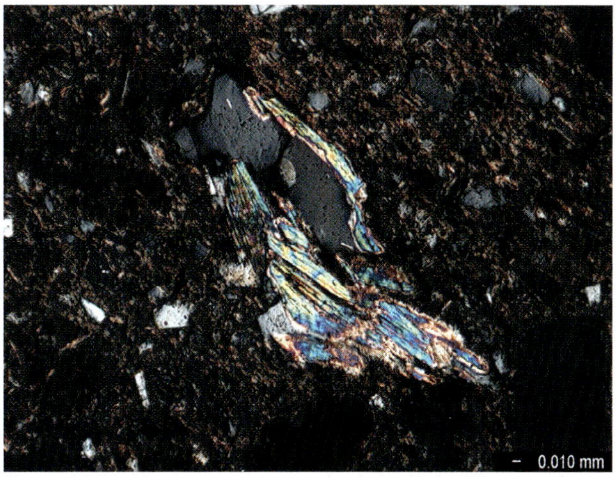

Figure 6.2.32 Mica in ceramic, crossed polarized light. Micas are often found as natural inclusions, probably deliberately selected for their properties. This example shows highly birefringent muscovite mica plates surrounding low birefringent quartz grains. The thin section is from a sherd of Blue Rock Incised Ware from Lancaster County, Pennsylvania, 1000–1300.

Woods (1985a) illustrated salt voids obtained from experimental work in which she mixed seawater into clay. She noted that few of the voids were cubic or square, instead nearly all were rounded, and difficult to distinguish from other voids that tend to appear in ceramic materials. However, the presence of salt could be recognized by yellow reaction rims that formed around shell or other calcareous fragments or along the edges of the salt crystal voids. These yellow reaction products were most visible in plane polarized light, although in very dark fabrics they were difficult to detect.

Crushed bone is sometimes added as a temper material. As mentioned in section 6.1.3 above, phosphatic, or "bone china," popular in England in the eighteenth century, can have as much as 50% bone ash added to serve as the flux. The bone ash imparts a pure white color and produces a translucent body. However, crushed or ground bone has also been added to prehistoric low-temperature ceramics, sometimes in the form of coarse fragments. The primary constituents of the bone fragments are calcium and phosphorus, making a good fluxing agent. In some cases, abundant and relatively large bone fragments were added as a temper material (facilitating water removal during drying) rather than a flux (Velde and Druc 1999).

Fragments of slag have also been used as temper material. "Slag" refers to the byproducts of metal smelting. In one example, slag temper was seen in ceramics from the site of Timna in the Negev, Israel. Timna was a major Bronze Age copper-smelting site, so the presence of slag in ceramics at the site is logical, and serves as an indicator of local production (Freestone 1995).

Philpotts and Wilson (1994) studied a Late Woodland vessel from the Butternut Knoll site in South Windsor, Connecticut that was tempered with stone knapping debitage in addition to shell fragments and grass (which burned out in the firing process). The stone debitage was composed of quartz that was identified as being of the clear vein variety, with many fluid inclusions; some fragments had small fibers of rutile. The authors found that the temper additive was primarily quartz, whereas the local sands contained a variety of other minerals in addition to quartz (such as a high percentage of feldspar). In addition, the shape of the quartz additive was unusual: particles were very elongated, with sharp terminations, and detrital quartz in sand and quartz temper that was prepared by crushing large pieces had a very different morphology. Transported grain from sand tends to be equidimensional, and is often rounded. Crushed quartz would have sharp boundaries, but would again tend to be relatively equidimensional. It was concluded that the most likely source of the ceramic temper was debitage from stone knapping. Indeed, vein quartz constituted about one-quarter of the lithic assemblage at the site. The authors proposed that because the knapping debitage grains were elongated, they had a tendency to become aligned during slab construction of the vessel's wall, with a preferred orientation parallel to the wall. This orientation would decrease the likelihood of a crack propagating through the wall, so this specially shaped quartz may have been selected to increase vessel strength.

Others (e.g. Gibson and Woods 1997) suggested that a knapping debitage material like chert or flint would be problematic as a temper additive because it usually requires much preparation. They claimed that the sharp, highly angular fragments are unsuitable for mixing with clay, because the clay would then be uncomfortable for the potter to handle, especially with the abundance of temper needed when open firing is done. They also noted that during firing, water in the structure of flint is released at about 400 °C, causing the flint to crack and disintegrate, creating problems in the ceramic body. To alleviate these problems, flint was calcined, as is done with shell, to drive off the chemically combined water, and to facilitate crushing the flint into chunkier and less sharp fragments.

Ryan (1978) noted that flint has been a popular form of silica for pottery production, especially with more recent high-temperature wares, because the small crystal size facilitates conversion to cristobalite upon firing. Flint pebbles from coastal or chalk deposits are calcined at about 900 °C, causing them to disintegrate into small pieces and change color from black to white. The calcined flint is then ground in water to a fine and uniform size.

Linné (1965) described the highly siliceous burnt bark temper used throughout the Amazon basin. He reported that in this region there were often no good sources of sand or lithics available, and crushing sherds for grog was not practical with the native wooden pestles. Instead, bark from the many available tropical siliceous plants was collected and burned; the resulting material resembled sand, and was added to the clay at ratios of about one part bark silica to three parts clay. Linné also reported that in Brazil, potters mixed burned and crushed freshwater sponges into their clay. The sponges have a skeleton of siliceous spicules that, when dried, consist almost entirely of silica. In thin section, silica spicules near the exterior surface are aligned horizontally due to shaping of the rim with a tool that exerts pressure on the clay; in contrast, on the inner side of the vessel the spicules are oriented at an angle of 90 degrees to the others.

Potters may also alter the plastic and drying qualities of their material by combining clay from two different sources, so that a coarser clay with a higher nonplastic content serves as the temper material for the finer clay. Each clay may have a different suite of nonplastics, and may also have some inclusions (such as quartz and feldspar) in common. It can be difficult to identify the presence of two clay sources, as differences between them may be very subtle. Velde and Druc (1999) illustrated some examples.

Isphording and Wilson (1974) discussed a material cited in the literature as "volcanic ash" temper used in pre-Hispanic ceramics in northern Yucatan. The presence of this temper material in northern Yucatan Maya ceramics had posed a puzzling problem because no deposits of volcanic material were known to exist in the area. However, the authors proposed that no volcanic temper actually existed; instead, the material in question is the mineral palygorskite (also known as attapulgite), which is widespread in northern Yucatan, and was known to the Maya as *sak lu'um* (white earth). In this area it was often found in association with a similar mineral, sepiolite, also identified in many of the ceramics containing palygorskite. Thus, this temper material (which strengthens the ceramic product), found along with other clay minerals, is essentially a case of natural or deliberate clay mixing.

Arnold (1971) and Arnold et al. (2000) mentioned this material in a discussion of contemporary pottery and raw materials from workshops in Ticul, Yucatan, where they found that temper was obtained at a small group of temper mines. This temper was a mixture of calcite, dolomite, attapulgite, and sometimes a bit of montmorillonite. The potters distinguished attapulgite (which they called "white earth", or *sak lu'um)*, understood it had some unique properties, and identified it as a crucial ingredient in pottery temper.

However, L. D. Jones (1986, 1991) disagrees with the idea that volcanic ash temper was absent and had been misidentified. She supported Shepard's identification (1942, 1968) of ash tempers in Lowland Maya pottery. Ash is described as being very fine (about 0.1 mm), slightly fibrous and usually colorless, occurring in sherds of varying shapes. Pumice is more globular and often opaque and white or pale colored, and is the same fine size as ash. She illustrated both of these tempering materials with photomicrographs of thin sections. At first glance, sherds with an ash temper appeared to be untempered; however, in the Preclassic and Classic, untempered sherds were very rare, so Jones noted that this appearance should cause one to look more closely at the material. She proposed that volcanic ash could have been imported into the Lowlands from elsewhere, and then used as a raw material for pottery production. However it was obtained, she suggested that the fineness of the material and the qualities it imparted to the clay and to the finished products (such as facilitating the use of unslipped vessels with uniform coloration and texture) were sought after by Lowland Maya potters. Some minerals tend to co-occur with the ash temper, including biotite (often in the form of lepidomelane), hornblende, clinopyroxene (mainly augite), and plagioclase feldspar.

Shepard (1968) noted that volcanic ash temper as free from phenocrysts as possible would be selected by potters making fine-textured wares. Many examples of ash do contain crystals in various amounts, however. These may include incipient crystals, rock fragments, and material associated with the environment where the ash originated.

7

Pottery Provenance Studies

Provenance studies usually involve identifying the region, site, or quarry that is the geological and/or geographic origin of the raw material used to make an artifact. Therefore, provenance studies using thin-section petrography of ceramic materials focus on nonplastic inclusions that retain characteristics of their original clay or temper sources. This tool works best for low-fired ceramics, since high-fired ceramics tend to show severe alteration of some original inclusions and the formation of new, high-temperature minerals as a result of the breakdown of some nonplastic and clay minerals. Successful provenance studies can answer a variety of anthropological questions regarding local production, imports and exotic goods, trade routes and exchange patterns, location and extent of raw material resources, and the social organization of production and distribution (Rapp and Hill 1998).

Identifying nonplastics and quantifying related data provides information for provenance interpretations. Classifying pottery by nonplastic materials and tempering technology can aid in initial ceramic classification schemes. Identifying what raw materials were selected by potters can help in the identification of production groups or workshops, or can link specific products to one production group or another. Temporal and spatial change and variation in use of nonplastics can provide clues about material culture manufacture and product variation and change. And, finally, in some cases, specific geological sources of nonplastics can be identified, providing even more concrete information about cultural contact, exchange relationships, and changes over time in materials selection.

7.1 Classifying Pottery Using Nonplastic Inclusions

One goal of characterization studies is to identify the type of nonplastic material present. The aim is often to classify sherds or wares according to material or technology, or into ceramic categories such as wares, series, and types; for example, sand-tempered ware versus grog-tempered ware (Figs. 7.1.1–7.1.4). This classification stage is often very important to archaeologists, who may wish to sort the entire ceramics suite from a site according to this criterion. The more precise information possible with thin-section analysis means that a subset of specimens can be analyzed to develop a set of standards that can be used to classify a larger number of specimens using only low-power binocular microscopy (Shepard 1939, 1942; Freestone 1995; Chandler 2001). Without the inclusion of thin-section petrography, errors are frequently made regarding temper identification, and thin-section work can often inform classification schemes and their interpretations (Porter 1962; Stoltman 2001).

Peacock (1968) examined 100 thin sections of Early Iron Age pottery from the Herefordshire-Cotswold region

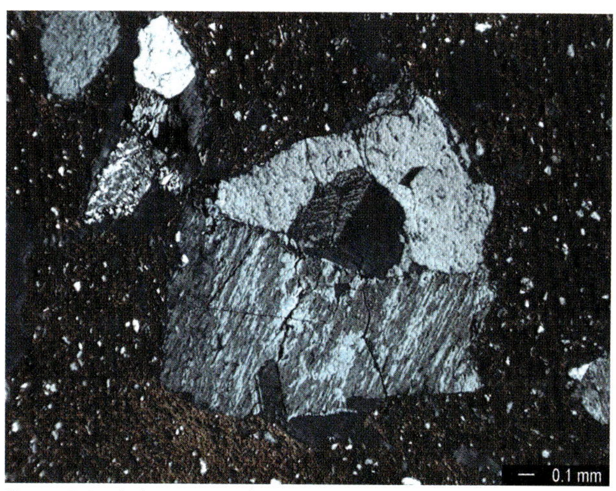

Figure 7.1.1 Lithic-tempered ceramic, crossed polarized light. This thin section was made from a sherd of Owasco Corded Ware from the Upper Susquehanna valley, Pennsylvania, 900–1200. The primary added temper consists of large angular grains of plutonic lithic fragments. There are also smaller mineral grains of a polymineralic mixture, derived from the flaking off of minerals from larger lithic chunks.

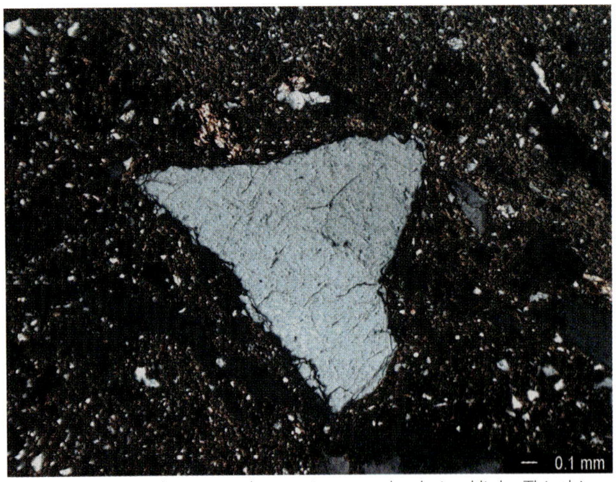

Figure 7.1.2 Sand-tempered ceramic, crossed polarized light. This thin section is from a sherd of Blue Rock Incised Ware from Lancaster County, Pennsylvania, 1000–1300. There are many large grains of sand, mainly quartz, that were added to a clay that contains very fine sand particles and micas.

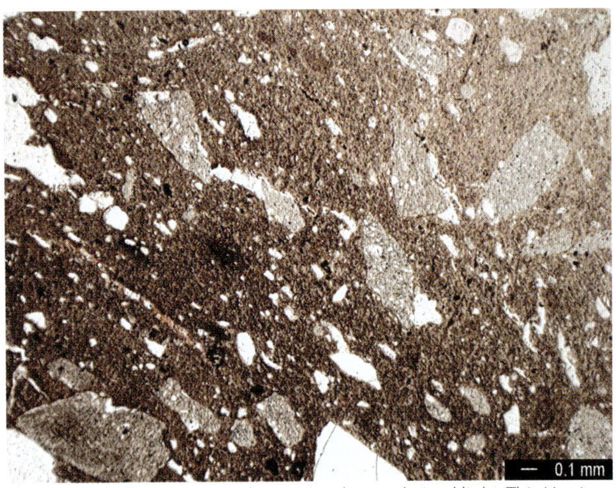

Figure 7.1.3 Grog-tempered ceramic, plane polarized light. This Hopi sherd (eighth–eleventh century) has many large, angular inclusions of grog, along with some sand grains (mainly of finer size).

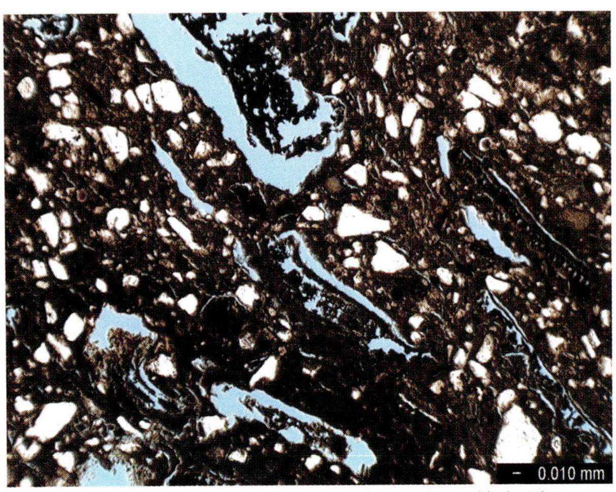

Figure 7.1.4 Organic-tempered ceramic, plane polarized light. This sherd from Ban Chiang, Thailand, has organic material (rice chaff) that is mostly carbonized in place with many areas of loss, creating a very porous ceramic. There is also a significant sand component present.

in Britain. Two wares were compared: Western Third B ("duck stamped") and Western Second B (linear tooled). Previous archaeological assumptions were that the linear-tooled wares were made by an indigenous population, while the duck-stamped ones represented the presence of an intrusive culture. However, on the basis of lithic tempers, Peacock defined three groups: one group had inclusions of igneous and metamorphic rock and mineral fragments; the second group had inclusions of limestone, some of which are probably Paleozoic and others Mesozoic; and a third group had inclusions of sandstone. Peacock identified the nearby Malvern hills as the probable origin for all three temper types. Thus there was in fact no distinction between the two wares—both were made locally with the same set of materials. Three separate pottery production centers, each utilizing a different set of raw materials but making both types of pottery, would probably account for the petrographic results.

Beynon et al. (1986) used petrographic analysis of temper materials to supplement other methods for classifying ceramic assemblages and understanding the cultural implications of their manufacture. They focused on temper types and sources for Early Bronze Age ceramics from two sites in Jordan, Bab edh-Dhra' and Numeira. A preliminary classification of wares was accomplished by macroscopic sorting at the site and further refined by limited laboratory analysis. Various morphological types and wares were selected for petrographic analysis (136 samples in all), in order to refine the ware categories by identifying temper types, ascertain any variations in temper usage over various chronological periods, specify any patterns in temper usage and vessel shape and function, identify local sources of temper, and compare use and sources of temper at the two sites to look for the possibility of trade or exchange between the two sites. Analysis was "blind" in the sense that the thin sections were studied without information about archaeological site, vessel form, or ware type, to prevent any bias from entering into the analytical results. Data were recorded on grain composition (quartz, limestone, basalt, and chert), grain size (measured on the longest diameter of grains within a sherd by using an ocular micrometer), grain shape (using three categories of rounded to subangular, subrounded to angular, or blocky and angular), and temper density. Four grit-tempering categories could be identified, which were then compared with vessel type and ware.

Beynon et al. found a good correlation between vessel type and ware and the four tempering categories. The most common tempering material was a sand that consisted of subangular grains of quartz, limestone, and chert, similar to the sands found at Wadi Kerak and Wadi Numeira. This wadi sand was used in fine and common ware bowls, jars, and jugs from Early Bronze Age I through IV at both sites. The second category was a temper comprised only of sand-sized quartz, with subrounded to subangular grains. It was identified as originating from weathered outcrops of Nubian Sandstone, which outcrops near Nemeira but not near Bab edh-Dhra'. This temper material was used in fine cream ware bowls, jars, and platter bowls made at Numeira during Early Bronze Age III. The third tempering material was a coarse, angular, tabular grit of fossiliferous and micritic limestone. The authors proposed that this material was prepared by intentionally crushing limestone cobbles and boulders collected near Bab edh-Dhra', where this type of limestone was more common than at Numeira. Larger crushed fragments of this limestone were used as temper for cooking pots and coarse hole-mouth jars at Bab edh-Dhra' during Early Bronze Age III and IV. Finally, the fourth category consisted of angular, blocky fragments of basalt with large crystals of plagioclase feldspar, amphibole, and pyroxene. As with the limestone fragments, the angular form of the basalt fragments and the presence of only basalt both suggested that the basalt was intentionally crushed. Similar basalts were found on the surface near Bab edh-Dhra'. Larger fragments of the crushed basalt were used as temper for bowls and platter bowls of burnished ware during Early Bronze Age III at both sites.

The authors also noted that there were some functional reasons as to why different tempers may have been used on different wares and vessel forms. For example, they found a significant correlation between temper size and wall thickness, with coarser tempers selected for thick-walled vessels. The largest vessels used limestone and basalt tempers, which also had the biggest size of all of the temper categories. The Wadi and Nubian sand tempers were probably chosen because of easily availability, and were found with a wide range of forms, including some larger vessels. This material also consistently appeared in vessel forms designed to hold liquid, possibly because of porosity properties. The thermal expansion property of limestone (similar to the clays) made it advantageous for cooking pots that needed to be repeatedly heated and cooled. A paddling procedure during vessel formation gave the tabular limestone grains a subparallel arrangement parallel to the vessel wall, which may strengthen the vessel and make it more suitable for large hole-mouth storage vessels. The basalt temper was found only on bowls that were slipped and/or burnished, and were probably used as serving and eating vessels.

Whitbread (1986a, 1995), in summarizing work done on Greek transport amphora fabrics, emphasized the usefulness of thin-section analysis as an adjunct to other archaeological data for characterizing fabrics and classifying vessels. The amphorae were made at a variety of different sites, with a wide range of coarse fabric inclusions, and functioned mainly for transport of liquids and semi-liquids. Many vessels have been classified by type and sometimes even provenance based on form and/or stamp marks. However, although nonplastic inclusions are coarse, they can only be identified through optical microscopy. Thus, thin-section analysis resulted in an improved description of fabrics, making fabric properties much more reliable for recognizing amphora types. It provided a test to confirm or refute fabric differences proposed by archaeologists, to help in assigning pieces with unusual stamps or typology to known series, and to give additional data useful for classifying fragments from amphora series with little extant information about stamps and typology. For example, Rhodian jars were found to have a fabric rich in small grains of quartz, serpentine, and orthopyroxene. In contrast, those from the neighboring Knidian peninsula were found to have a very different fabric with abundant quartz, quartzite, and phyllite, with some containing serpentine and rare volcanic glass. Compendia of the forms and features of Roman amphorae now also include a description of the fabric and a photomicrograph and/or detailed description of a typical example of that fabric in thin section (Peacock and Williams 1986; Peacock 1994; Tomber and Dore 2002).

MacSween et al. (1987) employed thin-section analysis to characterize and study Neolithic and Late Iron Age coarse wares from the eroding settlement mound at Pool in the Orkney islands off the northern tip of Scotland. Point counting (250 points using the multiple intercept approach), and grain-size analysis (for quartz grains) were performed in addition to mineral identifications. The main constituents, in addition to quartz, were sedimentary lithic fragments (sandstones, siltstones, and mudstones) that were locally available. The rock fragments were angular, suggesting that they were crushed and added as a temper. Size comparison of quartz in the ceramics versus that of the clay local to the site suggested that the quartz was a natural constituent of the clay and not added as a temper.

Thus, grain-size analysis was considered a way of characterizing and classifying vessels. To do this, 50 quartz grains were measured in each of 51 ceramic thin sections, and cumulative frequencies of size distributions were constructed. Each cumulative frequency was statistically compared with each other by tests of dissimilarity using the maximum difference between cumulative frequencies. The conclusion was that similarly textured clay sources were used throughout the period of study, with an additional source being used in the early part of the Late Iron Age. Comparative clay samples were collected near the site and fired, then analyzed in thin section. Grain-size analysis showed similarities to the pottery samples. It appeared that the clays close to the site could have been used without needing additional tempers, due to the presence of quartz. However, in practice, lithic fragments were often added by the potters. The additional, new, clay source in the Late Iron Age was often represented by burnished ceramics; that clay was finer, with fewer quartz inclusions, so might have been selected to facilitate the burnishing process.

Mason and Keall (1988) used thin-section analysis of surface collections to study the medieval pottery of Yemen. Kiln products such as wasters were characterized, as were imported fine ceramics, many of which have counterparts in excavated contexts elsewhere in the Near East so can be used in construction of a chronology for the Yemen assemblage. Petrographic work included study of grain size, shape, and sphericity; optical activity description for the clay matrix; and modal mineralogy (percentage content of mineral types) estimated using comparison charts. The mineral assemblage of local wares such as the kiln wasters was consistent with the lithology of the study area. Imports include *sgraffito* pottery and lusterwares. Some of the *sgraffito* examples collected in Yemen were found to be similar to examples from Iraq. Others had a fabric with a distinctive bimodal grain-size distribution and a slip of white clay with occasional fine quartz inclusions, similar to wares from

Siraf (in Iran); others matched fabrics from Egyptian wares (which include large angular grains of volcanic rock). The lusterwares found in Yemen were similar to ones originating in Iraq and Egypt. Later work by Mason (1991) looked at archaeological ceramic finds of Islamic ceramics from Yemen (where pottery was imported from sites in Iraq, Egypt, Syria, and Iran) in detail, including modal analysis and analysis of grain-size distributions. He noted that the ability to distinguish between various types, especially of lusterwares, enabled a more precise appraisal of the stylistic definition of each type.

Stoltman (1989, 1991, 2001) utilized thin-section petrography to resolve taxonomic issues regarding Upper Mississippi valley ceramics. For example, a Middle Woodland taxon, Linn Ware, included five ceramic types (Spring Hollow Incised, Spring Hollow Cordmarked, Spring Hollow Plain, Levsen Stamped, and Levsen Punctuated). However, distributional data emerged indicating that Spring Hollow Incised might not be associated with the other four types of Linn Ware. Thin-section analysis was carried out with the null hypothesis that the compositions of the Linn Ware types should be very similar. In fact, Spring Hollow Incised turned out to be sand tempered, whereas the other Linn Ware types were lithic tempered. Contextual evidence ruled out functional considerations as accounting for the different choices. As a result of this work, Spring Hollow Incised was removed from identification as Linn Ware, and has been renamed Prairie Incise, a type of Prairie Ware, a new Early Woodland ceramic series.

In his work, Stoltman emphasized quantitative thin-section petrography that incorporated point-counting procedures to collect data on both natural and human-added mineral inclusions. The quantitative approach proved extremely useful in discriminating local from nonlocal vessels. This approach was applied to the problem of identifying the extent of culture contact among prehistoric communities in the Upper Mississippi valley region (Hartley Fort in northeast Iowa and the Fred Edwards site in southwest Wisconsin) (Stoltman 1991).

A 1-mm counting interval was used over the entire area of each thin section; if fewer than 100 countable points were found, then the thin section was rotated 180 degrees and was counted a second time. Thin sections were counted "blind" as to which site they were from, to avoid bias. The 1-mm intervals tended to give more than 100 points per thin section, yet minimized repeat counting of large grains. The observations were categorized into clay matrix, silt, sand (natural component), temper, or void. To distinguish between added temper and natural larger mineral inclusions in the clay, Stoltman advocated collecting clay from the local area to identify natural constituents, when possible. Otherwise, the added temper may be identified as present when the grains are larger, polymineralic, and more angular than the natural inclusions. Individual minerals or lithic types could also be tabulated where that appeared to be useful. Sand and temper grains were assigned to size classes (fine, medium, coarse, very coarse, or gravel). Four basic thin-section properties resulted from these analyses: kind of temper, temper size, temper amount, and paste characteristics.

Stoltman distinguished the clay "body" as the bulk composition of a ceramic vessel, including clays, larger natural mineral inclusions in the silt, sand, and gravel size ranges, and temper. The term "paste" he then reserved for the aggregate of natural materials (clays and larger mineral inclusions) to which temper was added to form the clay body from which a vessel was made. Quantitative data could then be collected separately for the body, the temper, and/or the paste. Classifications could be done separately on the clay sources, based on the characteristics of their natural inclusions, and on the temper additives. Ternary diagrams were used to compare the quantitative data from ceramics from the two prehistoric Mississippian sites. Both body and paste data suggested that vessels from the Fred Edwards site were all of local manufacture and were different from those of the Hartley Fort site.

Vaughan (1990) analyzed 67 thin sections of Early Cycladic sherds from Akrotiri, Thera, with a first goal of characterizing the range of materials used at the site; then, she hoped that provenance assessments could follow. Twelve ware groups were analyzed in the study, including coarse storage vessels and finer decorated wares that may have been imported. Within each ware group, an assessment was made of the uniformity/variability; it was considered likely that multiple fabrics within a ware group might represent multiple workshops or sources of raw materials for that group. For each sherd, a horizontal section was prepared, labeled only by a number to avoid inadvertent bias during the analysis. Visual estimation charts were used to record shape and percentages, and grain sizes were measured by an eyepiece micrometer.

A wide variety of raw materials and sources was found, including clays derived from fine-grained pyroclastic/tuffite and acid lavas, micaceous-phyllite and schistose clays, siliceous sediments, and fine-grained sedimentary fabrics. Since geological data regarding Thera before the Late Bronze Age eruption is sketchy, what is known about locally available sources is somewhat tentative. However, some conclusions could still be made, even though matches could not be made with specific geologic sources. The range of materials found did suggest contact with other contemporary cultures in the immediate Cycladic region and perhaps even farther afield, where similar materials were also being used. For example, micaceous ware samples pointed to contacts with Naxos, while a talc ware and glaucophane schist ware suggested contacts with a variety of other islands. Some technological conclusions were also possible, such as the fact that most of the fabrics appear to constitute the use of raw materials prepared without any deliberately added tempering materials.

For example, Vaughan proposed that carbonates in one of the fabrics probably did not represent the addition of temper by the potter. This conclusion was reached because the carbonate content within the ware (and sometimes within a single sample) varied considerably in type, relative abundance, size, and distribution. Calcite was also present in the sherd matrix, discrete fossils were well preserved, which would not be likely if the carbonate fraction were crushed to produce a temper additive, and the age of the fossils was incompatible with the locally available limestone of Thera. Instead, the carbonates and fossils most likely originated with the parent rocks of the volcanic clays.

Barlow (1991) presented a good example of a pottery classification project that employed thin-section petrography. The project focused on constructing a classification system for Red Polished ware, the predominant pottery (handbuilt, low fired, with a red surface that may or may not be polished) found in Cyprus from the Early and Middle Bronze Age periods. Previous classification systems were not satisfactory, as there appeared to be a wide variety of fabrics, shapes, and surface finishes, and it was not clear where the subdividing lines for classification should fall, or what, if any, underlying rationale existed for those classifications.

Technical studies were initiated to provide additional data and insights to aid the classification efforts. Analysis included thin-section petrography of rock and mineral inclusions, and elemental analysis of pastes and slips. The thin-section work proved to be crucial in solving the classification dilemma. After studying thin sections of sherds representing the full range of Red Polished fabrics found at the site of Alhambra, a habitation site near Nicosia, two different clay pastes could be observed. An earlier geologist had also observed this fact, but had concluded they represented two separate production centers. In contrast, the Alhambra study found that the two clay pastes were used at the site both separately and as a mixture, and that both types were locally available.

One clay type was a product of the breakdown of chalk and limestones of the Lefkara formation. This calcareous clay was highly plastic and produced a soft, light-colored product (buff or pink in hue). It had natural inclusions ranging from fine to medium sized (0.125–0.5 mm), including calcium carbonates and microfossils. The second clay type was dark red and noncalcareous, a product of the breakdown of nearby pillow lavas. It had natural inclusions ranging from medium (0.5–0.25 mm) up to very coarse (2.0–1.0 mm), including volcanic lithics. It was not very plastic, and resulted in a moderately hard, dark red ceramic. Both clay types were accessible to ancient potters throughout much of Cyprus. The mixed-paste ceramics often visually resembled the red clay products. However, using thin-section analysis of a small subset of sherds, the researchers were able to develop a rapid field test to identify the presence of the calcareous clay (a microchemical test using a 10% solution of hydrochloric acid). The field test allowed large numbers of sherds to be tested.

Barlow discovered that the fine calcareous clay was used for the more decorative incised pots, while the red lithic-rich clay was used for cooling vessels and large bowls. The mixed clay was used for a particular category of ceramics: those with closed shapes (jugs, jars, and pots of liquids). She concluded that potters were selecting clays based upon the properties they desired for their products. The clay rich in volcanic lithics would be resistant to thermal shock, so made a good cooking vessel. The calcareous clay has low shrinkage and high plasticity, so would be good for decorative wares. Adding some of the lithic-rich clay would increase porosity and produce a closed vessel capable of keeping stored liquids cool.

Stoltman et al. (1992) combined petrographic and elemental analysis to look at vessels from two sites in the Kayenta region of north central Arizona (1250–1300). The two sites are located on opposite sides of a 14-km gap in settlement, so the level of interaction between them was of interest. Two types of vessels were found at each site, one a black-on-white painted ware and the other a plain ware. Thin-section petrography involved the point-counting procedure outlined in Stoltman (1989, 1991). Ranges and means were calculated for percentages of matrix, sand, and temper, the main type of temper was recorded, and a Temper Size Index was calculated by adding the scores for all measured temper grains and then dividing by the total number of such grains. Most black-on-white vessels, regardless of site location, were characterized by angular inclusions of volcanic glass. None of the plain ware vessels contained volcanic glass—those wares were tempered with sand. The ash-tempered sherds had extremely fine inclusions, whereas the sand-tempered ones were larger overall. Even the minority of black-on-white sherds not tempered with volcanic ash were finer grained, with finer sand temper than was seen in the plain wares. The authors concluded that, regardless of site location, the potters chose volcanic ash as the preferred temper for black-on-white wares, and sand as temper for the plain vessels. The plain vessels were also made with larger temper grains than the black-on-white ones, regardless of temper constituents. In addition, a significant difference in sand size indices between plain wares from the two sites was found, indicating that plain wares were locally made following slightly different recipes. Elemental analysis of the clays supported these conclusions, finding a difference between the two wares and between the two sites.

Joyner (1997) examined Byzantine and Frankish cooking wares dating from the eleventh to the fourteenth centuries excavated at Corinth, Greece. The two wares had distinct typological differences, with the early Byzantine ones being thick walled with a spherical body and the later Frankish ones being thin walled with a globular body. The design differences were thought to be due to differences in diet, with the Byzantines eating pulses and the Franks eating meat stews. In thin section, six major fabric classes were observed, with nonplastic inclusions that suggested a local manufacture: (1) angular radiolarian chert fragments; (2) monocrystalline quartz with less dominant chert; (3) mudstone fragments in a well-fired matrix that was optically inactive; (4) mudstone fragments in a poorly fired matrix that was optically active; (5) sandstone fragments; and (6) a fabric with a complex mix of inclusions of quartz, mudstone, and chert. A variety of fabrics, including all six on this list, occurred with the Byzantine and Early Frankish materials, probably representing the products of several workshops that were exploiting different clay sources. However, by the Middle and Late Frankish periods, only a single fabric appeared (no. 6 in the above list). There appeared to be a transition from the Byzantine form with many fabric classes to the Frankish form with one major fabric class, suggesting a change in more specialized pottery production associated with the incoming Frankish Crusaders. Clearly, pottery was still produced locally, but perhaps within a single workshop, or from multiple workshops all using the same clay recipe.

7.2 Identifying Production Groups or Workshops

A second goal of characterization studies is to provide a thorough description of the nonplastics present or added to the clay selected for use by potters, as a means of characterizing workshops or production centers. Beyond just "sand" or "grog," this level of analysis aims to fingerprint the materials of a set of specimens well enough to identify groups that may represent a production source, distinguish them from specimens that may come from different sources, and even identify workshops within a regional source. When there is knowledge of the geology at a site, identifying the existence of a production center may be achieved by analyzing the proportion of local versus foreign tempering material in a site or region (Vincent 1988). The approach taken to achieve success of this goal is partially dependent on the nature of the nonplastic material present; some materials may be too homogeneous or nondistinctive to allow for very precise discrimination. For example, with sand tempers, the assumption may be that pottery made in a specific restricted geological area should contain sand grains of approximately the same mineralogical composition (from the same rock formations) and similar size and morphology, as they are subject to the same processes of sorting and weathering (D.F. Williams 1983).

Freestone and Hughes (1989) pointed out that thin-section analysis, like other analytical methods applied to pottery sherds, can often be used to identify subgroups within a ceramic assemblage, but understanding the significance of those groups may be difficult. Groupings might reflect natural variations in a single source of raw material collected by one workshop or by multiple workshops, separate sources of raw materials within a single workshop or between different workshops, within-workshop variation in production methods over time or among potters, or the presence of different production methods found in a variety of workshops. To aid in interpretation of groupings identified by thin sections, additional methods are helpful (which may include careful examination and macroscopic study of sherds and vessels, comparison with other archaeological data, and elemental analysis of clay pastes).

Success also often depends upon the use of more detailed characterization of a thin section than simply identifying the constituents present; instead, at this stage the petrographer also needs to identify the volumetric proportions of the constituents as percentages (modal analysis), quantify the grain-size distributions, and perhaps gather other quantitative information such as degree of rounding of quartz grains. Obtaining quantitative information permits the use of multivariate statistical analysis, thus improving the possibility of characterizing a ceramic source and distinguishing it from other sources. In addition to identification of nonplastics and quantification of certain variables, a thorough thin-section analysis may also include a more detailed mineralogical analysis and geological interpretation of results. For example, although many ceramics have a feldspar component, feldspars can be further specified in thin section as to phase composition, and that information may be used to deduce a volcanic, plutonic, or metamorphic source (Fig. 7.2.1) (Kerr 1977). However, if initial characterization work indicates that the materials under study may be quite homogenous and lack distinction in their nonplastic components, trace element analysis to characterize clay deposits may be required. Freestone (1995) recommended a pilot study analyzing up to 20 thin sections before embarking upon a major research project, in order to assess whether or not the particular ceramic materials under study are likely to be useful in answering the research questions of interest. Vaughan (1995) also emphasized the need to ensure that the samples taken for a pilot study include the full range of vessel sizes and functions within each pottery class or ware, and that the area sampled is representative of the fabric of the vessel.

Peacock (1977) focused on the so-called "Pompeian Red Ware," Roman bowls and dishes with a thick internal slip, found all over the Roman empire from the first century B.C. to the first century A.D. (the name refers to a similarity of the color of the red slip to red pigments used to decorate walls of houses in Pompeii). Thin-section analysis had the goal of identifying various production centers for this ware so that issues of distribution and interrelationships could be addressed. Sherds from a wide variety of sites were analyzed and the fabrics classified, in order to propose how many production centers might have been involved in making vessels of this ware. Seven main fabrics were identified, and Peacock was able to make some initial suggestions regarding possible origins and distribution areas for Pompeian Red.

Focusing on interpretation of a single object, MacKenna (1978) compared a thin section from an amphora excavated

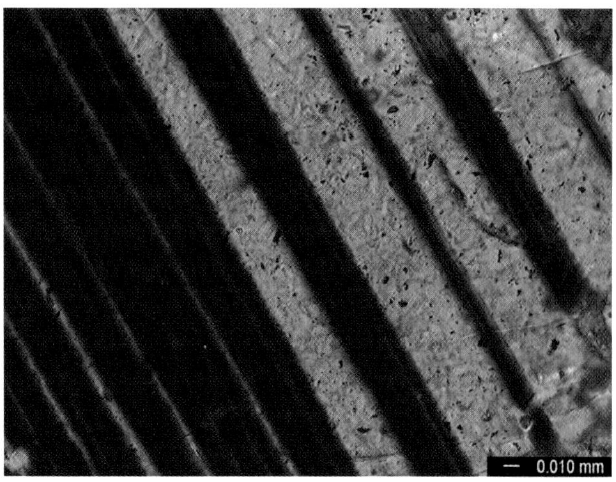

Figure 7.2.1 Plagioclase feldspar (labradorite), crossed polarized light. One method of determining phase composition (on the albite to anorthite scale) of a plagioclase feldspar grain is to measure the extinction positions of combined Carlsbad-Albite twins (the Albite twinning is visible as the alternating strips of dark and light areas; the Carlsbad twinning is seen in the division of the grain into two halves that differ in their extinction pattern).

at Brockley Hill in England with five reference samples from the site. Fifty quartz grains from each thin section were measured and grain size/frequency curves were constructed. The proportion of quartz grains exhibiting undulatory extinction was also calculated for each thin section by measuring the number of degrees of rotation of the microscope stage required to bring the entire grain to extinction, and a mean value calculated. Two statistical tests, Student's t and the Kolmogorov-Smirnov test, indicated that there was no significant difference between the amphora sherd and the five reference samples. Thus, it could be concluded that the amphora was a product of the potteries of Brockley Hill.

Lazzarini and Calogero (1989) studied *sgraffito* ceramics in Venice. Commerce between Venice and the Oriental empire led to the development of Venetian *sgraffito* around 1250. Early *sgraffito* in Venice was copied from imported Byzantine *sgraffito*, and distinguishing between the two production areas can be difficult. *Sgraffito* eventually became the most popular ceramic produced in Venice. A combination of petrographic (thin section and x-ray diffraction analysis) and chemical analysis was carried out on imported *sgraffito* and presumed local examples of the twelfth–thirteenth century. The data were compared with data from Byzantine sherds of known provenance and date (mainly kiln wasters). The primary goal was to clarify the presence and extent of local production of *sgraffito* in Venice.

The early (archaic) and later Venetian examples were rather similar in materials, indicating a similar origin. Both were characterized by the prevalence of angular or subangular quartz grains, and the frequent presence of polycrystalline quartz and chert. Small feldspar grains were also present, along with calcite and muscovite (sericite). Illustrated thin sections showed abundant quartz and sericite/illite flakes. Differences were seen mainly in the amount of quartz and the average grain size.

The Byzantine examples were quite different, and, as to be expected from wares with many sources, also quite variable. Illustrated thin sections showed a probable imported sherd with abundant metamorphic lithics and fine-grained quartz within an illite-containing paste. Comparison sherds of known sources included one from Corinth with large calcite and metamorphic particles within a paste containing only scarce inclusions, and one from Constantinople with scarce but larger metamorphic lithics within an illitic paste. Lazzarini and Calogero concluded that the local Venetian archaic *sgraffito* could be distinguished from examples of imports from Byzantine centers, but it was not yet possible to identify the origin of all of the imported *sgraffito*, although sources for some of the sherds could be suggested.

Rigby et al. (1989) focused on a group of seven Celtic painted vessels from the La Tène burials of the Champagne region in France. Only two of the seven vessels could be sampled safely, but thin sections were available from 130 unpainted comparative vessels from the same region and period. Thin-section analysis showed that one large group of the vessels was tempered with grog, another group was made with clay that was naturally tempered with glauconite (Fig. 7.2.2), and in some cases both grog and glauconite were present. A smaller group of vessels had calcite tempers

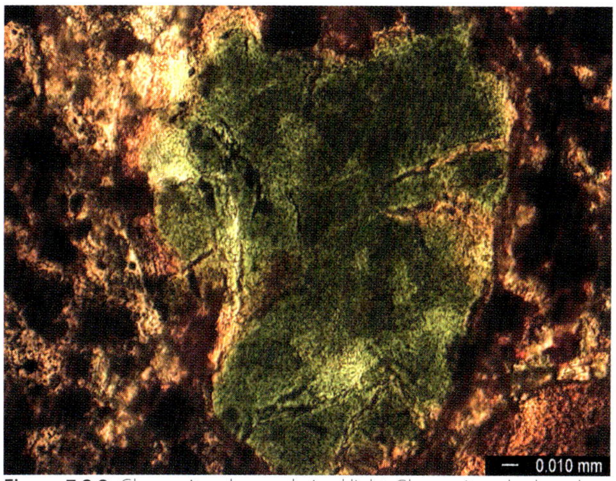

Figure 7.2.2 Glauconite, plane polarized light. Glauconite, a hydrated silicate of potassium or iron or magnesium, can occur as a natural constituent of clays. It is green in plane polarized light, and occurs in grains or pellets that are in part aggregates of minute crystals and in part single crystals.

(including shell and limestone fragments). A final group was made with clay that was rich in quartz sand or silt. In the fine-grain examples, the quartz was a natural component of the clay, but in others it may have been added as temper. Both painted pots fit into this last group.

Thin sections of objects in this small group showed a pale brown birefringent clay matrix with 30–40% well-sorted, subangular monocrystalline quartz grains about 0.1 mm in diameter, some red-brown iron oxides, and occasional calcite (Rigby et al. noted that calcite survival would depend on the burial environment). The group consisted of two comparative wheel-formed unpainted vessels, and two painted ones. A textural analysis was carried out to test the integrity of the group: the sizes and proportions of quartz grains were measured and compared with those of quartz-tempered pottery from later excavations (in a plot of mean grain diameter in millimeters versus percent quartz). The four vessels in question formed a tight cluster, similar in spread to clusters formed by other typologically well-defined groups. These results suggested that all four vessels may have been made with the same raw materials and techniques, and therefore may all originate in the same workshop.

Peña (1990) employed thin-section petrography to analyze red-slip cookware (Pompeian Red Ware, third century B.C. to the second century A.D.) from an Etruscan/Roman settlement in the Chianti Mountains of Italy. He was able to identify three distinct fabric classes. The most common fabric class at the site contained clastic materials whose origin is most likely somewhere in northern Etruria, in contrast to previous assumptions that the early centers of production were located in the Bay of Naples area. The other two fabric classes were rare at the site and contained volcanic materials, with multiple possible production centers.

Olcese (1991) analyzed Roman coarse ceramics from the site of Albintilum in Italy, in an attempt to see if ceramic production groups identified by typological and macroscopic criteria could be supported by thin-section data. The author was also interested in determining the extent of local production. Comparison of ceramic sherds to wasters found at the site was helpful, as was a study of the local geology. Olcese found that most of the Albintilium ceramics were characterized by the presence of quartz, feldspar, and fragments of schist and limestone. This assemblage fits with the geological characteristics of the area around the site, thus supporting local production. However, some of the production groups (as identified by typological and macroscopic criteria) had a mineralogical composition that did not correspond with the local geology. Pottery excavated at levels dating from the first century B.C. to the first century A.D. contained volcanic material (augite and sanidine). This assemblage suggested a source in central or southern Italy (south Etruria, Latium and Compania). Olcese concluded that coarse wares from this site during the first century B.C. to the first century A.D. have various origins, some nonlocal. In later centuries, an intensive local production met the local needs for coarse wares, so few such ceramics were imported. Other archaeological evidence showed that these coarse ceramics were even a source of export to the surrounding area during this period.

Peacock and Tomber (1991) studied sherds from Roman and early Byzantine period amphora excavated at 17 kiln sites in Tunisia. There were two main groups of kiln sites: those in coastal areas and those in inland areas. The coastal ones have urban locations and date mainly to the late second/early third to the late fourth/early fifth centuries. The inland ones are more rural and date mainly to the early fifth to mid-sixth century. Thus, it was proposed that during the fifth century a reorganization of amphora production took place, as production shifted from coastal areas to inland sites. Thin sections were taken from sherds representing all the kiln sites to confirm whether or not this proposal could be supported.

Since most of Tunisia has a uniform sedimentary geology, the same range of inclusions was found at all the kiln sites (mainly fine quartz and calcareous material). However, Peacock and Tomber found that three distinct fabric groups could be identified on the basis of clear differences in proportions of quartz to limestone. Fabric 1 had an abundance of calcareous material (limestone, fossiliferous limestone, and shell) that occurred in equal or greater proportion to quartz. Inclusions were very well sorted and densely packed. Fabric 2 had abundant moderately to poorly sorted quartz and lesser, but still moderate, amounts of limestone. There was a higher proportion of quartz to limestone than in Fabric 1. Fabric 3 had sparse amounts of limestone, with abundant quartz, again with some differences between kilns in size and sorting.

Fabrics 1 and 2 occurred mainly at the early, more urban, coastal kiln sites. In contrast, Fabric 3 was found only at the later, more rural, inland kiln sites. The differences identified by thin-section petrography indicated a change in technology concurrent with a change in location of production sites. The authors proposed that Fabric 1 had an added limestone temper, and Fabric 2 was probably a naturally occurring clay still containing a fair amount of limestone inclusions. However, Fabric 3, with a decreased amount of limestone, would have produced an improved product less likely to crack during firing. This fabric may have been a deliberate attempt to solve firing problems. It was noted that with pottery from sedimentary environments that contain only common mineral inclusions in abundance, fabric type may be determined less by geological source than by manufacturing technology.

Morzadec (1992) used thin-section petrography combined with geological field studies to investigate Iron Age pottery workshops in Brittany, France. He studied both temper and matrix materials, and found that the raw materials originated from alteration products of mafic and ultramafic rocks on the Ty Lann formation (where traces of early quarries and clay mines are still visible). He was able to identify several distinct workshops based on materials identification and comparison with locations where those materials were available.

In an examination of four classes of pottery recovered at 16 sites in the Central Mississippi valley (Riley et al. 1994), it was found that these objects were similar in form and function to ceramics recovered at the mound site of Cahokia in the St. Louis, Illinois area. Extensive trade networks existed during the period 1150–1400 between Cahokia and other sites in the Mississippi floodplain. Thus, the question under investigation was whether objects were "tradewares" or whether they were actually locally made copies of forms recovered at Cahokia. Thin-section petrography was done on 126 sherds. The technical approach taken was to calculate the percentage of temper material present and the grain size (given as the largest grain measured in a thin section and expressed as abbreviated Wentworth scale categories). Silt content was estimated as high, medium, or low, and the presence of biogenic opal, indicative of an alluvial backwater mud, was noted.

The results suggested that most of the wares examined were manufactured locally at the site where they were found. Vessel pastes usually derived from backswamp muds from floodplains, as indicated by varying amounts of biogenic opal. In most cases, the vessels were tempered with shell; in other cases, grog was used, although shell was visible in all of the grog tempering materials. Both utilitarian forms such as the Cahokia cordmarked jar and plates, and more specialized forms such as the beaker and bottle forms, were apparently all local productions rather than "tradeware" objects, according to the results of thin-section petrography.

Bedal et al. (1995) analyzed pottery from two sites in northwestern Iran, Hasanlu Tepe and Dinkha Tepe, from the Late Bronze Age and Early Iron Age. An abrupt stylistic change from one period to the next at these sites had been attributed to the migration of new people into the region. However, thin-section petrography (as well as elemental analysis) revealed that although a cultural change may have influenced the style of pottery, no major change in clay sources or tempering practices took place. In the Late Bronze Age, painted wares of forms typical at each site differed in mineralogy of their tempering materials, but not in the amounts of materials that were added. The painted bowls typical of Hasanlu had a calcareous matrix with inclusions of calcite, amphibole, and quartz (grains were both large and small). Dinkha painted jars had a finer, less calcareous matrix with inclusions of fine-grained quartz and iron-rich nodules. The conclusion was that each site had its own pottery workshops, each of which followed their own procedures regarding clay tempering.

In the Iron Age at both sites, the painted wares disappear. In thin section, an internal difference at Hasanlu is seen between cooking pots and gray ware vessels. The cooking pots are heavily tempered with large quartz, feldspar, and calcareous grains, similar to the technology used for the earlier painted wares, whereas the gray ware was made of finely levigated clay that contains only minute or small inclusions natural to the clay. Apparently, a new specialized workshop at Hasanlu used a new process specifically for the newly introduced monochrome gray wares. At Dinkha, however, even though a new assemblage of pottery types also appeared, the earlier process for paste processing and tempering continued to be used. Thus technological differences between the two sites suggested the presence of some workshops that specialized in certain ware types.

MacSween (1995) used thin-section analysis of coarse ware assemblages (hand-thrown, often poorly fired, and mostly roughly finished and undecorated) from sites in the Northern Isles of Scotland to identify workshop groups related to chronological periods. For example, at the site of Kebister in Shetland, about 4,000 sherds representing about 400 vessels were recovered. Most were undecorated body sherds that were chronologically undiagnostic. The fabrics could be divided into four main groups (untempered sandy clay, sandy clay with rock temper, grass-tempered clay, and talc-rich clay with rock temper) which were then compared with the few morphologically diagnostic sherds excavated at the site. MacSween was able to conclude that the rock-tempered pottery was associated with early phases on the site, while untempered clays dominated the later phases of the site. Since there were so few diagnostic sherds available, these conclusions were tested by plotting each fabric type on a site grid to look at distribution, with the hypothesis that rock-tempered sherds should be concentrated in the areas of earlier settlement while sandy and grass-tempered sherds should be concentrated in the areas of later settlement. The hypothesis was supported by distribution data.

MacSween noted that sometimes the creation of too many fabric groups can interfere with the ability to see trends in a larger dataset, as trends may be masked by small differences in parameters such as size or angularity of a particular type of temper. She recommended focusing on the effect that a particular temper addition might have on the pottery in grouping data. For example, grouping together all fabrics with coarse rock temper, whether those lithics are siltstone, mudstone, or sandstone, may be useful because all the lithic tempers would have had a similar effect on the properties of the pottery. In contrast, a more finely crushed igneous rock temper may have created a different effect on the pottery properties so might be better kept as a separate grouping factor.

Marcus (1995) conducted a petrographic analysis of storage jars from the site of Tel Nami, a small port in Israel that had substantial eastern Mediterranean contacts in the Middle Bronze IIA period (ca. 2000–1759 B.C.). Her interest was in delineating production centers, with the goal of identifying zones of interaction and trade networks during this period. Some large storage jars found at the site contained imported Aegean legumes, so she investigated whether they might have been made locally or if the jars, too, were likely imports. In thin section, they were seen to consist of a calcareous clay with rounded to subrounded quartz-feldspathic sand inclusions and silt-sized quartz and feldspar within the clay matrix. Micrite and fossil bioclasts had a bimodal distribution of rounded silt to coarse sand size; there was also secondary calcite in the clay, in the form of silt and fine sand-sized rhombs. Most importantly, aeolianite (quartz and feldspar in a calcareous cement matrix) sandstone detritus was present (Fig. 7.2.3), along with amphiboles, clinopyroxenes, and chert. This fabric type predominated at the site and was found with most pottery forms (except for cooking pots, which had olivine basalt fragments). Local geological deposits contained aeolianite sandstone, so, although an exact geological source was not identifiable, she concluded that the storage jars were consistent with local production.

Morter and Iceland (1995) analyzed Middle Neolithic sherds from Calabria, Italy (fifth millennium B.C.). Many had a decorative style known as *Stentinello*, comprised of a sand-tempered fineware with impressed and incised designs that were often filled with colored pastes. These contrasted with other wares having mainly painted decoration and more crudely executed impressed designs. The thin-section analysis was conducted to try to determine if the *Stentinello* wares might have been products of local workshops, or if they were imports. The nonplastic component was found to be mainly quartz with some feldspars (potassium and plagioclase), and in some cases significant amounts of lithics (schists, granites, and sandstones). The authors suggested that all these materials were available in local stream and beach sands. In contrast, examples of painted sherds, found in the later upper stratum, did not seem to have been made with local materials, so may have been imported.

Material from two strata at the excavated site showed a similar suite of clays and tempering material used over

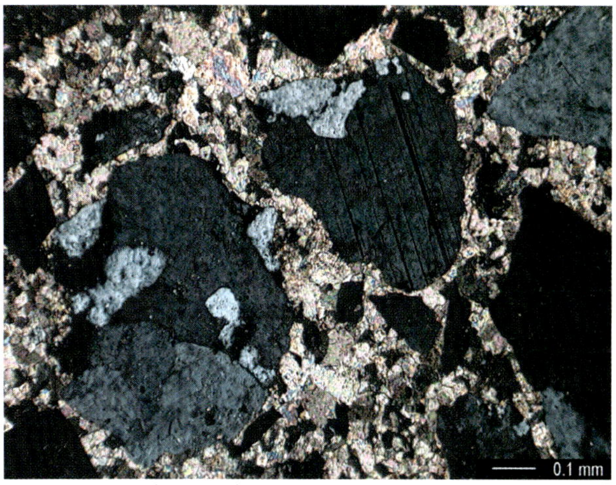

Figure 7.2.3 Aeolinite sandstone, crossed polarized light. Aeolinite sandstone consists of quartz and feldspar in a calcareous cement matrix. The matrix shows very high birefringence, with pastel-colored interference colors, in contrast to the low birefringence of the quartz and feldspar grains.

time. Point counting was utilized to quantify a number of the constituents: fine quartz (monocrystalline, less than 0.25 mm), other quartz (polycrystalline or monocrystalline greater than 0.25 mm), feldspars, pisolites (large ferruginous inclusions that are usually rounded with sharp boundaries, Fig. 7.2.4), micas (biotite and muscovite), lithics, clay matrix, voids (or pores), and "other" (opaques, heavy minerals, foraminifera and other micritic inclusions, and chert). In a discriminant analysis between strata, the lower stratum had a consistently higher concentration of pisolites, and micas were more frequent in the upper layer. The authors noted that these differences could be due to the use of slightly differing temper sources, or to different firing processes. The hematite of the pisolites should have burned out if firing temperatures were much above 800 °C. Yet, in spite of great similarity in materials, very different vessels were made in the earlier and later periods, even within the overall *Stentinello* style. The earlier products had a uniformly darker finish (probably due to smudging), not seen in the later pieces.

As part of the process of characterizing and identifying production groups and workshops, it is possible that thin-section petrography can aid in identifying the presence of specialization. Rice (1996) proposed that product uniformity co-varies with the numbers of producers within a community. Following that idea, Stoltman (2001) noted that paste and body properties can be measured quantitatively with thin-section petrography, so identification of specialized ceramic production may be possible. As an example, he developed a variability index, to contrast the relative amount of variability in the body and paste indices of a ceramic class from a site with those of comparable ceramics from other sites.

Sherds from Greek transport amphorae found in a warehouse of Archaic Kommos in southern Crete had been assigned on the basis of typology to production centers in Samos, Lesbos, and Chios in the eastern Aegean. Thin-section petrography was conducted (De Domingo and Johnston 1997) to see if there were any petrographic differences between these types and their proposed production sites; sherds from transport amphorae of Miletus, in western Anatolia, were also included. The Samian fabric was found to contain silty metasediments, having very low amounts of nonplastic inclusions (5–15%), the Lesbian fabric contained large fragments of volcanic glass with inclusions of feldspars, and the amphorae from Chios had a fabric containing sedimentary and medisedimentary inclusions. Milesian amphorae fabrics contained abundant inclusions from sandy metasediments (about 35%). The Samian and Milesian fabrics were similar mineralogically, but had very distinct textures. The authors concluded that the fabrics of these four types of amphorae could be distinguished in thin section, confirming their origins in different production centers and thus the presence of commercial links between eastern and southern Greece and Anatolia during the Archaic period.

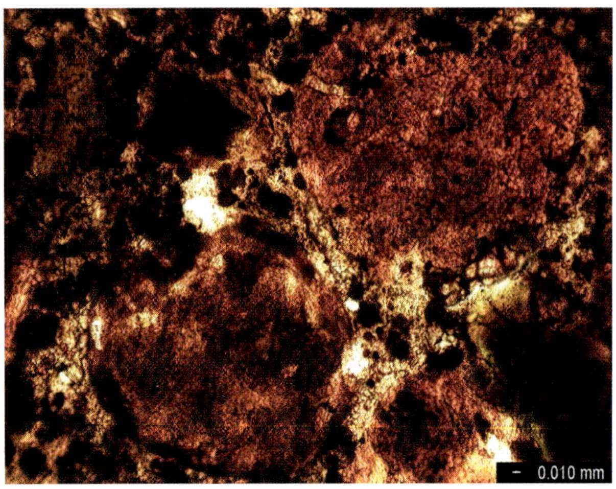

Figure 7.2.4 Pisolite, plane polarized light. Pisolites may be found in soils or clays, and are ferruginous inclusions that are usually rounded with sharp boundaries. The iron oxides impart a reddish color to this structure.

Buxeda I Garrigós et al. (2003) characterized ceramic products from a contemporary pottery-making community in Spain (at Pereruela), where they were able to observe the entire production sequence. Unused pots were obtained from four potters, and their clays were also sampled. The potters were mixing red clays with kaolin, and there was much geochemical variation in the red clays found near the village. Thin-section petrography helped to shed light on the problem that products from different potters at this one production center differed so much that they formed distinctly different compositional groups; and even within a single pot large variations in composition were found from one sample location to another. As it turned out, some of the red clays had monazite inclusions, sometimes as large as 100 μm in size. This mineral can occur as an accessory in granitic igneous rocks, and is a rare earth phosphate mineral. The variations in elemental composition caused by sampling from areas within the ceramic that could randomly contain some of the scattered large monazite grains (or small ones, or none) could have serious effects on provenance studies. Thus, identifying certain minerals in thin section can help to construct and interpret pottery studies based on elemental analysis.

Montana et al. (2003) studied Roman thin-walled (0.5–3.0 mm) ceramics (of the type used mainly for drinking) from northwestern Sicily. The main goal of the petrographic analysis was to test the hypothesis that Roman fine tablewares were part of a local production center at the site of Segesta. Archaeologists had proposed that numerous centers of production probably existed in the western Mediterranean, rather than this brittle ceramic being an item for long-distance trade. Forty samples plus seven kiln wasters (of other wares) from the Segesta site and three local silty-clay specimens were thin sectioned for modal analysis, in an effort to determine whether or not a production center might have been located there. Thirty-five of the samples were found to be petrographically similar enough that they could be considered one group—there was sufficient similarity between them and one of the silty-clay specimens, in terms of nonplastic inclusions, that they could all be considered one petrographic group. The observed similarity was both qualitative (type of nonplastic inclusions present) and quantitative (relative abundance of mineral and lithic components). The kiln wasters also matched this clay source and the thin-walled specimens. Thus, the thin-section petrography supported the idea that there was a local center of production for thin-walled fine tablewares at Segesta, Sicily during the Roman period. The five nonmatching samples probably represented imported objects.

Maritan et al. (2005) analyzed 82 Bronze Age and Iron Age pottery specimens from Tell Mishrifeh/Qatna in Syria. The fact that kilns and separation basins were found indicated that pottery was produced at the site. However, it was unclear if all of the pottery remains originated in the local workshops, or if some represented imported luxury wares or containers for imported goods. In order to answer this question, thin-section petrography (as well as chemical and diffractometric analyses) was performed on sherds and on clay material found in a separation basin. Petrographic work included characterizing the fabric, identifying mineral and lithic inclusions, determining absolute and relative frequency of grains, characterizing porosity, and assessing grain size, shape, and roundness. Nine petrographic groups were identified, including those: (1) rich in basalt fragments; (2) rich in limestone fragments with scarce basalt fragments; (3) containing volcanic glass fragments; (4) rich in mollusk shell fragments; (5) having an optically inactive, isotropic groundmass with scarce pores and numerous inclusions of a variety of shells, minerals, limestone, chert, or quartz grains; (6) rich in sand-sized euhedral crystals of spathic calcite; (7) having a dark brown groundmass that is highly porous, containing many elongated channels and planar voids; (8) rich in rounded quartz and angular chert fragments, with micritic limestone and a variety of other minerals; and (9) having a fine-grained paste with few inclusions.

Maritan et al. noted that the lithology of the inclusions found in all the petrographic groups was compatible with rock types outcropping in the area. For example, there were nearby outcrops of olivine basalt, chert, micritic foraminerfer-rich limestones, limestones with spathic calcite; and mollusk shells were locally available. The hypothesis of ceramic production in local workshops was thus supported by the petrographic evidence, results that were further supported by the chemical data.

Quinn and Day (2007) examined Middle Bronze Age 'Dark Faced Incised Ware' (DFIW) excavated at Knossos on Crete. The coherence of the objects as a group had been questioned because of stylistic variations, even though petrographic and chemical analyses supported the presence of a distinctive group. Quinn and Day utilized their novel method for characterizing both microfossils and nannofossils by combining thin-section analysis with a smear-slide technique (using a small scraping from a sherd) under very high magnification with an oil-immersion lens. They were able to confirm the consistency of the group. Fifteen out of 16 thin sections studied had raw materials that could be assigned to the Early Pliocene on the basis of a detailed taxonomic analysis of the nannofossil assemblage, meaning that the same raw material source was apparently used in the production of the DFIW ceramics.

7.3 Examining Temper Changes and Variation Over Time and Space

Fitzpatrick et al. (2006) noted that a necessary first step towards eventual provenance determination is the analysis of ceramic sherds to see if those from one locale exhibit mineralogical homogeneity, while being different from those of other locales. Only then can one begin to try to look at possible trade and exchange relationships or to identify specific geological sources for temper. Thus, clarifying where homogeneity, variation, and changes occur is one step in the study of ceramic tempers, and interpretations of those findings may in themselves address a number of archaeological questions.

One of the earliest and most successful petrographic thin-section studies of pottery was of material excavated at the site of ancient Troy (Felts 1942). This study remains a model of the potential utility of thin-section analysis of ceramics and sherds. Among the research questions addressed was whether or not ceramics produced in different time periods differ in thin section. Felts compared thin sections of ceramics with soil samples taken in and around ancient Troy. Most ceramics were found to have rock and mineral inclusions native to the local region and seen in the local soils (such as quartzite, a green hornblende trachyte porphyry, siltstone, and a strongly foliated mica schist). In a small percentage of the objects, the inclusions consisted of minerals and rock fragments not seen locally (such as basaltic lava). Therefore, these objects were hypothesized to be imported. The imported wares also seemed to differ from local ones in shaping and firing techniques, which served as supporting evidence for their exotic origins. Pottery and soils from various archaeological strata were examined for temporal changes. The earliest sherds were very porous (up to 20% of a typical thin section consisted of spherical and elliptical vesicles). Through time there was a clear change in technology toward smaller grain size for tempering materials, less porosity, and an increasing percentage of objects produced on a wheel. However, only a few mineralogical differences were found among the different strata.

Shepard (1942) performed thin-section petrography on ceramics from the Rio Grande, examining tempers and surface treatments. Whereas the regional chronology had previously been constructed on the basis of changes in ceramic styles, she used the additional thin-section data to better understand temporal change in the region as well as to identify probable pottery production centers. Shepard (1953) also used petrographic analysis of 33 thin sections from Anasazi and Mogollon pottery to identify chronological changes in paste composition. Brown ware from the Petrified Forest in Arizona was found useful for establishing chronological seriation. Four classes of brown paste were identified: micaceous, sand tempered, grog tempered, and lithic tempered. The lithic-tempered sherds occurred in minor amounts and Shepard classified them as trade pottery. The other three paste types were found to mark a chronological sequence.

Kolb (1977) looked at a large number of thin sections of Thin Orange Ware recovered in the Teotihuacán Valley

Project. He identified a Type Alpha that consists of mainly angular and subangular quartzite, angular calcite, talc, angular and rounded quartz, and muscovite, with smaller amounts of various other minerals and lithics sometimes present. A Type Beta had generally the same inclusions but in different percentages (with muscovite being present in especially greater amounts). Kolb concluded that Type Alpha was the most important ware and was mainly produced outside of the basin of Mexico from 150 B.C. to A.D. 805 and was imported into the Teotihuacán valley. It was used with a variety of vessel forms and types. In contrast, Type Beta was probably locally produced at Teotihuacán during 700–850 and was confined to two vessel forms with rare decoration.

Kamilli and Lamberg-Karlovsky (1979) carried out thin-section petrography of ceramics from Tepe Yahya, Iran, with the main goals of looking at continuity or discontinuity in ceramic production and of clarifying the role of foreign elements in technology and production during the period of about 3400 to ca. 3000 B.C. These questions were intriguing because in the earlier period the authors found clear ceramic parallels with ceramics of Tal-i Iblis, while the later period was characterized by a Proto-Elamite settlement and ceramics that had their parallels to the west at sites such as Tal-i Ghazir, Malyan, and Susa. The thin-section petrography was intended to determine if the distinct ware styles found at Tepe Yahya also reflected material variation and technological differences, and to find out whether the abundant, probably locally made, buff and red wares had a common paste and paint mineral assemblage. Less common gray wares and Ubaid-style wares were to be contrasted with these, and matches to other localities would be sought.

Standard thin-section petrography was supplemented by examination of the polished sections in reflected light and by electron microprobe. The thin-section analysis included detailed studies of twinned plagioclase grains, using the Michèl-Levy method (Figs. 7.3.1–7.3.4). In this procedure, the extinction angles of albite twins (polysynthetic twins) of multiple grains are measured and the results used to identify the specific plagioclase feldspar present (along the albite to anorthite range) (Kerr 1977). Kamilli and Lamberg-Karlovsky concluded that for the common buff and red wares, the mineral assemblage (abundant biotite, chlorite, quartz, untwinned feldspar, traces of twinned calcic plagioclase, pyroxene and chert) suggested that the potters used materials derived mostly from metamorphic biotite- and chlorite-rich schists. This finding correlated with the occurrence of many examples of chlorite-rich stoneware at the site, and with the fact that the site lies near metamorphic rock. In contrast the gray wares differed (lacking biotite and containing a greater heterogeneity of feldspar types); the single Ubaid-style sherd examined also differed in mineralogy and texture from the local ceramics. The authors proposed that these wares were probably brought to the site.

The consistent mineralogy of the common wares indicated that the potters at Tepe Yahya probably chose their materials from a relatively consistent source or set of sources. These local sources were derived from the local mica and chlorite schists. The ubiquity of such material near the site was used to support the proposal that the gray and Ubaid-style wares were made elsewhere—not only was the differing material not available locally, but it would have been difficult for potters at Tepe Yahya to avoid clays rich in mica and calcic plagioclase. Differences were also found in paint preparation and firing technology that reinforced the conclusions.

Ferring and Pertulla (1987) looked at temper variability of a red slipped ware found at prehistoric sites in Texas and Oklahoma. A total of 28 vessels from 14 sites across six drainage basins were studied in thin section. The authors noted that in sedimentary geology, determining the provenance of sand or sandstone using petrographic analysis is an integral part of basin analysis. Techniques used in sandstone petrography are routinely applied to thin sections made of unconsolidated sediments of beaches, stream channels, or

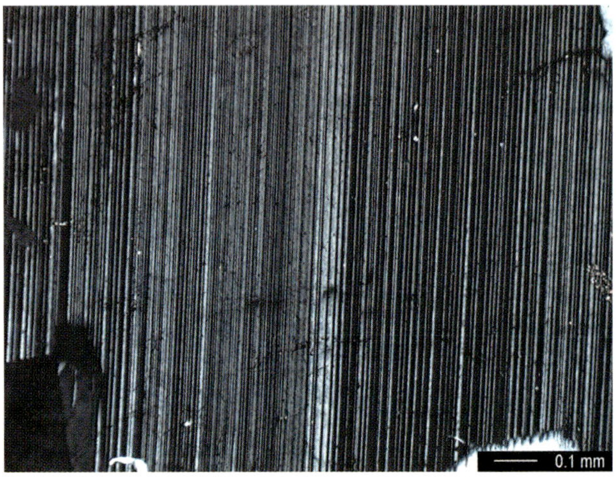

Figure 7.3.1 Determination of phase composition of plagioclase feldspar, crossed polarized light. With the Michel-Lévy method (Kerr 1977) for identifying phase composition of plagioclase feldspar, one set of polysynthetic twins is extinct at a stage rotation of five degrees (consistent with oligoclase). This method of phase determination requires measurement of 8–10 grains per thin section.

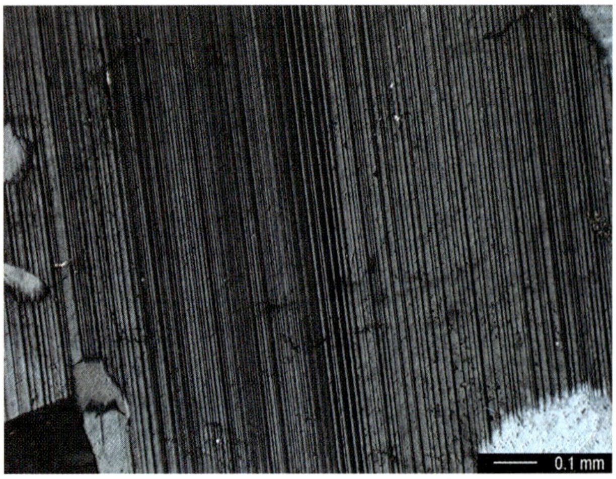

Figure 7.3.2 Determination of phase composition of plagioclase feldspar, crossed polarized light. The microscope stage was rotated five degrees from the center towards the other direction, so the other set of polysynthetic twins is now extinct.

soils (Basu 1985a, 1985b; Lucchi 1985; Morton 1985). In one approach, after point counting, grain types are grouped according to mineralogy or type of rock fragments into classes that have significance with regard to source areas. Triangular plots showing the relative frequencies of the different types of grains are constructed; these most often compare the frequencies of quartz, feldspars, and lithic grains. Quartz-rich, or mature, sands have undergone the greatest amount of weathering and/or long-distant transport. Immature sands have higher frequencies of feldspars and lithic grains (less resistant to weathering than quartz). Micas and other accessories are not used in the comparisons. Ferring and Pertulla thus constructed triangular compositional graphs using modal frequencies of different kinds of quartz, feldspars, and lithics for ceramics from each of the six drainage basins (using point counting of about 200 grains per thin section). The results showed that the sands in the ceramics were similar within each drainage basin, and were dissimilar between drainage basins. The conclusion was that local production most likely accounts for this observed patterning.

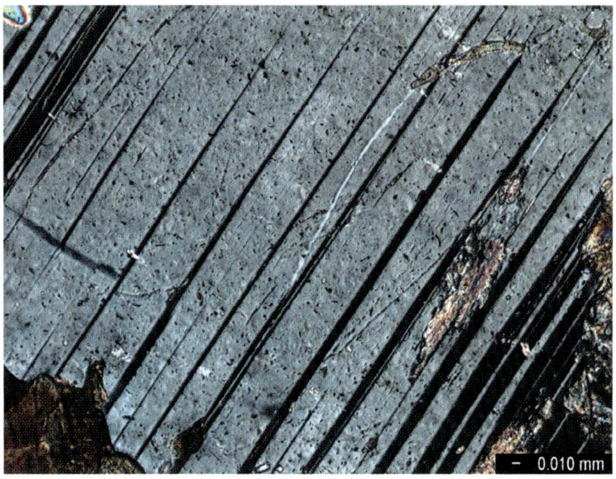

Figure 7.3.3 Determination of phase composition of plagioclase feldspar, crossed polarized light. For this example of plagioclase feldspar, the microscope stage was rotated 32 degrees from the center to bring one set of polysynthetic twins to extinction (consistent with labradorite). The high birefringent material is sericite, a weathering product of the feldspar.

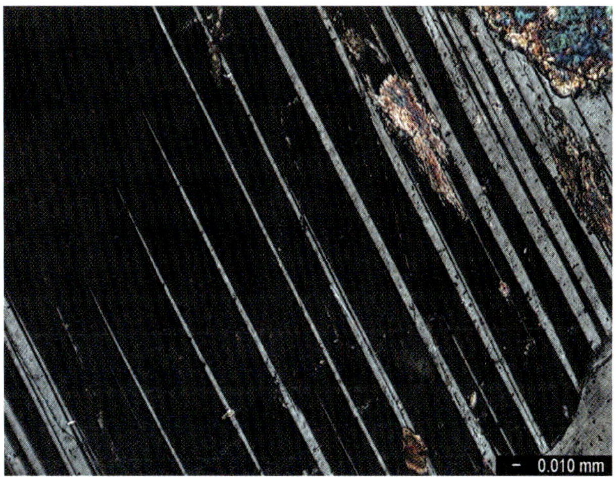

Figure 7.3.4 Determination of phase composition of plagioclase feldspar, crossed polarized light. This is the same grain as seen in Figure 7.3.3, with the microscope stage rotated in the other direction 32 degrees from the center. Thus, the other set of polysynthetic twins is now in extinction. The high birefringent material is sericite, a weathering product of the feldspar.

Zubrow et al. (1988) combined a mineralogical and textural study of temper materials with trace element analysis to examine local production and domestic exchange networks in the prehistoric Philippines. Since low-firing temperatures were used, the minerals present could be taken to represent geological sources. Because of the geological diversity in the Philippines, ceramics made from material collected in one area should be distinguishable from material in other areas. The project focused on two areas containing many prehistoric sites, the Bicol peninsula and the province of Sorsogon. Thin-section analysis included identification of minerals present and their relative proportions (reflecting local geology), and texture analysis (size and shape of inclusions, reflecting transport processes). Size and shape parameters included the proportion of coarse grains (larger than 0.062 mm) as determined through point counting, the degree of roundness, and grain-size distributions. Additionally, mineral composition (of plagioclase feldspars and pyroxene grains) was determined by electron microprobe analysis.

Most of the sherds were found to have a continuous distribution of nonplastics, indicating that the mineralogy did reflect the clay source. Modern potters in the region also do not add temper to their clay. Zubrow et al. noted that the relative proportion of various grain sizes in clay-rich sediments decreases with increasing grain size (with silt-sized particles more abundant than fine sand-sized ones which are more abundant than medium sand-sized ones, etc.). If temper had been added, then the proportion of coarse particles would be skewed (positive skewness), which was found not to be the case except in two instances. However, they pointed out that this analysis cannot detect the mixing of clays. The data were used to construct spatial networks where the movement of pottery may have occurred.

Day (1989) illustrated how ethnographic data can help in interpretation of the variation seen in ceramic temper technology. He proposed that thin-section analysis of archaeological materials, combined with ethnographic work, has significant potential to facilitate small-scale regional studies of artifact production and distribution. Current potters may help lead researchers to raw material sources not previously considered, and analysis of ceramics whose manufacturing process is known can be a valuable comparative source for understanding variation within and between kiln groups. Similarly, the effects of processes such as clay preparation, mixing, forming, and firing can be studied to improve interpretation of variation in thin sections of materials from the same and from different geological sources.

Day applied this approach to a study of ceramics in the Knossos/Archanes area of Crete. In the past, pottery from Knossos had been presumed to be locally made. Day focused on storage jars as an example of Bronze Age coarse ware pottery in the area, as these common products have large nonplastic inclusions, and good parallels (in both technology and function) with modern ceramic products of Crete.

Vessels from eight excavation sites in around the palace and town at Knossos were sampled. Over 30 modern ceramic production sites in Crete were also studied.

Analyses of archaeological samples had previously shown clear differences in technology of manufacture at various sites on the island, with recipes for Knossos/Archanes storage jars (use of complex clay mixing and crushed rock tempers) different from those seen in southern Crete (where sand tempering predominates). Detailed study of the lithic tempers in the Knossos/Archanes region showed a variety of distinctive materials. The villa site of Vathypetro had some ceramics with igneous rock fragments; others had siltstone. Other sites, too, had some distinctive ceramic fabrics, some of which were paralleled by those of modern storage jars. As a result, observation of the location of raw materials used for modern production helped in identifying past production materials and methods.

Feinman et al. (1989) were interested in changes over time in reduced gray ware pottery of the valley of Oaxaca, as it changed into a burnished black ware. They looked at the transition in serving bowls from a vessel type known as G-35 (a flat-based vessel of inverted conical shape dating to 200–900, mainly undecorated) and a type known as G3M (bowls dating to 900–Spanish contact which were thinner and highly burnished both inside and out). It was uncertain whether or not these two vessel types were similar mineralogically. Quantitative data were obtained through point counting of thin sections (89 G-3M samples and 9 G-35 samples).

The mean, range, and standard deviation for percentage of matrix, silt, and sand in the thin sections showed that the G-3M vessels generally contained finer inclusions, although there were overlaps. In addition, it appeared that in preparation of G-3M pastes, more than for earlier pastes, processes such as levigation, fine screening, or manually working the clay thoroughly were used to remove larger inclusions. The same mineral inclusions were used to temper the G-35 and G-3M vessels, but the G-3M pastes were more refined. One conclusion was that both wares were most likely produced locally.

Feinman et al. then examined petrographic variation within the category of G-3M bowls. That variation was analyzed in relation to a proposed distribution of Mixtec and Zapotec speakers in the sixteenth-century valley of Oaxaca, to see if variation could be explained by differences in ethnicity. Eighty-eight G-3M sherds from across the study region were used for qualitative thin-section petrography (identifying the minerals present) and for quantitative point counting (identifying the size, shape, and proportion of inclusions). For point counting, they followed the procedures described by Stoltman (1989), moving the microscope stage in 1-mm intervals across the entire area of the thin section. At each interval, the grain under the cross-hair was identified, and attributes of size and shape were recorded. The point at the cross-hair was first assigned to a category: clay matrix, silt (inclusions less than 0.0625 mm diameter), sand (inclusions greater than 0.0625 mm), grit temper, or void. Sand and grit were further divided as fine (0.0625–0.2500 mm), medium (0.26–0.50 mm), coarse (0.51–1.0 mm), very coarse (1.01–2.0 mm), or gravel (greater than 2.01 mm). They differentiated sand and grit temper (derived from lithics) by defining sand as having a tendency to be more rounded, smaller, and simpler in mineralogical composition while grit temper tended to be more angular, polymineralic in composition, and larger in size.

Using these data, the authors tried to identify discrete paste groups, following the assumption that standardization in ceramic production steps, such as exploiting the same clay source and temper additives, would reduce variation within a specific ceramic industry. Eighteen paste groups were identified. Subsequently, provenances of the sherds were plotted on a map, and spatial patterns regarding paste groups were revealed. For the G-3M samples studied, homogeneity in petrographic composition may indicate a locality or area of manufacture, but did not correlate with location of proposed Mixtec communities.

Maggetti (2005) studied ceramic remnants from about the eleventh to the sixth century B.C. representing the "Laugen-Melaun" culture in northern Italy, eastern Switzerland-Liechtenstein, and western Austria, for what they might reveal about contacts and exchange. He was interested in the extent to which the Alps served as a barrier or as a passage for ceramic trade, so was curious about the extent of variation in tempering technology along that route. Thin-section petrography was performed on 454 sherds. He found that many of those from northern Italy contained a volcanic temper (rhyolite) that could be linked to rocks of that region, and so represented local or regional production. Many ceramics containing that same temper material were imported into southeastern Switzerland (which did not have local sources of rhyolite), although the amount of imported ware versus pottery made using local temper materials decreased significantly from about 70% in the eleventh century B.C. down to about 10% by the sixth century B.C. None of the rhyolite tempering material, and thus apparently no imported pottery, was found north of the Alpine crest in Switzerland, in Liechtenstein, or in Austria. Instead, serpentinite temper was used in those regions.

7.4 Identifying Specific Geological Sources of Nonplastics

While elemental analysis potentially can be used to pinpoint clay sources of ceramic materials, if comparative clay samples are available, thin-section analysis is less likely to be useful in locating an exact site where raw clay materials were collected. However, thin-section petrography can sometimes establish the geological area of origin of natural or added nonplastic inclusions in the clay. Rock and mineral inclusions in clay provide information about the geology of the area in which the clay and temper materials were collected. If those inclusions are distinctive enough, it may be possible to deduce the nature of the source geology. Then, information about regional geology may help to locate and pinpoint a probable source. As Freestone (1995) pointed out, in some cases a single thin section will suffice, unlike elemental analysis to identify source clays, which requires a large number of analyses and reference groups for comparison. Unfortunately, such provenance studies are often not possible or are very difficult, because there may be undiagnostic temper or the geology of a region may be homogeneous across many production centers. The possible constraints for each particular study need to be well understood in order to determine whether or not thin-section petrography is likely to be a useful approach, and to aid in development of a good research design.

Once it has been established that the nonplastic components may be useful for identifying geological sources (e.g. presence of lithic inclusions, polymineralic sand tempers, or unusual additives that may be source distinctive), a knowledge of the geology of the relevant study area can provide an assessment of the variability that may be present (or absent) for different production centers and can help prepare the researcher for what local nonplastic materials to expect. Here the use of geological maps and reports, fieldwork to collect potential raw materials, and the availability of reference collections and/or publications of comparative materials can be very useful. For example, if knowledge of the local geology indicates that a particular nonplastic would not have been available locally, the vessel may be identified as an imported product, even if the original place of production is not known. And, if the source area of a particular nonplastic is fairly restricted within the region of interest, it may be possible to trace the source fairly precisely (Hodges 1963).

However, additional constraints to keep in mind include the following: potters may mix clays from two different sources, so naturally occurring nonplastic inclusions may represent two different source areas or two different locally available clays; in processing of clay, potters may refine their material by removing heavier or coarser components, so comparative materials gathered in a source area may not completely resemble any ceramic materials produced there; locally gathered clays and/or temper additives may be sorted and refined, crushed, or mixed; temper additives may come from exotic materials quite different in source than the clay deposit itself; and unusual materials not representative of the source area may be chosen as temper additives, such as boulders that result from river transport.

Marie Farnsworth (1964) used a mineralogical analysis of clay fabrics to study Greek coarse wares from the Agora, particularly cooking and other household wares used in Athens during the sixth–fourth centuries B.C. Her main goal was a provenance study, identifying tempers that had likely sources in Attica (a metamorphic region), Corinth (a sedimentary region), and Aegina (a volcanic region). Through thin-section petrography of temper materials in coarse wares, combined with a knowledge of the pertinent geology, she was able to determine that most of the cooking ware found in Athens was made locally, but a few other fabrics were present, particularly those from Aegina. Most reddish-firing cooking wares used in Athens were from Aegina, while the reddish-firing cooking and household wares used in Athens in the fourth century B.C. originated from Attica.

By combining petrographic analysis of excavated sherds with knowledge of pertinent geological source regions, Hays and Hassan (1974) were able to identify local sources of temper for Neolithic ceramics from Sudan and confirm stylistic studies of decorative motifs that suggested local origins. Three different tempering regimes were found, all with local sources. One group contained basalt lithic fragments and augite grains, indicating a source from the Abyssinian basalts in eastern Sudan. The second group contained quartz, orthoclase feldspar, microcline feldspar, sodic plagioclase feldspar, muscovite, biotite, and polymineralic fragments of biotite gneiss and chlorite schist. The source was identified as the Precambrian Basement Complex, which consisted of granites, gneisses, and schists; the archaeological sites investigated were near exposures of that complex. The third group contained subrounded to rounded quartz grains and hematite inclusions. These were identified as originating from the Nubia Sandstone, which had extensive exposures in the area where the archaeological sites are found. The authors noted that the three different local methods of tempering ceramics may indicate subgroups of pottery traditions maintained by different social or workshop groups.

Southward (1979) employed thin-section analysis of ceramics from a Jornada Mogollon village in New Mexico to identify probable sources for temper additives. She examined six ceramic types, and found that all were tempered with either granite or granite-gneiss. She concluded that the angularity of the particles and their multi-mineralic tendency suggests that the granites were crushed and then deliberately added to the clay paste as a temper. Most of the granite samples were medium grained with a mineral suite consisting of 55–70% alkali feldspar (primarily microperthite) (Fig. 7.4.1), 25–35% quartz, 2–5% plagioclase feldspar (often slightly sericitized), and minor amounts of hornblende, epidote, hematite, opaque minerals, and granophyre (a fine-grained granite having a groundmass of irregular intergrowths of quartz and feldspar with larger crystals embedded within it). The second type of granite was identical except it also had inclusions of a rare mineral,

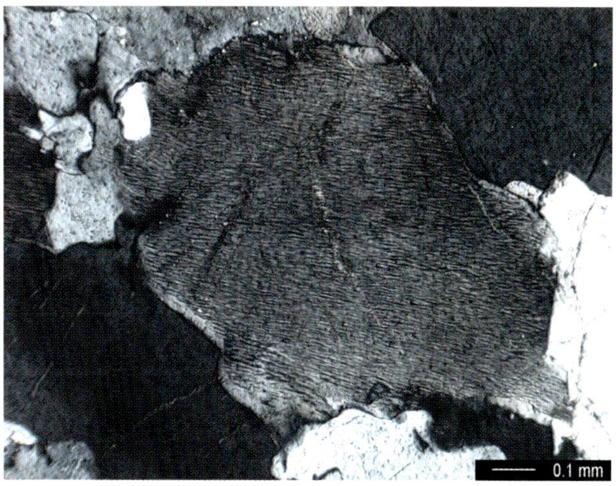

Figure 7.4.1 Microperthitic texture, crossed polarized light. This potassium-rich alkali feldspar (dark areas) is intergrown with a sodium-rich feldspar (albite, light areas) in the form of narrow lamellae, forming microperthite. The grain is within a granite lithic, and is surrounded by quartz grains.

xenotime (a phosphate mineral). The granite-gneiss is fine to medium grained, and contains less alkali feldspar (50–60%), 20–40% quartz (of variable types), and about 10% plagioclase feldspar (mostly sericitized), with minor amounts of hornblende, epidote, hematite, and opaque minerals. She found that possible source outcrops existed some distance away along the east escarpment of the San Andres range; none was found locally.

Southward suggested that since the granite was crushed before use, it was most likely chosen as the temper material for a particular reason. It, or the finished ceramic products, could have been transported. The behavioral models she proposed included (1) that the village inhabitants traveled to the granite source area and quarried it for use, then took the raw material back to their village for use or trade; (2) that the granite temper was obtained through trade with other villages near the source area; (3) that finished vessels with granitic temper were trade items and were imported into the village; or (4) that itinerant potters carrying their own materials visited the village and made vessels there from time to time.

Riley (1982) noted that in some regions, such as the Mediterranean, in practice, petrographic analysis of ceramics usually cannot point to a specific origin of a material. Instead, it more often points to a general origin. Thus classification and comparison of ceramic fabric groupings is important. He mentioned that usually not all the possible geological origins are being considered, but instead the focus is on the geological origin most compatible with the archaeological evidence. Sometimes petrographic analysis can support or refute an origin suggested by the archaeological data; alternatively, petrographic analysis may guide an archaeologist to a possible origin not previously considered. He provided examples for both of these situations, and also emphasized that projects to identify the provenance of pottery or sherds using thin-section petrography would benefit greatly from the establishment of more comprehensive databases of materials of known origin. Such controls could include vessels stamped or marked with their origin, a full range of kiln materials, and perhaps modern pottery made in a traditional manner at known locations and with known source materials.

Maniatis et al. (1984) included thin-section petrography in their study of amphorae found at Corinth, Greece. The data allowed them to postulate an origin along the Atlantic coast close to the Straits of Gibraltar, either in Morocco or Spain. Other types of studies had already established that there were two general groups of amphorae, distinguished by slightly different chemical compositions and manufacturing methods. The major goals of the thin-section work were to determine whether the two groups were mineralogically distinct, and if the mineralogical data could be matched to possible geological sources. Mineral identifications were supplemented by point counting to give relative proportions and sizes of the constituents.

One group had metamorphic inclusions. Lithic fragments were most often biotite schist and muscovite schist, sometimes amphibole schist. The polymineralic suite included garnet, chlorite, amphibole, kyanite (some with sillimanite inclusions), and chloritoid, along with quartz and orthoclase feldspar. The second group had sedimentary inclusions. This group could be further subdivided as containing: sand with lime (minute calcium minerals) scattered throughout the matrix; sand with some limestone; sand with limestone and fossils; sand, limestone and fossils, but with some rounded metamorphic grains; and sandstone fragments with rounded low-grade metamorphic rock fragments (phyllite with microfolds) and no limestone or fossils.

Maniatis et al. concluded that both groups could have been produced in the same general region, with angular temper materials for the first group collected closer to the metamorphic source. The rock types represented were relatively common, so it was not possible to choose between the proposed sources of northern Morocco or southern Spain. However, the Atlantic coast was proposed as a more likely source for the sedimentary group than for the metamorphic one.

A geological study of the sediments underlying the archaeological site of Deir el-Balah in the Gaza Strip provided comparisons to Late Bronze Age Egyptian and Canaanite ceramics excavated at the site. The primary goal was to see if the ceramics might have been made locally (Goldberg et al. 1986). Samples of different pottery types from several locations at the site were examined in thin section, and found to be similar. Nonplastics included a coarse, angular quartz silt ranging from 25 to 50%, and 50–60 μm in size, and subrounded quartz sand (1–3%, 300–400 μm) within a calcareous, fine silty clay matrix with traces of calcium carbonates, feldspars, chert, and/or mica (muscovite and biotite). There were also some heavy minerals present (hornblende, epidote, pyroxenes, zircon, and/or garnet). The mineralogy and texture of the sherd samples closely matched that of the Late Pleistocene loamy sediments underlying the site, so the conclusion was that both the Egyptian and Canaanite pottery vessels were made from the same material that was quarried at the site. The thin-section analysis results corresponded well to clay studies undertaken with neutron activation analysis.

Whitbread (1986a) emphasized that using petrographic data for provenance studies in isolation could lead to very misleading results. He noted that the fabric of a ceramic vessel from one region is sometimes indistinguishable from that of a very different region. Thus, it is important to start a provenance study by utilizing all archaeological evidence available for identifying the possible source localities of an object. Then, by studying the geology of possible source regions, thin-section petrography can be used to further limit the range of likely localities for production and to exclude some possibilities.

Blackman et al. (1989) studied ceramic materials from the site of Hili on the Oman peninsula, in order to investigate contacts and interactions of protohistoric site inhabitants within and outside of the peninsula. Fabrics and ware types were initially classified using macroscopic and low-power microscopic examination. Three of the wares were the focus of further analysis: "domestic" ware, considered to be a local product used for everyday functions and found in abundance in the Hili area; "black-on-red" fine ware, found throughout the Oman peninsula, mainly in funerary contexts, and stylistically similar to wares from southern and eastern Iran; and Emir gray ware, consisting mainly of miniature jars also found in funerary contexts in Oman, and in settlements and graves in eastern Iran.

A total of 75 sherds from the three ware types was analyzed by thin-section petrography, along with ten samples of earth, sand, and rock from the area around Hili. The "domestic" ware specimens were all very similar in thin section, indicating a single source. The many nonplastic inclusions ranged from 10 μm to 0.5 mm in size, and included detrital quartz, some granito-gneissic rock fragments, and small quantities of basic and ultrabasic grains (pyroxenes, amphiboles, olivines, and serpentines). In contrast, the geology of the area around Hili was dominated by sedimentary rocks with carbonates such as limestones, dolomites, and marls. In addition, many of the grains seen in the thin sections were angular or splintered, including the harder lithic components such as quartz and feldspars which would become rounded and altered by depositional transport. The nearest areas mineralogically consistent with the thin-section findings were located about 60 km from Hili, and because of the lack of transport indicated by grain morphology, production centers were most likely located near the parent rock sources.

Blackman et al. found that the "black-on-red" ware specimens had a paste that contained a nonplastic silt fraction (ranging from 5 to 100 μm in size) that was mainly tiny angular quartz with some biotite, fractured amphiboles, and rare lithic fragments, possibly basalt. As with the "domestic" ware, carbonates were totally absent from this silt fraction and so the area around Hili could be excluded as a raw material source. However, differences in mineralogy indicated a source different from that of the "domestic" ware, although exact location of either of these sources was not identified.

The Emir gray ware was seen in thin section to be quite different from either of the other two wares. It exhibited an argillaceous matrix with silt-sized, fractured angular quartz, plagioclase, and microcline. The angularity of the quartz, as well as the absence of many other minerals or lithics, indicated this material is not the product of long-range transport. The lack of carbonate, basic, and ultrabasic components made the Oman peninsula an unlikely raw material source.

Neutron activation analysis was also done on samples from the three wares, and three separate compositional groups were found, supporting the results of the thin-section petrography. The authors concluded that three ceramic types were produced from different raw materials (and not, for example, simply variants created by differing levigation, tempering, or firing regimes). The "domestic" and "black-on-red" wares were more abundant, very homogeneous (thus indicating standardized production), and did not change very much over time. Blackman et al. hypothesized that these wares were probably produced somewhere on the Oman peninsula where there were compatible geological environments. However, the Emir gray ware was not geologically compatible with the peninsula, and may have originated from eastern Iran where the ware was found in high concentrations and from an early date. "Black-on-red" fine wares were a local product on the Oman peninsula throughout the third millennium B.C.; in mid-millennium "domestic" ware appeared there as a new specialized product and indicated development of a local ceramic industry. The first ceramic evidence of contact with Iran appeared in the mid-third millennium B.C. with the importation of Emir gray wares. This work demonstrated how thin-section petrography, combined with geological study, can contribute to addressing complex questions of regional craft production and outside exchange contacts.

Mason et al. (1989) employed both archaeological materials and geological information to propose the probable place of production for sherds collected on the surface at medieval Islamic sites in Yemen. Wasters and kiln furniture such as trivets, which indicate ceramic production at a particular site, were studied in thin section along with surface-collected sherds. Mineral composition of the nonplastic components was determined, and modal frequencies (abundance) of major and minor inclusions estimated (using visual comparison charts), as were sphericity, roundness, and grain size of each mineral phase; texture and mineralogy of lithic inclusions were also recorded. The results were compared with known lithology of the study area, and found to be compatible with local ceramic production. Five main groups could be distinguished on the basis of the modal mineralogy. Kiln furniture and wasters, supported by geological evidence, were used to identify probable production locations for the pottery sherds.

Panjwani (1989) compared nonplastic inclusions found in sherds with local clay deposits, to identify whether or not the pottery found at Lothal might have been locally produced. Lothal was a major Harappan port town located in Gujarat in western India. The alluvial plain around Lothal is surrounded by raised tidal flats filled with microfossils that have been abraded by the surf action of the tides. Thin sections of samples taken from the mud flats showed abraded pellets of microfossils and quartz, along with a few grains of feldspars and heavy minerals. Comparative samples for thin-section analysis were taken from several different wares at the site, including both Harappan and non-Harappan (Red Ware, Buff Ware, Black and Red Ware, and Micaceous Red Ware). A characteristic feature of all Lothal pottery was that it contained a considerable amount of altered microfossil fragments of foraminifera; there were also rounded or subrounded quartz grains, feldspars, micas, and heavy minerals similar to that seen in the mud flats. Thus, Panjwani concluded that both the Harappan and the non-Harappan potters at Lothal appear to have exploited the locally available clay.

Porat et al. (1991) looked at ceramic materials excavated at two Early Bronze Age sites in Canaan where Egyptian-style pottery was found (an Egyptian house in En Sesor and from excavations in Tel Erani). Their main focus was whether the Egyptian-style pottery was imported from Egypt, or whether it was made locally. They compared the ceramic materials with Nile alluvium and with the local loess soil. They were also interested in whether or not the same material was used in En Besor as was used in Erani. Other analytical techniques were then used to confirm the results. The petrographic criteria that the authors found to be the most useful in this case included the sorting and volume (percentage) of silt-sized quartz, the heavy minerals associated with the quartz, and the amount of carbonate present. Using these criteria, they could distinguish three groups among the sherd specimens.

Group 1 had poorly sorted, rounded, sand-sized quartz grains that graded into the quartz present in the clay matrix. Feldspars were abundant, along with heavy minerals such as clinopyroxene, hornblende, mica, and epidote. Straw molds were often present. The clay matrix included small mica laths and poorly sorted silt-sized quartz. They also found phytoliths in the clay. Volume percentage of quartz was generally between 5 and 7%, mainly as silt. The authors concluded that Nile muds and sands, mixed in different proportions, were used for the objects in this group. They noted that the petrography of Nile alluvium was already well characterized. The silty quartz of Nile sediments is often not very abundant, is poorly sorted, and has fine grain sizes. Pyroxene is a common component of the heavy mineral assembly in the Nile silts, and biotite micas are also abundant, whereas carbonates are rare.

Group 2 included jars made of a marly clay containing rounded, sand-sized quartz grains and a decomposed limestone temper. The clay matrix was very calcareous, with silty quartz, rare mica laths, and iron-filled foraminifers. Porat et al. concluded that the texture of the sand- and silt-sized quartz and the heavy mineral assemblage indicated that this group of objects was made with a Nile mud component that was mixed with marls (as temper addition). Weathered limestone outcrops occurred on both sides of the Nile valley and could have served as the source.

For their Group 3, they found some differences between the two sites. At En Besor, samples showed 7–10% angular, well-sorted, coarse silt to fine sand-sized quartz grains, often coated with clay. Accessories included feldspars and heavy minerals such as hornblende, epidote, zircon, and muscovite mica. Some silt-sized rounded carbonate fragments also occurred, possibly from poorly preserved foraminifers. There were some finely dispersed carbonate particles seen only at high magnification. Straw molds were also common. Rounded sand-sized mud balls made of pure clay stood out in the thin sections. At Erani, the clay matrix appeared similar to that of En Besor, with well-sorted silt to fine sand-sized quartz and a heavy mineral assemblage dominated by hornblende, epidote, zircon, and mica. However, the material differed from that of En Besor in that silt-sized micritic carbonate nodules were abundant, and there were also phytoliths present. The mud balls so visible in the En Besor samples were absent. Based on the size, sorting, and angularity of the quartz, the authors concluded that locally available loess soil was used for this group of objects. The differences between the two sites suggested that each collected their material from different sources within this region of loess sediments.

However, the technology used in the manufacture of the pottery that falls in this group more closely resembled that used for the Egyptian Nile clay pottery than that used for Early Bronze Canaanite pottery. For the Canaanite pottery, a silty clay was chosen and little mineral temper was used; instead, vegetable matter was used as a temper additive. The authors proposed that while some jars were imported from Egypt, other objects, especially household utensils, were made locally, with Egyptian potters producing them for a permanent Egyptian population which inhabited these sites.

Kilka (1992) conducted both a geological study and sand sampling of streams around a Bronze Age site at Fiave, Italy, for comparison with ceramics excavated at the site. Mineralogical and chemical analyses were performed on 48 sherds. The elemental analysis, even of a form considered to be local, was so variable that no local reference group could be constructed. The thin-section work showed two primary groups, one with mainly calcareous temper, and one with a quartz and feldspar temper. Some inclusions were fine grained (less than 0.07 mm) and were thought to be naturally present in the clay. Others were coarser, and might have been deliberately added by the potter. Each of the primary groups could be further subdivided on the basis of additional features (such as presence of fossils, diorite, chert, rhyolite, granite, basalt, or amphibole). Hypotheses about origins were made based on geological maps. Comparison of thin- section data with analyses of sands collected in the region refined the probable places of temper origins for many of the sherds.

Douglass and Schaller (1993) used thin-section petrography as their primary tool in a provenance study of prehistoric ceramics in the American Southwest. They focused on Little Colorado White Ware, a decorated black-on-white ceramic fabricated by the Anasazi from about 1050 to 1250. Sherds from this ware are found at archaeological sites throughout the Little Colorado River valley of northern Arizona. The authors were interested in understanding the regional extent and magnitude of exchange networks, so wanted to be able to identify production sources for this ware. A total of 269 sherds from 11 archaeological sites in the region was studied in thin section (supplemented by some XRD and electron microprobe analyses of augite grains). Modal abundances of sand-sized particles were determined by point counting. Geological studies of potential source areas were analyzed for comparison. Two major volcanic fields located on opposite sides of the Little Colorado River were proposed as probable source areas: the San Francisco Peaks in the west and the Hopi Buttes in the east.

For maximum sample area within each sherd, thin sections were cut parallel to the surface of the vessel instead of perpendicular. These sections were cut from the interior to reduce the chance that weathering processes had altered the original mineralogy. Point counting (300 points for each sample at 1.0 mm intervals horizontally and vertically) was supplemented by visual examination to identify rare constituents that might be diagnostic but may have been missed in the point-counting procedure. Grains of less than 0.06 mm maximum diameter were considered to be part of the clay matrix. Many of the constituents are widely distributed minerals such as quartz and feldspars, so the volcanic lithics were the most diagnostic for identifying original production locality. Basaltic rock fragments were found, as well as pyroxene grains that originated in sand deposits derived from the weathering of volcanic rocks (and many of these grains were found in a matrix of volcanic rock). Most were clinopyroxene, probably augite (Fig. 7.4.2).

The volcanic lithics found in the ceramic samples matched the volcanic rocks of the Hopi Buttes in texture and mineralogy. In contrast, the San Francisco Peaks volcanic rocks contained plagioclase feldspar, distinctly absent in the sherd lithics. Douglass and Schaller found that the mineral suite of the sand-sized fraction of mineral grains in the sherds appeared to have derived from weathering of fairly mature sandstones (accessory minerals dominated by ultra-stable minerals such as zircon, tourmaline, and rutile and a relatively low feldspar/quartz ratio). They identified the Wingate Sandstone in the Hopi Buttes area as having a mineralogical composition identical to that of the ceramic sand component. Sand of this composition, along with volcanic rock fragments and pyroxene crystals derived from weathering of volcanic rocks, was present in many of the dry washes traversing the Hopi Buttes area. Thus, the evidence pointed to this area as being the source for the ware, while none of the samples studied contained sand of a composition that could be shown to have originated west of the Little Colorado River.

Bray (1995) used thin-section data as one piece of evidence in a provenance study of a distinctive, thin-walled ceramic ware known as Panzaleo. It was found in many locations in the northern highlands of Ecuador, although rarely as the sole or primary ware. Thin sections from Panzaleo examples found at País Caranqui (in a wide variety of domestic and mortuary contexts) were compared with local Caranqui wares. The main difference was that the Caranqui wares contained minerals and rock fragments of volcanic origin, while the Panzaleo thin sections showed metamorphic components. Bray noted that in Ecuador there are two main Andean mountain ranges; only the one in the eastern part of Ecuador has metamorphic rocks, so eroded metamorphic materials would be found mainly in the foothills and drainages of the eastern mountains. Finding Panzaleo pottery at País Caranqui thus supported the idea that there was a significant level of contact between the highlands and the eastern lowlands throughout much of the pre-Columbian period. Combining the thin section data with other information, Bray hypothesized that this contact was in the form of exchange relationships that may have been connected with ritual and festival activities.

Williams and Jenkins (1997) found that the local geology of northern Wales was conducive to matching nonplastic inclusions of ceramic sherds to a specific source and to identifying the presence of foreign sources. They analyzed Neolithic, Bronze Age, and Roman sherds from a variety of sites, and found that "clast petrography," or the analysis of rock fragments, was the most useful way to identify sources. Rounded fragments suggested a natural inclusion in the clay, angular ones a deliberately added temper. The geology of northern Wales is quite varied, making the matching of sherd inclusions to geological outcrops possible. For example, sherd inclusions from one Neolithic site were found to be serpentine (Fig. 7.4.3), which has limited outcrops close to the site, therefore it was possible to propose a local origin for the sherd temper.

In addition to the many different types of lithics found in sherds that could be matched to specific, limited geological outcrops, some sherds contained siliceous bioliths (such as phytoliths, spicules, and diatoms) that could also help in identifying provenance. To facilitate the identification of such materials, a novel "re-ignition" procedure was used in preparing thin sections. Williams and Jenkins noted that unoxidized sherds are often dark and opaque, with details obscured. To counteract this problem, the sherds were ignited overnight at 500 °C and re-oxidized, clarifying the components in a thin section.

Schmitt (1998) used digital image analysis to obtain quantitative data on thin sections of 47 amphorae found in the Lyons area of France, with the goal of assessing whether

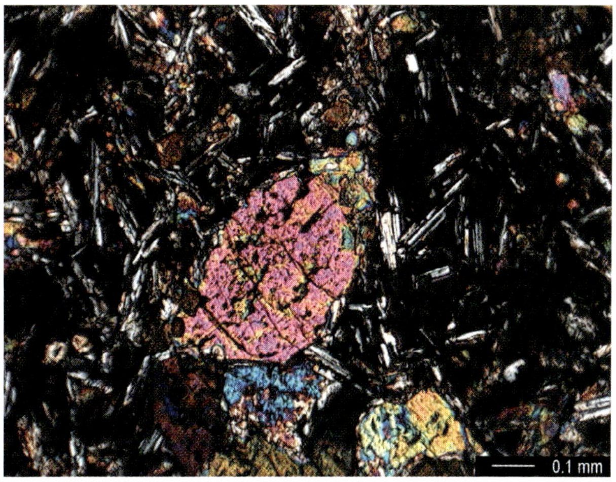

Figure 7.4.2 Augite within basalt lithic fragment, crossed polarized light. In ceramic materials, augite is often found in the form of small prismatic crystals. Birefringence is moderate, with maximum interference colors seen when the grain is oriented at about a 45-degree position. Small laths of plagioclase feldspar surround the augite, along with opaque iron oxides and low birefringent chlorite.

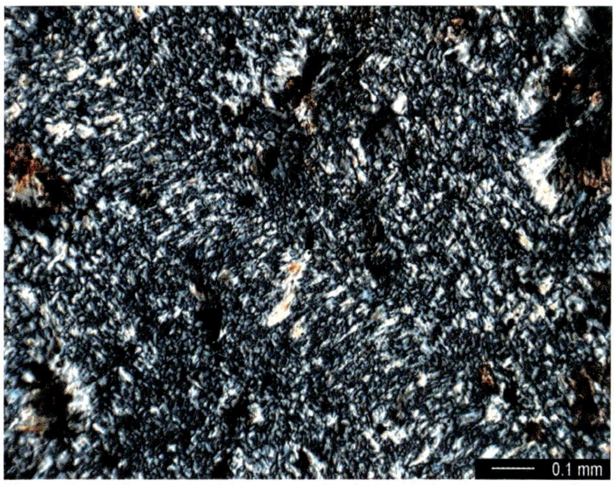

Figure 7.4.3 Serpentine lithic, crossed polarized light. Serpentine is fine grained with low interference colors in crossed polarized light, occurring in fibrous, platy, or flaky forms. Here there are also accessory iron oxides present.

or not they might be of local production. Intentionally added temper in these vessels was coarse and abundant, comprised of sand determined to be of granitic type (with quartz, alkali feldspars, plagioclase feldspars, micas, and granite lithics). There were also examples of the use of volcanic lithics (rhyolite). Data collected through image analysis (adding the results obtained from ten contiguous fields of view in each thin section) included overall abundance of temper (the ratio of the surface occupied by temper grains to the total surface), granulometry (calculated by measuring at the major axis of each grain and then summing the surfaces of the grains for each class of major axis), and percentages by grain types. Schmitt noted that the image analysis program could not identify minerals or lithics, so that work still had to be done by the petrographer. Still, the new approach was not only more rapid but also more precise than the traditional technique of point counting, because it was the total area of the grain which was measured and not the number of occurrences.

It was concluded that the temper characteristics were consistent with the local geology of the Lyons region. However, the results were also consistent with the geology of other regions where the amphorae might have originated (such as the Rhône valley). To further narrow down the possibilities, samples of alluvial sands were taken along the Azergues, the Saône, and the Rhône rivers, and compared with the thin-section sand temper. Geochemical data augmented the petrographic work. All the data taken together supported the hypothesis of local production in Lyons.

Heidke and Miksa (2000) introduced a sophisticated and thorough approach for provenance studies of sand-tempered ceramics, focusing on sherds excavated at sites (dating between 100 and 1350) in the Tonto basin of central Arizona. Their procedure began with geological work to locate and map the pertinent main waterway (trunk stream) and its tributaries. "Low-order" streams were identified as those at or near their headwaters, with small drainage basin areas; "high-order" streams are far from their headwaters and have many smaller tributaries and large drainage basin areas. Sands were then collected, some from the main waterway, but with the primary focus on collecting from tributaries of similar stream order close to the main waterways (the archaeological sites included in the study are all located within a kilometer from a sand sample location).The purpose of the sand collection was to petrographically characterize the potential sand tempers within the study area.

Petrographic analysis was done using point counting by the Gazzi-Dickinson technique. This method differs from traditional point counting in that sand-sized monomineralic crystals (larger than 0.0625 mm) occurring within lithic fragments are classified in the category of the crystal, rather than in the category of the larger rock fragment. The advantage of this approach is that it gives more consistent results for all grain types recorded, and reduces the effects of grain size on composition (since sand-sized grains can potentially be broken away from the lithic fragments), so frequency of grain types between samples is based mainly on composition. Fine-grained lithic fragments were identified using information about mineral composition and internal texture and fabric. Staining of alkali feldspars helped reduce errors in distinguishing them from quartz.

The quantitative point-counting data were transformed to log-ratio data, and used in a sequence of statistical analyses. Heidke and Miksa suggested that frequently in archaeological provenance studies, interpretation of quantitative data has been limited by reliance on a visual comparative approach using ternary diagrams. Instead, they began their statistical work with an iterative series of correspondence analyses, as an exploratory step to refine the boundaries of petrofacies (sand temper resource procurement zones) and to assign the sampled sands to a particular petrofacies. Two types of compositions were seen among the petrofacies: those that were rich in minerals and those rich in rock fragments. No pure quartz sands were found in the Tonto basin area.

Once the petrofacies were identified, a series of "nested" discriminant analyses was performed. The first step was to test the assignment of sands to either the mineral-rich group of petrofacies or the lithics-rich group, a second discriminant analysis tested the assignment of mineral-rich sands to specific petrofacies, and a third tested the assignment of rock-fragment-rich sands to specific petrofacies. The final classification functions were then used to classify each ceramic sand temper to a petrofacies, running the samples as "unknowns." The authors noted that the data collected in point counting (composition) were chosen because in this region, sand tempers are compositionally heterogeneous and texturally similar; in other regions, where there is more textural variability, other parameters (such as grain size, roundness, and sphericity) could be measured and included along with or instead of mineralogical and lithic composition data.

Stoltman (2001) proposed that the first step upon recognition or identification of a ceramic type should be to demonstrate that it was locally produced, by checking whether or not the paste and temper materials were available locally. He suggested four possible approaches: (1) comparing mineral inclusions in ceramic bodies to locally available sands, minerals, and rocks; (2) comparing ceramic pastes to local sediments; (3) comparing ceramic bodies, tempers, and/or pastes to those of ceramics that are accepted as having been locally produced; and (4) examining between-site variation in ceramic bodies, tempers and/or pastes of a single class of vessels. He pointed out that for the first three procedures, the null hypothesis is that if locally produced, the products under investigation will not differ significantly (qualitatively or quantitatively) from local materials or products. The fourth procedure, which he terms the "spatial pattern postulate" assumes local production of a ceramic class if vessel compositions display a pattern of within-site homogeneity in contrast to between-site heterogeneity.

Stoltman (1991, 2001) reported on an application of the first two of these approaches in a study of three ceramic classes at the Fred Edwards site in southwestern Wisconsin. The goal was to verify that the ceramics were locally produced. Quantitative thin-section petrography (with point counting using a 1-mm interval) was performed and the results were compared with locally available raw materials (one class was grit tempered, with hematite, another was shell tempered, and the third had hematite temper). The data were presented in two forms. The first was composition of the body, with mean values for percent matrix (clay and silt), sand (natural sand sized and larger inclusions), and temper (deliberately added materials of any size); temper type; and temper size index (on an ordinal scale of size classes ranging from 1 to 5). The second was composition of the paste with temper additives excluded, with mean values for percent matrix (clay only), silt, and sand, and sand size index.

Hematite and shell were found on the site itself, so local temper materials were clearly used (Figs. 7.4.4 and 7.4.5). Local sediments were collected for comparison on and around the site. A local subsoil clay was found to have a paste index almost identical to the mean for the 12 vessels studied, in terms of percent matrix, silt, and sand, indicating that the paste was probably local as well. Some of the shell-tempered vessels turned out to be likely imports—they did not have local pastes, and they also had a higher percentage of temper additives than the local products. The size of the temper also differed.

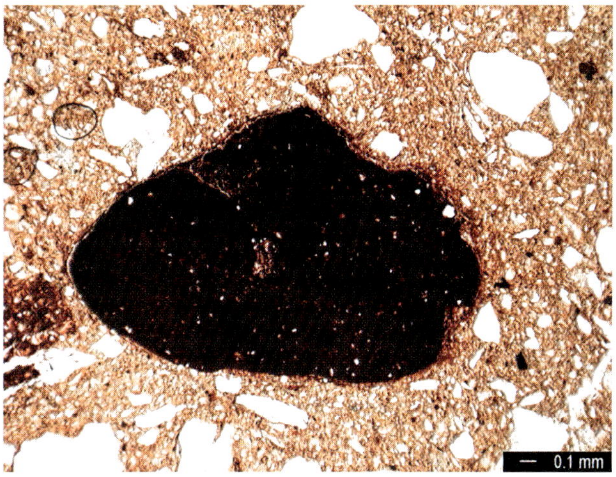

Figure 7.4.4 Hematite temper, plane polarized light. The large chunks of hematite in this sherd, contrasting with the smaller colorless sand grains in the clay matrix, may have been a deliberately added temper.

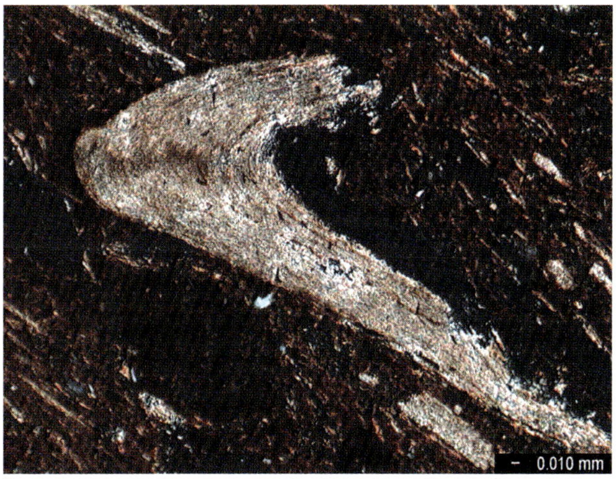

Figure 7.4.5 Shell temper, crossed polarized light. Some of the original structure of the shell temper additive in this sherd is still discernable, and the calcium carbonate displays a typical high birefringence (with pastel tinges). Small fragments of shell make up most of the non-clay material.

One particularly interesting example of pottery temper research was work carried out on sand-tempered Pacific island ceramics. This research helped to reconstruct the history of the settling of the islands by providing evidence for the movement of people from one island or island group to another (Dickinson and Shutler 1974, 1979; Dickinson et al. 1990; Dickinson and Shutler 2000; Dickinson et al. 2001). Island potters were relatively restricted in their choice of where to obtain sand tempers, generally to those available within their island or local island cluster. After petrographically characterizing the sands of 50 islands and 15 major island groups, the researchers were easily able to identify local versus imported ceramics. When objects found on one island were clearly made elsewhere, the patterns of past human movement could be retraced. Since the differently tempered ceramics did not vary significantly in visual appearance or quality, and materials for manufacturing ceramics were readily available on all the islands, the authors concluded that the addition of foreign-tempered examples was more likely due to movement of people than to a ceramics trade. In addition to the presence, size, and shape of diagnostic minerals, inclusions within mineral grains were very important in characterization. For example, glass inclusions within plagioclase feldspar and pyroxene grains indicated that they derived from volcanic rocks.

Dickinson and Shutler (2000) summarized some of the implications of petrographic temper analysis for Oceanian prehistory. Thin-section analyses (from over 1,500 thin sections) revealed major distinctions in sand found from one island to another, as local island geology is controlled by systematic and well-known geotectonic patterns that led to relatively homogeneous, unmixed sand within an island but variations between islands. The only caveat was that calcareous temper sands of reef detritus were undiagnostic in origin. Quartz, which tends to be the most abundant temper material in continental ceramics, and is difficult to trace to source, tends to be absent or less abundant in Pacific island tempers while other, more diagnostic types of sand grains predominate. Since islands are so distinctive in their terrigenous sands, ceramic transfer or movement of raw materials from island to island can be detected clearly. Prehistoric pottery was made locally on multiple islands, rather than being distributed from a few ceramic centers; however, some limited ceramic transfer did take place within nearly all island groups. Combining data from sherd temper analysis with data from other portable artifacts such as obsidian provided detailed physical evidence for migration, trade, or exchange within and between island groups.

For example, detailed information was available about the geologic sources and geographic distribution of sand tempers in prehistoric sherds from the Mariana islands (Dickinson et al. 2001). Temper compositions were characterized in terms of frequency counts of the relative percentages of different types of sand grains visible in thin section. A minimum of 400 grains was counted in each thin section whenever possible. Areal, ribbon, or traverse counts were performed depending upon the number of sand grains present in a given sherd. Dickinson et al. noted that frequency counts do not give the same results as grid-point counts that are standard in sedimentary petrology, but were advantageous here because the clay paste, which dominated the sample, was ignored and only temper grains were counted. A clear size difference was seen between the largest silt grains of the clay paste and the smallest temper grains, indicating that the Mariana temper sands were deliberately added by potters to the clay paste.

In this dataset, calcareous-sand tempers were not diagnostic as to source. However, volcanic sands and the terrigenous component of hybrid sands of mixed terrigenous and calcareous grain types could be traced to sources on Saipan and Guam, where volcanic bedrock was extensively exposed. Quartz-rich tempers were traced to dacitic volcanic rocks on Saipan. Quartz-poor tempers derived from andesite came from both Saipan and Guam, but abundance of orthopyroxene and clinopyroxene was diagnostic of Saipan, olivine of Guam, and placer temper sands rich in heavy ferromagnesian minerals occurred only in sherds on Guam. The temper analysis was used to document widespread ceramic transfer from Saipan to other islands throughout Mariana prehistory.

In contrast to the success achieved with sourcing of sand-tempered ceramics on the small islands of Pacific Oceania, temper analysis was not successful at identifying whether sources of pottery dating to the Early Yayoi-Heian period (around 300 B.C.–A. D. 300) from archaeological sites in Okinawa were located on that island or on mainland Japan (Fitzpatrick et al. 2006). Elemental analysis of the pastes was equally unsuccessful. The basic problem was that, geologically, few if any fundamental differences can be expected in temper materials. Fitzpatrick et al. pointed out that Japanese ceramics were made on islands of varied geology tapped by rivers that mixed sediment derived from multiple bedrock sources, so, as expected, temper from those ceramics showed much variation and complexity at single localities, making provenance studies quite difficult.

Fowles et al. (2007) examined pre-Classic ceramics from Taos, New Mexico. In this case, most of the locally produced black-on-white pottery was untempered. Although abundant nonplastics were found, these were naturally present in the raw clays available to prehistoric potters. In addition, there was little clay processing (such as levigation or grinding), so variability in ceramic materials was a function of variability in the geological distribution of raw clays. Since the geology of the Taos district is rather complex, a combination of thin-section petrography and elemental analysis proved useful for identifying clay sources in the Rio Grande del Rancho drainage system. The ceramic sourcing studies revealed that settlement aggregation was accompanied by a reduction in diversity and quality of clay sources used by local potters, probably because the formation of buffer zones between village aggregates led to restricted procurement strategies.

Quinn and Day (2007) used their technique, as mentioned above, for characterizing both calcareous microfossils and nannofossils (by combining thin-section analysis with a smear-slide technique and very high magnification under an oil-immersion lens) in provenance studies of Bronze Age Aegean archaeological ceramics from Crete. On the island, similar geological deposits over wide areas meant that source areas identified through thin-section and compositional studies were rather broad. Quinn and Day used micropaleontology to narrow the source definitions more precisely and to make "micro-provenance" determinations. Their work involved sampling pertinent sediments and identifying their microfossil assemblages to compare with sherd analyses. The success of this work demonstrates the utility of detailed studies of calcareous microfossils and nannofossils when they are found in archaeological clays and tempers.

8

Pottery Fabrication, Use, and Deterioration

Although much work involving thin-section petrography of pottery focuses on identifying and characterizing nonplastics—and on using that information to answer questions relating to pottery classification, identification of production groups and workshops, examination of changes and variation in temper additives over time and space, and identification of specific geological sources of nonplastics—there are also many other research questions for which thin-section analysis can be useful. Thin sections can reveal information regarding clay selection and processing choices, pottery forming methods, firing conditions, pottery function and history of use, porosity patterns and their implications, surface decoration methods, issues of technological choice, and deterioration mechanisms.

8.1 Identifying Clay Processing Choices and Paste Characteristics

Raw clays are often processed in various ways before being used, to improve working properties or performance characteristics. For example, two or more clays may be mixed together, clays may be ground to break up larger particles or pellets, coarse particles may be removed through levigation (water settling) and decanting, or through sieving, and nonplastic temper materials may be added. Some clues to these processes may be visible in thin sections.

Strictly speaking, the clay paste component refers to the clay matrix and any inclusions naturally associated with the clay deposit selected by the potter. While clays are usually difficult to identify in thin section, it is sometimes possible. For example, Morter and Iceland (1995) concluded that illite was used for most of the eastern Calabrian (Italy) sherds that they examined, based on an orange birefringence under crossed polarizers; x-ray diffraction (XRD) analysis arrived at the same conclusion.

Although individual clay grains cannot be distinguished in thin section, due to their very small size, most clays contain natural coarser inclusions which can be identified and characterized by thin-section analysis. Based on a study of natural inclusions, it may be possible to deduce that two clays were used, or that different parts of a vessel were formed using different clay batches. Maggetti (1982) discussed a case where potters at a single production site used different clays: a marly clay (very calcareous), a moderately calcareous clay, and a noncalcareous, siliceous clay.

A key problem is distinguishing between natural impurities and added temper material, which is important in understanding preparation technologies and for identifying geological sources of clays or deliberate additives. Once the temper can be excluded, the paste can be compared to sediment samples to try to match ceramic composition with raw material sources (Stoltman 2001). However, sometimes clays are modified for use by removing material by sedimentation or levigation to alter their properties; it is much more difficult to determine whether material was removed than that it was added (Woods 1982).

Clues as to whether inclusions found are natural or not include the variation in size and shape, as well as the overall profile of materials present. In addition to focusing on the naturally occurring coarser nonplastic components, there is also a variety of other attributes of the clay matrix itself that can be used to systematically describe and classify sherds in thin section. A paste with a very homogeneous appearance in terms of grain size, content, and distribution, for example, may represent the considerable effort that went into processing the clay (Fig. 8.1.1). Lumps of clay distributed here and there within a matrix of uneven temper distribution indicate that less effort was expended on grinding, sieving and mixing of the clay (Fig. 8.1.2). Wedging and kneading eliminate large air bubbles and produce a uniform distribution of the finer air bubbles. In thin section, the presence of many voids of varying sizes suggests that little kneading was done (Fig. 8.1.3). In contrast, the presence

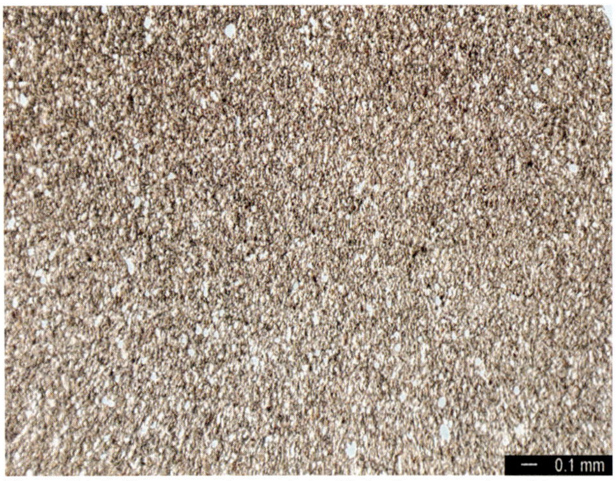

Figure 8.1.1 Clay paste, plane polarized light. A paste such as this one (an English Ironstone) that has a very homogeneous appearance (in grain size, content, and distribution) may represent the extended effort put into processing the clay.

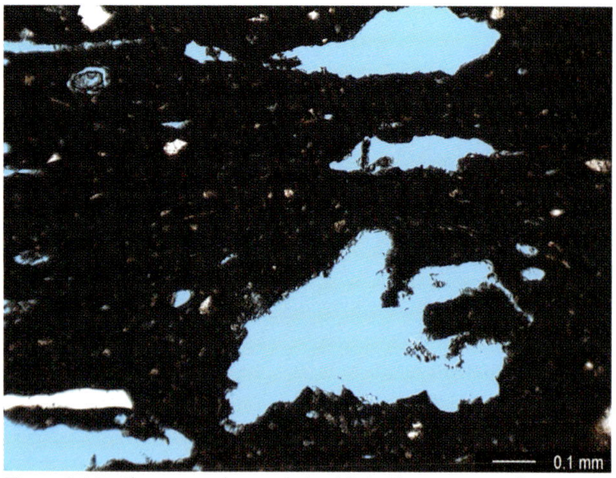

Figure 8.1.3 Clay paste, plane polarized light. The presence of many voids (blue) of varying sizes suggests that little kneading was done.

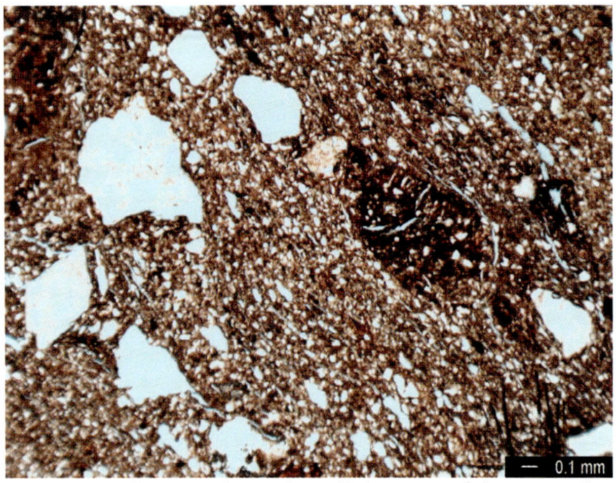

Figure 8.1.2 Clay paste, plane polarized light. Lumps of clay, scattered streaks and lumps of hematite, and an uneven temper distribution indicate less effort may have been spent in grinding, sieving, and mixing of the clay.

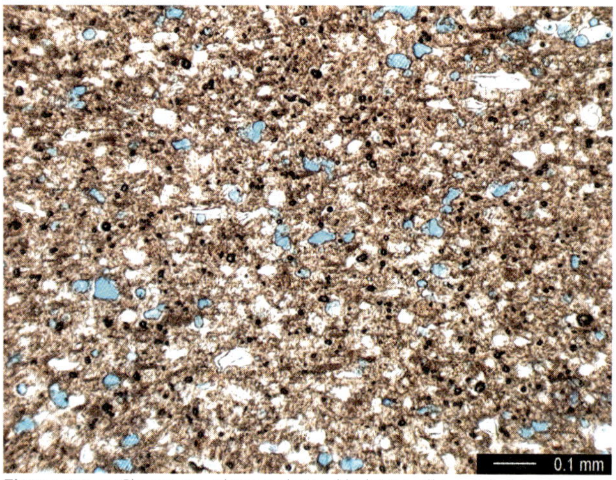

Figure 8.1.4 Clay paste, plane polarized light. Smaller voids (blue) of uniform size and distribution may indicate that more thorough kneading or processing of the clay was accomplished.

of voids of uniform size and distribution may indicate that more thorough clay processing or kneading was accomplished (Fig. 8.1.4) (Rye 1981).

Sometimes the presence of streaks of different color may indicate the mixing of two different clays (Fig. 8.1.5). However, as Woods (1982) cautioned, that may not always be the case. Many clay deposits contain clays of different color, and even kneading well may not easily disperse and mix the colors, so color streaking is not necessarily a sign of clay mixing and/or insufficient kneading by the potter.

In an early use of thin-section petrography to study clay materials and processing methods, Felts (1942), one of the pioneers in thin-section petrography of ceramics, looked at materials excavated at the site of ancient Troy. He was able to identify a variety of clay selection and processing approaches that changed over time at the site. The earliest clays showed fine, stream-rounded natural inclusions of quartz, feldspars, micas, and calcium carbonate minerals, indicating that detrital silts were collected with little or no purification. The

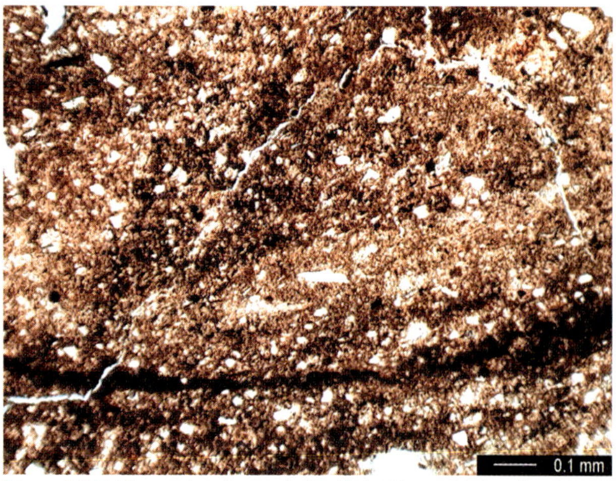

Figure 8.1.5 Clay paste, plane polarized light. The presence of streaks of different colors in the clay (here light and dark brown) may indicate that two different clays were mixed in the production of this ancient Chinese cord-marked vessel from Jiangxi province.

purity of the clay increased over time and was accompanied by an increase in the use of added tempers.

Arnold (1972) analyzed materials from clay and temper mines and from potters' collections in Quinua, Peru, a modern village located in the southern Andean Highlands near the ancient site of Huari. He found that the potters distinguished between clay to use for untempered paste and that for tempered paste by the presence of "gold-like" particles of mica (mainly phlogopite) in the former. The gold-colored phlogopite reflected light and thus was easily visible to the potter. The clay for tempered paste sometimes had very small amounts of mica, but the quantities were too small to be easily detectable by eye. In addition, the clay for untempered paste had more than 50% naturally occurring nonplastics, so adding a temper was unnecessary. In contrast, the clay for tempered pastes had only about 5% so did need the added temper for improved working properties. Arnold's work demonstrated the usefulness of analyzing raw materials from ethnographic settings where potters are still using traditional methods and materials that may have been available to historic or archaeological villages in the same region. New ideas can be gained that can aid research designs and data interpretations for analyzing archaeological sherds.

Hegde (1979) studied three ancient deluxe wares from India: Painted Grey Ware, Black Slipped Ware, and Northern Black Polished Ware. These wares are important archaeologically because of their frequency and wide distribution, and also because they represent high technological achievements in ancient Indian ceramic history. Sherds from three excavated sites (Hatingapura, Mathura, and Kausambi) were analyzed. Thin sections showed that the clay was carefully sifted and elutriated; it was revealed to be fine-grained, mature clay that was well elutriated to remove heavier particles.

Kilmurry (1982) studied Stamford ware, ceramics made at Stamford in England between 850 and 1250. The goals of thin-section petrography were to characterize the ceramic fabrics, compare them with characteristics of local clays, and assess any variations in clay selection. Thin sections showed that the ceramics were lacking in distinctive inclusions, so quantitative analyses of the quartz component were undertaken. A minimum of 200 quartz grains per thin section was counted and measured—grains less than 0.03 mm across were classified as part of the clay matrix. Measurements included shape (using visual estimation charts), with most found to be either angular or subangular, ratio of quartz to clay matrix background, and size (with distributions plotted as the percentage of quartz grains of each size from 0.04 to 0.28 mm across) with those of larger sizes combined into the last column. As a summary measure of the sandiness of fabrics, the percentage of quartz present was compared with the mean size of the grains.

Kilmurry concluded that the data suggested four clay sources may have been used at Stamford, in a variety of workshops throughout the town. Local clays were collected and fired to various temperatures, then compared with the ceramic thin sections to locate possible sources and identify probable firing temperatures. A chronology of clay usage was then proposed. Some sources were used throughout the period studied, others were used only early or late. Some of the shifts in clay sources selected corresponded to changes in form, decoration, and manufacturing technique. Some clays were carefully sieved to remove larger quartz grains, while others were less carefully processed; one fabric was found to be a mixture of two clays.

Russell (1982) used grain-size analysis of naturally occurring quartz in Early Saxon pottery from East Anglia to try to discern if all the pottery might have been made with one clay source, or if multiple clays were used at the site. Cumulative frequencies were compared using the Kolmogorov-Smirnov test for deviation between two cumulatives, at a significance level of 0.05. He was able to conclude that more than one clay source might have been exploited for the site analyzed.

Jornet et al. (1984) studied sherds from sites near Valencia (Spain). They focused on tin-glazed majolica (stonepaste) wares from Paterna and Manises, two small towns on the outskirts of the city of Valencia. Thin-section petrography showed an uneven distribution of fine-grained calcite and isotopic matrix, leading the authors to conclude that the potters were using blended clays (a calcareous clay and a non-calcareous, quartz-rich clay).

A study of Koan and Corinthian amphorae with similar clay pastes that may have been processed differently found that the Koan jars all had the same stamp and a similar fabric mineralogy, but upon refiring at 1100 °C, one fragment was red while three were yellow (Whitbread 1986a). The hypothesis was that the same clay sources were used for the red and yellow clays, but three had seawater added, which tends to make red clays turn yellow. However, the Corinthian jars showed some difference in mineralogy indicating that red and yellow clays may actually have originated from different sources. To test that hypothesis, a grain-size analysis was done, which confirmed that the red clay was extracted from *terra rosa* soils by levigation, and the yellow clays were excavated from beds of Neogene calcareous clays.

Vincent (1988) noted that interpretation of distinct clay pellets within the clay paste of a ceramic requires great care. With hand-formed fabrics, localized variations in moisture content may create small dry lumps within the moister clay paste during preparation processes. Upon firing, the lumps may look like natural inclusions of clay pellets, so it can be difficult to determine if they are natural or the result of clay-processing techniques.

Whitbread (1989, 1995) illustrated the use of a descriptive system derived from soil micromorphology methods that provided the means for characterizing and classifying clay matrix components of a sherd thin section (this approach later was discussed also by Josephs (2005)). For example, it included a section on microstructure where the focus was on describing the shapes and arrangement of voids, micromass (<0.01 mm in size), and inclusions. Voids (trapped air or pores that result from them) were defined and characterized in a variety of ways (such as irregular with wavy edges, or regular in shape with smooth edges). Also included in his microstructure description scheme was the distribution of coarse and fine particles (such as porphyric when coarse particles occurred in a dense groundmass of fine material and there were few pores), with modifiers to further describe the distances between coarse particles on the basis of mean grain diameters.

Whitbread also pointed out that the optical state of the micromass is an indication of the extent to which the vessel had been fired, so he included a variety of descriptors for this feature (inactive, slightly active, or active) (Figs. 8.1.6–8.1.8). If it is active (anisotropic or birefringent), there are additional descriptors that can be applied (such as stipple speckled or mosaic speckled, striated, etc.) (Fig. 8.1.9). Other suggested types of features to aid in description and classification included mottling of the micromass, defined as local variations in color and optical density, appearing as patches and coatings with clear boundaries around voids.

How to identify whether or not a clay was refined by levigation in order to remove coarse particles was discussed by Freestone and Hughes (1989) in their study of ceramics from the Late Assyrian site of Qasrij Cliff and the post-Assyrian site of Khirbet Qasrij. They found that most of the clays were very fine, having few inclusions greater than 0.1 mm, with calcite abundant in the lower fired wares. Since the higher fired wares had high lime contents, the authors concluded that calcareous clays were used for all the wares. Some of the local clays contained very coarse particles of calcite, which would have been problematic in firing due to spalling. Thus it was quite probable that the potters chose to refine the clays to remove the coarse calcite fraction; this procedure would account for the very fine-grained appearance of the clay bodies as seen in thin section.

Panjwani (1989) noted that micas are tiny, platy crystals that tend to become concentrated along with the clay when it is repeatedly elutriated. He illustrated as an example a thin section prepared from Micaceous Red Ware from the site of Lothal in Gujarat, India. Angular to subangular grains of quartz and feldspars appeared, along with a large number of tiny mica grains; Panjwani interpreted this finding as evidence suggesting that the clay used in preparing the pottery was repeatedly elutriated.

Figure 8.1.6 Clay paste, crossed polarized light. This well-fired ceramic (a hard-paste porcelain dinnerware) exhibits an inactive micromass. Only the quartz and feldspar grains are visible in crossed polarized light, with the clay matrix completely black.

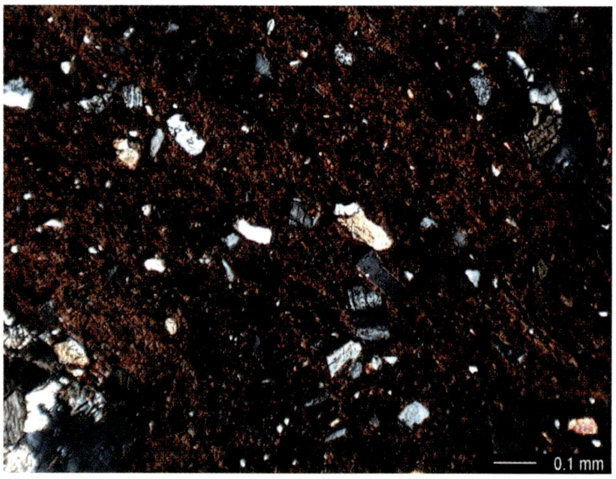

Figure 8.1.8 Clay paste, crossed polarized light. This Iroquoian ceramic from southeastern New York (1300–1500) exhibits an active micromass. The silt- and sand-sized particles (quartz, feldspar, amphibole, and lithics) are set within a matrix that has a reddish-brown birefringence.

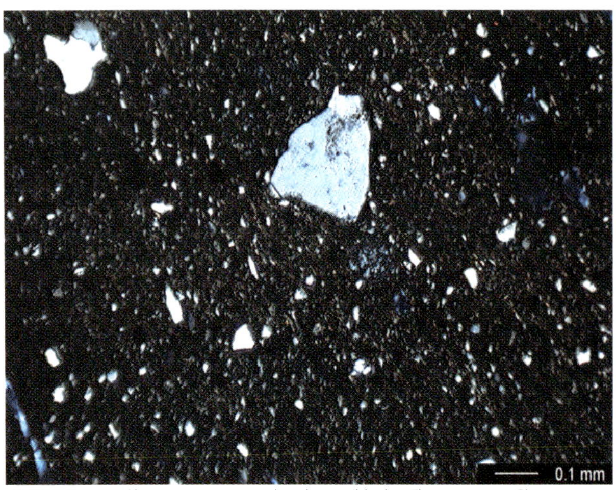

Figure 8.1.7 Clay paste, crossed polarized light. This ancient Chinese sherd exhibits a slightly active micromass. Although sand- and silt-sized quartz grains are the most visible, the finer grained matrix shows some birefringence.

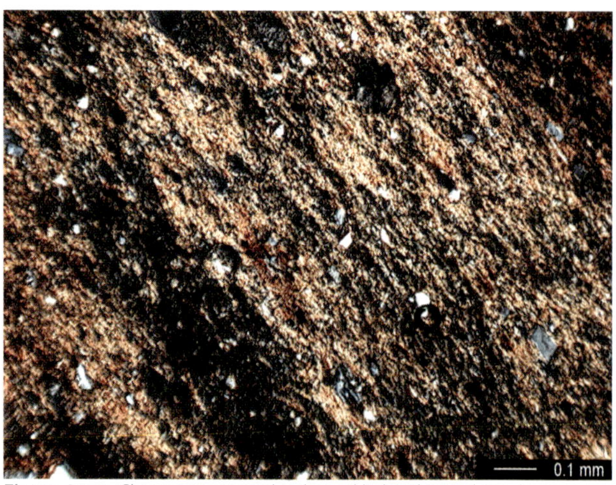

Figure 8.1.9 Clay paste, crossed polarized light. This prehistoric ceramic from southern Delaware (300–800, Mockley Ware) has an active micromass that can be described as striated.

A methodology for using thin-section petrography to identify the possibility of modification of clays prehistorically was developed by Barnett (1991). He started with quantitative point counting (a 200-observation line count) of coarse grains (larger than 0.06 mm) in archaeological ceramics and of local clays, to try to match the two and account for possible past modifications. Size was organized for analysis by Wentworth scale units; a mineral taxonomy was used, and angularity of each grain was assigned to one of four qualitative classes. Barnett pointed out that three basic types of modification result in changes to size, angularity, and mineralogy of coarse grains in clay: tempering, sieving, and working by either crushing or grinding. Size information was used to identify sieving and the addition of coarse temper grains. Angularity measurements indicated whether the temper was crushed, added material or a natural component of the clay. Voids might be evidence of addition of organic material, if found in sufficient amounts. Angularity and size ranges for each mineralogical group were first measured in the clay samples, and then compared with the vessel samples. This procedure was followed with an assemblage of Early Neolithic pottery from the site of Balma Margineda in the Andorran Pyrenees. One vessel was identified as nonlocal, and the modification of local clay deposits was demonstrated for two local vessels and suggested for others.

Peña (1991) used thin-section mineralogy studies (along with XRD) to investigate raw material acquisition and paste preparation practices among Roman-period potters in the Orte-Vasanello area of the central Tiber valley of Italy. Ethnographic research among modern potters at Vasanello indicated a complex pattern of clay utilization. Local clays, fired test tiles made from those clays, pottery from the modern Vasanello workshops, and ceramics from two Roman ceramic production sites were analyzed and compared. It was found that similar clays and tempering materials were apparently used in both Roman and modern workshops. There was a high degree of mineralogical similarity between functionally equivalent groups of pottery produced in the modern workshops of the Orte-Vasanello area and by the ancient Roman potters.

Vaughan (1991a, 1991b) described thin sections of distinctive sherds of Base Ring ware in Cyprus. By eye it was apparent that there were strongly contrasting colors: in thin section it could be seen that the matrix was ferruginous, but there were patches of non-ferruginous clay (later identified as kaolinite). She found local South Coast Pakhna Formation clays associated with shale outcrops that were iron rich, and some kaolinitic clay was known to occur in some of the soils of southeastern Cyprus. It was unclear whether or not the mixture of these two clays was natural, or if the kaolinite clay had been deliberately added to the ferruginous clay.

Methods of micromorphological analysis used in soil studies were included by Adan-Bayewitz and Wieder (1992) as part of their fabric characterization of ceramics from Roman Galilee and Golan. Sherds from 11 different sites were studied; all samples came from sherds that had already been assigned a compositional group through neutron activation analysis. Thin-section studies were done blindly and randomly with regard to compositional group, to determine whether the same groups could be discerned with thin sections. The authors found that micromorphological analysis was useful in classifying pottery into major fabric groups; however, this approach requires a basic knowledge of the nature of the clays available to the potters and the clay formation processes involved. In addition to classifying sherd fabrics, they were able to correlate fabric specimens with defined clay types, thus improving provenance determinations, as well as clarifying fabrication issues.

For example, one group (the Golan Group) was made with basalt-derived clays. Similarity between the microfabrics of the Golan pottery and the microfabric of the basalt-derived clay material of the Golan indicated that no temper was added—the basalt component was a natural constituent of the clay from which the pottery was made. Naturally occurring rock fragments and silt-sized particles fulfilled the same function as added temper would have served; indeed, even cooking vessels did not contain added temper. In addition to mineralogical differences, criteria based on characteristics of the clay paste were useful in characterizing groups. For example, the Golan Group specimens tended to have a more isotropic or opaque appearance, which Adan-Bayewitz and Wieder attributed to release of iron by the primary minerals in the basalt rock component.

Vaughan et al. (1995) incorporated thin-section petrography in investigations of prehistoric white ware pottery from the settlement of Akrotiri on the island of Thera in the southern Cyclades of Greece. Forty thin sections of Burnished, Bichrome, and Dark-on-Light wares were analyzed. The groundmass of all fabrics contained finely disseminated micritic carbonate with some microspar or pseudospar mosaics. Abundant discrete microfossils had calcite replacement. Igneous rock fragments in the fabric included vitric clastics and crystalline ashes. Although specific geological sources were not found, the general type of clay deposit that would match the pottery fabrics was locally available. At first glance, all the pottery fabrics appeared similar, however, careful examination revealed that the Dark-on-Light wares tended to have carbonate inclusions uniformly less than 1 mm in diameter, while most bichrome wares and all burnished wares had carbonate inclusions with a greater maximum grain size (1.0–3.35 mm). Although the Dark-on-Light wares had a fine-grained carbonate fraction, they also had larger inclusions of volcanic lithics. Since these lithics would have been harder and more resistant to crushing, their presence was not seen as contradicting a fine-preparation procedure, but water-settling methods should have removed these larger fragments. Their presence, though, would not pose the same technical problems as those created by large carbonates, so would not be unacceptable to the potters. Vaughan et al. concluded that potters of the Middle Cycladic period in Akrotiri were not preparing their clays by water settling, a method common in traditional Cycladic workshops today. Instead, clays were prepared by crushing and grinding, probably in a dry state, followed by hand sorting or crude sieving.

Velde and Druc (1999) contrasted the paste preparation evidence seen in thin sections of a fine-grained ceramic *sigillata* ware and coarse Neolithic wares of central France. The *sigillata* ware had a homogeneous mix of clays and small temper grains; the paste preparation steps of crushing, sieving, and impurity extraction were done carefully. The result was an even distribution of fine grains with a regular texture. In contrast, the Neolithic wares showed a high irregularity of grain-size distribution. The paste was made from an irregular clay source that was not homogenized after extraction from its source. Lumps of clay were joined to form the paste, giving a very irregular texture. Larger temper grains (probably a natural constituent of the clay) had an irregular distribution in a clay matrix that did not appear to have been ground, sieved, and mixed; there was a "zoning" effect with patches of fine grains interspersed with areas of larger grains.

In a study of thin sections from 56 kiln samples from Kommos, a Minoan Crete site by Day and Kilikoglou (2001), many of the observations could be related to clay-processing technology. For example, one fine-grained fabric had few nonplastic inclusions. The micromass had streaks of calcareous clay within a darker matrix, which the authors interpreted as indicating the mixing of clays in the preparation of the clay body. A medium-coarse fabric, which displayed a similar micromass, also showed evidence of clay mixing. There was also evidence of clay mixing in a medium-to-coarse calcareous fabric, although the matrix was overall more calcareous. A coarse fabric with sand inclusions displayed less evidence of clay mixing than the medium and fine equivalents of this material. The authors incorporated the petrographic data into a study of vessel size and shape, and concluded that the petrographic analysis clearly showed that different clay paste formulations were selected for different vessel sizes, with the larger ones having a coarser fabric. Slight differences were discernable in clay matrix appearance due to differential mixing of clays to form the body. There were also clear differences in the proportion and size of nonplastics according to vessel size. Thus, thin sections revealed differences in recipes including both the amount and type of nonplastic tempers added and the presence and degree of mixing of separate clays.

Miksa (2001) carried out thin-section petrography of prehistoric Hohokam buff ware pottery from Arizona using the Gazzi-Dickinson method of point counting (emphasizing the mineralogical composition of rock fragments). She conducted this petrographic modal analysis mainly to quantify the proportion of different components in the paste, rather than as a provenance tool. Experimental work helped her develop criteria to distinguish crushed schist temper from natural schist-bearing sands. The two sources of schist in the pottery were found to vary in grain-size distribution, rock homogeneity, grain angularity, crushed grain features, finely disseminated mica, and with large volcanic rock or quartzite fragments. For sand tempers, criteria for identifying an intentional addition or a natural constituent of the clay included analysis of grain-size distribution, mineralogical composition (including textural characteristics of grains), and grain angularity. She concluded that in most cases the clay and the sand originated from the same deposit. Clay lumps and caliche were also incorporated into many of the clay pastes. Using her criteria for differentiating components in complex pastes, Miksa identified several paste recipes, or standard paste mixtures, among 130 thin sections of buff ware pottery from nine sites in central Arizona.

Amadori et al. (2002) discussed thin-section analysis of Bronze Age pottery from Imola, Italy, with the petrographic work informing a study of technological style and the working methods of potters. Most of the samples studied contained argillaceous rock fragments with a microstructure similar to that of the clay matrix, but differing in color and having clear boundary discontinuities. Accompanying these fragments were rounded to subrounded particles that were reddish to brown, sometimes surrounded by shrinkage boundaries, identified as "bohnerz," or nodules rich in iron and present as natural aggregates in the clay. In addition, there were also some carbonized vegetal fragments with part of the original structure often still preserved, as well as isotopic nodules with a concentric structure ranging from reddish brown to dark brown in color (cylindrical vegetal residues sectioned transversely with respect to the longitudinal axis). Taken together, the authors proposed that the raw material included paedogenised alluvial sediments (clays from soils). Finding original structures partially preserved in the ceramics was seen as evidence for the use of poorly worked clays and a complete lack of purification or preparation of the clay material. They further suggested that the clays were collected from a shallow depth, and then used immediately with no preliminary homogenizing or purification operations.

In a review by Krishnan and Shah (2005), they illustrated how thin sections can be used to investigate clay-processing choices in the production of ancient Indian pottery. They noted that refinement of the clay paste, a result of levigation/elutriation allowing the settling of coarser particles, was reflected in thin sections by the presence of a finer paste with a reduction in both relative abundance and in size range of nonplastic inclusions compared with the original clay. In contrast, the addition of a sand temper to the clay created an increase in frequency and proportion of larger inclusions. Thus, variability seen in thin sections could be attributed to human alterations of the material, not just to the use of different geological sources. They then considered the process of kneading, reporting that ethnographic work and thin-section studies have shown that at successive stages of kneading, proportions of very fine sand, fine sand, medium sand, and coarse sand increase while that of silt decreases, as the clay becomes better mixed and homogenized, with the nonplastic inclusions more evenly distributed. This point was illustrated with photomicrographs of perfectly sorted and strongly unimodal inclusions. The process of kneading may leave unmixed clay lumps which show up in the thin section as having distinct shapes and trailing into the surrounding matrix, but differing in color and texture (Fig. 8.1.10).

Maritan et. al. (2005) performed thin-section petrography on Bronze Age and Iron Age pottery from Tell Mishrifeh/Qatna in Syria, and found that the shape, the grain-size distribution, and the rock type of the inclusions provided

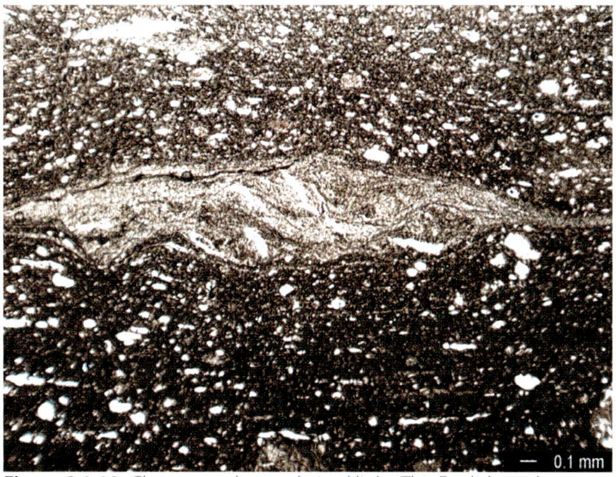

Figure 8.1.10 Clay paste, plane polarized light. This English earthenware (late seventeenth to early eighteenth century) has some unmixed clay lumps left from the kneading process. These differ in color and texture from the surrounding matrix, and, like this large one in the center, often trail into the surrounding matrix.

information about clay-processing techniques used in ceramic production. The presence of many clay pellets and argillaceous rock fragments suggested that the clay was not seasoned, or at least not for a long period of time. A wide variety of petrofabrics was found at this one production site (nine distinctive fabrics, all made with locally available materials), which indicated the use of many different recipes at one production site. An analysis of mineral assemblage and textural features helped to identify four main types of ceramic paste. The first was a paste with abundant, well-rounded inclusions and seriate grain size, which the authors suggested was consistent with pottery prepared without levigating or tempering the raw clay material. The second had a fine-grained matrix with scarce small inclusions, perhaps representing levigated clay. The third had abundant, angular rock fragments, indicating that this paste was probably tempered with ground rock fragments. The fourth was tempered with mollusk shells or fragments of spathic calcite, and had a gap in the grain-size distribution with other well-rounded inclusions—this was possibly a starting material of the first paste type, to which the temper was added.

Thin sections of Middle and Late Neolithic pottery from the rock shelter site of Abri Pendimoun in France, which occupied a key position along communication routes between France and Italy, were examined by Basso et al. (2006). After a preliminary study of 1,245 ceramic finds with a binocular microscope, in which three major fabrics were identified, 46 representative fragments were chosen for thin-section analysis (supplemented by other analytical data). Another 15 thin sections made from local raw materials that might be potential sources were also examined. The thin-section characterization included textural and compositional parameters (grain size and shape, mineralogy of nonplastic inclusions, and calculation of the ratio of nonplastic inclusions to clay matrix). The three fabric groups were identified as three basic recipes, with subgroups within each. Comparison with the raw materials indicated that all groups could have derived from local clays. In some cases, the characteristics of nonplastics suggested that an unprocessed material was used, with no temper additions. For other subgroups, one of the same local raw materials was used, but temper additives such as crushed limestone or aplite fragments were mixed into that material.

Capone (2006) investigated pottery from the final periods of glazeware manufacture in the Rio Grande region, looking at technological change over time through thin-section petrography. One aim was to identify clues to processing of raw materials—she found that although raw materials remained constant, they underwent less processing through time. For example, average grain size of the large fraction increased from 400 μm in the earliest period up to 700–900 μm by the latest period. In addition to larger overall grain size, there was also a change over time leading to a higher frequency of large grains.

Neolithic pottery from the Apulia region in Italy was examined by Laviano and Muntoni (2006); 375 thin sections were analyzed (among other analytical techniques) along with 134 clay samples from the region, the technology of pottery production being one of their interests. They identified different choices made by ancient potters in the preparation of their raw materials. The clays were usually at least somewhat refined, and in some cases a mineral temper (such as fossiliferous sand, alabastrine limestone and calcite), grog, or vegetal material (as evidenced by curvilinear and long pores where organic material burned out during firing) were added. At one of the settlement areas, different choices could be seen in the preparation of the pastes, with the main variation seen in grain size (from 50 μm up to 600 μm).

Shoval et al. (2006) employed thin-section petrography to characterize the ceramic technology used in manufacture of Iron Age functional pottery (cooking pots, storage jars, and tableware vessels) excavated at Tel-Hadar on the eastern shore of the Sea of Galilee. The thin-section data identified a variety of petrographic groups based on characteristics of the ceramic matrix and of the temper particles; for some of these groups, evidence existed that the potters mixed two different clays to make their paste. The cooking pots included three major petrographic groups: (1) basaltic soil tempered with calcite crystals; (2) *terra rosa* tempered with calcite crystals; and (3) basaltic soil with no temper additive. The storage jars and tableware vessels comprised the other petrographic groups, some of which were mixed clays. These groups included ceramics manufactured with one of the following: (1) calcareous alluvium (unsorted and rounded grains of basalt, chalk, limestone, and sand within a calcareous matrix); (2) calcareous alluvium soil (richer in iron oxide and quartz silt than the previous group); (3) calcareous rendzina soil (includes chalk fragments within a calcareous matrix rich in quartz silt) that was mixed with non-calcareous basaltic soil (as evidenced by mud balls of basaltic soil and a few basalt fragments); (4) calcareous marl (only a small amount of quartz silt present in the calcareous matrix, plus foraminifera fossils) mixed with *terra rosa* (as evidenced by mud balls of *terra rosa*); or (5) calcareous sediment of the coastal plain (some rounded quartz and skeletal fragments of Mediterranean fauna within a calcareous matrix rich in quartz silt).

Quinn and Day (2007) used micropaleontological techniques (combining thin-section analysis and smear-slide studies, both with very high magnification under an oil-immersion lens) to characterize microfossils and nannofossils to identify the presence or absence of clay mixing in Bronze Age ceramics from Crete. They were also able to discern differing degrees of mixing of raw materials. It was noted that a biostratigraphic analysis of nannofossils indicates that on the south coast of Crete and at Knossos, light-colored highly calcareous marls of Pliocene age were exploited preferentially, but that on their own such sediments are often unsuitable for ceramic production. Mixing them with a low-calcareous clay-rich material would produce a better paste for fabrication of ceramics. This micropaleontological approach can identify such choices made by potters.

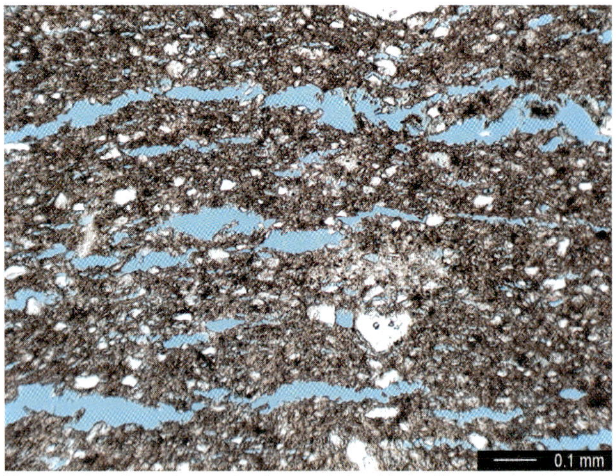

Figure 8.2.1 Ceramic matrix, plane polarized light. Elongated voids or pores (blue areas), oriented in parallel alignment, indicates a forming method that involves strongly applied pressure. This is a seventeenth-century German stoneware.

8.2 Reconstructing Forming Methods

Some ceramic petrographers are especially interested in understanding ceramic technology and fabrication methods (Hodges 1963; Freestone et al. 1982; Middleton and Freestone 1991). Although they draw on methods of sedimentary petrography and soil micromorphology, they note that since ceramics are made by humans, the methods used by geologists are not sufficient for understanding ceramic fabric variations caused by human interactions with raw materials. Thus ceramic petrographers are working to develop their own methods to describe ceramic fabrics in such a way as to highlight both the nature of the raw materials and the effects of ceramic technology. Interpreting thin-section evidence reliably requires careful study of all aspects of clay processing, forming methods, firing regimes, and vessel function and use through ethnographic studies and replication experiments combined with thin-section analysis (Whitbread 1989, 1995). Of course, the information about possible forming techniques that is gleaned from thin-section analysis should supplement a careful study of sherds and any existing whole vessels (Matson 1970; Rice 1987, 1996). As with most research questions in ceramic analysis, thin-section petrography is one line of evidence but rarely the only one.

Forming methods may include techniques such as pinching, press molding, slab building, hammer and anvil, coil building, and wheel throwing. One vessel may involve the use of multiple techniques (Rye 1981; Rice 1987, 1996; Whitbread 2001). Thin-section analysis may reveal clues not visible by a surface examination as to which forming methods were used. Forming a vessel may require a complex sequence of activities, each involving different techniques (Courty and Roux 1995; Vandiver et al. 2002). For example, forming techniques that involve strongly applied pressure can result in parallel alignment of platy particles such as micas; other elongated inclusions and voids or pores can also become oriented during forming (Fig. 8.2.1). When pressure is applied to the plastic clay, elongated voids and particles may become aligned, usually with their long axes

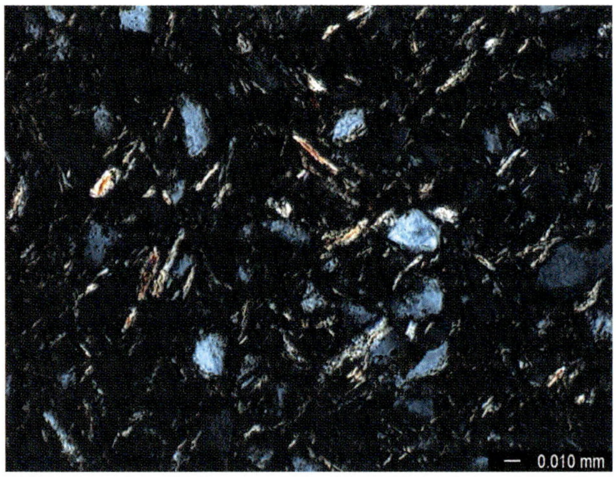

Figure 8.2.2 Ceramic matrix, crossed polarized light. In this early Chinese ceramic, the platy particles (sericite, showing higher birefringence than the surrounding sand grains) are oriented randomly rather than in a parallel arrangement.

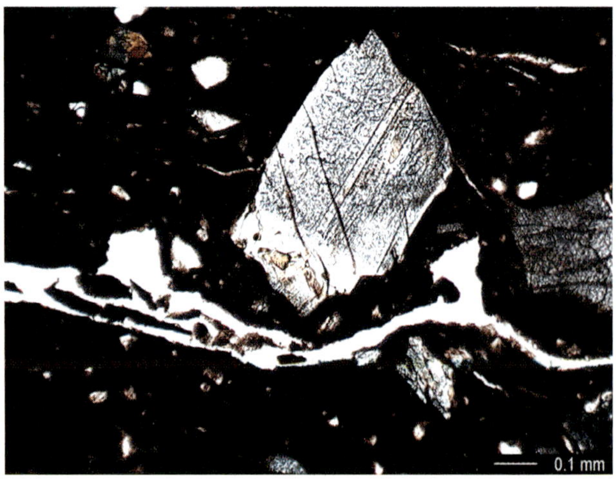

Figure 8.2.3 Ceramic matrix, plane polarized light. Evidence of coil- or ring-building is the presence of join voids indicating poor bonding, so the two sections separated more as the clay dried. This example is from an Iroquoian sherd from New York; the clay has been pulled apart at the join. Coarse-grained nonplastics include very weathered plagioclase feldspar grains, and smaller clear quartz grains and bioite flakes.

at right angles to the direction of the force; equidimensional particles will not be affected (Woods 1982).

Vessels thrown on a wheel may tend to have particles that are oriented with their long axes lying more horizontally (parallel to the rim and base) but at a diagonal—the faster the wheel rotates the steeper the angle of orientation away from the horizontal (Gibson and Woods 1997). In contrast, vessels formed by other methods tend to have particles oriented vertically, in circles, or randomly (Fig. 8.2.2). However, this analysis is not straightforward, because a more parallel alignment can also be an indication of other techniques such as paddle and anvil, and coiling, depending upon the direction of shaping and smoothing activities (Hodges 1963; Velde and Druc 1999). Other thin-section evidence of coiling, ring building, or building in segments may include join voids indicating poor bonding; the two sections separate more as the clay dries (Fig. 8.2.3). Other evidence for wheel throwing is the presence of a very fine body with few inclusions (since it is difficult to throw a pot on a wheel if there are inclusions greater than about 3 mm or if the inclusions are angular), and voids should also be less common because clay for throwing needs thorough preparation and kneading (Woods 1982; Gibson and Woods 1997). Whitbread (1989) emphasized that the use of preferred orientations of inclusions and micromass features such as voids and channels can only be used to deduce forming methods if the thin sections represent specific and known orientations (vertical, horizontal, and/or tangential) with respect to the vessel wall. In addition, later forming actions may obliterate or alter the evidence of the primary forming method, some methods of manufacture may not leave visible patterns of particle orientation, and some orientation patterns might result from more than one fabrication process. Finally, evidence of burnishing or polishing may be visible in thin section. Fine particles, especially the platy clay and mica minerals, may show a clear parallel orientation near the treated surface while having a more random orientation further in the vessel (Vaughan 1995).

In the study by Felts (1942), mentioned above, of materials from the site of ancient Troy, he hypothesized that a lack of internal preferred orientation of grains and the haphazardly distributed porosity seen in thin sections of sherds from the site indicated that a minimum of working was done during the shaping process. Shaping techniques identified through his thin section examinations included the use of a potter's wheel, indicated by strong internal alignment of pores, platy minerals, and rod-like mineral fragments.

Freestone and Hughes (1989) employed thin-section petrography to help reconstruct the fabrication method of tripod bowls from the post-Assyrian site of Khirbet Qasrij, especially to identify procedures for adding the legs to the vessels. Organic temper was added to the clay, and burnt out during firing (identified as dung by the fine particle size and the mixture of ragged stalk-like and irregular impressions left behind). The bowls themselves were wheel thrown (as evidenced by corrugations on the interior of the body sherds and spiral patterns on the bases, defined by voids), allowed to dry to the leather-hard state, then turned down on the wheel using a sharp tool to form footrings and well-defined rims. The separate legs were then added as wet clay. Consequently, the clay material used for the legs was heavily tempered with dung in order to reduce the drying shrinkage of the legs so that they would better fit the partially dried body of the vessel without pulling apart from it due to differential shrinkage as they dried.

The use of thin-section analysis to provide evidence for fabrication in a mold was illustrated by Whitbread (1989) using an analysis of the distribution of coarse versus fine particles in a thin section taken from a Corinthian washbasin. Three types of fabrics were found: the first, which formed the core of the vessel, had quartz grains in an open-spaced porphyric distribution, and no mudstone inclusions; the second was a layer about 1 cm thick, with dominant mudstone inclusions—this layer had a vug-filled microstructure, and coarse particles were linked by sections of finer material; the third was a thin surface layer only 1–3 mm thick without mudstone. From these data, he concluded that the layer with dominant mudstone was probably pressed into the mold before the core clay was placed in, as a release agent to aid extraction of the finished piece. After the piece was removed from the mold, it was then covered with the fine surface layer.

Vaughan (1991a, 1991b) used thin sections in her study of Cypriot Base Ring ware. Although she was also trying to characterize the raw materials and possible provenance, one of her goals was to define the ware in general in terms of technical attributes, including those related to fabrication methods. She began with a macroscopic study under a binocular microscope, using sherds with fresh fractures, then moved into thin-section study of a subset of sherds. Thin sections made from fired samples of over 40 Cypriot clays provided comparisons to potential raw materials. Variables recorded in the macroscopic study included those related to raw materials and preparation (nature and distribution of inclusions and voids), as well as those related to primary and secondary forming procedures (evidence of assembly, scraping, paring, degree of compaction, wall width, smoothing and trimming), and finishing techniques (evidence of wet-hand slurry or slip application, and burnish or polish evidence). The subsequent thin-section work included identifying inclusions. The technological features visible in thin section included the presence of very thin slip layers. She also detected orientation of fine particles near the surface that might have resulted from scraping, smoothing, or wet-hand slurry and burnishing activities.

She noted that some surface treatments were difficult to discern in hand specimen, but could be corroborated in thin sections, for example, slips were sometimes very thin (30 μm). A strong orientation of the smallest micas associated with the groundmass, orientation of some inclusions and voids, and a grading of inclusions with the finest located near the exterior surfaces, were considered to be consistent with vigorous secondary forming techniques such as scraping and smoothing. These techniques were applied to pare the vessel walls to very thin widths, and to prepare the surfaces for burnishing or polishing of the slip or slurry finishing of the exterior. The resulting regular alignment of particles on the surface produced a subtle reflective luster that was characteristic of the Metallic Slipped and Red Burnished categories of ware. In thin section these appeared as a graded sediment.

Philpotts and Wilson (1994) studied a Late Woodland vessel from South Windsor, Connecticut, that was tempered with shell fragments that have mostly dissolved, leaving large cavities, elongated quartz from stone-knapping debitage, and grass that burned out, leaving small holes. The elongated quartz and the cavities from the original shell and grass temper were aligned parallel to the vessel walls. This alignment was interpreted as resulting from vessel fabrication, as the clay was either rolled into coils or flattened into sheets for slab construction. They noted that the more the clay was deformed, the greater would be the resulting preferred orientation. Orientations were measured for 1,280 elongate quartz grains and cavities left from temper, using a thin section cut parallel to the vessel rim, and excluding grains near the surfaces as their orientations might have been disturbed during surface decoration. The authors concluded that the clay was stretched during working by a factor of about two, and proposed that such a small degree of stretching is not consistent with the coiling method (which might have stretched the clay by a factor of three). Thus, slab construction was deemed more likely, a hypothesis that was confirmed by x-radiography.

Courty and Roux (1995) worked with two traditional potters in India, one who specialized in coil building and one in wheel throwing, to test how vessels of known fabrication method appeared in technical analyses. In one procedure, clay from a single source and prepared in a standard way was used, and in another, a variety of clays was included. Analysis was done by examination of surface features and of microfabrics using thin-section petrography and scanning electron microscopy (SEM). Wheel-thrown vessels were found to have a dense fabric, random orientation and distribution of the coarse-grain fraction, and slight elongation of the few vesicular pores. The pores were predominantly parallel to the clay walls near the rim, as were asymmetrically shaped grains. In crossed polarized light under medium magnification, the fabric had a speckled birefringence. Under higher magnification, two sets of small birefringent streaks of oriented clay, regularly distributed, could be seen when the microscope stage was rotated. Throwing a vessel rapidly on the wheel produced an irregular discontinuous network of fine fissures, random orientation of coarse grains, and a heterogeneous appearance of the fine mass. Under higher magnification, this fine mass showed random orientation of clay domains and silt-sized inclusions, with a random birefringent fabric. Beating a wheel-thrown object produced an outer region with an increase in density of the fine mass and vesicular pores aligned almost parallel to the clay walls; the internal structure was similar to that of non-beaten wheel-thrown objects.

The authors described the thin-section appearance of coil-built vessels as a dense homogeneous fabric having clearly discernable elongated vesicular pores, with structural discontinuities at regular intervals, corresponding to joins between coils. The discontinuities appeared as a network of fine fissures and elongated pores at an abrupt contact of diagonal direction between zones of dense packing of large aggregates of clay domains and coarse inclusions in random orientation. It was noted that press-molded ceramics differed in thin section from both wheel-thrown and coil-formed vessels, by the common occurrence of large aggregates, abundant fine fissures, and some elongated vesicular pores with subparallel orientation. At higher magnifications, the fine mass was seen to have a regular subparallel orientation of clay domains, silt-sized inclusions, and micas, with a subparallel striated birefringence to the fabric.

Courty and Roux used the information gained by their experimental study to identify fabrication methods of a group of third-millennium archaeological ceramics from the Near East. They concluded that the vessels were initially formed by coiling followed by shaping on a wheel; this information implied that wheel shaping may have developed as an evolutionary step before wheel throwing. Thinning or shaping a coil-built pot requires less rotational kinetic energy than wheel throwing, which corresponds to the type of early rotative devices that have been found in the Near East.

Whitbread (1996) carried out replication experiments to study how forming methods affected preferred orientation in ceramic fabrics. He noted that preferred orientation in elongate inclusions and voids can provide information about manufacturing method. For example, features in coil-built pottery may show horizontal alignment due to horizontal stacking of coils, whereas wheel-thrown vessels may show a slightly raking angle of alignment due to the potter pulling up the clay wall while the pot is spinning on the wheel. However, such analyses are difficult when grains or voids are equidimensional rather than elongate or are present in very small amounts, and, by focusing on these relatively large features, the fine detail needed for distinguishing between different forming processes is absent.

Thus Whitbread developed a method for relatively low-fired ceramics (below 900 °C), focusing on the fact that the birefringent properties of the fired clay matrix often survive to some degree under these low-temperature firing regimes, depending upon clay composition. These birefringent properties highlight regions of preferred orientation. He used crossed polarizers with a gypsum plate to identify optical orientation in the clay matrix and to see what is termed "relic clay mineral domains," or clusters of clay minerals packed in such a way that they appear as an optical unit. These features highlight specific regions of preferred orientation. Using the gypsum plate helps to map these areas within the fabric, because domain structures in most clay minerals will be length slow. Birefringent domains in the same optical orientation as the gypsum plate will appear greenish blue, whereas domains in different orientations will be yellow-pink. So, for example, where a section is almost all yellow-pink or all greenish blue, it indicates strong preferred orientation (Fig. 8.2.4); a mixture of yellow-pink and greenish blue areas indicates little preferred orientation (Fig. 8.2.5). Using a gray-scale conversion to create a binary image, with the yellow areas represented as white and the greenish-blue areas as black, image analysis was performed to look at the proportion of black to white areas as a measure of the development of preferred orientation in a given region. A predominance of either black or white indicated well-developed preferred orientation; in contrast, an image closer to equal in proportion of black and white had poorly developed preferred orientation.

Whitbread prepared test materials using four forming methods (wheel-thrown pots, thumb-built pots, coil-built pots, and slabs). He then cut vertical, horizontal, and tangential sections from each for thin-section preparation, and found that horizontal sections from all the pots had a high degree of preferred orientation. For the wheel-thrown pots, this was parallel with the vessel wall; for the thumb-built pots, strong in the center and strong near vessel surfaces (with the inner wall having poorly developed preferred orientation reflecting the original working of the ball of clay); and for the coil-built pots, parallel to the vessel wall. The slab showed a poorly developed preferred orientation.

Sections taken from the vertical and tangential orientations were more useful. In the vertical sections, wheel-thrown pots showed a tendency for preferred orientation, depending on composition; thumb-built pots had weakly developed preferred orientation except at the wall surface; coil-built pots showed patchy preferred orientation and were similar in appearance to the thumb-built ones; and the slab had well-developed preferred orientation which was sometimes at an angle to the horizontal. In tangential sections the wheel-thrown pot showed strong preferred orientation; the thumb-built pot showed poorly developed preferred orientation in the center of the wall (but might be well developed on the outer part of the wall); the coil-built pot showed very well-developed preferred orientation aligned to the horizontal plane; and the slab had very poorly developed preferred orientation.

Whitbread concluded that this method of analysis is a useful way of testing hypotheses about methods of vessel formation based on visual examination. A thumb-built vessel can be distinguished by poorly developed preferred orientation in both vertical and tangential sections, while a slab will show well-developed preferred orientation in a vertical section but poor preferred orientation in both horizontal and tangential sections. Both a wheel-thrown and a coil-thrown vessel will have well-developed preferred orientation in horizontal and tangential sections, but in vertical sections the wheel-thrown pot will have preferred orientation parallel to the vessel wall while the coil-built one will have poorly developed preferred orientation but with patchy localized areas of preferred orientation if some areas of the original coils were preserved.

Roux and Courty (1998) used thin-section analysis to identify vessel-forming methods for very early ceramics made in Oriental Asia (Mesopotamia, Iran, and the Indus valley) during the fourth–third millennium B.C. In addition to coiling, vessels could be shaped on a wheel, but without being thrown. Four methods were outlined by which vessels could be formed by making different choices in combining coiling and wheel shaping. The authors hypothesized that the choice between the four methods was related to cultural behavior, and understanding those choices could help in understanding why the potter's wheel was adopted in the region by the second half of the third millennium B.C. An experimental study was designed in which all four methods were used to produce vessels and sherds. The material used was an illitic clay with about 30% fine silt-sized micaceous particles, along with about 9% silt-sized calcite particles and about 5% subangular medium-sized quartzitic sands. After firing at ca. 800 °C, these objects made by known methods were studied to identify diagnostic surface features and microfabrics in thin sections. Results were then compared with archaeological materials to demonstrate that a variety of wheel-shaping methods has been used in the past.

Specimens for thin sections were cut perpendicular to wall surfaces. Clay fabrics were correlated to mechanical deformations brought about by the different vessel

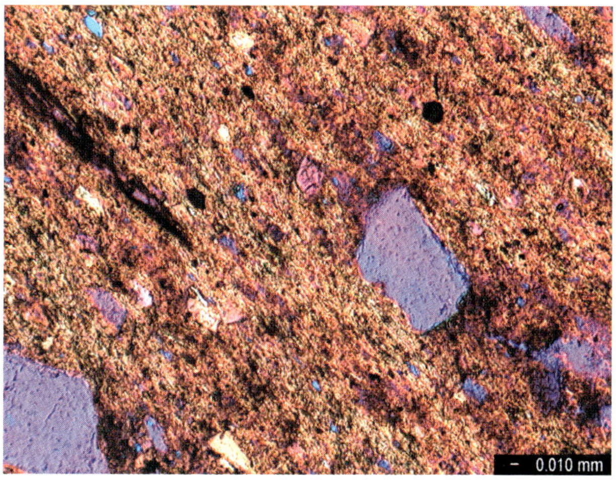

Figure 8.2.4 Ceramic matrix, crossed polarized light with gypsum plate inserted. When the gypsum plate is inserted into the light path under cross polarization, birefringent domains in the same optical orientation as that plate will appear greenish blue, whereas domains in different orientations will be yellow-pink. In this case, the resulting appearance of the clay matrix is almost all yellow-pink with little greenish blue (some of the silt and sand-sized minerals are bluer). This relatively uniform color indicates strong preferred orientation of the clay matrix.

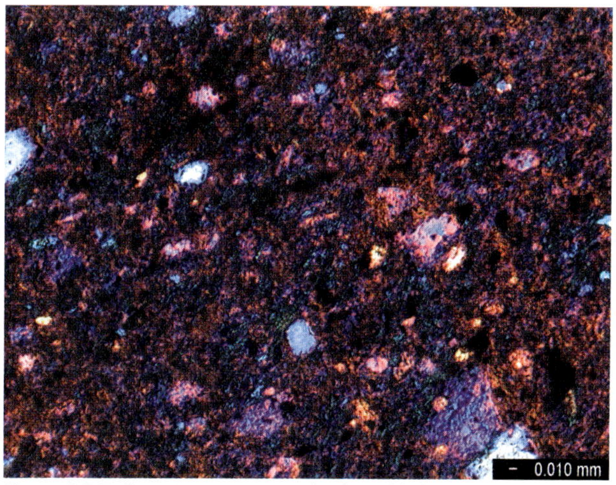

Figure 8.2.5 Ceramic matrix, crossed polarized light with gypsum plate inserted. When the gypsum plate is inserted into the light path under crossed polarization, birefringent domains in the same optical orientation as that plate will appear greenish blue, whereas domains in different orientations will be yellow-pink. In this case, the resulting appearance of the clay matrix is a mixture of greenish-blue and yellow-pink. This non-uniform color pattern indicates a more random rather than preferred orientation of the clay matrix.

formation processes. Their procedure, as in their earlier work, was to start at low magnification to examine air void patterns, coarse particle distribution, and general aspects of the fine-grained matrix, and then move to higher magnification to examine morphology and arrangement of clay domains formed by packing of clay-sized and fine silt-sized particles and their birefringent fabrics under crossed polarized light.

Low-magnification observations showed a similar appearance for all fabrication methods, with a dense, homogenous fabric and random distribution of the coarse inclusions and of the rare, fine vesicles. However, at higher magnification, the clay domain arrangement and the birefringent fabrics showed diagnostic features that could be used to characterize a forming method. For example, one method resulted in two distinct microfabrics, one with a subparallel striated birefringent fabric and another with random orientation showing a speckled birefringence. Another method resulted in a regular arrangement of clay domains with a cross-striated birefringence making a herringbone pattern. A third formation method gave a regularly spaced alteration of two types of birefringence representing a marked contrast between the central part of the coils and the joins between coils. For the fourth formation method, a dense packing of strongly compressed clay domains gave an overall homogeneity to the clay microfabric, with subparallel striated microfabrics at the periphery and the herringbone pattern to the inner part. Strong pressures for joining coils resulted in differential displacement of coarser grains, which were expelled at the periphery of clay domains. Thinned ceramics showed an elongated morphology of clay domains, and those most thinned lost their diagnostic attributes.

The findings were employed in an examination of ceramics from a wide variety of sites. The microfabric study was combined with a study of surface features, to determine whether wheel-forming methods could be identified. The same types of microfeatures were visible, and facilitated identification of vessel-fashioning methods with a high degree of precision. This information, in turn, permitted Roux and Courty to look at technological methods and their possible sociocultural implications.

Vandiver et al. (2002) studied Longshan pottery from Liangchengzhen in southeast Shandong province, China (ca. 2600–1900 B.C.). The most famous Longshan ware is a black pottery, such as black cups, but there were also other objects such as red and white tripod pitchers. Longshan objects are particularly renowned for having very thin walls, elaborate forms, and a shiny black surface. Among a variety of other analytical techniques, the authors included optical microscopy of sherds to reconstruct complex fabrication sequences. The earthenware vessels had nonplastics consisting of quartz, feldspars, micas, and a weathered, high-iron mudstone. As evidence that the cups were thrown on a wheel, they cited the alignment of porosity (horizontal voids) parallel to the surface. The pores were elongated 10–40 times their width. X-radiography and xeroradiography showed a diagonal alignment at an angle of 30–45 degrees from the horizontal throwing ridges and/or base of the vessel. It was noted that this diagonal alignment suggested that the deformation of the ceramic body was done by a combination of centrifugal wheel rotation and raising of the vessel wall by compression and upward movement using hands and/or tools (as in the potting process of throwing).

In Capone's (2006) investigation of pottery from the final periods of glaze ware manufacture in the Rio Grande region already mentioned, she concluded that vessel construction became more expedient over time, with less compaction of the body coils. Evidence for a decreasing amount of compaction during vessel fabrication was a change in the presence of large voids: the average frequency of voids increased from 4% in the earliest period up to 7–10% in the later period.

8.3 Inferring Firing Conditions

Thin-section analyses aimed at identifying probable firing conditions most often focus on the relationship between firing temperature and changes that occur in the mineralogy and/or the microstructure of pottery. Examples of such changes include the breakdown of clay minerals and micas, decomposition of calcite, changes in quartz, and the formation of high-temperature minerals such as spinel, gehlenite, wollastonite, and mullite. Examples of microstructural changes include progressive sintering and vitrification of the clay matrix (which also creates changes in porosity) (Tite 1995). Most studies of these changes have been done with analytical methods such as XRD, infrared spectroscopy, differential thermal analysis, SEM, or physical testing procedures such as hardness tests. However, some of these features are also visible in thin sections of sherds.

Investigating firing conditions can be complicated, because the function and appearance of a fired ceramic depend upon a number of variables including the mix of raw materials, maximum temperature attained, rate of temperature rise, rate of cooling, total duration of firing, time maintained at peak temperature, and atmospheric conditions during firing (oxidizing, reducing, or a combination) (Hodges 1963). A typical research design will involve refiring fragments from a group of sherds in the laboratory, and noting any changes in mineralogy and microstructure occurring at progressively higher temperatures (al-Saa'd and Roussan 1999). The presence or absence of diagnostic minerals such as calcite, mica, and calcium silicates can be used to propose a probable firing temperature. The upper range of firing temperatures for the group of sherds can be estimated, although a precise temperature is usually not possible because of the wide range of variation found in maximum temperature within a single kiln, whether open or closed. If clay sources are known, samples of that clay can be test fired as well and compared with the archaeological pottery. Usually, the choice of an oxidizing or reducing atmosphere is based upon the color of the sherds under study (Maggetti 1982; R.E. Jones 1986; Tite 1995).

Tite (1995) recommended focusing on trying to identify "equivalent firing temperatures" rather than maximum temperature. He defined this term as the temperature which, if maintained for one hour, would produce the observed mineralogy or microstructure. He made this recommendation because the changes observed actually reflect overall heat input during firing, which will be a combination of firing temperature plus firing time. He noted that because open firings usually involve very fast heating and very short times maintained at maximum temperature (compared with kiln firings which involve slower heating rates and longer periods of maximum temperature), equivalent firing temperatures for open firings generally range between 550 and 750 °C versus 750–950 °C for kiln firings. He further recommended combining information on equivalent firing temperature with that on firing atmosphere (based on the color of the pottery), and texture to distinguish between the use of an open firing or kiln firing.

It is not always possible to observe all relevant minerals in thin section (due to the very small size of some of the high-temperature forms), but thin sections can be a useful adjunct to other studies such as those with SEM. For example, original mica and calcite concentrations and appearances can change according to firing conditions (Echallier and Mery 1992). At high temperatures, these components react to form very fine-grained silicates. If the vessel is fired at high temperatures for a long enough period, quartz will recrystallize into the higher temperature forms of silica, tridymite (at 867 °C) and cristobalite (at 1250 °C) (Fig. 8.3.1) if the temperature in the kiln is held that high long enough for them to form (Freestone and Hughes 1989; Henderson 2000). Temperatures above about 1000 °C can create separation along cleavage planes within plagioclase feldspars and biotite (Chandler 2001). Tite (1995) noted that achieving a firing temperature up to about 1000 °C is relatively easy, especially with use of a kiln, but that achieving temperatures above about 1100–1150 °C requires much more technological sophistication (such as better thermal insulation of the kiln and/or higher throughput of hot gases).

Other high-temperature minerals formed when clay is heated to about 1000 °C or above include anorthite (a calcium-rich feldspar), mullite, cordierite, spinel, and wollastonite (a calcium silicate). Which minerals will form will depend upon which clay minerals one starts with, the presence of minerals that can serve as fluxes, the temperature achieved, firing atmosphere, length of firing time, and the length of time allowed during the cooling phase. For example, mullite (Fig. 8.3.2) is formed with non-calcareous clays that reach about 1000 °C in a reducing atmosphere, and hercynite (iron spinel) will also form in a reducing atmosphere at about 850–900 °C (Maggetti 1981; González-García et al. 1990; Broekmans et al. 2006).

Layering in clay color from the surface to the interior of a sherd can sometimes provide information about raw materials and firing conditions. Most clays contain some iron-oxide minerals; if a clay source is naturally oxidized, it will stay red or brown when fired in oxidizing conditions (colored by pigmenting ferric oxide), and will turn black or gray under reducing conditions (due to the formation of iron spinels). However, often a red or tan center will remain while the outside will be black or gray. Organic-rich clay deposits often have naturally reduced iron oxides, and a gray or black color. When fired in an oxidizing atmosphere, the clay will turn red or brown, but sometimes a gray or black core will remain and only the outer edges will be red or brown. The presence or absence of these layering effects is also influenced by length of firing time, type of fuel, placement in the furnace, positioning of the vessel, amount of air ingress, and presence of a protective glaze (Letsch and Noll 1983; Velde and Druc 1999).

Organic material, when present, can also provide evidence about firing conditions. In oxidizing conditions, organic matter will partially or completely burn out, leaving only a pore behind (although it may leave a darkened ring of reduced iron oxides around the borders of the pore) (Fig. 8.3.3). In reducing conditions (with a low oxygen supply) organic matter will char in place and may burn out only slightly or not at all (Fig. 8.3.4). Tite (1995) noted that open-fired (versus kiln-fired) pottery is rarely fully oxidized because of contact with smoky and sooty fuel, as well as

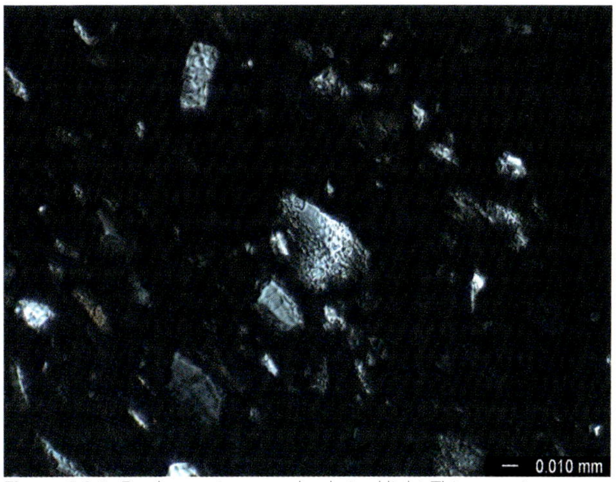

Figure 8.3.1 Fired ceramic, crossed polarized light. This ceramic (Sèvres soft-paste porcelain) was fired at a high enough temperature for a sufficient period of time to initiate breakdown of many of the quartz grains, such as the one in the center of the field of view, and recrystallization into minute crystals and aggregates of trydimite and cristobalite. These appear as tiny wedges or tile shapes over portions of the quartz grains.

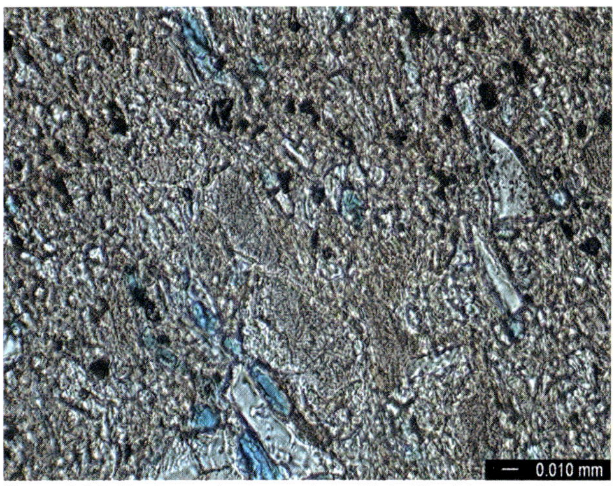

Figure 8.3.2 Fired ceramic, plane polarized light. The firing conditions of this ceramic (Sèvres soft-paste porcelain) were conducive to formation of mullite (the fine needle-like crystals crisscrossing the matrix in random orientations, as in the center of the field of view).

insufficient firing time for the organic material to completely burn out. In contrast, kiln firing involves separation of fuel and pottery, so organic material can completely burn out and fully oxidized pottery can be produced if the kiln is kept unsealed.

In the presence of oxygen, the organic material is first converted to carbon, which will burn out if conditions remain stable for a long enough period of time. If kiln conditions lead to slow burnout of the carbon, residual carbon will diffuse throughout the ceramic body; under conditions of rapid burnout there may be a sharp line of carbon produced, giving the visual appearance of a slip (Hodges 1963).

Dissolution of calcium carbonate occurs at temperatures of 600–800 °C, with release of carbon dioxide, leaving behind calcium oxide (lime). Therefore, the presence of primary calcium carbonate (such as calcite that has not undergone recrystallization) is one indicator of low firing temperature (Fig. 8.3.5). However, primary carbonates can still occur with firing temperatures as high as 1000 °C, depending upon a variety of factors including size of the carbonate crystals, size and mix of other inclusions, composition of the clay matrix, degree of vitrification, and location of the carbonates within the vessel. The free lime that forms as calcium carbonate dissolves can react with clays such as illite to form higher temperature minerals (plagioclase, gehlenite, and diopside) starting at about 700–800 °C (Peters and Iberg 1978; Letsch and Noll 1983; Vaughan 1995).

Another indication of firing temperature is the presence and degree of birefringence of the clay matrix. If a temperature of 800–850 °C is attained, some of the clay minerals will begin to break down to form spinel and glass, giving the matrix an isotropic appearance under crossed polarized light (see Fig. 8.1.6) (MacKenna 1978). If the ceramic was held at this temperature only briefly, only the surface will show this alteration while the inner areas remain unaltered (giving the appearance of a slip or surface coating) (Hodges 1963; Vaughan 1995). For example, Vaughan (1990) concluded that most of the thin sections from Early Cycladic wares of Akrotiri, Thera, represented pottery fired to temperatures no higher than 850 °C because in most cases the clay matrix remained optically active, or with an anisotropic appearance (see Figs. 8.1.8 and 8.1.9).

In an examination by Buxeda I Garrigós et al. (2001) of thin sections of ceramics from a Late Bronze Age kiln at Kommos, Crete, six major fabric groups were identified. For some sherds, the matrix was optically slightly active to inactive, indicating a high firing temperature for those objects. For others, the micromass of the matrix was highly optically active, showing a generally lower firing temperature.

A combination of color and birefringence of the matrix can provide information about both the atmosphere and the extent of firing. As the firing process initiates removal of volatiles (water, organics, and carbonaceous material) and decomposition of some components (carbonates, sulfides, and sulfates) it also initiates development of color by the oxidation of ferrous compounds to ferric oxides. However, iron compounds cannot be oxidized fully until the carbonaceous components have been given off as carbon dioxide. As a result, the appearance of a birefringent, evenly colored

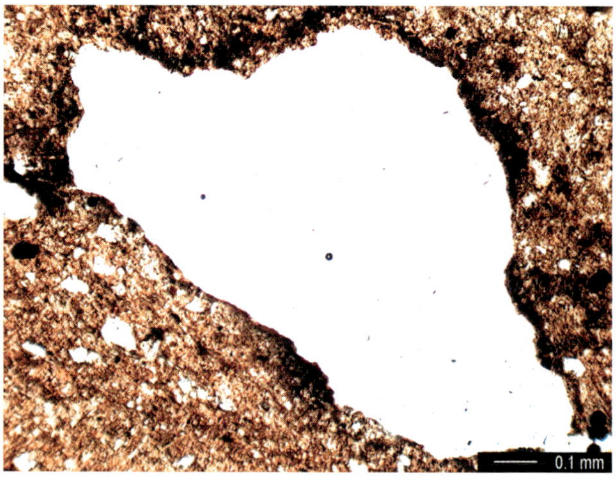

Figure 8.3.3 Fired ceramic, plane polarized light. This prehistoric sherd from southern Delaware exhibits pores with a darkened ring of reduced iron oxides around the borders, which formed as organic matter burned out during oxidizing firing conditions.

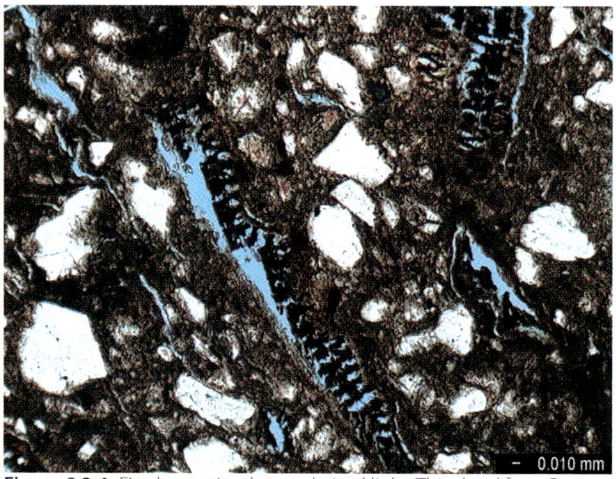

Figure 8.3.4 Fired ceramic, plane polarized light. This sherd from Ban Chiang, Thailand, was fired in reducing conditions so that the abundant organic matter charred in place.

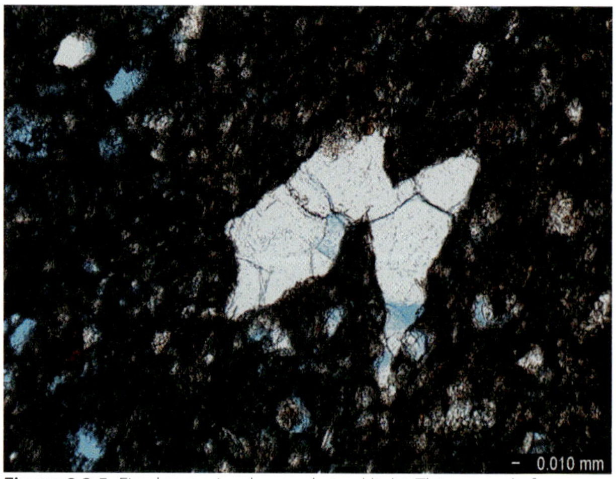

Figure 8.3.5 Fired ceramic, plane polarized light. This example from Iznith was fired at a low enough temperature to retain primary calcium carbonate grains. These grains are colorless in plane polarized light, and show traces of the rhombohedral cleavage. The clay is dark brown with small clear quartz grains (smaller than the carbonate grains), bits of reddish-black iron oxides, and small pores (blue).

red or yellow matrix indicates that one or more conditions conducive to complete oxidation was present in the firing process (high temperatures, a gradual rate of temperature rise, prolonged firing, and/or an oxidizing kiln atmosphere). In contrast, a dark, optically inactive matrix indicates that one or more conditions were present during firing that would inhibit the burnout of carbon (low temperatures or a reducing kiln atmosphere) (Krishnan and Shah 2005).

Intermediate conditions resulting in partial but incomplete oxidation (such as a rapid firing to high temperatures that are not held for long) can lead to oxidation of iron oxide and alteration of clay minerals only on the surface of the vessel. A progressive variation in color from the inside to the outside of the ceramic body may indicate a firing environment that lacked a uniform atmosphere, or which contained insufficient oxidizing capacity due to poor exchange of air (Amadori et al. 2002). However, interpreting these data can be complex, as other factors affect the ease with which a ceramic material will achieve complete oxidation (such as the presence of fluxing agents and the density, thickness, and porosity of the clay body) (Krishnan and Shah 2005).

Interpreting firing conditions can be a complex matter. One has to keep in mind the iron content of the clay, ceramic porosity, composition of nonplastics, and variations that can occur in temperature and atmosphere within a single kiln. Thus the safest approach is to refrain from generalizing about firing conditions based upon analysis of one or a few sherds; instead, analysis of a sufficient sample size to account for the full range of variation within a ware/site/region/period of interest should be the goal.

One of the research questions raised by Felts (1942) in his study of ceramics excavated at the site of ancient Troy (see above) was whether or not it would be possible to infer firing conditions by examining and interpreting ceramic thin sections. He estimated the degree of firing qualitatively by observing whether the clay matrix had been partially vitrified, whether calcite had been thermally dissociated, and if there were thermal changes appearing in minerals such as quartz. He also concluded that firing occurred in a reducing atmosphere, producing gray pottery with reduced iron oxides and small amounts of carbon from organic matter. However, frequent patches of oxidized materials showed that the firing conditions were not well controlled. Very few sherds reached temperatures high enough to cause calcite to disintegrate.

In studies of ceramics from Tepe Yahya, Kamilli and Lamberg-Karlovsky (1979) employed thin-section petrography to compare the common locally made red and buff wares with the less common, probably imported, gray wares. They concluded that the common wares were produced with unsophisticated firing techniques; an unfused paste matrix suggested a firing temperature of 600 °C or less. In contrast, an imported "Ubaid-style" sherd appeared to have been fired at a temperature above 900 °C because the groundmass was completely fused and any fine mica or clay minerals had been destroyed by firing. Traces of a fine-grained green silicate (probably pyroxene or olivine) might have been a result of high-temperature matrix fusion. The fused, glassy paint texture also suggested a very high firing temperature.

Maggetti (1982) studied the mineralogy of Roman pottery from four workshops at Lousonna (Switzerland). Calcite-rich clays were used to make a calcareous *terra sigillata* ware at one of the workshops. Minerals were classified as primary (relic minerals from the original clay, not destroyed during firing) such as quartz, firing minerals (formed during firing) such as diopside, and secondary minerals (formed after firing, mainly during burial) such as precipitated calcite. A marly clay at the site was used for firing experiments, and led to estimates of firing temperatures for the pottery of 850–1050 °C. Minerals that could form during firing of the high-lime clays included plagioclase feldspar (bytownite), diopside, gehlenite, and hematite in an oxidizing kiln and diopside, plagioclase feldspar, and spinel in a reducing kiln. The low-lime clays could form plagioclase feldspar, sanidine, spinel, and hematite in an oxidizing kiln and plagioclase, sanidine, and spinel in a reducing one.

Palmiter and Johnson (1988) reviewed a variety of methods used for estimation of firing temperature of ancient ceramics, and ranked them in order of least effective to most successful as follows:

- colorimetric analysis
- measurement of apparent porosity and hardness
- measurement of elastic properties
- measurement of electronic and magnetic properties
- Mossbauer and electron spin resonance spectroscopy
- measurement of thermoluminescence sensitivity
- use of SEM to assess internal morphology
- mineralogical analysis through thin-section petrography
- differential thermal analysis.

They noted that thin-section analysis is relatively inexpensive so can be used on a large number of specimens. In addition to identifying the minerals present, one can assess their size, shape, orientation, and interrelationships; these characteristics can be compared with a mineralogical temperature scale, developed by observing temperatures at which minerals transform and decompose in the ceramic body. However, since the thermal stability of a mineral is a function of the atmosphere, the amount of molten phase present, and the concentration of minerals in the ceramic body, simple mineralogical temperature scales need to be used with caution.

De Andres and Balcazar (1989) studied firing temperatures and atmospheres of ancient (Early Bronze Age) ceramics from the shelters of Peña Corva in Spain, using a variety of analytical methods. In addition to studying the original samples, some samples were refired at temperatures of 600, 700, and 850 °C and these heat-treated specimens were studied by the same analytical methods. Quartz grains were generally rounded, indicating the presence of alluvial sand which is available near the archaeological site; fresh rhombohedric, angular calcite fragments of sharp outline were seen in most of the samples, indicating an addition of crushed calcite by the potters. In most refired samples, calcite disassociated at 700 °C (and at 850 °C in one sample), no alkaline feldspars were observed after 700 °C, and plagioclase feldspar, infrequent to begin with, showed a slight increase following the heat treatment. No minerals that tend to appear as a result of mineralogical transformations of ceramics at firing temperatures above 750 °C were seen (e.g. gehlenite, wollastonite, diopside, and spinels), suggesting an upper limit of firing temperatures used in the study. The presence of large calcite fragments excluded firing temperatures above 700–750 °C.

Freestone and Hughes (1989) found two types of ceramic matrix in their study of pottery (made with calcareous clays) from the Late Assyrian site of Qasrij Cliff and the post-Assyrian site of Khirbet Qasrij. One type contained relic clay minerals and the other did not. They noted that illitic clay minerals remain in diminishing amounts in calcareous clays at firing temperatures up to about 900 °C but disappear by about 950 °C. Thus they concluded that the first ceramic fabric type probably represents firing temperatures up to 900 °C while the second type represents higher firing temperatures.

Kolb (1989) studied "Hard Buff Limestone Tempered Ware" excavated at prehistoric sites in northern Afghanistan. He noted that calcite, a major temper constituent which appeared in all sherds, was found in its calcined form, and that calcination begins at >620 °C and is completed at 898 °C (as reported by Shepard 1968). At the same time, orthoclase feldspar is stable at <900 °C and melts at 1150 ± 20 °C; quartz grains crack and become altered to a high-temperature form of silica at >579 °C. Based on his observation of these particular minerals, Kolb deduced that the "Hard Buff Limestone Tempered Ware" was probably fired in the range of 850–900 °C, with some specimens attaining slightly higher temperatures (up to 950 °C).

Schneider (1989a) studied thin sections (augmented by XRD analysis) of stonewares from northern Mesopotamia, ca. 2700–2200 B.C. The vessels were made from a very fine clay that naturally contained mineral and rock inclusions of about 30–40% by volume. There was abundant quartz and plagioclase feldspar, the latter partly melted in the more vitrified samples. Small amounts of micas and mullite were also found, along with microcline feldspar in one of the two groups. The other group had a few sherds with cristobalite and hercynite (an iron aluminum oxide member of the spinel group). Since the source of the clay was not known, Schneider studied mineral changes during firing by refiring sherds at different temperatures up to 1250 °C. The mineralogical information was then used to estimate the original firing temperatures of the sherds.

It was noted that micas and potassium feldspars can be stable at temperatures above 1000 °C, depending on duration of firing. At about 850 °C, as quartz begins to decay, illitic calcareous clays can form diopside, gehlenite, and anorthite; in noncalcareous clays mullite will form at about 950 °C. The cristobalite, which as Schneider pointed out is abundant in modern stonewares, was found only in sherds also containing mullite. Hercynite will form under reducing conditions from clays rich in iron and aluminum. Based on this analysis, he was able to deduce that firing temperatures for the wares ranged mainly between 950 and 1050 °C. The partly melted quartz and presence of feldspars and mullite were evidence for the upper part of this range.

Martin-Socas et al. (1989) included thin-section petrography in work that used mineralogical composition to evaluate firing temperatures of Purchena pottery from Almeria, Spain. Primary minerals in the clay included quartz, feldspars, micas, and calcite. The authors concluded that firing occurred at temperatures below 800 °C, and probably below 700 °C. As a result, calcium carbonate appeared as relatively large crystals, and there was no apparent vitrification of the matrix. They pointed out that archaeologists often link firing regimes (oxidizing or reducing) to the color of the vessel surface, but this is often not valid because color is more strongly related to mineralogical composition in conjunction with firing temperature.

Thin-section petrography was employed by Porat et al. (1991) to estimate the range of firing temperatures found with Egyptian-style ceramics excavated at two Early Bronze Age Canaanite sites, En Besor and Tel Erani, using evidence of mineralogical changes that had occurred in firing to estimate the temperatures reached. Most importantly, the authors needed to know whether decomposition of calcite and the reaction of the products with the clay minerals, which occur between 750 and 850 °C, were causing vitrification within the clay matrix. Another visible change seen was oxidation of green hornblende to deep red oxyhornblende, which occurs at 800–850 °C. Three groups were identified based on materials analysis: Group 1 was made with Nile muds and sands (so imported to the sites), Group 2 was made with the same material but also with a limestone temper (so also imported to the sites), and Group 3 was made of local loess soil, with each site utilizing different sources within the region. It was concluded that Group 1 (storage jars) attained firing temperatures higher than 800 °C because oxyhornblende was present, Group 2 jars had firing temperatures ranging from 850 to 950 °C as indicated by completely decomposed limestone and a highly vitrified, milky matrix, and Group 3 firing temperatures exceeded 800–850 °C. With the En Bensor samples it could be seen that some of the fine carbonate particles present had reacted with the surrounding clay and the matrix was vitrified; the Erani samples showed oxyhornblende. All of these examples, then, showed firing temperatures consistent with Egyptian techniques, rather than the firing temperatures of common Canaanite vessels (about 750 °C, so about 100 °C lower).

Echallier and Mery (1992) studied pottery from Hili in the United Arab Emirates from the third millennium B.C., made as copies of Iranian wares. Once they had identified a local marl as having a composition very close to the objects of interest, that material was used to conduct a series of firing experiments. Experiments also included other calcareous fabrics, to determine their reaction when fired at incremental temperatures. The thin-section illustrations included changes in carbonates from the marl and shell temper materials, and in green hornblende and olivine grains as firing temperature was gradually raised.

Venkatachalapathy and Ramaswamy (1992) examined thin sections from an archaeological site in Tamil Nadu (Narayanakuppam in Sambhuvarayar district), India, with a wide variety of wares from different historic periods of unknown dates. A representative sample of pottery was

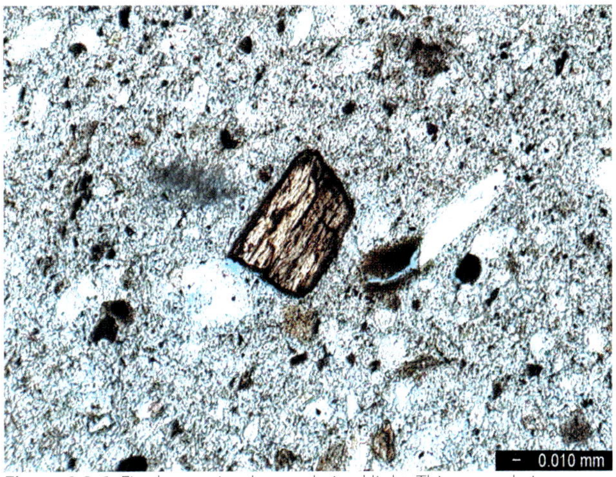

Figure 8.3.6 Fired ceramic, plane polarized light. This example is from a medium-fired jar body from Heijo Palace site, Nara, Japan. The development of ferruginous iron oxide around the rim of this biotite grain may be an effect of firing.

studied. The main nonplastics observed were quartz, sodic oligoclase, orthoclase, clinopyroxene, hornblende, biotite, and magnetite; about 45% of the total volume was clay matrix. The authors suggested that the development of ferruginous rims around the clinopyroxene, hornblende, and biotite (Fig. 8.3.6) was an effect of firing. Similar changes in the properties (color, structure, and optics) of black mica (biotite) due to heating during the firing process had already been documented (Sabine 1958).

In the discussion by Kapur et al. (1993) of fibrous minerals in sixteenth-century Iznik ceramics, they proposed that the presence of uniform patterns (weblike and radial clusters) of pseudowollastonite in the glaze suggested a uniform and slow increase of firing temperature in the kiln. The pseudowollastonite was found as an infilling within rounded pores in the glaze, so was thought to indicate formation of vapor phases in pores leading to crystallization. The uniform orientation differed from wollastonite. It was noted that wollastonite would require a temperature of 1125 °C for formation, whereas pseudowollastonite has a minimum formation temperature of 600 °C. The angularity of quartz in these ceramics also indicated temperatures below 1000 °C.

Bartl et al. (1995) studied "brittle wares," red or red-black pottery wares from the Hellenistic to early Islamic periods in Syria and Palestine. Thin-section petrography showed that these ceramics were made of noncalcareous clays with a high amount of temper (fine-grained or coarse mineral inclusions or lithic fragments). The authors concluded that firing temperatures were relatively low because of the remains of primary calcite from the temper, as well as volcanic inclusions with no indication of melting. Thus the upper limits for firing temperatures would have been no more than 850–900 °C. At least the last stage of firing was oxidizing, resulting in brick-red colors on the exterior of the sherds (due to the use of high iron and low calcium clays). Where black interiors were seen, earlier stages of firing were apparently done with a reducing atmosphere in the kiln.

One of the goals set by Day and Kilikoglou (2001) in their study of thin sections of ceramic samples taken at the Kommos kiln site of Minoan Crete was to collect information on the firing temperature and atmosphere used in the kiln. They noted that the optical activity of a vessel's micromass is a function of the degree of vitrification, which in turn depends on firing temperature. One of the fabric groups examined showed a micromass that varied from being optically highly active to optically slightly active. The analysts interpreted this difference as indicating the use of a range of quite different firing temperatures to fire essentially the same clay body type. Most of the examples in another fabric group contained a fabric that was optically only slightly active, suggesting a generally higher firing temperature than the previous group. Yet another group had an optically active matrix that suggested a relatively low firing temperature. The authors concluded that at this single production site there were significant differences in firing temperatures. These differences did not seem to be random, but may indicate different regimes for different vessel types and whether the vessels were unpainted (resulting in lower firing temperatures) or painted.

Fabbri et al. (2002b) conducted experimental work on the firing of coarse ceramics, of the type found in Roman archaeological excavations in northern Italy, in an attempt to improve their understanding of the behavior of the material during firing, as well as to monitor microstructural and mechanical characteristics. They found that the firing atmosphere (reducing, as in a traditional pit-kiln, or oxidizing, done in an electric laboratory kiln) did not affect the characteristics of samples containing grog inclusions, but did with those containing calcite inclusions. The reducing conditions of the pit firing allowed temperatures to rise as high as 800 °C without calcite decomposition, while the same was not true of an oxidizing environment. They therefore concluded that for the ancient potter, firing in a reducing atmosphere allowed better management of the firing process and functionality of the finished products.

8.4 Deducing Intended Function

A number of archaeological scientists have conducted experimental work aimed at better understanding the relationship between the kinds, amounts, and sizes of temper and the performance characteristics of ceramics (such as thermal shock resistance, impact resistance, and strength). The goal is to facilitate understanding of the functional reasons why potters may have made the technological choices that they did, and to identify the ways in which deliberate product design, engineering, or fabrication choices were made so that particular functional needs could be met (Bronitsky 1986; Bronitsky and Hamer 1986).

Anna Shepard (1939) conducted a famous petrographic study on La Plata valley ceramics of southwestern Colorado and New Mexico. She characterized temper materials in order to examine differences between two pottery types with differing functions: culinary (with igneous rock temper) and non-culinary wares (with grog temper). Ultimately, she concluded that the temper differences were due to exchange from specialized production centers. However, her findings—that functionally distinct ceramic types within a single community or site may be tempered with different materials—were pioneering; where they were produced was not as important as the functions they were engineered to perform (Stoltman 2001).

In addition to looking for different types of temper material (such as sand versus grog), other data sought in identifying functional classes include differing amounts and sizes of temper. In a study of functional variation within a ceramic assemblage (shell-tempered ceramics from Moundville, Alabama), Steponaitis (1984) included some thin-section analyses to determine the size and percentage of shell by volume for jars (cooking vessels) versus bowls/bottles. He found that the jars tended to have coarser and more abundant temper, and concluded that perhaps this helped to maximize resistance to crack propagation that would result from thermal shock. Bronitsky and Hamer's experiments (1986) supported this idea. Steponaitis (1984) and Stoltman (2001) pointed out that shell has clay-like thermal expansion properties and also fractures easily into long plates with large surface area (Fig. 8.4.1), so has greater resistance to crack propagation than many other potential temper materials. Thus, Steponaitis proposed a major technological change occurred in Alabama over the period of 1000 B.C. to A.D. 1500, as the type of temper used in cooking vessels changed from plant fiber to coarse quartz sand, fine quartz sand, grog, and eventually shell.

Other studies have found that different functional classes of ceramics involving non-shell tempers can also exhibit different sizes and amounts of temper. Garrett (1986) analyzed Black Mesa ceramics from northern Arizona, and found that the main utilitarian gray ware type of Pueblo II times had coarser and more abundant sand temper than the three main black-on-white painted types of the same period. Stoltman (1996) found that on average, utilitarian brown wares of Wind Mountain in southwestern New Mexico had more abundant and coarser grit temper than painted vessels during the Mimbres occupation of the site.

Casadio et al. (1991) focused on identification of fire-resistant pottery (from Emilia Romagna in Italy, sixth–fifteenth centuries). Two technological approaches were taken to attain fire-resistant wares suitable for cooking. In most of the samples, the clay bodies contained coarse calcite grains added to a non-carbonatic, iron-rich clay in order to achieve specific technological properties. The authors noted that the thermal conductivity of calcite is at least three times greater than that of clay, so the addition of this temper material enhanced the thermal shock resistance of the ceramic. These earthenware vessels with calcite temper would be eminently suitable for cooking purposes, with good thermal shock resistance and heat transmission properties. Of the 15 sherds studied, most had abundant (30–35%) coarse (on average 0.5–1 mm) inclusions of calcite. The presence of many sharp edges and clean-cut outlines, with no interactions with the surrounding clay matrix, indicated a deliberate addition. Quartz, feldspar, and micas were also present, although in very low proportions. It was suggested that the structural changes these additional minerals caused (increasing the mean pore diameter) also led to enhanced thermal shock resistance. Firing temperatures were probably in the range of 600–700 °C, to prevent decomposition of the calcite. One example showed partial calcite decomposition and the presence of wollastonite, forming from the reaction of calcite and clay; that sample may have had a slightly higher firing temperature.

Another strategy for production of cooking pots was discussed by Bartl et al. (1995) in the study mentioned above. Thin-section petrography (along with elemental analysis by x-ray fluorescence (XRF) spectrometry) was performed on 54 samples. The ceramics were made of noncalcareous clays with a high amount of temper (fine-grained or coarse mineral inclusions or lithic fragments), which the authors noted would make the vessels suitable as cooking pots, their proposed function. The combination of clays low in calcium and the addition of high amounts of mineral temper would reduce cracking from thermal shock. The temper composition (as well as the clay composition) differed significantly from other pottery found in the region—geological characteristics of the rock fragments in the temper suggested a provenance in western Syria or Palestine. More commonly, wares from the Neolithic to Islamic periods in the region were made with local calcareous clays or marls, and were fired at higher temperatures, thus they would have been much less suitable for cooking than the "brittle wares."

It has been observed (Velde and Druc 1999) that even within a single ceramic tradition, thin-walled vessels differ from thick-walled ones in temper technology. For example, with Gallo-Roman common kitchenware pottery from the site of St. Marcel in France, the mica-rich clays had quartz and feldspar temper particles, whether the vessels were thin- or thick-walled. However, a quantitative study showed that the thick-walled pieces had a larger percentage of nonplastics of greater diameter than in the thin-walled pieces.

Heat is transmitted more easily along the sheet structure of a clay mineral than across it. Thus, worked clay, which

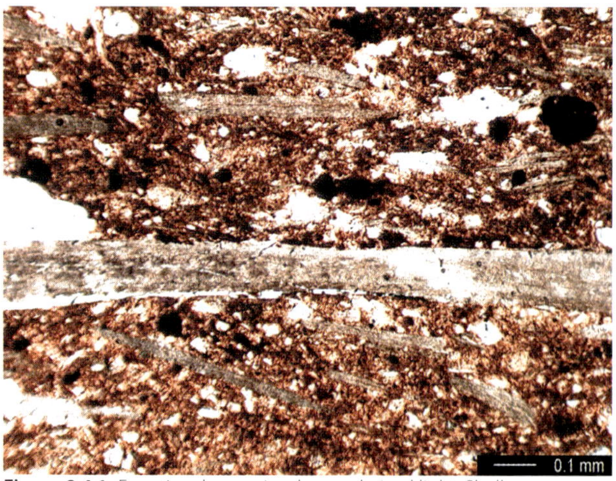

Figure 8.4.1 Functional ceramic, plane polarized light. Shell temper tends to fracture easily into long plates with large surface area, so has greater resistance to crack propagation than many other potential temper materials. This example (a prehistoric sherd from Lancaster County, Pennsylvania) shows a large, long plate of shell in the center of the field of view, surrounded by a number of smaller shell pieces.

tends to gain parallel alignment, does not transfer heat as well as particles in random orientation. Nonplastic components not only disrupt the parallel structure of the clay, but also have different thermal properties. Abundant temper, then, helps to create a more homogeneous heat flow and increases heat retention of the vessel, improving cooking qualities. For a vessel to function well as a cooking pot for faster cooking foods, abundant nonplastics with random orientations might be expected (Fig. 8.4.2). In addition, highly porous ceramics will also have decreased thermal conductivity, so a good cooking pot should also be somewhat dense with few pores. A pot for simmering, with heat well retained, should have little temper, be clay-rich with highly oriented particles, and have high porosity (Velde and Druc 1999).

Tite and Kilikoglou (2002) also examined the issue of pottery technology and ability to withstand thermal shock. They noted that thermal shock resistance is a measure of the ability of the vessel to undergo rapid changes in temperature without cracking. They also pointed out that the main cause of thermal shock is differential expansion or contraction of inner and outer surfaces of the vessel as one surface undergoes rapid thermal change, bringing a temperature gradient through the vessel wall and resulting stresses. The magnitude of the stress is affected by factors such as thermal conductivity of the vessel body, bulk thermal expansion coefficient, thickness of the walls, and shape of the vessel. The bulk thermal expansions of shell- or limestone-tempered fabrics are lower than quartz at temperatures up to 600 °C, resulting in less stress. They also note, as did Steponaitis (1984), that a platy temper such as shell helps to dissipate crack propagation energy, minimizing crack propagation, compared with angular or rounded temper grains. Thus shell is more effective than limestone, quartz, or grog temper in minimizing crack propagation from stresses due to rapid thermal change.

So to deduce, for example, that a sherd could have come from a vessel well suited to function as a cooking pot, control of mineral inclusions must ensure that they have thermal expansions very close to that of the fired clay matrix, that they should not have crystal lattice structure changes that occur within the range of heat used for cooking, or that the mineral inclusion content should be very low and consist of small particle sizes (Rye 1976). However, because of the complexities that can be involved when all factors, including the clay composition, are considered, it is helpful if experiments can be conducted preparing and firing pottery test pieces of known material and firing regimes to measure the actual effects of technological parameters found with specific archaeological vessels.

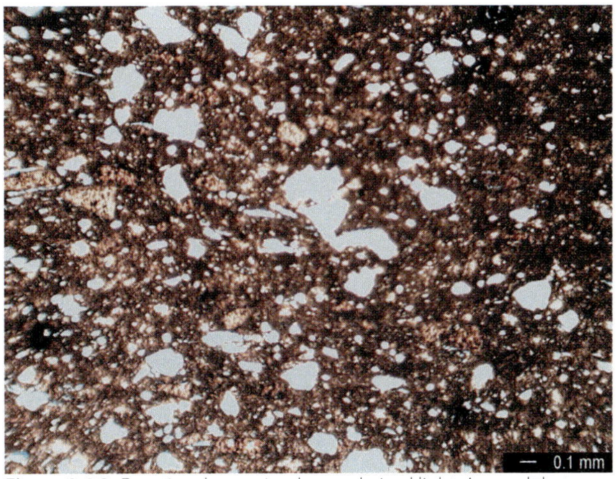

Figure 8.4.2 Functional ceramic, plane polarized light. A vessel that functions well as a cooking pot for faster cooking foods might have abundant temper additives with a random orientation, as with the sand (mainly quartz) and argillaceous rock fragments seen here. In addition, highly porous ceramics will also have decreased thermal conductivity, so a fast cooking pot may also be somewhat dense with few pores, as here (the thin section is mounted in blue epoxy so pores would appear blue).

8.5 Interpreting Voids and Porosity Data

The pores or voids (holes or empty spaces) often present in a ceramic material can result from a variety of causes. The term "pores" is sometimes used to refer specifically to the very fine interstices formed by the packing of materials within the ceramic body, with the term "voids" referring to hollow spaces that are large enough to be seen without magnification (Rye 1981). However, since "pores" and "voids" are all empty, hollow spaces, and can originate by a variety of mechanisms, these terms are often used interchangeably.

Air bubbles can become trapped in the clay during clay processing and fabrication, and more pores often result from the drying and firing processes, although other factors can affect porosity (such as the chemistry and mineralogy of the raw materials). Organic materials such as chopped grass, straw, seeds, or dung tempers often burn out partially or wholly during firing, increasing porosity. Careful selection of the organic material can be used to control pore size, but it is important to keep in mind that porosity can be increased during use, or during burial of the vessel by depositional processes such as secondary calcite formation around pores, dissolution of calcite grains or other soluble minerals within the ceramic, and crystallization of gypsum or various salts. Field or conservation laboratory treatments of ceramics can also affect the porosity (Rye 1976; Maggetti 1982; Freestone and Hughes 1989).

Porosity can provide information about both pottery production methods and about intended function. For example, a decorative porcelain vessel designed for drinking hot tea must be impermeable to water, sturdy enough to last for a long period of time, but does not need to be able to withstand the heat of a cooking fire. A dense, hard, glassy non-porous material is a good choice (Fig. 8.5.1). However, water storage vessels that must keep contents cool (by a pore structure that allows for evaporation), and cooking vessels that must be able to expand and contract without cracking over a cooking fire require a more porous material (Fig. 8.5.2). They may take less time investment for making, firing and decorating, so do not necessarily have to last as long as a glazed decorative porcelain object.

Primary pores are the interstitial spaces between clay minerals where water was driven off and replaced by air during drying. Clay that was not highly worked through kneading and other clay-preparation procedures designed to more closely pack the clay minerals together has a higher amount of primary porosity. Secondary pores appear during firing through shrinkage of the clay substrate as water is released, release of gases (such as carbon dioxide released as calcium carbonate is heated) producing pockets of trapped gas, or from the burning out of organic temper materials in an oxidizing atmosphere. Large nonplastic particles often have a porous rim around them from shrinkage of the surrounding clay matrix (see Fig. 6.2.15). Secondary pores tend to be irregular in shape and somewhat oriented. An overfired fabric can show many elongated pores resulting from shrinkage (Herz and Garrison 1998). Vitrification at higher firing temperatures can also create secondary pores as the clay and silica minerals melt, bubble, and shrink; these pores are often very rounded (Fig. 8.5.3).

Pores can also be characterized by their forms, which can have varying impacts on the properties of the ceramic material. A variety of these pore forms was illustrated by Rice (1987). Her examples included closed pores (sealed rather than open to the exterior surface), channel pores (connecting two pores or a pore and the surface), "blind-alley" pores (extending from the surface without other connections), loop pores (two interconnected ones), pocket pores (having a narrow aperture or neck), and micropores.

Long, linear pores with parallel alignment may indicate shrinkage of the clay as excess water is released during firing. The parallel pores tend not to be connected as a network,

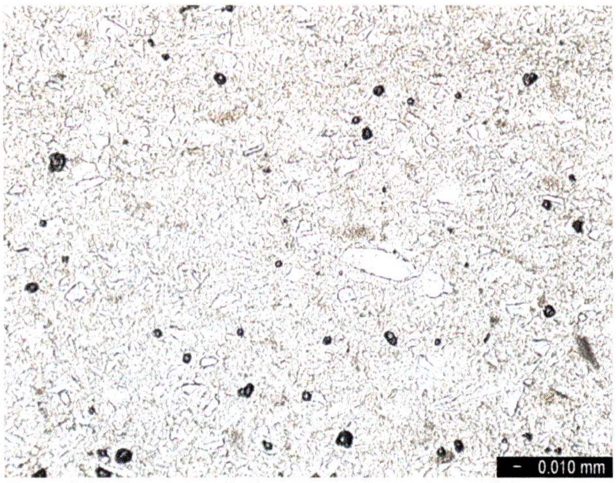

Figure 8.5.1 Porosity studies in ceramics, plane polarized light. Decorative porcelain designed for drinking hot tea must be impermeable to water and sturdy, but does not need to withstand the heat of a cooking fire. A dense, hard, glassy non-porous material as used with this Japanese tea bowl is a good choice. Much of the clay has vitrified, but not to the extent that shrinkage pores have developed. There is some fine-grained quartz and feldspar component.

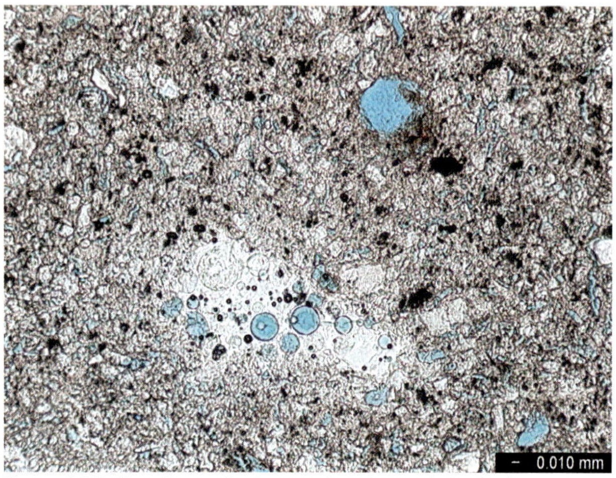

Figure 8.5.3 Porosity studies in ceramics, plane polarized light. Vitrification at higher firing temperatures can create secondary pores as the clay matrix and silica minerals begin to melt, bubble, and shrink; these pores are often very rounded. In this early Chinese example (early Western Zhou, Yazhou), rounded voids (blue areas) have formed on and around melting quartz grains and within the vitrifying clay matrix.

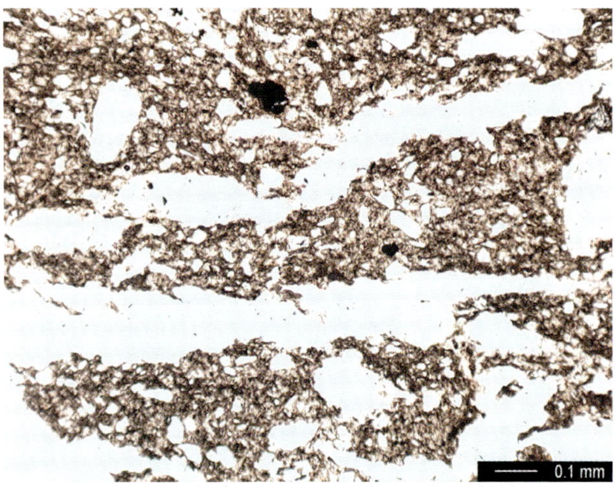

Figure 8.5.2 Porosity studies in ceramics, plane polarized light. In contrast to the tea bowl in Figure 8.5.1, this example (a prehistoric sherd from the Upper Susquehanna valley, Pennsylvania), is extremely porous. Most of the smaller, rounded clear areas are sand grains, while the larger, often elongated clear areas are pore structures.

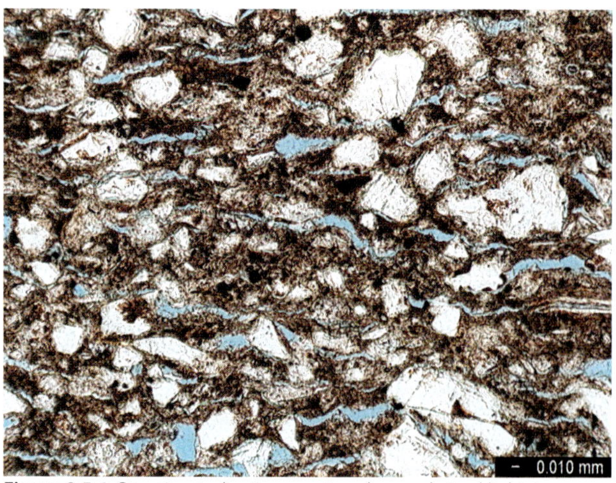

Figure 8.5.4 Porosity studies in ceramics, plane polarized light. Long linear pores with parallel alignment can appear as a result of shrinkage of the clay as excess water is released during firing. These pores tend not to be connected as a network, and are often wavy and may taper at the ends. This early Chinese sherd has many fine, separate wavy elongated pores (blue) that developed throughout the clay matrix of this very sandy ceramic.

are wavy, and may taper out at the ends (Fig. 8.5.4). The presence of long linear pores without tapered ends may also provide clues about working methods. In areas where the clay was more compressed, such as in the lip of a vessel, the clay minerals have greater parallel alignment, accentuated by shrinkage cracking (Velde and Druc 1999) (see Fig. 8.2.1). Longer elongated pores may represent a join void between two separately made sections of the vessel, for example, where rings, coils, or other separate sections were joined and separated slightly during drying and firing stages (join voids) (Gibson and Woods 1997) (see Fig. 8.2.3).

Pores that are deliberately produced to allow for the permeability needed to facilitate evaporation of water through the pot tend to appear as a network of connected pores. However, this structure transfers heat out as well, making for a less functional cooking pot. A better structure for cooking would be isolated pores that conserve heat but allow for expansion and contraction without cracking of the material.

Quantitative measures of ceramic porosity data connected with firing temperatures (for a given clay composition) have also been used to infer firing conditions. This approach requires studies that show how pore size and structure change as the temperature of firing rises. Rye (1976) looked at shell-tempered cooking pots in Oceania and noted that low-temperature ceramics (fired at 600–900 °C), with grain boundaries simply sintered, exhibited mainly open and interconnected pores (resulting in a ware that is permeable to water). With firing at higher temperatures, and the beginnings of vitrification, some of the pores become closed off. At very high temperatures, the open and interconnected pores disappear completely, leaving closed pores that contain gases. He further noted that the relatively high porosity seen in many low-temperature wares can be advantageous for cooking vessels. If the pores are large, they can substitute for temper additives such as calcite that are used to improve thermal shock resistance and prevent crack propagation through the vessel walls. According to Rye, the ideal pore size is between 7 and 9 mm, and the range of advantageous pore sizes somewhere between 5 mm and 1 cm depending upon cooking fire temperatures. If a crack starts, it will be halted when it reaches a large pore, in the same way it would have when it reached a mineral inclusion.

Maggetti (1982) demonstrated how pore size can change with a study of kaolinitic clays, looking at the effects of firing temperatures for clays of known mineralogy and chemistry. He found that the fine-grained clays have much finer pores than coarse-grained clays, and their pore size distribution curves are much steeper. He originally proposed that for a given clay type, an increase in firing temperatures can increase the mean diameter of pores, while also closing micropores, due to sintering of the clay and the crystallization of new minerals. He tested this idea with firing studies of illitic calcareous clays. He had assumed that once the mean pore radii was known for a sherd, segments could be refired at increments, and when the mean pore radii started to increase, it would indicate that the original firing temperature had been exceeded. However, he found that this approach did not work; no reliable original firing temperature estimation could be made by refiring and porosity analysis. Thus, it is useful to remember that although a sherd or a ware can be characterized as to porosity, interpretation of how that porosity came about is quite complex.

Douglass and Schaller (1993) recorded the presence of pore spaces while doing point counting of prehistoric Southwestern ceramics, although pores were not included in the 300-point total for each sherd (used for modal analysis). A quantitative measure of porosity was obtained by dividing the number of observed pore spaces by 300 plus the number of observed pore spaces. It was concluded that the presence of diagenetic minerals lining many of the pores indicated that the true porosity of the samples was close to the reported porosity and thus the observed pores did not result from mechanical removal of temper grains during sample preparation.

In the study by Philpotts and Wilson (1994), previously mentioned, of a Late Woodland vessel from South Windsor, Connecticut (tempered with shell, quartz, and grass), three thin sections were cut, each at a different orientation: one slice was oriented perpendicular to the rim, another parallel to the rim, and the third at an oblique angle to it. The thin sections revealed that 22% by volume of the sherd consisted of pores (cavities) (69% was clay and 9% was a quartz temper). In the section cut perpendicular to the rim, the cavities formed concentric and swirled patterns; in that parallel to the rim, the cavities appeared elongated parallel to the walls of the vessel. The authors attributed the alignment patterns as indicating either the presence of coils or the boundaries of slabs, and thus reflecting the way in which the clay was deformed during fabrication. It was pointed out that although the cavities were elongated in a form that bubbles of air in clay would tend to take as a result of deformation during fabrication, their terminations were irregular and jagged, unlike the rounded ends that would be expected for air bubbles. In addition, it was noted that the walls of the cavities had markings indicating the former presence of shell fragments and pieces of grass, both of which were probably added as a temper to the clay paste. The shells dissolved and the grass was burned off during firing. The shell fragments left very large holes, while the vegetal temper left small holes.

Williams and Jenkins (1997), in a study of Neolithic, Bronze Age, and Roman pottery in northern Wales, found that it was useful to divide voids into those resulting from the process of vessel production and those due to dissolution of original constituents. They attributed irregular longitudinal crevices meandering through the fabrics parallel to the outer surfaces of the vessel (Fig. 8.5.5) to production processes, caused by insufficient wedging of the clay in the manufacturing stage. They also attributed voids surrounding inclusions to production, caused by differential shrinkage of matrix and inclusions in the firing process. Voids arising from dissolution of original constituents they found directly relevant to provenance studies. For example, Neolithic wares from some of the sites of northwestern Wales had a "corky" fabric with high porosity. The shapes of voids often corresponded with those of dissolved sponge spicules, shell fragments, rhomic crushed calcite grains, or long and arching grass blades.

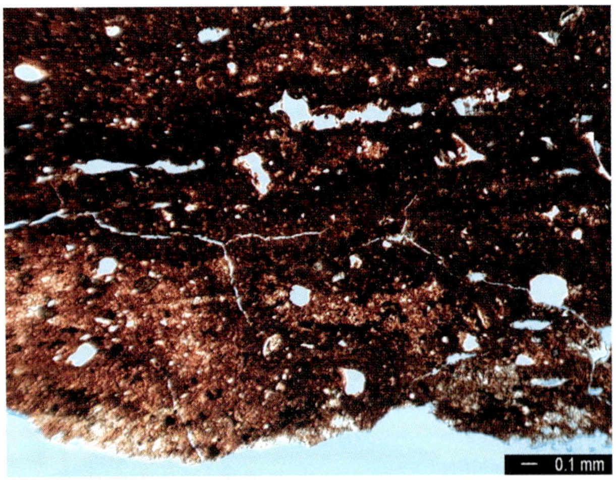

Figure 8.5.5 Porosity studies in ceramics, plane polarized light. This early Japanese sherd (from Heijo Palace, Nara) has, in addition to larger pores, irregular longitudinal crevices meandering through the fabric parallel to the outer surface of the vessel. It has been suggested these might be due to insufficient wedging of the clay.

Newer methods of digital image analysis can provide quantitative data on porosity in thin sections of ceramic materials. Schmitt (1998) used typically prepared thin sections to measure overall porosity in amphorae found in the Lyons area of France. He obtained two images for each field of view: one under plane polarized light (so the holes in the ceramics appeared as light zones with a high gray level) and the other under crossed polarized light (the holes appeared as dark zones with a low gray level). The second image was then subtracted from the first, and for each pixel a new gray level was calculated corresponding to the difference in value of the two images. The ratio of surface occupied by the holes to the total surface provided a measure of total porosity of the ceramic.

Another faster and less complex approach is to prepare the thin sections with a dyed resin (both blue and red are commercially available, or a fluorescent dye can be used if a fluorescent microscope is available). Pores will then be easily visible in plane polarized light, as they will be the color of the dyed epoxy and therefore easily distinguished from quartz. (As an added benefit, the blue dye tends to make quartz and alkali feldspars easier to distinguish from each other in plane polarized light.) Using an image analysis package such as Image-Pro Plus, the dyed pore areas can be quantified rapidly (Reedy 2006a).

Barone et al. (2002) noted that there are five main factors that can affect pore structure in ceramic materials: firing atmosphere, firing temperature, amount of temper in the sherds, initial grain-size distribution of the clay, and clay mineralogy and chemistry. Given these complexities, determining the factors leading to a given pore structure can be difficult. However, differences in pore-size distribution may help in differentiating ceramics from different periods, locations, or workshop groups.

Koh Choo et al. (2004) studied the microstructure of celadon and whiteware sherds excavated at Sori kiln in Korea. They examined pore size, shape, and number within the ceramic body as an indication of degree of vitrification. Experimental refiring of sherds created visible changes in the pore structure of some—pores became rounder and fewer in number, indicating that the vitrification process continued as temperatures rose to 1150 °C. Further vitrification, however, at a refiring temperature of 1200 °C caused the pores to become slightly bloated, an indication of overfiring. The bloating was more obvious at 1225 °C. Similar microstructural changes in the celadon and white ware bodies suggested that both were originally fired at similar temperatures, probably in the same kiln, but since the celadon recipe would have a higher amount of flux, those ceramic materials were appropriately vitrified or even overfired, while the whiteware products from the same kiln appear underfired.

Smith and Trinkley (2006) examined thin sections from 34 sherds of Late Archaic to Early Woodland pottery recovered from the Crescent site in South Carolina. Point counting included over 300 counts per thin section at a point step of 0.1 mm. In addition to identifying minerals and grain sizes, secondary porosity voids containing some carbonized remnants were studied as evidence of the presence and abundance of fiber temper. Point-count categories included paste, quartz, fiber, feldspar, opaque minerals, other minerals, and argillaceous fragments. The fiber temper was most often only visible in the form of pores, which included both elongated and oval to round voids. For the fiber-tempered examples, these voids comprised 35–40% by volume of the thin sections.

The authors found that cross-sections of the stem (represented by oval voids) were more abundant than stem fragments (represented by elongate voids) and had a specific orientation within the sherds. The round voids were concentrated in the reduced core while the elongate voids were found near the interior and exterior surfaces. They suggested that this orientation resulted from the manufacturing process. Outer surfaces were devoid of any fiber, perhaps from floating or smoothing practices intended to bring up fine clay to cover the surface fibers. Some of the pores were partially filled or rimmed with calcite, probably from post-depositional groundwater interaction.

8.6 Studying Decoration and Surface/Clay Body Interfaces

8.6.1 Slips

A slip is comprised of a fluid suspension of finely processed clay in water. It is applied to the ceramic surface either as decoration, to provide a smooth surface texture (often increased by burnishing), to serve as an undercoat for paint or glaze, or to decrease the permeability of the vessel walls (by filling pores with fine material). Various materials such as wine, vinegar, and salt might be added to the clay and water suspension to improve the settling properties (Maish et al. 2006). The slip is usually applied to an unfired but shaped and partially dried ceramic. A leather-hard vessel could be dipped in the slip, or the slip might be poured over the surface of the vessel or even painted on. A fluid suspension can be processed using the same clay as that used to form the body, but often a different raw clay is used for the slip, chosen for aesthetic or functional purposes. The application of a separate slip material differs from what is often called a "self-slip," achieved by fabricating the vessel in such a way as to concentrate micas or other fine particles from the clay body itself onto the surface. Alternatively, the two processes can be combined. Wet smoothing of the clay surface can concentrate finer clay particles at the surface, and they become more intensely colored upon firing in comparison with the inner surface where coarse particles dilute the color (Matson 1970).

A slip—which can be a mixture of clay and nonplastics, and even organic compounds—can be colored by the addition of oxides. Arnold (1972) noted that the clay mineral montmorillonite has excellent film-forming properties, so improves spreading quality and helps prevent flaking of the slip after the vessel has dried. He also pointed out that montmorillonite, kaolinite, and halloysite give a white color, while hematite or goethite can be added to impart a dark red color, and goethite alone can impart a black color. Blues are made by adding ground cobalt mineral pigments. For example, in Germany the cobalt pigment smalt, mixed into a clay slurry with water, was applied to the ceramic surface with a brush (Gaimster 1997a, 1997b). Stoneware from some regions of Germany (sixteenth–seventeenth century) was known for a very thick chocolate brown slip made from an iron-rich clay. It was used on stoneware bodies not fired high enough to hold liquids well, so the thick slip helped to create waterproof ceramics (Gaimster 1997a, 1997b).

The slip is applied while somewhat viscous, so little is absorbed by the clay body (a very thin slip is sometimes called a "wash"). Upon firing, the fine-grained slip will harden and adhere to the ceramic surface, and in thin section it appears as a distinct line on the ceramic surface (Fig. 8.6.1) (Woods 1982; Velde and Druc 1999). In thin section the slip is seen to be a fine-grained clay-rich layer, often with similar petrographic characteristics to the fine fraction of the ceramic fabric itself (Katsarou et al. 2002) (Fig. 8.6.2).

Betancourt (1979) noted that viewing the thin sections in reflected light can highlight the structural nature of a slip. He studied Early Bronze Age pottery from Crete covered with reddish yellow (goethite, lepidocrocite, and iron oxide hydroxide) and black (magnetite and other iron oxides) slips. The slip layers were opaque with transparent inclusions of quartz and calcic plagioclase feldspar. Since the same inclusions were seen in the clay body, with the only difference being in the degree of fineness, Betancourt proposed that a geologically similar clay was used for both slip and clay paste (no tempering material appeared to have been added to the clay fabric). On the exterior it measured 0.02 mm thickness, while the interior slip ranged in thickness from 0.02 to 0.05 mm.

Myer and Betancourt (1981) employed thin-section analysis to study mottled slips used for Vailike Ware, the principal type-ware for Early Minoan IIB in Crete. The ceramic bodies contained plagioclase feldspar, hornblende, biotite, quartz, and metamorphic lithic fragments. Lepidocrocite was also present, possibly as an alteration product. The slip and clay body were similar in composition, differing mainly in fineness. Viewing the thin sections in reflected light aided in discerning application layers of the slips; there was usually a single uniformly thick application of slip. In some areas the surface of the slip appeared pale, with large lenses of black (which was magnetic, indicating it was magnetite) beneath the surface; in others the slip was black, with pale brown (reduced then reoxidized) or red areas (oxidized). The only coloring agents appeared to be iron oxide, and the mottled color of the slip seemed to have been formed through selective reduction of the surface by selectively oxidizing and reducing to produce ferrous and ferric oxides.

A complex decorative scheme is seen in Figure 8.6.3 in plane polarized light, in Figure 8.6.4 in crossed polarized

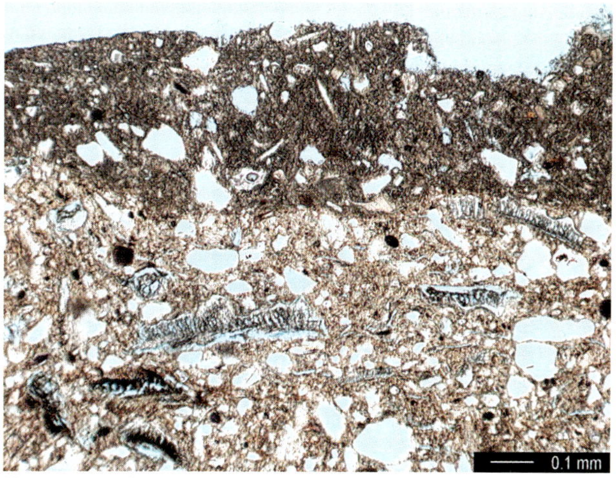

Figure 8.6.1 Slip, plane polarized light. The slip on this sherd from Ban Chiang, Thailand, appears white macroscopically. In thin section the slip is a distinct layer on the ceramic surface. The slip clay appears to be different from the clay used in the body, and, while the slip contains a significant number of nonplastics, these are mostly finer grained than those of the body and are restricted to silt and sand particles (the abundant organic temper in the ceramic body is missing in the slip layer).

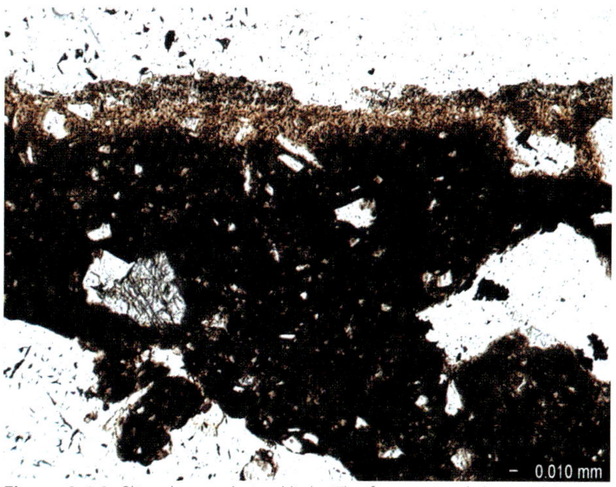

Figure 8.6.2 Slip, plane polarized light. The fine-grained slip on this Iroquoian ceramic from southeastern New York appears in thin section as a distinct line on the ceramic surface. The iron-oxide rich clay of the slip is very similar in appearance to that of the body, but has been processed to be finer and the largest size fraction of sand grains present in the body does not appear in the slip. The body itself is very porous and contains some large nonplastics.

light, and in Figure 8.6.5 in reflected light. The thin section is from an English slip-decorated, lead-glazed earthenware piece dated 1675–1725. The clay is dark brown in plane polarized light, and has a moderate amount of sand temper ranging from small to large grains, along with pores. Particles (some rather large) that are black in both plane polarized and crossed polarized light are hematite, confirmed by using reflected light (where they are steel gray to black with many thinner areas having a reddish tinge). The slip layer that serves as a ground layer between the clay body and the glaze layer is quartz rich, consisting of clay plus quartz (with much more quartz than is seen in the clay body, but also more fine grained). In reflected light, it appears that the clay of the slip is the same as the clay of the body. The slip also contains some hematite, most easily visible in reflected light. Much of that is on the upper edges, indicating it may have entered from glaze. The lead glaze over that slip is mainly glass, with black particles (lead crystals). The outer portion of the glaze has a red iron-oxide colorant, also quite visible in plane polarized light and enhanced in reflected light. Much of it is in solution within the glaze, but some small dendritic crystals of hematite are also visible, especially along the upper edges of the glaze. That region may have cooled more quickly, allowing some of the dissolved colorant to begin to crystallize into dendrites. The glassy glaze is dark in crossed polarized light, while the many quartz grains of the crystalline slip are visible. This quartz-rich slip provides a transition between the porous clay body and the silica glass of the glaze—without it, the difference in co-efficient of expansion would be much greater and the glaze would tend to spall off.

Types of slips can include metallic, burnished, or matte. Rather than painting or applying the slip to a cool vessel before or after firing, an alternative procedure is to sprinkle it on the surface of a hot vessel. This facilitates bonding between the slip and clay body. Descombes (1986) discussed how this procedure is used in Rwanda, where a whitish color is achieved by sprinkling a clay-water mixture onto the surface of hot vessels.

A "self-slip" or "floated surface" is made from the clay body itself. A fine fraction is brought up to the surface and concentrated there, by smoothing the clay under upward pressure. This process leads to a gradient of texture and composition from surface to body, rather than a distinct demarcation, with an absence of coarser particles near the surface. For example, in the stoneware from northern Mesopotamia ca. 2700–2200 B.C. examined by Schneider (1989a) any sherds had a surface coloring that appeared to be a thin (about 1 mm) slip, often a very thin darker slip on a red or brown body or a red slip on a gray body, described frequently as a self-slip or wash. However, under the microscope no deliberately applied surface layer was apparent—there was no structural boundary or change in microstructure, except that the thin surface layer seemed to be a bit more vitrified. Schneider noted that this thin self-slip can occur accidentally if the potter applies dirty and salty water while smoothing the surface by hand. The very fine clay fraction and salt content in the water can add potassium, calcium, magnesium, and sodium, which will increase the fluxing elements on the surface. Alternatively, efflorescence of salt from the wet clay during drying can also enrich fluxing elements on the surface, or there may be a surface enrichment of alkali elements during firing if hot wood ash is deposited on the surface of a vessel in the kiln. However the flux enrichment occurs, it causes the surface to be slightly more vitrified than the interior, so less permeable to oxygen, creating differences in color during oxidation or reoxidation.

Cotkin et al. (1999) used thin-section petrography to examine slips and other inorganic surface materials on 386 utilitarian pottery vessels from 18 Early Woodland through Early Fort Ancient period sites (1150–1300 B.C.) in south-central Ohio. They focused primarily on characterizing unpigmented slips and washes, and documenting their frequency and distribution through time. Previously,

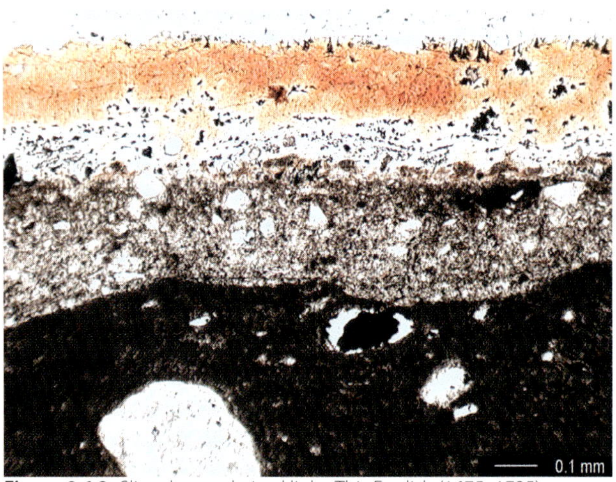

Figure 8.6.3 Slip, plane polarized light. This English (1675–1725) earthenware has a ceramic body with some sand component and chunks of hematite (black in plane polarized light). The body is covered with a slip layer made of clay plus abundant silt- and sand-sized quartz, with some hematite. The slip is covered by a glaze consisting of clear glass, some black lead particles, and hematite colorant. Some of the hematite is in solution in the glass, and some has crystallized in small dendritic structures.

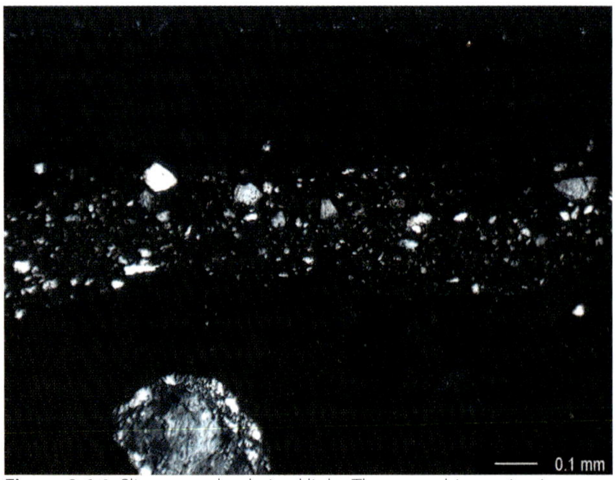

Figure 8.6.4 Slip, crossed polarized light. The same thin section is illustrated here as in Figure 8.6.3. In crossed polarized light, only the silt and sand particles are visible. These are much more abundant in the slip layer than in the clay body, but the particles are also much smaller.

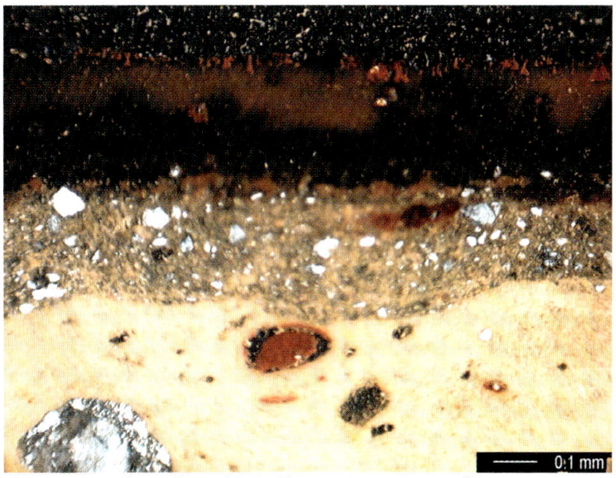

Figure 8.6.5 Slip, reflected light. The same specimen of Figures 8.6.3 and 8.6.4 is seen here in reflected light. In this view, the hematite has a reddish tinge. A large hematite particle is seen in the clay body, some is seen in the slip, mainly on the upper edges (so it may have entered from the glaze), and the glaze contains both smears of hematite in solution and dendritic particles (especially on the upper edge, where it may have cooled quickly enough to begin to crystallize). In reflected light the clay components of the slip and of the body appear to be similar.

uncolored slips and washes had not been identified on pre-Mississippian utilitarian ceramics of the Eastern Woodlands—the use of slips in general had been considered to be a Southeastern rather than Midwestern prehistoric ceramic feature. This research revealed that many of these Midwestern vessels actually have unpigmented pottery slips and washes on their surfaces. Additional electron microprobe work on a small selection of the slips showed that they were probably made from the same raw clay as the ceramic bodies, but with little or no added rock temper and/or with sieving of the slip clay. The authors suggested that slips and washes were used for decoration and to decrease vessel wall permeability.

Cotkin et al. defined the term "slip" to refer to a surface material that exhibited three primary characteristics: (1) discontinuity from the ceramic body; (2) fine texture (mainly clay and silt-sized particles; and (3) having few or no temper particles like those present in the paste of the ceramic body. They defined three additional characteristics that sometimes helped in distinguishing a slip from the ceramic body: a distinctive color, different clay composition, and orientation of particles. The term "wash" was used to mean a very thin slip, achieved by extra fluidity of the applied suspension or by the method of application.

The thin sections were examined in both plane and crossed polarized light, as well as in reflected light. The slips could be distinguished from the clay body in texture and/or color, with the two phases always separated by a visible discontinuity. The fine-grained slips had clay and silt-sized quartz, with some feldspar and mica; most grains less than 0.1 mm were often rounded, and appeared to be natural constituents of the slip material. Coarse lithic temper particles (1–3 mm in size) seen in the ceramic body were generally absent in the slips. The conclusion was that the slips were probably made from the same clay as the ceramic body, but without the addition of temper; both clays were similar in color and in fine-grained mineralogy. It was considered possible that the slip clays were also sieved to remove any particles larger than 1 mm, however, the fine fraction of the ceramic body was similar to the slip so long-term water-settling processes were probably not performed. The similarity in appearance made it difficult to identify the slips macroscopically, so thin-section petrography was very useful in this case. Maximum thickness of the slip was measured at 0.6 mm. A few samples were identified as "washes" instead of slips—they had similar characteristics but were much thinner than slips. Since these examples had flat exterior surfaces, the presence of eroded or deteriorated slips could be ruled out.

Gomez and Doherty (2000) looked at Late Bronze Age White Slip II wares from Cyprus. The slip had a granular texture, different from the usual fine clay nature of most slips. In thin section, this slip exhibited two main components: grains that appeared to have originated as aggregates of clay minerals, and a fibrous chlorite. Chlorite slips were earlier identified on samples of comparable White Slip ware by Courtois and Velde (1980), so this finding was consistent. There were also rare grains of sphene, magnetite, rutile, and albite. Some chlorite pseudomorphs had also formed after pyroxene grains, which still retained their original morphology. In thin section there was clear separation between the body and the slip, so previous suggestions were that the slip had been applied after a biscuit firing. However, Gomez and Doherty pointed out that some of the coarser particles were visibly impressed into the outer surface of the clay body, with the body–slip interface being very irregular. They thus proposed that the slip must have been applied to a moist, plastic surface before the vessels were fired. The granular nature of the slip meant that the potters would have had to press it onto the body to attain enough adhesion to keep it into place until firing, resulting in the pushing into the surface of coarser grains and the irregular interface. No evidence was seen that the coarse grains were dragged along the surface, ruling out wiping operations.

Chandler (2001) analyzed thin sections of ceramics from a number of Early Harappan sites in the northwestern Indus valley, including slipped wares. He observed compression or concentration of mica inclusions near the outer vessel surfaces of some of the thin sections, and suggested that this might be an indication of burnishing that was done after drying but before firing. The burnishing would compact and reorient the fine clay particles. Micaceous slips were visible, but it was not clear if these were applied before or after the burnishing. However, he noted that it is more usual to burnish after slip application to compact and orient the particles and impart a luster.

Some firing residues can give the visual appearance of a slip, but in thin section can be seen to have formed by other processes. For example, if a low-fired ware was held at maximum temperature for just a short period of time, only the surface will be isotropic and altered, whereas the inner areas will remain birefringent and porous. Rapid burn-out of carbon can leave a thin dark layer on the surface. And, if a clay body has a high salt content, the salt may concentrate on the surface during the drying phase and then react with the clay body during firing to produce a crust that superficially may appear to be a slip (Hodges 1963).

8.6.2 Paint and enamel

Ceramic paints are often slips or washes that are deliberately pigmented, and serve primarily decorative purposes. They can be applied to form design elements either before or after firing. In the first case, a paint slurry is applied to the surface of a dried but unfired ceramic, and both are then fired together. The pigments will often fully form during the firing process. In the second case, a finished pigment is applied to the fired ceramic and fixed with the aid of binders (Noll 1981).

Pigments are often colored by finely ground metal oxides (such as hematite, graphite, or manganese), and frequently burnished onto the surface to ensure adhesion; alternatively, they may be applied as very thin aqueous suspensions. An organic vehicle or medium may also be used to improve the spreading quality and help bind the paint to the surface before firing. For example, Shepard (1968) described the use of juice of baked mescal, syrup of yucca fruit, and other plant preparations that were mixed with the ground pigment and water. Variations in hues can be affected by particle size, the combination or mix of metal oxides selected, the presence of a white ground or other background preparation, the amount (or thickness) of the applied pigment, the firing environment (oxidizing or reducing), alterations in a burial environment, and the presence of restoration treatments (Habicht-Mauche et al. 2006; Maish et al. 2006).

Paints applied directly to the clay surface and not covered by a protective glaze layer often include iron oxides (red, brown, or black) and manganese oxides (brown or black). Paints may also be covered by a glaze layer, or sandwiched between two glaze layers, providing protection during and after firing, examples being cobalt oxide (blue), copper oxides (red or green), and lead-tin oxides (white) (Velde and Druc 1999). Inglaze painting is accomplished by dipping a biscuit-fired ceramic into a glaze (such as a tin glaze) and then, onto the dry but unfired glaze, applying the decoration (such as with a blue cobalt pigment) followed by the second firing (Keblow Bernsted 2003). Thin sections can be used to examine the layer structure of paints and enamels, and whether or not they were applied over a thicker slip layer or glaze, or are protected by glaze layers (Figs. 8.6.6 and 8.6.7). Thin sections are also useful for measuring the thickness of the layers, and of any interaction layers.

Enamel paints are essentially colorant-containing glazes (see section 8.6.4 below) of low melting temperature that are applied with a brush onto discrete design areas, rather than to a larger or overall surface area, so they consist of a pigmented glass (Fig. 8.6.8) (Harrison-Hall 1997a). Sometimes an opacifier is added to the enamel to make it opaque rather than transparent (Fig. 8.6.9). A fired porcelain would be painted when cold with the crushed enamel recipe, then refired to melt the enamel. Sometimes the enamel is fused to a glaze ground (overglaze enamel), or is covered by one (underglaze enamel).

Kamilli and Lamberg-Karlovsky (1979) employed a combination of transmitted and reflected light to examine the paint layers on the edges of thin sections from sherds excavated at Tepe Yahya. Transmitted light was used to identify fused or remnant silicate minerals and translucent oxides, and reflected light helped them to characterize the mineralogy of opaque minerals and their textural relationships. Their photomicrographs illustrate that it is possible to distinguish between sintered and glass-rich fused paints. Silicate minerals crystallized from the paint melt are also visible. Electron microprobe analysis of some of the individual grains in the same polished thin sections established the elemental composition of some of the components.

The authors found that the paint used for the common locally made buff and red wares contained mainly thickly applied iron oxide fired to a partly fused, sintered texture of hematite. Bright metallic grains appeared disseminated within a spongy, matte ground. Grains of unfused original quartz were also scattered throughout the paint. Some secondary silicates produced in the firing process (such as anorthite and fayalite) were present but the groundmass retained a red color because of abundant hematite. They

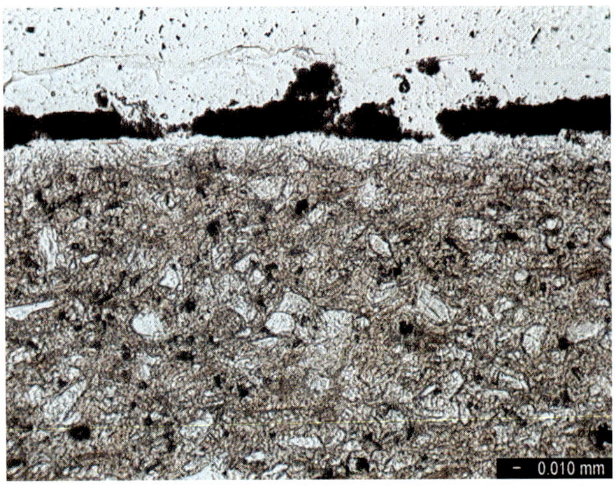

Figure 8.6.6 Paint, plane polarized light. This hard-paste dinnerware has a paint layer that is protected by a thin, clear overglaze. In reflected light the paint has a deep red tinge, indicating it may be hematite.

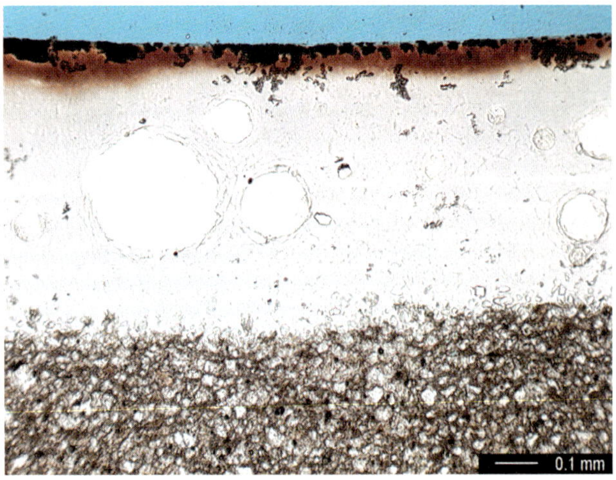

Figure 8.6.7 This enamel paint layer sits atop a very thick clear glaze (with some bubbles). The sample is from a sherd from Jingdezhen, China, mounted in blue epoxy.

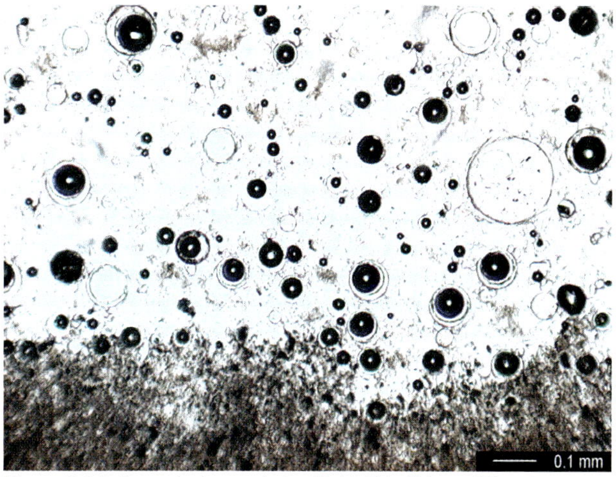

Figure 8.6.8 Enamel, plane polarized light. This Chinese Ming dynasty blue and white porcelain has a very thick enamel paint, or pigmented glass, layer. The clear glass is pigmented in the blue areas with dispersed droplets of cobalt oxide. There are also some clear bubbles visible in plane polarized light.

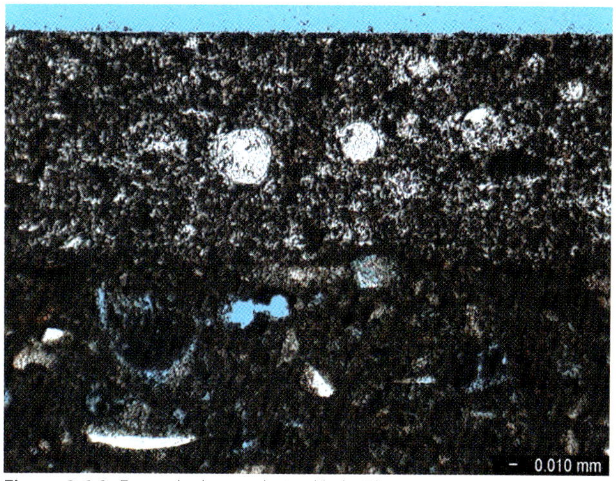

Figure 8.6.9 Enamel, plane polarized light. This Japanese opaque blue and white glazed sherd (1630–1720, mounted in blue epoxy) has an opacifier added to the glass in the blue areas, visible in plane polarized light.

proposed that the original paint mixture was probably an iron mineral such as goethite (as in ochre pigment) with naturally occurring magnesium oxide. Clay minerals and quartz may have been present naturally or added as a firing binder. The sherd was probably fired in an oxidizing atmosphere that ranged between 700 and 900 °C, resulting in the partially fused, sintered state of the paint layers. Some black paints also appeared as a result of oxidized manganese dioxide which masked the red color of the hematite.

In contrast, less abundant gray wares, which were probably brought to the site, used a different paint technology. On two gray wares, a fused black (reduced iron oxide) paint was shown in reflected light to contain glass, bright metallic grains (perhaps ilmenite), gray metallic grains (possibly hematite) and dendrites. The microprobe work showed that in one specimen the glass had a high potassium content, which probably acted as a flux to facilitate the formation of glass and indicated that the firing temperature might not have been as high as had been thought previously. In the other case, where there were only traces of glass and dendrites, potassium was only present in small amounts. Metallic oxides and fusion silicates predominated and created a dull, matte surface. Another example, a red-painted gray ware, appeared in reflected light as a barely sintered red mass. The translucent red material was probably hematite. Electron microprobe work showed an abundance of aluminum and little potassium, which meant that the paint would be inhibited from fusing if fired at a relatively low temperature. The authors concluded that this was an early, unsophisticated (unfused and unreduced) version of the later higher fired examples. Transformation to the later version could have been accomplished by switching the paint mix to include a clay mineral with less aluminum to serve as a binder, or by using less binder.

Kamilli and Steinberg (1985) examined the paint layer on a thin section taken from an "Ubaid-style" sherd from Ur in Mesopotamia. No iron-oxide minerals were found; instead, the paint contained hedenbergite, fayalite, pseudobrookite, and chromite. In thin section, the paint layer appeared to be melted and recrystallized to form an interlocking mass of greenish crystals of hedenbergite and fayalite, and long brown needles of pseudobrookite (an iron- and titanium-bearing mineral). There were also round to subangular, nearly opaque grains (which Kamilli and Steinberg hypothesized were chromite relict from the original paint layer).

Under reflected light, the texture of interlocking crystals was easier to see. The green crystals were now pale gray rosettes, the brown needles of pseudobrookite were white, and the chromite relics were subangular white grains that were zoned (dark gray in the center and white on the rims, probably a result of high firing temperatures). A microprobe analysis confirmed their identifications. The reflected light view also clarified the extent of the paint-to-paste transition zone, which in this case was quite wide.

The mineral assemblage also indicated high firing temperatures (above 900 °C) and reducing conditions. Other samples from the site that were fired to a lower temperature had only a slightly sintered paste composed of ilmenite, chromite, clay minerals, micas, and some quartz, feldspar, and iron oxide. At higher temperatures, all but the chromite melted and recrystallized into the green silicates and pseudobrookite giving a brownish-to-greenish black color that would be difficult to distinguish with the unaided eye from an iron-oxide paint that had been fired to high temperatures under reducing conditions.

Fukang (1987) used thin sections to study the underglaze paint technology on Changsha stonewares made at Tongguan in Hunan province of China (late eight to early tenth century). Changsha vessels are known for their opaque white or yellow-buff glaze with green or brown underglaze or overglaze decoration (painted or splashed on). They are significant because they may represent the first high-fired polychrome-glazed ceramics known, fired at stoneware temperatures too high for lead glazes; they were exported widely, and have been found throughout Asia and the Middle East (Chui-mei and Bronson 1987). Fukang characterized the underglaze methods used for finely painted designs on Changsha ceramics, and found that the methods were quite different from those used later in China. For example, the paint colors are more intense near the surface than at the body–glaze interface, and transparent glaze layers over the colored ones were either absent or very thin.

Green and brown pigments were used for coloring the glaze and painted designs. It was noted that the conventional process for underglaze painting entails grinding the mineral pigments (containing copper, iron, or cobalt) with water to form a fine paste. The paste is then painted or sprayed onto the surface of an unfired ceramic body. Then, a layer of transparent or translucent glaze material, several times thicker than the pigment layer and of uniform thickness, is applied, and the piece is fired. In the kiln, pigments may partly or completely dissolve in the glaze. The pigment painted onto the surface of the clay body or over a slip layer often does not dissolve completely in the glaze, so residual particles of undissolved pigment remain at the body–glaze or slip–glaze interface. Pigment layers are usually very thin, generally less than 1 mm in thickness, in contrast to the thicker glaze layer. The overglaze is colored by dissolved pigments through diffusion and movement of bubbles; under the microscope gradual changes in color strength can be seen from the lower layer (strong) to weak or very weak in the upper part of the glaze.

The Changsha thin sections revealed that in some of the wares there were distinct differences from the conventional underglaze appearance. No residual particles of undissolved glaze were seen at the body–glaze interface (a reflection of the composition of the coloring compound used and its flux). Furthermore, many of the sherds had no transparent glaze layer over the colored areas. And, in some, color strength changed from strong at the glaze surface to very weak at the body–glaze interface; some pieces had only the upper part of the glaze layer colored—the lower part was colorless.

Fukang proposed that the thin section observations could be explained by the presence of a variety of decorative techniques used by Changsha potters. In some cases, the coloring compounds were painted onto a colorless base glaze, and upon firing the pigment would diffuse into the base, giving a gradual color change from strong at the surface to weak at the body–glaze interface. In other cases, a transparent glaze layer may not have been used, so those examples were not actually underglaze wares. The colored layer was much thicker than the base glaze, giving painted areas a raised appearance. In yet other cases, coloring compounds were applied to the body and a very thin layer of uncolored glaze added on top. Because of the thinness, it became diffused with color during firing, so would no longer be discernable. However, the glazed surface over the painted areas tended to be depressed, providing evidence for an underglaze technique. An electrolyte and binder might have been added to the coloring compound, as was done later in China, to improve its workability. These additives made the coloring compound less porous, so less glaze adhered to the colored areas than to the undecorated areas, creating depressions in the glaze surface above the colored areas.

Using thin-section analysis to examine decorative surface treatments of Neolithic ceramics from the island of Yaiali, Greece, Katsarou et al. (2002) deduced that a thick (30 μm) layer of fine-grained dark red pigment was applied after the vessel had dried. As a result, well-defined limits were visible between the pigment layer and the underlying clay layer.

8.6.3 Organic and salt coatings and "glazing"

Organic coatings are a useful and decorative solution to the need to decrease the permeability of vessels to be used for the storage of liquids. Such coatings are sometimes achieved by rubbing the vessel surface with plant leaves or extracts from plant preparations. Oil or other organic material, such as animal fat or butter, may also be applied onto the surface of a vessel to make it impermeable; burnishing or polishing along with this application compresses the surface and aligns clay platelets, and improves impermeability as well as helping to achieve a shiny and lustrous surface (Vandiver et al. 2002). Sometimes these coatings (such as gums and resinous substances) are intended to improve the luster of a slipped or painted surface, or to protect the decorated surface, especially where a fugitive paint is used (Shepard 1968).

Some organic coating materials, dyes, or decorative designs are purely for aesthetic purposes and not functional (Rye 1976). For example, some Late Western Zhou ceramics from China appear to have been coated with lacquer and/or inlays of shell pieces as a decorative device (Valenstein 1989). Pottery may also be deliberately blackened by deposition of carbon from a smoky fire into the pores of the clay, in a process called "smudging." If only the surface is blackened, in thin section, this process will appear as a wavy indistinct line caused by differential penetration of carbon into the pores, versus the distinct line produced by a black slip or a carbon paint. The "smudged" deposited carbon also tends to be an extremely thin layer, with a great deal of variation in measured width from one spot to another, while a black slip or paint is thicker and more evenly applied (Woods 1982; Gibson and Woods 1997).

Gosselain and Livingstone-Smith (1995) noted that post-firing treatments, including the application of organic coatings, or smoking or soaking, were widespread in sub-Saharan Africa. Examples included organic mixtures prepared by boiling, soaking, or infusing parts of at least 26 plants or trees. These substances might be rubbed on the surface of the cooled vessel after firing was complete, or they could be sprinkled onto the surface of the still-hot vessel during firing. The authors found that in addition to being decorative, organic coatings were used for waterproofing a ceramic. Ethnographic studies also showed that symbolic functions were often very important in the selection of compounds to use.

Natural ash-glazing (Harris 1997) takes place in wood-fired kilns with strong air flow, allowing ash from the fuel to move into the firing chamber and settle onto the surface of the pottery. The hot ash is rich in lime and alkalis, and will react with the hot ceramic surface. The ash serves as a flux for the clay surface, forming a glaze that is often patchy and uneven (Fig. 8.6.10), as it depends on the patterns in which the ash happens to fall onto the surface. The natural formation and settling of fuel ash can be facilitated and manipulated by packing spaces between vessels with organic material.

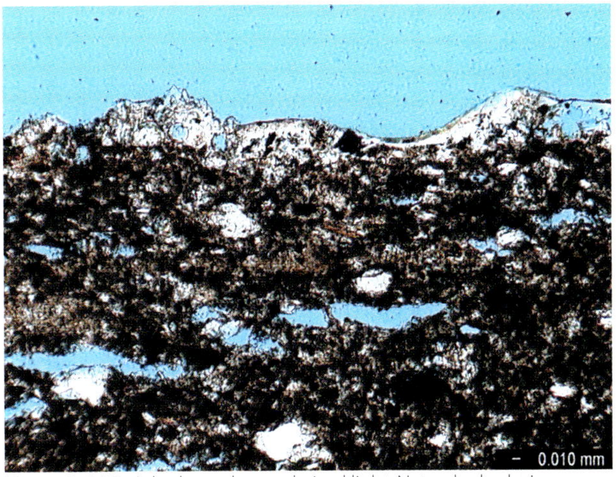

Figure 8.6.10 Ash glaze, plane polarized light. Natural ash glazing allows hot ash from wood fuel to move into the firing chamber and react with the hot ceramic surface to form a thin glaze. Lime and alkalis within the ash serve as flux with the clay surface, and with a high enough temperature a patchy and uneven glaze will form in patterns where the ash happens to fall on the surface.

Deliberate ash glazing (Kerr and Wood 2004) is accomplished by applying a mixture of ash and clay or glaze preparation onto the surface of the vessel before firing. This deliberate glazing has a more uniform thickness and controlled appearance. Prior to their use as a glaze additive, wood ashes were usually processed to separate the more soluble components from the insoluble ones. For example, ashes may be steeped in water and then the top water removed while leaving the ash-rich sediment. After multiple steepings, the most caustic materials, potentially damaging to ceramics, are removed.

White "skins" can form on calcareous clay bodies when there is some salt present in the drying process. Sometimes these "skins" are relatively slight, and may form as a result of some salt being naturally present in the clay, rather than being deliberately added (Freestone and Hughes 1989). In other cases, the salt is deliberately added to the clay in order to enhance this surface decorative effect upon drying (Rye 1976).

Salt glazing during the firing, rather than the drying, process is another decorative technique often used for stoneware, and was especially popular in historic America, Europe, and Japan. It requires a special kiln design with openings along the top. At the height of firing, salt (usually sodium chloride) is shoveled into the openings; the sodium chloride vaporizes in the heat of the kiln, forming sodium oxide, which reacts with the clay surface to form a thin glassy glaze. A byproduct is hydrochloric acid, which vents out of the kiln in great white acid clouds. Salt glazing is a good way of achieving an overall glaze layer, without having to paint or dip each individual piece. This process also produces an interesting surface texture, with the appearance of an orange peel. The textured surface is partially due to the fact that the sodium oxide forms a glaze of uneven thickness, but also because some of the hydrochloric acid attacks the clay surface and the glaze as it forms (without good ventilation major damage can be caused to the ceramics) (Gaimster 1997a, 1997b). In thin section, under plane polarized light, the unevenness of the deposition of the glaze, with peaks and missing areas, and surface imperfections created by the hydrochloric acid vapor in the kiln environment are all visible (Fig. 8.6.11). Salt glazes typically vary in thickness from 20 to 100 μm (Freestone and Tite 1997).

Salt glazing in Europe first appeared in the Cologne area of Germany in the fifteenth century (Lowenstein 1985). The salt glaze was formed on top of gray and blue stoneware bodies, used mainly for storage and drinking functions. The earliest examples have thin and matte glazes, probably a result of dipping the clay objects into saltwater or salt slurries before firing. Later, ordinary cooking salt was shoveled into the furnace during firing, giving a heavy, tough, glossy glaze. Sometimes a pigment was mixed with the salt, the fine powder coating the ceramics as the vaporized salt formed a glaze surface, creating a colored glaze (Gaimster 1997a) (Fig. 8.6.12).

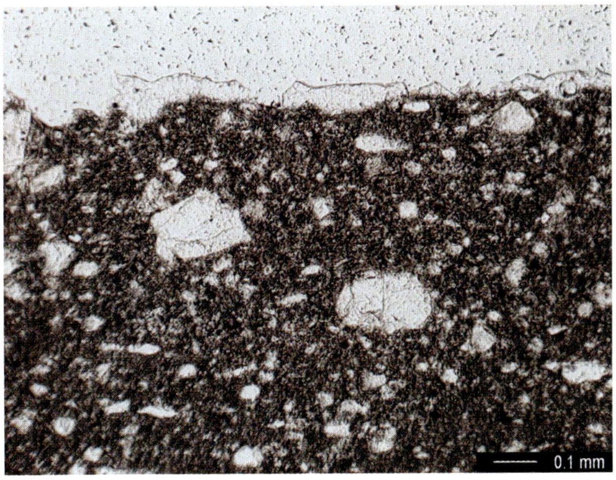

Figure 8.6.11 Salt glazing, plane polarized light. This European (ca. 1800) salt-glazed stoneware exhibits a typical uneven deposition of glaze and a pitting of the surface of the ceramic caused by hydrochloric acid in the kiln environment.

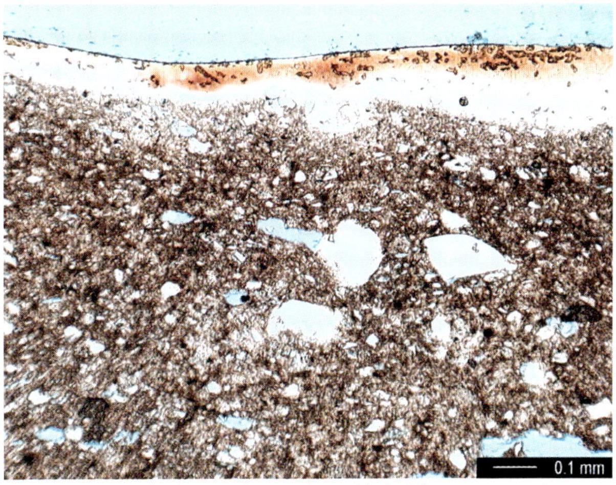

Figure 8.6.12 Salt glazing, plane polarized light. This seventeenth-century German salt-glazed stoneware had pigment mixed with the salt. The fine pigment powder then coated the ceramic as the vaporized salt formed a glaze surface, creating a colored glaze. As is typical of salt-glazed wares, the glaze is of uneven thickness and the ceramic surface is pitted by hydrochloric acid in the kiln.

8.6.4 Glazes

Glaze is made by mixing powdered glass (mostly silica with some alkali flux, with or without metal oxide colorants), a binder (such as animal glue) and a wetting agent (such as water) so the glass powder is suspended in liquid form. It is then usually applied over the entire ceramic surface by dipping or pouring. If fired under high enough temperatures to melt the material and form a glass, it will fuse onto the clay body and provide a long-lasting, impermeable decorative surface. The clay body as well as firing temperature, atmosphere, and stability of the kiln can greatly affect the appearance of a glaze color, thickness, uniformity of coverage, and stability of adhesion.

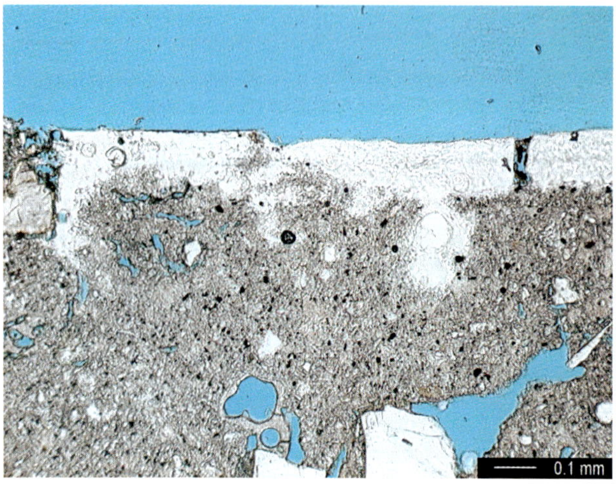

Figure 8.6.13 Glaze, plane polarized light. In the Shang-Zhou period in China (1600–220 B.C.), the transition to porcelain occurred and true glazes appeared. They tended to be thin (this example, Early Western Zhou, averages 115 μm) with small cracks, transparent but with small bubbles, and poor bonding.

Jiazhi (1984) reviewed some of the problems that ancient potters had to overcome in order to successfully produce a vessel with a transparent, glossy, and impermeable glaze. He illustrated a thin section from a Neolithic Jiangsu painted vessel. The pigmented coating contained about 10% flux, but since the firing temperature was limited to about 1000 °C, only a small amount of glassy phase was formed. There were large amounts of mica and quartz visible in the decorative coating, similar to the composition of the ceramic body except for their smaller particle size. A later, pre-Shang Jianshan ware showed evolution closer to a true glaze: there were higher amounts of glassy phase, but also many bubbles, quartz remnants, iron precipitates, and small particulates, creating a rough and permeable glaze. In the Shang-Zhou period (1600–220 B.C.) the transition to porcelain occurred, and true glazes appeared. They tended to be thin with small cracks, transparent but with small bubbles and poor bonding with the ceramic body (Fig. 8.6.13). The mature stage of Chinese glazes was described as beginning in the Han dynasty (206 B.C.). His thin sections of some of these glazes illustrate thin glaze layers with fewer quartz remnants or other crystals. Some bubbles appeared in the glaze, but a reaction layer forming at the body–glaze interface—with many calcium feldspar (anorthite) crystals growing from the glaze into the body—gave better bonding between the two layers (Fig. 8.6.14). Beginning in the Song dynasty (A.D. 960) celadon wares were seen, with thicker glazes that incorporated many small bubbles, small crystalline precipitates, and quartz remnants to create specific optical effects (semitransparent, lustrous depth, and jade-like qualities) (Fig. 8.6.15).

A variety of optical effects can be achieved by controlling the amount of crystallization within the glaze (and thus opacity versus translucency), the size and structure of crystals, and development of interaction layers at the

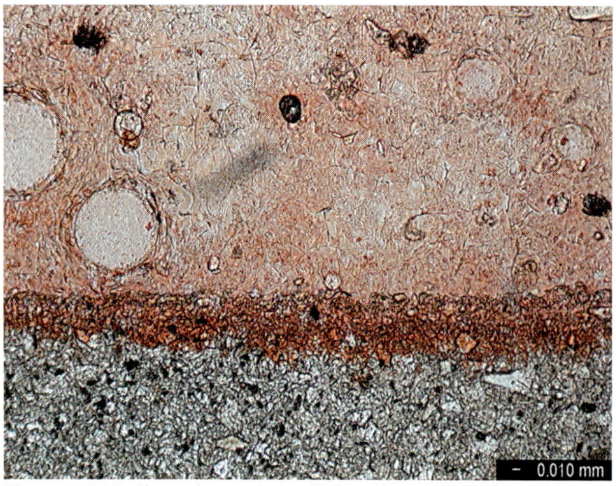

Figure 8.6.14 Glaze, plane polarized light with red plagioclase feldspar stain (barium chloride plus rhodizonate). A reaction layer, primarily consisting of tiny acicular anorthite (calcium-rich end member of plagioclase), has formed at the glaze–body interface in this porcelain. The stain helps to identify the boundaries of the layer. The glaze at the top of the field of view shows a low concentration, the ceramic body in the lower edge shows none, and the interaction layer is especially rich (with a deeper intensity to the stain). An interaction layer such as this improves bonding between the ceramic body and the glaze.

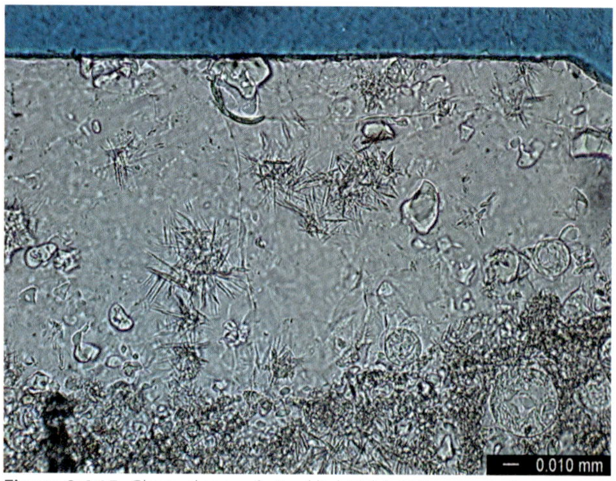

Figure 8.6.15 Glaze, plane polarized light. This Chinese celadon ware (a green jar lid) exhibits a typical thick glaze that incorporates many small bubbles, small crystalline precipitates, and quartz remnants, to help create the optical effects of semitransparency, lustrous depth, and jade-like qualities. The small acicular crystals are anorthite. Small blocky and rounded precipitates are the calcium silicate mineral wollastonite (actually pseudowollastonite). The edge of the interaction layer in the lower part of the field of view is enriched in the small crystalline precipitates.

Figure 8.6.16 Glaze, plane polarized light. This sixteenth-century English earthenware mug has a lead glaze. Due to the poor fit between the sand-rich clay body and the glassy glaze, most of the glaze has spalled off.

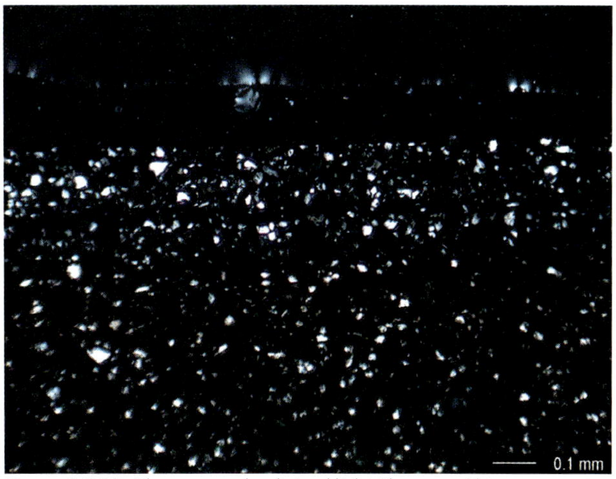

Figure 8.6.19 Glaze, crossed polarized light. The same Chinese export porcelain from Figure 8.6.18 is shown here in crossed polarized light. This example illustrates the "true Chinese porcelain" appearance with many small grains of quartz in both the clay body and in the glaze layer. Firing the glaze at fairly low temperatures under well-controlled conditions is necessary to retain such a large crystalline component. The glaze and ceramic body almost appear to be one material, with only the enamel layer being isotropic.

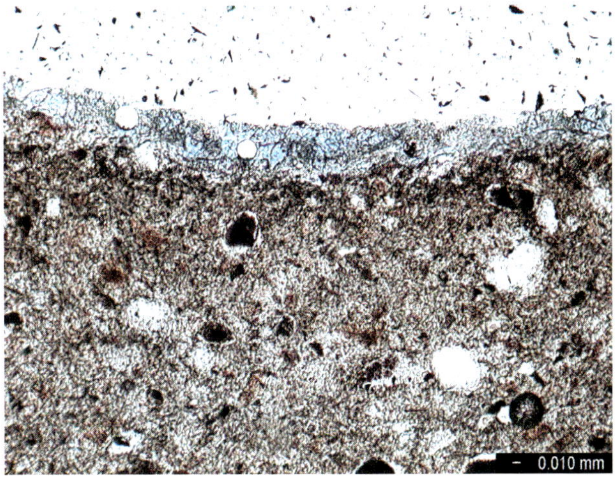

Figure 8.6.17 Glaze, plane polarized light. This English earthenware (1820–40) has a thin blue glaze of uneven thickness, which is deteriorating and spalling off on many places on the surface. The completely glassy glaze is a poor fit for the porous, sandy, iron-oxide-rich ceramic body. Carbonized organic material still in place in some of the pores indicates a mainly reducing firing atmosphere.

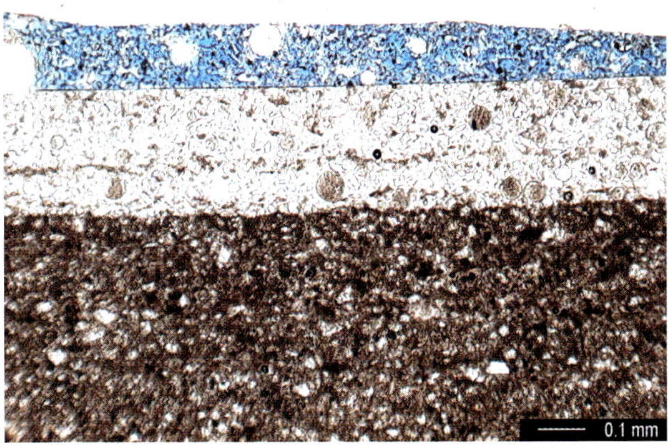

Figure 8.6.18 Glaze, plane polarized light. This Chinese export porcelain (1790–1820) has a very even, well-preserved glaze layer atop a homogeneous clay body. Both ceramic body and glaze contain abundant quartz grains. Blue enamel decoration was added on top of the clear glaze.

ceramic/glaze interface (De Benedetto et al. 2004). Other manipulations can include smoothing the wet clay surface to push the larger temper grains away from the surface, so that a clear glaze will adhere better to the less porous surface, giving a whiter and more homogeneous appearance to the object (Velde and Druc 1999). Although much work with glazes requires analytical methods beyond thin-section petrography, some information about raw materials, firing regimes, and ceramic properties can often be gleaned by thin-section analysis of minerals and textures of the ceramic body; thickness, porosity, and mineralogy of the glaze layer; and characterization of the glaze–body interaction layer.

The thin section of an English earthenware mug with a lead glaze from the sixteenth century, seen in Figure 8.6.16, illustrates one technology. There is a significant sand content in the clay body, yet the glaze is completely glassy. The glaze layer has severely deteriorated and flaked off completely in many parts, indicating a poor fit between the glaze and the ceramic body. In crossed polarized light the glaze is seen to be completely glassy (black, isotropic) with no crystalline component, while many sand grains are visible in the clay body. The difference in coefficient of expansion between the glass and the porous, sand-rich clay body has created difficulties in preservation of the glaze.

Similarly, the English lead- or tin-glazed earthenware, dated to 1820–40, seen in Figure 8.6.17, exhibits a poor fit between ceramic body and glaze. The body consists of a brown clay with many pores, as well as some quartz and opaque iron oxides. The carbonized organic material still remaining in some, but not all, of the pores indicates a mainly reducing atmosphere for firing. The glaze has a cobalt blue colorant, and is quite deteriorated with very uneven layer thickness.

The opposite situation is seen in Figure 8.6.18, illustrating a thin section of Chinese export porcelain made between 1790 and 1820, with a blue overglaze enamel. In plane

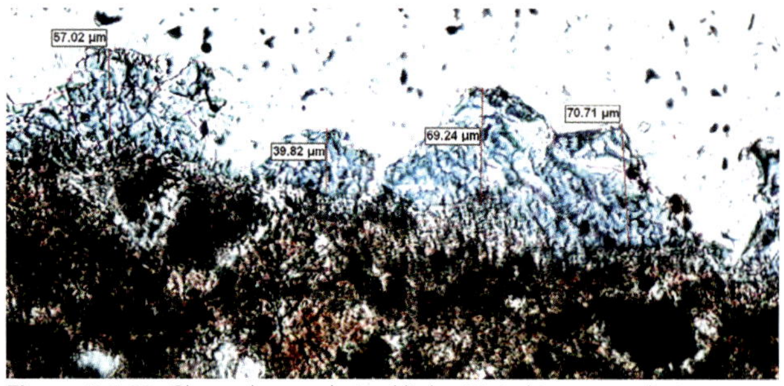

Figure 8.6.20 Glaze, plane polarized light. Digital image analysis was used to measure the thickness of this glaze layer, as it facilitates fast, multiple measurements. Since the glaze has seriously spalled, only the greatest thickness measurement may be of interest (here 70.71 µm); alternatively, the entire layer can be measured and the minimum, maximum, and average thicknesses calculated.

polarized light, a very even, well-preserved clear glaze layer sits atop a homogeneous clay body (containing abundant quartz grains). The cobalt blue decorative area was clearly achieved by the addition of a second glaze or enamel over the base clear glaze. Such decorative areas were usually painted on after the first glaze firing, then the piece was refired. In cross polarized light (Fig. 8.6.19), we see the usual "true Chinese porcelain" appearance of the underlying glaze layer, with the many small grains of quartz in both the clay body and in the primary glaze. The glaze must have been fired at a fairly low temperature and under well-controlled conditions to retain a large crystalline component; the long firing times would ensure that the porous body would absorb much of the glaze. In crossed polarized light the glaze and ceramic body almost appear to be one material.

Song dynasty celadon glazes made near Longquan in China are famous for their jade-like translucency. At one time, this translucency was assumed to result from the presence of residual unmelted quartz particles and bubbles. However, the translucency is now thought to be a result of the development of anorthite needles and wollastonite precipitates (Vandiver and Kingery 1986). Vandiver and Kingery mentioned that inclusions in the glaze, ranging up to 100 µm were visible, and included unmelted quartz, agglomerates of multiple quartz particles, and agglomerates of iron-bearing clay. They noted that no cristobalite reaction rims around quartz particles were visible (as are typically seen around residual quartz particles in some other Chinese glazes such as Jun ware), suggesting that the glazes were not held at high temperatures long enough for cristobalite to form at the quartz–glass interface. However, the authors did observe a large number of anorthite needles and fine spherical precipitates of wollastonite. The quartz particles were large enough that they would not create strong light scattering and translucency. Instead, the precipitation process that formed the acicular anorthite and the wollastonite (actually pseudowollastonite in the form of spherical particles as well as in blocky forms) was present in relatively high amounts in the most translucent jade-like glazes.

Sakarya et al. (1990) found a highly vitrified layer between the glaze and body in twelfth- and thirteenth-century ceramics from Samsat in southeastern Turkey. The glaze itself had a thickness of about 250 µm, and contained large spherical pores probably caused by reaction between the alkaline glaze and the ceramic body. The glassy layer underneath the glaze, overlaying the clay body, was thicker, about 500 µm. This interlayer was more highly vitrified than the clay body, also probably due to reaction with the highly alkaline glazes used on Anatolian ceramics of this period. The authors suggested that the presence and thickness of this layer might account for the high durability of the ceramics.

Vendrell-Saz et al. (2006) studied Islamic and Hispano-Moresque lead glazes from the tenth to the fifteenth century, from glazed ceramics found at several archaeological sites in Spain. Although much of their study involved use of SEM and analysis of chemical composition of the glazes, they also used thin sections to examine and illustrate development of the glaze–ceramic body interface. Inclusions in the glassy glaze itself (which ranged from 60 to 180 µm in thickness) included incompletely reacted quartz grains, trapped bubbles, and crystals that formed at the glaze–clay interface and became detached during firing. Occasionally, grains from temper in the paste were dragged into the glaze and appeared as coarse inclusions. The glaze–paste interface consisted of a layer of varying thickness (5–60 µm) formed by a network of crystals, formed as reaction products during firing. Nucleation and growth of alkali feldspars containing lead occurred throughout this interface region, while calcium pyroxenes developed only around calcium-rich grains. When the glaze was applied to an unfired body, the interface developed more thickly than the body when the glaze was applied to a biscuit-fired body.

Digital image analysis procedures can be used to measure layers that mark technological processes, such as glazes or slips (Reedy 2006a). Measurements can be made at individual points along a layer (Fig. 8.6.20). However, perhaps the most useful approach is automatic measurement of the minimum, maximum, and average width of a glaze layer. Information about the thickness and uniformity of that thickness may be helpful in characterizing a decorative glaze technique.

8.6.5 Transfer print

Figure 8.6.21shows a thin section of an earthenware plate with ink transfer decoration, from the Staffordshire Pottery in England, early nineteenth century. To improve on hand-painting of decoration, primarily to make the decorative process speedier and more economical, manufacturers used a design transferred from an inked copper plate to an intermediary surface (Atterbury 1982). The intermediate was usually a piece of paper that, while still wet, was pressed or rubbed onto the ceramic surface, leaving the inked design. Colorants that can stand the heat of firing were selected. Refiring at a lower temperature would fix the design permanently and designs could be repeated easily, with uniformity, allowing mass production. Transfer prints may be done over a very thin glaze layer, to give a more even and less porous surface for the ink. After the print was applied the vessel was refired at a low temperature to dry out the oils in the printing medium and to harden on the color. The vessel

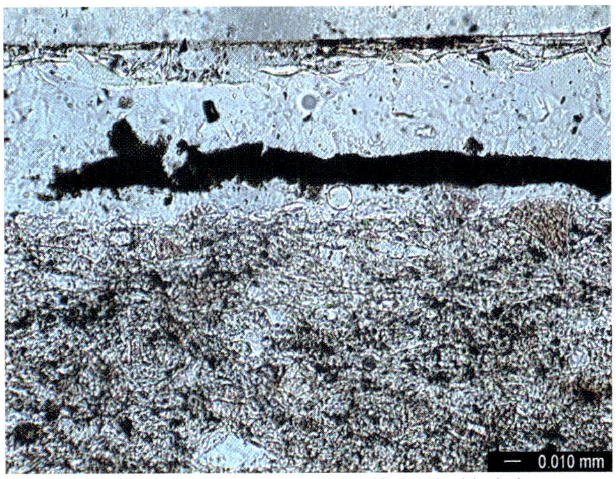

Figure 8.6.21 Transfer print, plane polarized light. The black decorative area here is ink, transferred onto the surface of this Staffordshire plate from the early nineteenth century, using a paper intermediary from an inked copper plate. It was placed over a thin clear glaze that provided a more even and less porous surface for the ink than the ceramic itself. After a low-temperature refiring to dry out the printing medium, the plate was dipped into a thicker clear glaze and refired at a higher temperature.

was then dipped into glaze prior to being fired at higher temperatures, producing the appearance in thin section seen in Figure 8.6.21, with the ink layer on top of a very thin clear layer then covered by a thicker clear glaze layer.

8.6.6 Incised, impressed, and applied decoration

Incised decoration is carried out by dragging a sharp tool over the surface of the clay (usually when it is leather hard). Identification and characterization of incised surface decorations can be enhanced by thin-section analysis (Fig. 8.6.22). For example, Gibson and Woods (1997) illustrated a thin section of an Iron Age sherd from Namibia that has incised lines. They noted that the presence of rounded grooves along the surface, giving a very rough edge, indicated that the clay was in the leather-hard state at the time the incision was made. Qualities such as the depth and evenness of incised decorations can be measured, and, in cases where the presence of slips may obscure the presence of incised work, thin sections can clarify their presence (Barlow 1991).

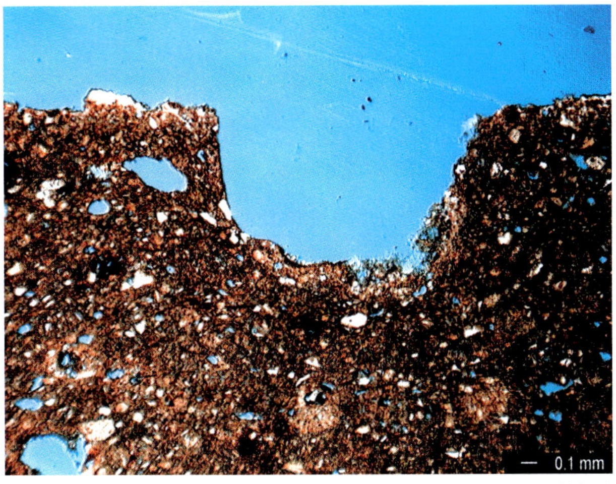

Figure 8.6.22 Incised decoration, plane polarized light. This ribbed black jar from the Song dynasty (Hebiji kiln) includes incised decoration.

Impressed decoration can be achieved with the use of a stamp or similar tool, generally used while the clay is still wet. Oil is often applied as a release agent to ensure that the clay does not stick to the stamp or patterned tool. Close examination of impressed decorations can reveal information about the state of the clay when the potter carried out the work. For example, if the clay was rather dry, then impressions are shallower; if, instead, the clay was rather plastic, then flow structure may be seen on the clay just below the surface. If the clay was too soft, then decorations will be poorly defined and may show tear marks and raised edges where the wet clay stuck to the decorating tool (Woods 1982).

Applied decorations can include features such as lines or blobs. In some cases, as with some Greek vases, applied clay can form a more complex decorative surface; for example, to indicate curls in the hair or on a beard. Early ideas that these were made from thickened black gloss (Noble 1988) has given way to identification of these areas as added clay, pressed as clay dots onto the leather-hard vessel surface then sometimes coated with levigated slip that turned lustrous black during firing (Cohen 2006). Cohen discussed mold-made figural reliefs also seen on some Athenian vases, with adhesion probably aided by a slip coat. Similarly, applied clay decorations, ranging from the very simple to the very elaborate, made by hand or in a mold, and simply pressed onto the clay body and adhered with a slip, are seen on many types of ceramic pottery from all over the world. For example, German stoneware of the thirteenth–nineteenth century often has applied decorations. Relief ornaments were made with the use of molds, and the stoneware clay is reported to be very plastic and especially suitable for use in such molds. Decorative elements such as medallions containing figural images and botanical ornamentation appear (Gaimster 1997a, 1997b).

In thin section, one can identify whether or not the applied features are made of the same material as the rest of the vessel. For example, even if the applied piece is of the same type of clay as the vessel itself, it will often have a different particle orientation and a thin line demarcating the original vessel surface (Gibson and Woods 1997). Furthermore, if the decoration was added when the vessel clay was still plastic, then some distortion of the clay body may be visible where pressure was applied to ensure adhesion. If, instead, the decoration was applied after the vessel was partly dry, then adhesion would be achieved by wetting the surface or applying a thin layer of slip to aid in adhesion (leaving a distinct line along the original surface) (Woods 1982).

Maish et al. (2006) discussed the presence of gilding (mechanically or chemically applying a thin layer of gold) on appliquéd areas of Athenian vases. They noted that identification of examples of this decorative technique is rare, in part because gilding tends not to be well preserved on ceramic surfaces. They reported that gold leaf on Athenian vases is most often found on areas not covered with the typical Athenian black gloss, and is more often seen on raised clay additions used to accentuate details of decorative paintings. Occasionally, small funerary vessels were completely gilded, probably to mimic the appearance of solid gold vessels. The authors conducted experiments to determine if gilding was applied before or after firing, and concluded that post-firing gilding was more likely as otherwise damage to the gold leaf was possible. On most of the objects they studied, no ground layer was found and no adhesives seen. However, modern restoration and preservation attempts can interfere with identification of original materials.

Gold "inlay" work and gilding are also seen on some Chinese ceramics (Qingzheng 2002). In some cases, a design is incised with a sharp tool on the surface of a vessel with a dark background then the lines are filled in with gold pigment. In other wares, gold leaf was cut into the required shapes, glued onto the ceramic body and the vessel was then fired at a low temperature. Sometimes the glazed rim of a vessel was mounted with gold or silver. Overglaze decoration can also be drawn using gold (or silver) powder as the pigment.

8.7 Explaining Technological Choice, Variation, and Change

Beyond identifying the raw materials available to a potter locally or through trading relationships, and the various functions that certain technologies may serve, thin-section petrography of ceramic materials is also used to investigate cultural choices that a potter may have made. For example, in some cases, grog was deliberately chosen as a temper additive, not because of specific technological qualities, but because the reuse of an older pot in the fabric of a newer one may have been a way of demonstrating a link in traditions between one generation and another (Sterner 1989).

Rye (1976), working with cooking pots in Oceania, took the approach that it is useful first to try to identify technological reasons for use of a material or technology. If some materials are found to have logical reasons or functions for their choice, then he could hypothesize that potters were deliberately choosing those materials as a result of long-term cultural learning. He believed that this approach would work better than simply being descriptive regarding the materials and technology found, determining that they were used and then trying to find other related places where they were used. For example, if potters were using only one material to make a wide range of vessels including cooking pots, then it could be expected that over time they would adjust their clay body to give the best thermal shock resistance characteristics. Then, post-firing techniques might be used to enable vessels to serve other functions (such as adding an organic coating to decrease permeability and allow for better storage of liquids).

Howard (1982) studied pottery from the Late Bronze Age site of Dainton in Devon, England. Thin-section analysis of archaeological sherds indicated that there was a variety of distinct fabrics used in pottery production at the site. After collecting samples of nearby clay deposits, she processed and fired them, then prepared thin sections for comparative purposes. She deduced that although the local clay would have been functional for eating vessels and small storage jars, it was not a good cookware material because of its high plasticity and lack of tensile strength. The clay contained only a small amount of subangular quartz grains and rare biotite; this fabric matched the matrix of two of the pottery fabrics found at the site. In one of the two, angular lithic fragments were also present, identified as weathered dolerite. Dolerite is a synonym for diabase, an igneous rock similar to basalt except that it is medium grained (Fig. 8.7.1) and typically ophitic or subophitic (having laths of plagioclase within a matrix of coarse clinopyroxene which largely encloses the laths; also similar to but finer grained than a gabbro). Weathering alteration typically includes altered feldspars and pyroxenes. Howard collected clay and rock samples from several south Devon dolerite outcrops, and conducted some experiments including adding the crushed rock to the clay previously collected. She concluded that the potters were deliberately selecting this rock temper to improve the thermal shock resistance of the locally available clay material.

The other pottery fabric was also made with the local clay, but in this case angular grog was added instead of dolerite. The functional differences between the two temper additives were not identified. A third pottery fabric from the site was characterized by the presence of elongated plates of weathered chlorite schist within a sandy, micaceous matrix containing laths of plagioclase feldspar. The nearest outcrop of chlorite schist was 35 km from the site, and in thin section matched the excavated material. Again, no functional difference between pottery with the schist additive and that with the dolerite or grog additives was found, so reasons other than functional criteria for selecting dolerite versus grog versus schist may exist.

Mainman (1982) sampled 150 sherds from Anglo-Saxon pottery (cooking vessels) found in Canterbury, England. The type of inclusions was identified through normal petro-

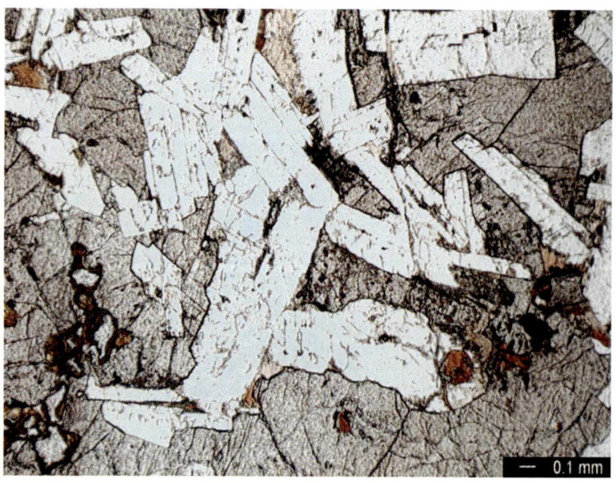

Figure 8.7.1 Dolerite lithic, plane polarized light. Dolerite (diabase) is a medium-grained igneous rock that typically has an ophitic texture (laths of plagioclase enclosed within clinopyroxene). The white laths here are the plagioclase, within the brown clinopyroxenes. Weathering products include clay, chlorite, and iron oxides.

graphic methods, and the relative quantity and size were ascertained through point counting. There was a variety of archaeological goals, with a focus on characterizing raw material selection and technological change through time. Five different fabric groups could be identified: three were tempered with chalk and quartz sand (but at different ratios of one to the other), one was tempered with organic material, and one had a sand temper (without any chalk). The number of each type found at each stratigraphic group was compared to examine any change through time in relative importance of one material over the other. Mainman also compared the materials with locally available clays. The first three fabric types were found to predominate in the earliest levels. They were then superseded by the fourth (organically tempered wares), and even later by the last type that was free of chalk. The first four types could have come from local materials, but the last type was inconsistent with local materials (although it had a coarser sand temper, the clay was free of the fine quartz sand, flint, and calcareous inclusions that made up the matrix of the other groups). This last type was also used for harder ceramic materials, fired to a higher temperature than the earlier wares.

The author interpreted these results as indicating that technological change occurring at the later levels of the site accounted for the change in choice of materials. The first evidence for use of a wheel to complete the rim occurred at this time; all the other fabric groups were used for vessels that were made entirely by hand. Kiln technology had probably also advanced, and potters were interested in firing to higher temperatures to achieve harder and less porous wares. However, the presence of significant amounts of calcium carbonate found in the local clays would have caused crumbling and spalling of the surface, as the higher temperatures would have led to decomposition of the calcium carbonate, moisture absorption, and formation of calcium hydroxide. Potters would have sought out clays better suited for high-temperature firing.

Accompanying this change in choice of materials was a move towards greater standardization, as could be seen both in mineralogical and textural data in thin section, and in greater standardization in vessel form. The change in clay source not only allowed for better high-temperature results, but provided more consistency, indicating a shift to professional potting. In contrast, the earlier fabric types showed considerable variation in textural parameters even within the same range of inclusion types. Mainman proposed that this textural variability can be explained by vessels being produced in a domestic context by multiple single households undertaking production as the necessity arose.

Williams (1982) noted that choice in materials selection and preparation may be a better reflection of tradition and culture than pottery form and decoration. The latter might simply represent local copying of external trade products, while the reasons why certain materials and preparation methods are chosen by potters may be deeply rooted in local traditions and cultural style. Williams used as a case study the analyses of a large number of thin sections from prehistoric (Neolithic and Iron Age) pottery from the Orkneys. The data showed the possibility of the existence of a "technological recipe" for making pottery that was widespread in the islands during the Neolithic and into the Iron Age.

Most of the prehistoric Orkney pottery contained a distinctive temper made of crushed dike rock (mainly lamprophyre, a group of intermediate igneous rocks with feldspars and ferromagnesium minerals that occur as dikes and minor intrusions). The resulting temper inclusions were mainly camptonite (a lamprophyric rock composed of labradorite feldspar, pyroxene, sodic hornblende, and olivine), with some bostonite (rich in alkali feldspars), monchiquite (a lamprophyre composed of olivine, pyroxene, and usually mica or amphibole phenocrysts embedded in a glass or analcime groundmass), and olivine basalt.

In thin section, it was clear that the fragments of lamprophyre dike rock were sharply angular. Thus it was postulated that the temper did not originate from drift material, which would tend to have rounded fragments along with more variety of inclusions. Yet, while some sites were located near dikes of the selected mineralogy, others were not. He ruled out the idea of centralized production at workshops located near the dikes with products traded to sites farther away, because the nonplastic non-dike materials in the sherds were extremely varied, making centralized production appear very unlikely. He concluded that fabric differences suggest that local potters at a variety of sites throughout the Orkneys were choosing their own particular clay sources, but often added crushed dike rock to the clay. One reason might be the fairly low thermal expansion of the crushed dike rock, a useful property for cooking vessels. An implication would be that there was a high level of communication between the prehistoric Orkney sites, and chronologically, the tradition of a particular type of temper being used was passed on through generations. Thus, beyond characterizing the ceramic material and identifying a source of the temper additive, Williams was able to characterize a technological style tradition, and argue for its functionality and cultural implications.

Braun (1983) studied Woodland-period cooking pots from Illinois and Missouri, and found that around A.D. 400–500, a change took place: vessel walls became thinner, and the average size and concentration of temper decreased. He proposed that this was due to a change in diet, with the utilization of more starchy seed foods that required longer cooking times, thus subjecting vessels to more stressful thermal conditions for which the new pottery technology was better suited.

Alvarez Arza et al. (2002) conducted thin-section studies of Bronze Age pottery from the site of Pixarelles Cave in Catalonia. Although all the vessels analyzed were made with materials that appeared to be consistent with a broadly local provenance, there was much variation, with ten fabric types identified. One distinction was that some of the raw materials used for ceramics contained carbonates while others did not (the sediments of the nearby River Ter contain abundant limestone fragments). The carbonate-free fabrics included some fabrics with altered granitic inclusions and others with monocrystalline inclusions of quartz and alkali feldspar. The carbonate fabrics also varied in the types of additional inclusions found.

Technological variability was apparent as well as variability in the choice of local raw materials. Variations included tempering with granitic rocks, with differences in grain size and quantity of tempering material added, and carbonate tempering in both carbonate-rich and non-carbonate-rich clay matrices. Some fabrics suggested the use of unprocessed clays, while some had coarse particles removed. Fabric variation did not seem to be related to type of vessel.

It was concluded that pottery at the site may have been produced by multiple individuals or groups, each making somewhat different technological choices. The authors found that some chronological differences did emerge from the data. In the Early Bronze Age, when the cave site may have acted as a temporary shelter for mobile groups of people, a number of different ceramic traditions were present. The fabric variety from this period suggested that many groups of people, each bringing their own ceramics, made in different manners, inhabited the cave. During the Middle Bronze Age the site was used on a more permanent basis by a specific group. Fewer ceramic fabrics were being used at the site during that period. In the Late Bronze Age, although variability still existed, one fabric group became dominant.

Faber et al. (2002) analyzed polychrome pottery of the Middle Bronze Age from Knossos, Crete. Thin-section petrography identified three distinct fabrics: one coarse, one fine, and one very fine. The coarse fabric had a calcareous clay with nonplastic inclusions consisting of altered basaltic rocks, biotite schist, and fine-grained sedimentary rocks, which together were interpreted as being a deliberately added sand temper. The fine-grained fabric also had a calcareous clay with a few inclusions of quartz, feldspar, and biotite. The very fine fabric was similar but with essentially no nonplastic inclusions. Interestingly, some vessels were found to have been made with both the coarse and the fine fabrics, with the fine one added as a surface layer or used for separate coils for different parts of the vessel. Thus it was clear that a single production group was preparing at least two pastes.

Tsirtsoni and Yiouni (2002) employed thin-section analysis to try to confirm the hypothesis that ceramics found at a Neolithic settlement site in Greece were probably cooking vessels. Findspots were next to hearths or ovens, and there were two forms: very large plates with elevated rims, and tripod open vessels. The authors made the underlying assumptions that (1) inclusions with low coefficients of thermal expansion or one similar to that of the fired clay (such as calcite, feldspars, or grog) should be more resistant to the effects of thermal shock in cooking; and (2) that large amounts of nonplastic inclusions, especially coarse ones, will increase the porosity of the ceramic fabric, thus helping to maximize thermal shock resistance by dissipating stress during the heating and cooling phases. The goal of their thin-section analysis was to see if the fabric of the potential cooking vessels was different in these parameters from the fabric used for other vessel forms. A total of 76 thin sections were prepared from the two vessel types in question and from a variety of other decorated and undecorated vessel types.

However, no sharp distinction was found between fabrics used for the two presumed cooking vessels and those used for other vessel types at the site. No material could be identified as solely a cooking vessel fabric. The fabrics of the presumed cooking vessels were different from those of some other vessel types, but not different from some of the other wares found at the site. The authors concluded that, although there did indeed seem to be some technological choices being made about selection of raw materials, those choices were most likely related in some way to the manufacture of vessels rather than to their intended function.

8.8 Studying Deterioration Products

There is a variety of mechanisms that can lead to the formation of deterioration products on a ceramic. Some of these were discussed in earlier sections on firing, evidence of use, and porosity studies. Preparation of the thin sections themselves can negatively impact preservation of the full suite of minerals within a ceramic sherd, therefore it is important to use preparation methods appropriate for the materials involved. Friable materials may require vacuum impregnation or consolidation, so that loose grains are not plucked out during grinding processes, and water-soluble minerals such as gypsum may require use of non-aqueous grinding procedures. Conservation treatments can affect the percentage and/or type of certain inclusions. For example, in Greece some pieces may be acid cleaned (Vaughan 1995). However, most deterioration products form after an object is no longer in use and has undergone exposure to the elements in a surface location or in burial.

The mineral composition and microstructures of ceramics have an effect on deterioration rates. Freestone (2001) pointed out that ceramics are usually fine-grained and somewhat porous materials with a high internal surface area available to react with environmental solutions. Highly porous ceramics, especially where pores are open or interlinked, provide good access to soil solutions. Thus a material such as a low-fired earthenware will be very unstable. In addition, ceramics often contain metastable minerals, which will react rapidly and accelerate the onset of deterioration. Examples are a high-temperature calcium aluminum silicate, gehlenite (Heimann and Maggetti 1981), and a calcium silicate, wollastonite. Freestone also noted that the particles within ceramic materials are often bound together in a matrix of fine grains and filaments of glass; as these corrode, the strength of the material deteriorates, leading eventually to complete disaggregation. He proposed that corrosion of low-fired ceramics can occur by the weakening and breakdown of particle contacts that depend on a glass matrix, the rehydration of fired clay minerals causing them to slowly revert back to clay. Dissolution of very soluble minerals that are found as natural components of the clay or as added temper is another serious problem. Calcite from limestone or shell temper often has large angular voids where the lime leached out in a moderately acidic environment (Fig. 8.8.1), and the dissolution of natural gypsum inclusions in clay

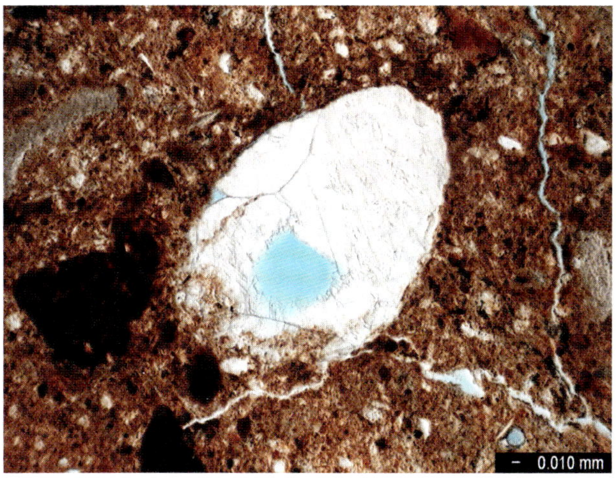

Figure 8.8.1 Deterioration in ceramic, plane polarized light. Dissolution of soluble minerals found as natural components of clay or as added temper is typical for buried ceramics. A void has been leached in this calcite grain, revealing the blue mounting medium.

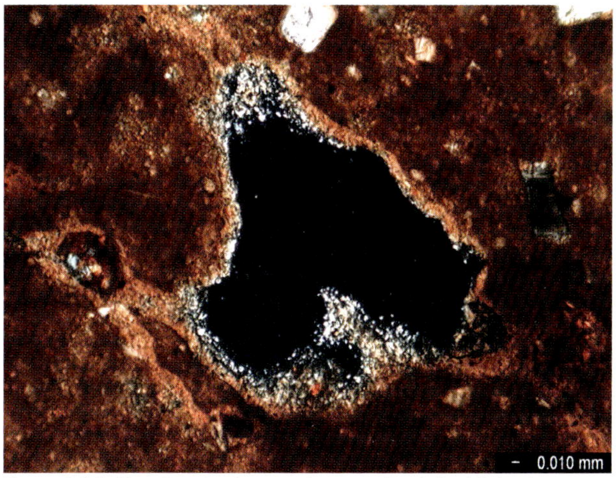

Figure 8.8.2 Deterioration in ceramic, crossed polarized light. A common alteration mechanism in ceramics is the formation after burial of secondary calcite and/or gypsum, deposited by groundwater. These deposits often appear as concentric rings around pores, and are usually microcrystalline.

may leave what are called "swallow-tail" void shapes within the clay.

One common alteration product that results from deterioration after burial is secondary calcite or gypsum deposited by groundwater during burial (often seen as concentric rings around pores or as patches scattered within the clay matrix) (Fig. 8.8.2) (Prag et al. 1974; Maggetti 1982; Vaughan 1995; Reedy 1996; Freestone 2001). It is usually fine grained, or even microcrystalline, and is distinct from carbonate concretions that often coat the surfaces of archaeological ceramics. Vincent (1988) observed extensive authigenic calcite deposits on prehistoric sherds from northeastern Thailand, and he pointed out that this post-depositional groundwater effect was probably accelerated in tropical conditions. There were both surface adhesions and networks of interstitial veins within the pottery fabrics. Calcite and silica-cemented sand appeared as surface deposits on some of the sherds.

Glazes, too, can deteriorate during burial (Freestone 2001). Alkaline glazes, common in the ancient world, are extremely susceptible to deterioration—they can weather to the point of being composed wholly of hydrated silica. Glossy clay slips can suffer physical abrasion in burial, as they tend to be composed of very thin layers.

De Andres and Balcazar (1989) studied ancient (Early Bronze Age) ceramics from the shelters of Peña Corva in Spain, looking for the effects of burial on mineralogy of sherds. In some samples, halite and gypsum crystals were seen, and the authors suggested that these products might have formed during burial as a result of periodic floods of the nearby Salado River. They noted that the low-fired sherds of the site would be very sensitive to mineralogical changes during burial; for example, rehydration of incompletely firing clay minerals could occur. Some secondary calcite formation was also proposed.

Owen and Day (1998) discussed the effects of weathering on soft-paste porcelain in their study of unglazed wasters from the Longton Hall factory site in Staffordshire, England. Although they focused primarily on elemental analysis and on changes in chemical composition, their conclusions were also applicable to thin-section analyses. For example, they found that prolonged exposure to subsurface water could cause dissolution of wollastonite in soft-paste porcelain (through decalcification). The glazed sherds better preserved their original mineralogy, apparently because the glaze protected the clay body from alteration. Other deterioration effects included an increase in porosity; and perhaps dissolution of the some of the high-temperature phases of silica (tridymite and cristobalite).

Cau Ontiveros et al. (2002) used thin-section analysis to study secondary calcite formation processes, types, and textures encountered with archaeological ceramics. They, along with earlier researchers, noted that not only recognizing the presence of secondary calcite, but also examining it closely, is important if one is doing elemental analysis of the ceramic materials. If the secondary calcite forms primarily by precipitation from external calcium carbonate-rich solutions, a range of elements can be introduced into the bulk composition of the clay, so merely ignoring calcium contents in statistical analyses is insufficient. However, if the calcium originates at least partly from within the sherd, the effects on elemental analysis will be different. A variety of secondary products and textures was described. "Micritic clots" were defined as crypto-microcrystalline masses of calcite that partially retained the form of an original carbonate grain or microfossil that was transformed during firing. Thermal decomposition of calcite could create lime, which might further react with the clay matrix to form minerals such as gehlenite, diopside, and anorthite. These minerals themselves could alter during burial in humid conditions (gehlenite to calcite and zeolites, plagioclase to calcite and smectite, and diopside to calcite) (Heinmann and Maggetti 1981). The authors noted that this process could result in lighter colored borders, or depletion features, around pores whose edges displayed the precipitation of secondary calcite as micrite and microsparite fringes.

Patches of microcrystalline calcite, when scattered in an irregular manner throughout the matrix, are often the result of the alteration of original raw materials and minerals formed during firing. In contrast, when the patches are concentrated in certain areas near the surface or around a pore that is filled with sparite or microsparite, they are more likely to be the result of penetration of aqueous solutions from outside the ceramic. Geodetic calcite, or crystals of sparite and microsparite filling pores and cracks in the ceramic, often having a "dog's tooth" texture, may be a product of precipitation from aqueous solution external to the ceramic material (in the absence of light-colored borders, micritic clots, or other evidence of primary calcite or calcium-rich firing minerals within the clay matrix) (Cau Ontiveros et al. 2002).

Other changes that can occur are that calcite grains within the original fabric may begin to dissolve, fine-grained gypsum may crystallize, especially around the edges of pores, and mineral coatings of various composition can be precipitated from groundwater onto the surface of the ceramic.

Ceramics that have been recovered from underwater shipwreck localities are likely to develop secondary minerals that are common precipitates in seawater, such as analcime (a white or light-colored zeolite), high-magnesium calcite, and aragonite. Sondi and Slovenec (2003) studied thin sections of Roman-period amphorae excavated at shipwrecks in the central Adriatic. Needle-like aragonite crystals were seen in microcracks and pores. The outer walls of the amphorae were encrusted with microfossils; in thin section this layer was revealed to be built up of calcareous micrite.

Maritan and Mazzoli (2004) discussed some of the processes that cause deterioration of buried ceramics, and some of the products that may form. They noted that while burial load, temperature variations, and frost can affect ceramics, circulating water is the biggest problem, dissolving some minerals and precipitating secondary minerals. They focused on some archeological finds from a Second Iron Age site in northeastern Italy, including a sherd that was studied in depth and was well illustrated in their paper. They found large polycrystalline aggregates of phosphate crystals, which ranged from 200 to 700 μm in size, and had a distinctive green to blue-green color in plane polarized light. The surrounding matrix of the sherd had a low porosity (less than 10% by volume). Most of the pores were less than 100 μm and were either irregular in shape (vugs) or elongated (channels).

Optical properties allowed the authors to distinguish the type of phosphate mineral present in the sherd. The mineral had prismatic crystals with good cleavage, was pleochroic from deep cobalt blue to pale yellow to gray, and was optically positive biaxially with a large optic axial angle. These characteristics identified the phosphate mineral vivianite (a hydrated iron phosphate). The mechanisms of formation of phosphate in buried ceramics were discussed, primarily related to precipitation of phosphates from aqueous solutions, depending upon environmental conditions in the soil, movement of water in the ground, and the presence of organic matter and bacterial activity. Crystals and thin films of phosphate may form or adsorb at the surface of primary minerals. The alteration of bones or bone fragments in iron-bearing solutions, or the presence of original phosphates in the soil, can also lead to precipitation of phosphates in buried ceramics.

9

Nonpottery Ceramic and Clay Materials

9.1 Clay Sculpture and Molded or Stamped Objects

The earliest known ceramics in the world are the small hand-formed objects from Dolni Věstonice, Czechoslovakia, found at sites dating to about 26,000 B.P., about 14,000 years before any other ceramic finds. Rather than being found at settlements, numerous fragments were recovered at sites located about 2 km away from settlements. Most of the finds were small pellets, fragments of female and animal figures, slab-like objects, and round balls. Most of these artifacts were fragments rather than whole, unbroken objects. Technical studies (Vandiver et al. 1989b, 1990) showed that the objects were made from the local soil, a loess, comprised of about 80% quartz. The loess was wetted, and preformed segments were joined to make the more complex objects. Firing was done in open kilns at temperatures ranging from 500 to 800 °C.

A combination of laboratory studies and experimental firings indicated that the objects were apparently made specifically to be exploded through thermal shock in the kiln. The researchers pointed out that loess has little shrinkage and low thermal expansion, so can be fired without drying and will not crack easily. The high porosity of the material meant that water could be removed efficiently from the interior during heating, preventing cracking. Since it was in fact difficult to produce thermal shock and cracking with this material, the fragmenting of the objects was probably deliberate. The still-wet objects were placed into the hottest part of the kiln, where they would then sizzle and steam, and the larger figurines would also pop and burst. The large accumulation of fragments at the kiln sites, with few intact objects, indicated that this "thermal shock party" was the goal, rather than a durable product. The reasons or functions for this activity, however, were not clear.

The fragments were comprised largely of quartz (angular to subangular) and micas (muscovite/sericite and biotite) (Figs. 9.1.1 and 9.1.2). Particle sizes ranged from about 5 to 150 μm, and these larger particles made up about 80% by volume of the loess body. There was also minor calcite, dolomite, anorthite, and chlorite, and a small amount of clay and glassy phase. According to Vandiver et al. (1989b, 1990), the paste was like a mixture of beach sand and a small amount of clay, so gritty when wet with little plasticity. Phosphorous in the soil and other alkaline impurities acted as a flux to help form incipient glass at quartz grain contact points, even at the low firing temperatures (Fig. 9.1.3). Porosity ranged from 20 to 40% (Fig. 9.1.4). One of the methods used to identify the firing temperature range was to fire new loess pellets and refire archaeological fragments, to look at changes seen at various temperature increments (such as glass formation at contact points, rounding edges of clay particles, and mineralogical changes such as disappearance of micas).

Other, later, small hand- or mold-formed clay objects, often called terracottas (since they tend to be made of a low-fired red, brown, or yellow unglazed porous clay material),

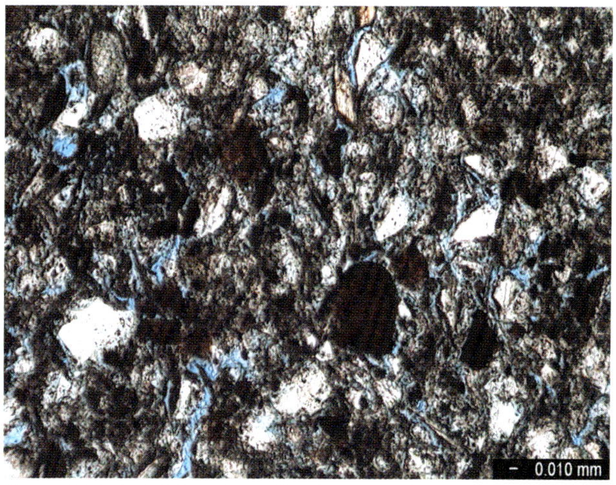

Figure 9.1.1 Dolni Věstonice artifact fragment, plane polarized light. The porous fragments from this site are comprised largely of quartz and micas. The larger clear minerals are angular to subangular quartz. Some brown biotite mica is also prominent, along with a few clear tabular grains of muscovite/sericite mica. There is also some chlorite and clay, together with minor calcium carbonates and plagioclase (anorthite).

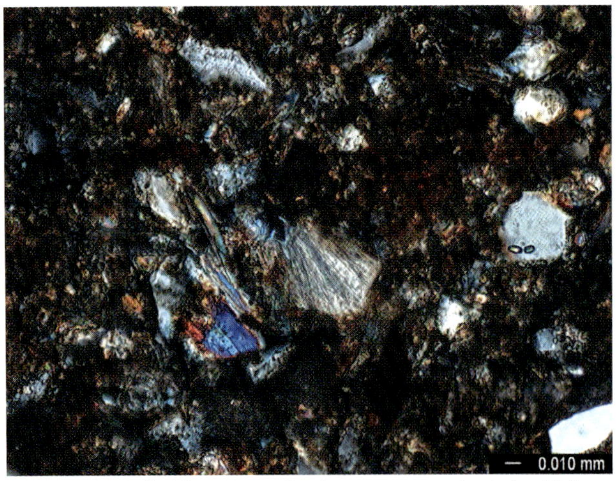

Figure 9.1.2 Dolni Věstonice artifact fragment, crossed polarized light. Some of the abundant muscovite mica particles are visible in the center of the field of view (with high interference colors, and in tabular form or scaly aggregates).

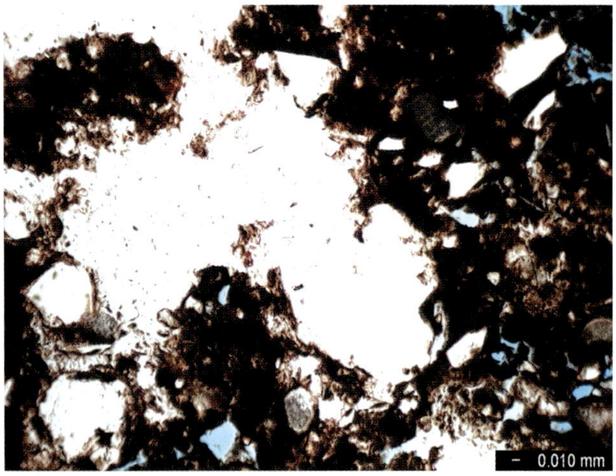

Figure 9.1.3 Dolni Věstonice artifact fragment, plane polarized light. A few patches of amorphous silica glass have begun to form, surrounded by angular quartz grains.

are found in many parts of the world. Such figurines may have some hollow parts made on a wheel (such as cylinder shapes), some components may be made in molds or dies, and some solid parts may be modeled by hand. Pieces are then assembled (perhaps with the use of pins and supports), dried, and often fired. Slips, polychrome paints, and/or appliqué decoration are common. Some objects are functional, others are decorative, and many serve various ritual functions.

Abbate-Zohar et al. (1985) studied terracotta ornaments found in the cloister of the Carthusian monastery in Pavia, Italy. Built between 1460 and 1470, it incorporated 50 arches decorated by terracotta ornaments. The authors identified two classes of ornaments: the first included pieces made by press-worked dies, featuring more generalized figures such as angels and putties, and architectural ornamentation; the second included unique life-like portraits of saints, prophets, or monks, attributed to major sculptors.

Thin sections were employed to identify minerals (along with x-ray diffraction analysis (XRD)) and to characterize the microtexture of the matrix (together with scanning electron microscopy (SEM)). Thin-section analysis identified the presence of about 5% medium to fine sand (≤ 400 μm), 20–35% coarse silt (30–60 μm), with fine silt, clay, and vitrified areas making up the remainder (60–80%) of the terracotta material. The authors concluded that the raw materials had been carefully sieved, and then the temper additives (mainly ground quartz and feldspar, with a small amount of small brick grog) carefully regulated. The quartz and feldspars were very angular, and homogeneously distributed within the silt-clay matrix.

It was noted that fabrication techniques could be observed in thin section. Hand-molded objects had a non-oriented texture, seen with the quartz and feldspar grain distributions as well as with the clay particles and micas. In contrast, press working produced a subparallel orientation of those grains, as well as preferred orientation of micropores and voids. Firing temperatures were suggested, by the range of original and newly formed minerals present, to be between 700 and 800 °C. The presence of goethite and hematite indicated oxidizing conditions. The main deterioration products were found to be gypsum crystals precipitated into micropores and fissures. These deposits have created stresses, which are producing new microcracks.

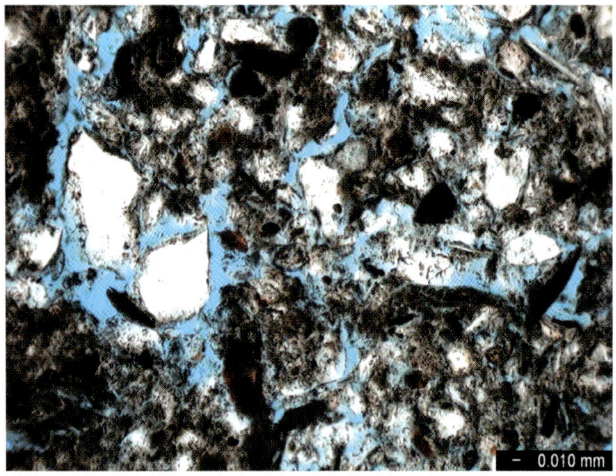

Figure 9.1.4 Dolni Věstonice artifact fragment, plane polarized light. The blue epoxy areas indicate the abundant porosity of this material.

Fragments of painted clay statues from Old Nisa, Turkmenistan, dating from the third–second century B.C., were studied in thin section (along with XRD analysis) as part of conservation and museum exhibition preparation efforts (Chiari et al. 1993). The statues were fabricated by modeling the core of the figures with a coarse greenish clay. Details of the figure were obtained by adding an outer layer in a finer yellow-brown clay (about 0.5 cm thick). The two layers had the same mineral inclusions in the clay (quartz, calcite, albite, orthoclase feldspar, chlorite, and muscovite); however, the green clay contained more chlorite. The particle size distribution also differed in the two layers, and they differed in amount of amorphous iron oxides and hydroxides which created the yellow-brown color in the fine layer.

While these two major layers were visible by eye, two other layers, on top of the fine layer, were discernable only in thin section. A thin white layer (of finely ground quartz mixed with kaolin and a small amount of ground calcite) was applied to the entire statue as a background color, which was then painted. On top of the paint colors was another layer of brown clay, probably produced as a deposited layer once the pieces had fallen to the ground. Separation lines between layers were very sharp, with little mixture of materials. The authors deduced from this that the layers were applied in succession, with each layer being allowed to dry completely to a perfectly smooth surface before the next layer was applied.

The earliest terracotta sculptures in South Asia, such as those made during the Harappan period (2400–1800 B.C.), were simply modeled by hand (Ardeleanu-Jansen 2002). Small pieces were often formed by pressing wet clay into molds, and the objects were sometimes unfired. Large sculptures were modeled over an inner core of straw that would carbonize during firing, leaving a hollow center and thus a much lighter piece. Later objects were made using a mixed technology: molds were employed in conjunction with hand-modeling techniques and a single figure could include several production methods. For example, the face might be made with a mold, the figure modeled by hand (and the two pieces joined with a tenon), and ornaments added by the appliqué technique. Sectional molds, made of terracotta themselves, have survived at Gandhara sites in Pakistan, and date to as early as the second century B.C. (Marshall 1951; Pal 1986; Poster 1986). It was only in the medieval period, however, that the potter's wheel came into use for terracotta sculpture production (Pal 1978).

Sometimes the clay was coarse enough and of the right working consistency to use alone, or temper might be added. A porous, coarse-grained material promoted rapid, even drying, leading to fewer problems with cracking and breaking. After the clay image was formed, the piece was thoroughly dried in the sun. Most objects were then fired, usually at low temperatures. Simple kilns would suffice, or the objects could be baked on an open fire. The final color of the clay was generally a red, brown, buff, or gray. Sometimes a slip was added, or, after drying or firing, a lime-based whitewash; either of these could serve as a ground for the application of decorative paints.

Two terracotta sculptures from Gandhara (an ancient region of eastern Afghanistan and northwestern Pakistan) were studied with thin-section petrography (Reedy 1992a). The first, a hooded female figure, was made with a hematite-rich clay containing a fine-grained sand component (Fig. 9.1.5). The quartz grains were angular to subangular, and often contained fine inclusions. Polygonized quartz (made up of two or three separate subcrystals) were common. Accessory minerals included muscovite, gypsum, anhydrite, carbonates, and clinopyroxene. This clay material was used to form a small hollow image (7 × 6 cm) that was made in three separate pieces (the head and two halves of the hood/bust), each pressed into a mold. The hood/bust pieces were pressed around and over the head and were then joined at the sides. Although fine details were completed on the front, the back was unfinished. The piece was covered with a white slip, with only remnants now remaining. Bits of red pigment remained in the hair, and traces of green, blue, and gray paint were also visible on the figure. This polychrome finish on the Gandhara piece was similar to polychrome finishes on Greek and Roman terracottas (major influences on Gandhara art), which were almost always painted (Higgins 1969).

The second female bust studied was made with a clay with much less hematite. Instead, it was marked by abundant carbonaceous material (Fig. 9.1.6) that was probably added to the clay. The quartz, as in the previous sculpture, was fine grained and subangular. In this case, accessory minerals included plagioclase feldspar, muscovite, chloritoid, and hypersthene. The front and back of the sculpture were molded in two separate pieces. Join lines were visible along the shoulders, down the center of the upper arms, and on the inside surfaces. The head was solid, but the torso was hollow. In contrast to the previous sculpture, this one was fully finished on the back, with a detailed long plait of hair running down its length.

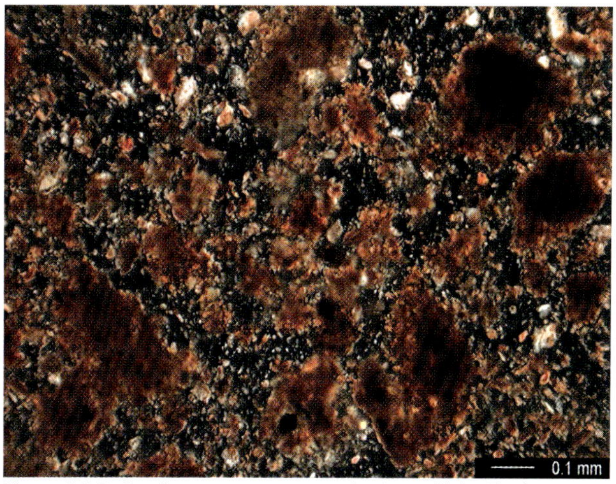

Figure 9.1.5 Gandhara terracotta, crossed polarized light. This loose-grain sample from a terracotta sculpture from ancient Gandhara is a hematite-rich clay containing a very fine-grained sand component. In crossed polarized light the hematite gives a reddish glow to the material.

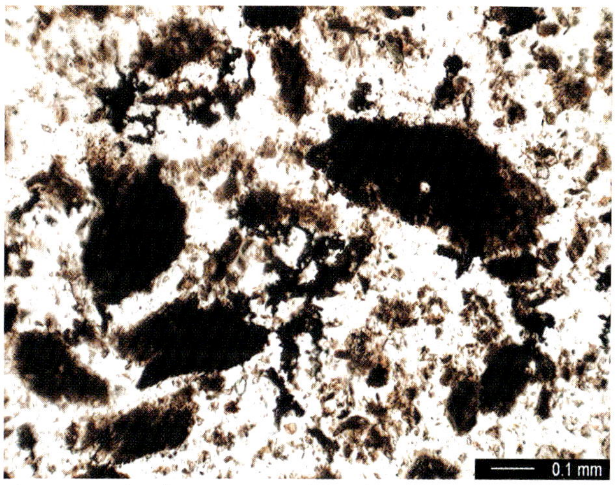

Figure 9.1.6 Gandhara terracotta, plane polarized light. The loose-grain sample from a second terracotta sculpture from ancient Gandhara is seen in this case to contain clay, with a very fine-grained sand component, and abundant carbonized organic material (that was probably added to the clay).

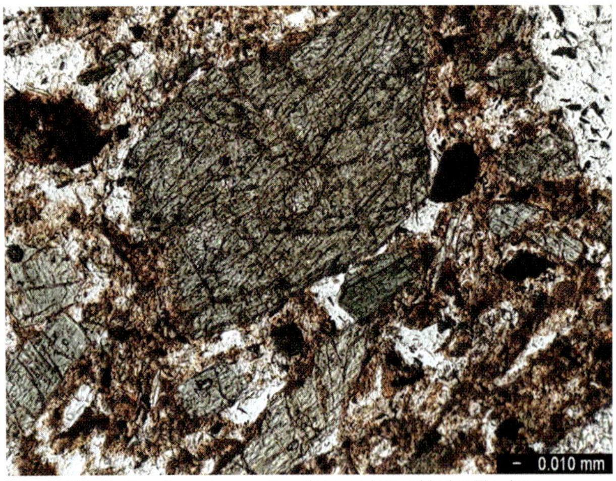

Figure 9.1.7 Gandhara terracotta, plane polarized light. The loose-grained sample from a third terracotta sculpture from ancient Gandhara shows a very different material than that seen in Figures 9.1.5 and 9.1.6. This material is more coarse grained, and contains abundant hornblende, green in plane polarized light.

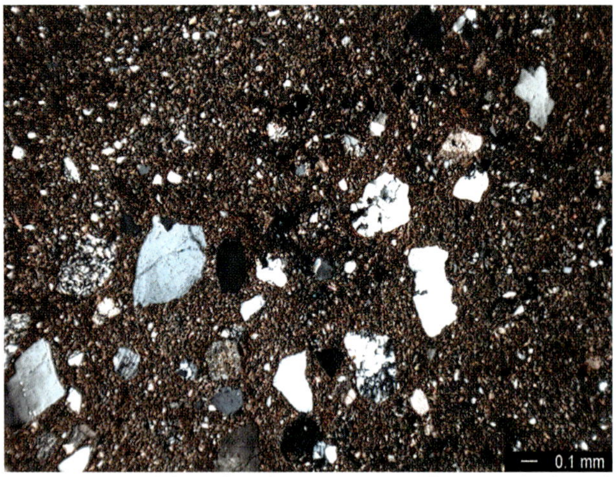

Figure 9.1.8 Mud plaster from large Bamiyan Buddha, crossed polarized light. The carved stone image of the large Buddha at Bamiyan was covered with a clay-sand-gypsum-lime plaster. The finer matrix contains relatively coarse sand particles.

A third Gandhara terracotta, depicting a man sticking out his tongue, was made with a very different, coarse-grained, material (Fig. 9.1.7). The main nonclay component was hornblende. This mineral has a deep green color in plane polarized light. It is interesting to see such a range of materials in objects made about the same time period in the same general region.

In Southeast Asia, small terracotta sculptures were made with similar materials. Three were analyzed by thin-section petrography (Reedy 1997c) including a Dvāravatī-period head of the Buddha, a standing Buddha in high relief attributed to Lamphun, twelfth century, and a high-relief head, probably of a disciple of the Buddha, dated to the fifteenth century. The sample taken from the Dvāravatī head was very high in quartz (more than 50% of the material). The quartz grains were both large and small—some were cracked, many had undulous extinction, and some were polycrystalline. Micas were abundant (including muscovite, chlorite, and biotite). Plagioclase feldspar was fairly common, and there were some scattered chunks of hematite. The sample taken from the standing Buddha from Lamphun had the least amount of quartz (20%) of the three terracottas. Some of those grains were polygonized, but the polycrystalline and undulous quartz so common in the other terracottas were absent here. Hematite was common, often in the form of large chunks. The only mica was chlorite, and there were also some scattered grains of zircon and garnet. Finally, the material of the third terracotta analyzed, a Sukothai-style head of a disciple, was somewhat like that used for the Dvāravatī head. There were cracked quartz grains and examples of undulous, polycrystalline, and microcrystalline quartz. The micas were muscovite and chlorite. For all three objects, the variety of minerals present was less than that usually seen in clay core materials (from inside cast bronzes) from Southeast Asia, indicating that perhaps the terracotta clays were more highly processed to remove impurities compared with the materials used for clay core materials (see section 9.4 below).

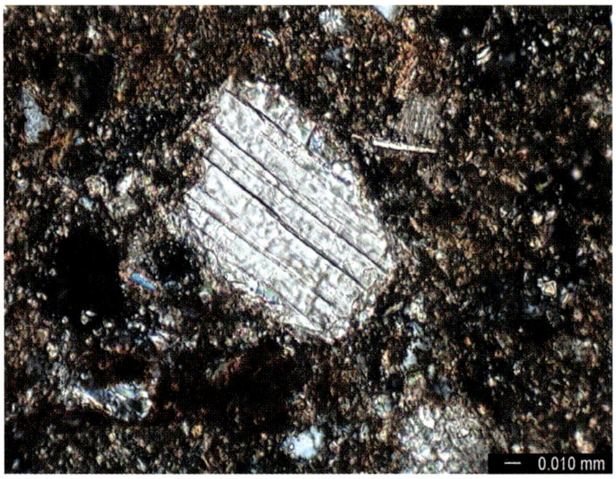

Figure 9.1.9 Mud plaster from large Bamiyan Buddha, crossed polarized light. Gypsum and its anhydrous form, the mineral anhydrite, often co-occur, and one can alter into the other. The grain in the center of the field of view is anhydrite, which has higher relief and higher interference colors than gypsum. It can be distinguished from calcium carbonates by the perfect cleavage in one direction, and imperfect cleavages at right angles. It is also distinguished by going to extinction parallel to the best cleavage direction; so here at a 45-degree angle it shows maximum interference colors.

Large clay sculptures are often unfired—they are simply formed and then dried in place. Life-sized or larger clay sculptures of great aesthetic sophistication and complex detail and symbolism are found in shrines and temples throughout Tibet and at other Buddhist sites in China, Afghanistan, Central Asia, and other Asian locations. In Tibet, for example, the material used for forming the statues is often prepared by mixing and kneading the clay with sand and jute or raw cotton, added to increase porosity and prevent cracking as the clay dries. To aid the drying process, the sculpture is built up in layers, with periods of drying between layers. The outermost layer is made with a finer clay, often mixed with Tibetan paper (which is handmade from a grass, is strong and fibrous and is considered to be a sacred material), for a smooth finish. Various medicinal substances are also added to the finer clay used on the outer surfaces, for their fragrant and protective qualities (Heath et al. 1991). Luczanits (2004) studied early Tibetan clay sculptures, finding clay mixed with fine sand, finely chopped straw, thin fibers (such as cotton or paper), coarse hair (black animal hair), vegetal fibers, or small pebbles. He noted that the use of hair in clay sculptures is characteristic for Central Asia.

The now-destroyed two colossal Buddha images (Dupree 1977; Gruen et al. 2002) at Bamiyan, Afghanistan, had cores that were carved from the sandstone and conglomerate of the cliff face overlooking the Bamiyan valley. The carved stone images were covered with a clay-sand-gypsum-lime plaster (Figs. 9.1.8–9.1.10) where finer details could be molded. A matrix of hemp rope and wooden pegs, or armatures, was used to support the plaster and attach it to the stone, and the plaster layer itself was bonded with organic fibers (wheat straw, ox hair, horse hair, and sheep's wool have all been reported) for tensile strength (Fig. 9.1.11). The statues were originally painted on top of a final fine gypsum plaster layer.

Smaller modeled objects, such as figurines and heads, made with a material that is often called "stucco" by art historians, are also found at and near Bamiyan. These are made (Sengupta 2002) by first modeling a core of clay. Over that, a fine layer of lime mixed with gypsum is added. Sometimes only lime was used; however, gypsum was locally available and was often the main component.

Molded figurines made of a very different material, soft-paste porcelain (see section 6.1.3), appeared in Europe in the eighteenth century. Kingery and Vandiver (1986) studied a soft-paste porcelain sculpture made in Vincennes, France in 1752, following a design of Françoise Boucher and titled *La Danseuse*. The piece was one of a series made with unglazed biscuit-fired figures, very precisely molded, having the white, grainy translucence of marble. It was found that the small sculpture was assembled with eight separately molded parts, probably made in plaster molds. Drying had to have been done slowly and carefully under controlled conditions, to ensure equal water content in each of the parts, otherwise cracking could occur at joins as differential shrinkage of various parts occurred after assembly. The details of the figure as a whole were then modeled and incised. Final drying and firing steps also had to be done with careful control.

The marble-like appearance of the surface was enhanced by polishing with a sandstone abrasive, to produce a fine roughness that scatters reflected light. This surface treatment prevented a glossy glazed ceramic appearance, and instead gave a more sculptural quality reminiscent of marble. Kingery and Vandiver pointed out that the French formulation used only 8–10% clay, for a low-alumina content. A high lime content was imparted by the use of limey clays from Argenteuil, which were very plastic so allowed successful molding even with such low-clay formulations. At firing temperatures of 1050 °C the main silica formed was cristobalite. The other main crystalline phase was wollastonite (calcium silicate), surrounded by a siliceous glass. The authors illustrated a microstructure with cristobalite grains, some still having an inner residual core of quartz, fine-grained blocky wollastonite grains, and a silicate glass matrix. It was noted how the quartz grains tended to fracture while changing into cristobalite during firing, so there were polycrystalline grains with fractures in and around them, rather than fractures through the glass matrix. This situation created greater susceptibility to breakage in conditions of sudden environmental change, and might have been one of the causes of the demise of the importance of this material compared with hard-paste porcelain (which has greater strength).

Of a smaller size, and often unfired, are impressed or stamped clay objects. Early examples include cuneiform tablets from the ancient Near East. These small clay tablets represent one of the earliest forms of written records.

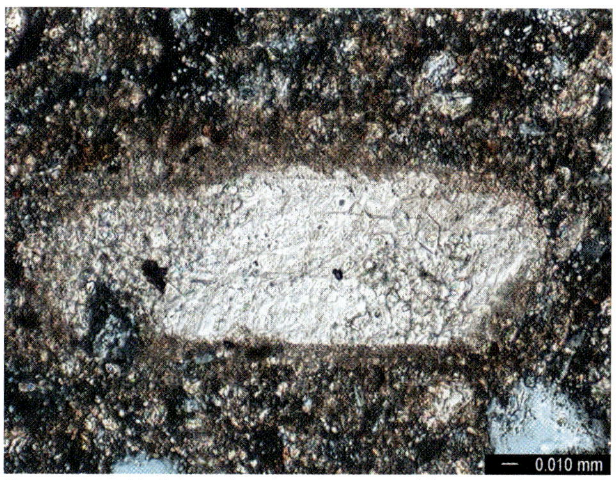

Figure 9.1.10 Mud plaster from large Bamiyan Buddha, crossed polarized light. Calcium carbonate grains can be distinguished in the Bamiyan Buddha plaster layer by presence of rhombohedral cleavage, as well as pastel to pearl gray or white interference colors. This grain is disassociating on both ends, with the formation of secondary precipitates.

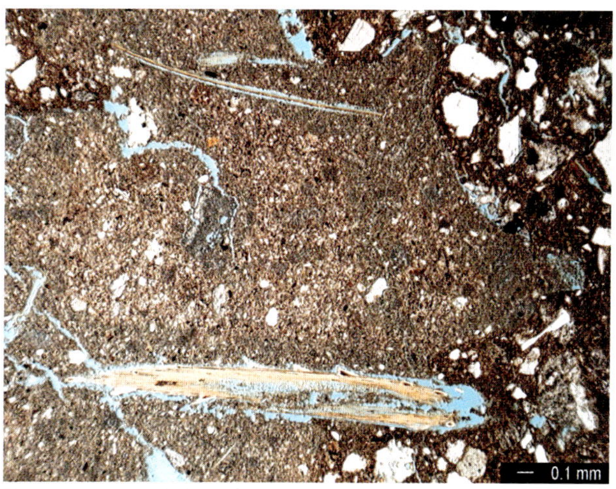

Figure 9.1.11 Mud plaster from small Buddha, Bamiyan, plane polarized light. This sample from an original part of the upper layer on the small Buddha at Bamiyan contains a piece of straw in the lower part of the field of view, and an animal hair in the upper part. It also illustrates variation between the finer layer on the left, and the coarse, sand-rich area on the right.

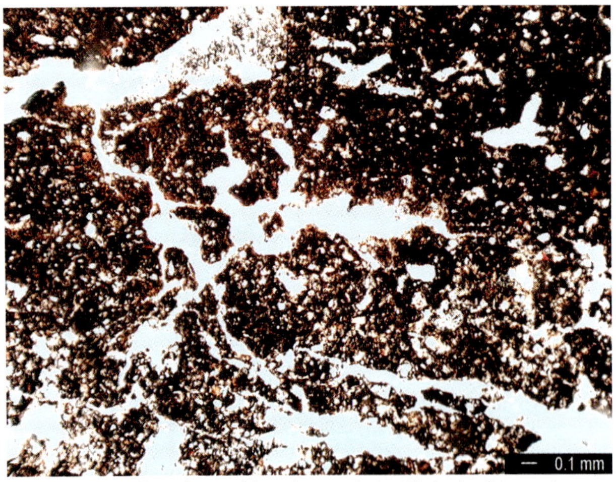

Figure 9.1.12 Cuneiform tablet, plane polarized light. In thin section, this cuneiform tablet is seen to have a network of interconnected pores.

Cuneiform, a wedge-shaped system of writing developed by the Sumerians before the fourth millennium B.C., was impressed into the wet clay with a stylus. Although some tablets were fired, many were simply dried. Many thousands of these tablets have been found at archaeological sites in Mesopotamia, Iran, and Turkey, documenting the political, legal, economic, and cultural activities of their times (Van de Mieroop 1999).

As unfired, or very low-temperature fired, clay objects that have been buried for millennia, cuneiform tablets often suffer from structural weakness and soluble salt damage. Many are close to disintegrating upon recovery, and cannot be handled by scholars. The script is often obscured by surface encrustation of lime and gypsum. The traditional preservation treatment approach is to first fire the tablets at temperatures of 750–900 °C. In addition to making the tablets more structurally sound for handling, firing also facilitates cleaning by hardening the tablets and limiting cracking, and softening the burial encrustation. Although firing is deemed a drastic and irreversible treatment, it meets the primary objective of preserving the tablet and script. After firing, many tablets are then immersed in water for 10–12 days to remove soluble salts. This procedure follows that originally developed at the beginning of the twentieth century by Friedrich Rathgen, the first director of the Royal Museums of Berlin (Rathgen 1901, 1908).

However, in thin section (Fig. 9.1.12) the extreme porosity of such tablets is visible, with large networks of interconnected pores permeating the material. Gypsum (hydrous calcium sulfate), occurring as small blocky aggregates with a somewhat fibrous structure, forms inside some of the pores (Fig. 9.1.13), as well as on the surface of some of the tablets (Fig. 9.1.14). One school of thought is that the gypsum in the pores and on the surface of tablets formed as a reaction with groundwater in burial, and should be removed by soaking the tablets. However, in thin section it appears that in some cases the clay itself is permeated with natural inclusions of anhydrite (calcium sulfate) (Fig. 9.1.15). Thus, the gypsum is forming as "inherent vice" from the calcium sulfate content of the tablet's clay itself, in the presence of moisture. Soaking the tablets will pull out the existing gypsum from pores and from the surface, but will also initiate formation of new gypsum crystals from the anhydrite present in the clay, and accelerate deterioration of the objects.

Another type of unfired or low-fired stamped clay object is the Tibetan *tsha-tsha* (Tucci 1988; Liu 2000). Usually ranging in size from 1 to 25 cm in height, flat ones are fabricated with metal stamps, and may depict a deity, multiple deities, shrine images, and sometimes letters or inscriptions (mantras). Another variety depicts small three-dimensional images of Buddhist stupas or Bon-pö "thousand-god" images, which are made by pressing wet clay into a hollow metal mold. After drying, the objects are sometimes whitewashed, gilded, or painted with a variety of colors. *Tsha-tsha*s serve a wide variety of functions in Tibetan society and religion (Reedy 2006b).

Thin-section petrography was carried out on five Tibeto-Chinese *tsha-tsha*s dating to the eighteenth century (Reedy 2006b). The objects were probably all from a single set depicting the correct iconography for a deity pantheon (the name of each deity appears on the backs in Tibetan,

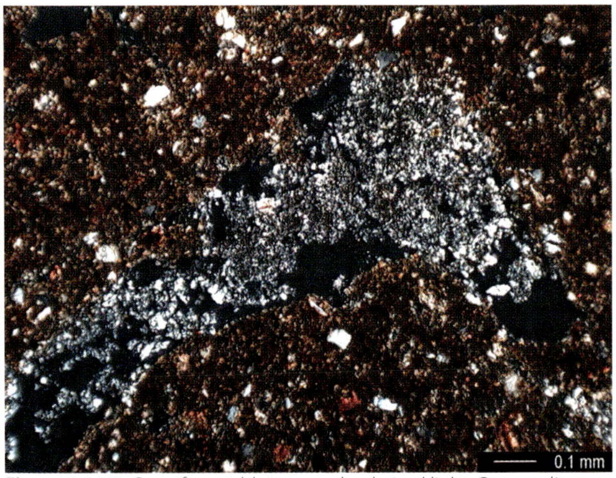

Figure 9.1.13 Cuneiform tablet, crossed polarized light. Gypsum lines many of the pores in this cuneiform tablet. Interference colors range from gray up to white or straw yellow. Anhedral to subhedral aggregates are common.

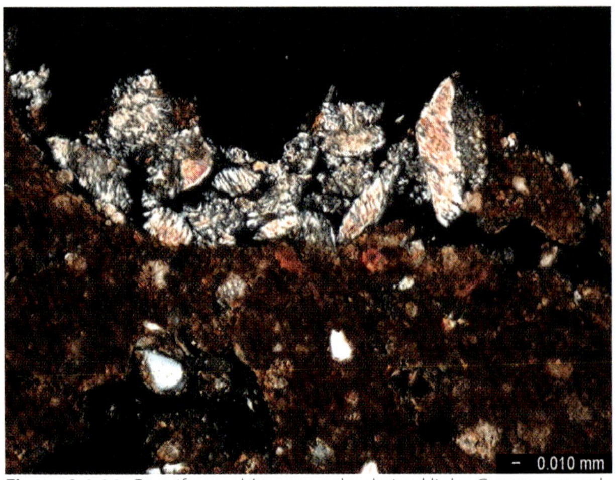

Figure 9.1.14 Cuneiform tablet, crossed polarized light. Gypsum crystals have formed on the surface of this cuneiform tablet.

Chinese, Mongolian, and Manchurian). The material used (Fig. 9.1.16) was a reddish clay containing chlorite and abundant hematite, along with biotite, muscovite, and sericite. A fine-grained sand component included quartz, microcrystalline quartz, a few larger chunks of polycrystalline quartz, and minor amounts of plagioclase feldspar. There were also some small linen fibers mixed into the material. The objects were covered in a red paint, and gold-colored paint was added to the front of each *tsha-tsha*. Thin sections revealed that the red pigment layer was applied over a fine-grained clay slip layer. The slip was a refined layer of the same iron-rich clay used for the body of the *tsha-tsha*, while the top pigment layer was hematite.

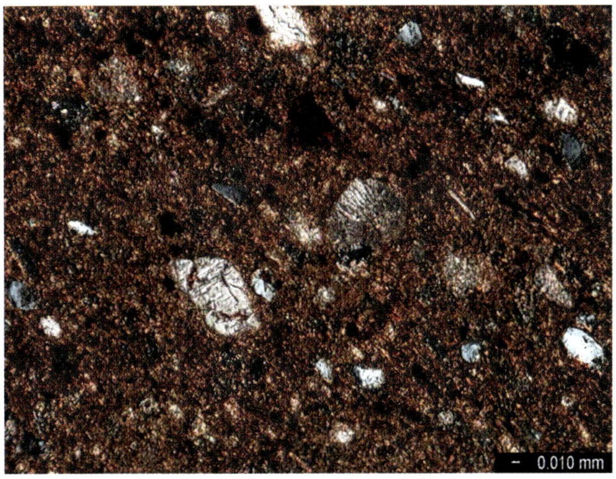

Figure 9.1.15 Cuneiform tablet, crossed polarized light. In thin section, this cuneiform tablet is seen to be permeated with many natural inclusions of anhydrite (calcium sulfate). These inclusions provide an internal source for gypsum (hydrated calcium sulfate) formation upon exposure to moisture from groundwater or from conservation "desalination" treatments. The large internal pore network will help to constantly replenish the surface gypsum encrustations.

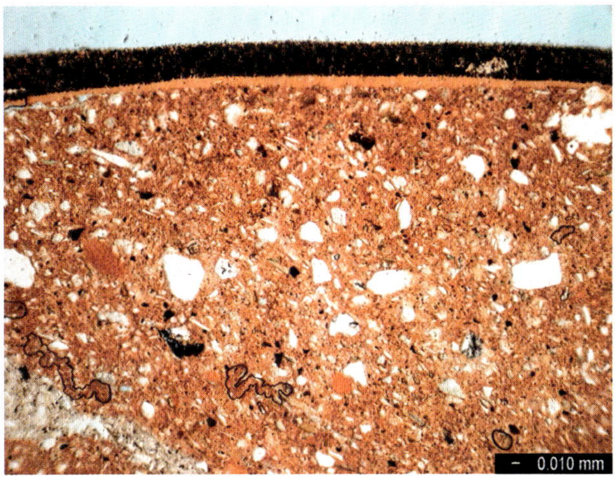

Figure 9.1.16 Tibetan *tsha-tsha*, plane polarized light. The material used for this Tibetan *tsha-tsha* was a hematite-rich clay and a fine-grained sand component, along with some small fibers that may be linen. The decorative layer is comprised of a fine slip that appears to be a refined layer of the same iron-rich clay as is found in the body of the object. The slip is covered with a red pigment, seen in reflected light to be hematite that in some areas was covered with an additional layer of gold paint.

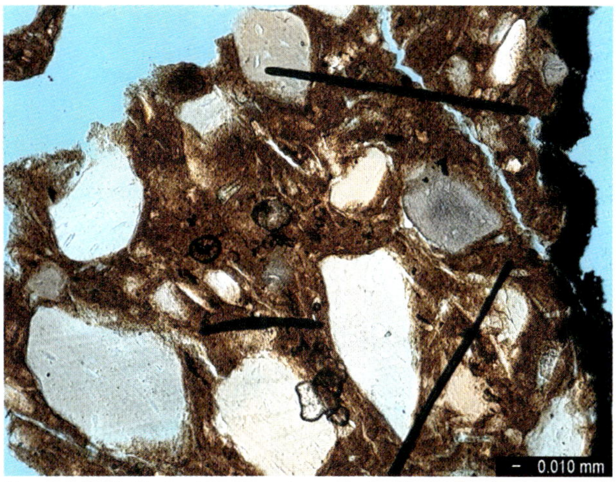

Figure 9.1.17 Tibetan *tsha-tsha*, plane polarized light. The clay mixture used for this Tibetan *tsha-tsha* includes abundant plagioclase that is very weathered, in a matrix of iron-rich clay that contains sand and some hairs.

Samples from two earlier *tsha-tsha*s found inside a portrait bronze were also studied in thin section (Reedy 2006b). One was a flat object depicting a seated Buddha, and the other was in the form of a small stupa. Both were made from a similar coarse, sandy, chlorite-rich clay (Fig. 9.1.17) that contained weathered plagioclase feldspar, quartz, microcrystalline and polycrystalline quartz, hypersthene, hornblende, uncarbonized organic matter, chlorite, and abundant hematite. Instead of linen fibers, these objects had some short, dark hairs mixed into the material.

9.2 Bricks

Bricks have been used as a building material since antiquity (McKee 1973; Woodforde 1976; Rosen 1986; Warren 1999; Trench 2000; Rapp 2002; Campbell and Pryce 2003). They are made by pressing wet clay into a mold; sandy clays, or mixtures of clay and quartz sand (Fig. 9.2.1) are used in the production of brick, often with the addition of vegetal matter such as straw (to increase resistance to cracking) (Fig. 9.2.2). Sand, silt, pebbles, grass, twigs, and other components may also enter into bricks as natural components of the clay. Preparation steps may include kneading, periods of exposure to the elements to break down large lumps, inclusion of additives to improve cohesion and consistency of the material (such as tree sap, fruit juice, or dung), addition of straw for strength, and, if needed, the addition of more silt or sand to help in reduction of shrinkage as the brick material dries and/or is fired.

A variety of methods has been used to form bricks from the finished clay material. The clay can be rolled onto a sand-covered surface and cut into slabs, or pushed into a mold (usually made of wood dusted with sand or ash to facilitate release of the brick). Bricks made after industrialization were formed by extruding the clay through a die and then slicing it with a wire, or by press molding (giving a relatively dense brick of consistent shape) (Fig. 9.2.3). A wide range of machines was invented for production of bricks.

Figure 9.2.1 Brick, crossed polarized light. This brick (from the Read House, Newcastle, Delaware) has a typical sandy clay, with grains ranging from fine to coarse in size.

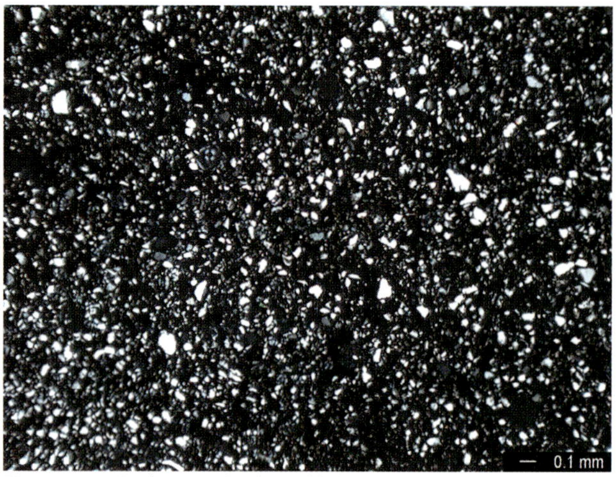

Figure 9.2.3 Brick, crossed polarized light. Post-industrialization bricks (such as this one from the Read House, Newcastle, Delaware) made by extruding clay through a die or by press molding, are relatively dense materials. This example is primarily clay plus a fine sand, with minimal porosity.

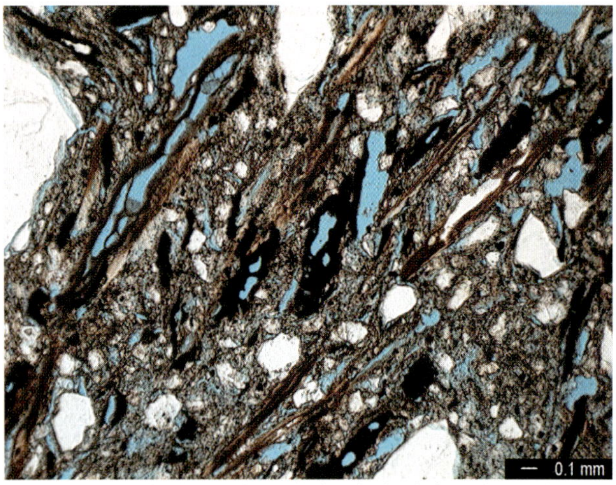

Figure 9.2.2 Brick, plane polarized light. This brick (from the Read House, Newcastle, Delaware) is an example of a brick with added vegetal matter (mainly straw), which functions to increase resistance to cracking. In this case, the straw additive is abundant (and partially carbonized), creating a very porous material.

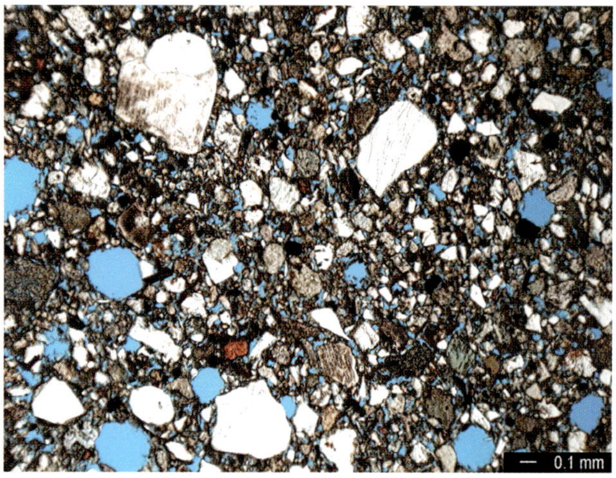

Figure 9.2.4 Brick, plane polarized light. This brick (from Arizona) is made from adobe. An unfired and very porous material, it consists of an abundant coarse fraction, predominantly sand.

Once fabricated, the bricks are then hardened—some of the earliest bricks were merely sun dried, like those found at Neolithic sites such as Jericho. Sun-dried bricks remain popular in areas of low rainfall where they maintain their durability for long periods of time (for example, in many parts of Africa, China, Latin America, and the southwestern area of the United States). In more rainy climates, unfired bricks may be used where they are protected from direct rainfall by a projecting roof and an external finish. Unfired brick is also called "adobe", "mud brick", or "clay lump."

Brown and Clifton (1978) reviewed the composition and properties of adobe (Fig. 9.2.4). They emphasized the fact that particle size distribution affects performance of adobe—too much gravel or too much clay will adversely affect durability. The clay and silt act as the primary binder, but since the clays tend to expand when they absorb moisture and shrink when they dry, an excess of clay creates dimensional instability. Too many large gravel particles may affect the binding forces of the silty clay matrix in the presence of moisture. Thus, a high sand-to-silt/clay ratio (70–80% sand to 10–15% silt/clay), with minimal gravel content, generally makes for dimensionally stable adobe. The sand is often added to materials that lack such a high ratio. Deterioration of adobe is usually due to exposure to excess moisture; cycles of swelling and shrinking of the clay minerals lead to cracking.

Brown et al. (1979) studied thin sections of adobe samples taken from three historic structures in Arizona (Tumacacori National Monument, a mission church of the late eighteenth century; Fort Bowie National Historic Site, a cavalry fort originally built in 1866; and Escalante Ruin, a prehistoric farming community dating from 900–1450). Their goals were to characterize the mineral assemblages, particle size distributions, soluble salt contents, and porosities. The data were compared with weathering observed on the structures.

Thin sections of samples from Tumacacori National Monument showed a coarse fraction consisting of rounded and angular grains of quartz, rounded fragments of quartzite, euhedral crystals of unaltered alkaline feldspar, and angular to rounded grains of calcic feldspars that were altering to clay. Chemical zoning of the feldspars was distinctive. Within the silt/clay matrix, chlorite, hematite, muscovite mica, amphibole, gypsum, and titanite were present. Twigs and grass were also found, along with some small pebbles (quartz, granites, and quartzite). The amount of clay was relatively small. Soil samples at the site were high in silt and clay, so the coarse fraction was probably added intentionally. Gypsum crystals were a common deterioration product, along with a variety of other soluble salts, indicating the movement of water within the body of the structure.

Thin sections taken from Fort Bowie National Historic Site revealed a high ratio of coarse grains to silt/clay matrix. The coarse fraction included polycrystalline quartz and feldspars (microcline, orthoclase, and plagioclase), with a wide size range. Minor amounts of micas, amphibole, and hematite were seen, over 25% of the coarse fraction was gravel, and many large pores were also present.

Thin sections of samples from Escalante Ruin showed a microcrystalline calcite-clay matrix cementing a coarse sand. Pebble fragments included quartz, granite, quartzite, and schist. Sand particles were primarily quartz and feldspar, with some biotite mica and calcite. The clay and silt contained much calcite, although it was not uniformly distributed—scattered areas were rich in clay and others were nearly pure calcite, very resistant to water erosion. The authors concluded that at the first site the action of soluble salts was the primary cause of deterioration. At the second site, particle size distribution contributed to rapid deterioration. Finally, at the third site, the adobe was well consolidated due to the presence of large amounts of calcite.

The primary purpose of firing is to create a stronger and more durable building material. Fired bricks date back to the third millennium in the Indus valley (Fig. 9.2.5) and Sumeria.

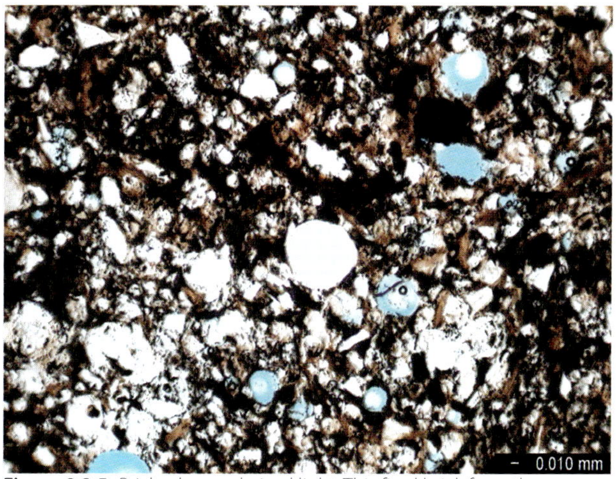

Figure 9.2.5 Brick, plane polarized light. This fired brick from the Indus valley (Harappa) has some quartz that has melted and formed amorphous silica glass (the clear, rounded areas).

The usual firing temperatures for brick range between 800 and 1200 °C. However, early brick kiln conditions often led to inconsistency and variation in firing, with some areas overfired (and distorted) while others are underfired (and softer). As with pottery, a variety of mineralogical changes occurs upon firing, as the clay and other minerals undergo high-temperature reactions (see section 8.3). The exterior of the brick, exposed to higher temperatures from the kiln, tends to vitrify and harden to a greater extent while the interior stays more porous (and softer).

Chemical and mineralogical composition of the raw material, firing temperature, and atmospheric conditions, as with pottery, will affect the final color of bricks. For example, clay rich in iron oxides and fired in an oxidizing atmosphere at relatively low temperatures will produce a red brick; higher temperatures can yield purple, brown, or gray. With clay that is rich in lime or chalk, and fired in an oxidizing environment, cream-colored, yellow, or greenish-yellow bricks result.

Decorative glazes are sometimes added to brick. These appeared as early as the sixth century B.C., at Babylon (Matson 1984; 1985), although glazed bricks were not widespread until the nineteenth century. A brown glaze can be produced by throwing salt into a hot kiln where bricks are being fired. Sometimes finely ground metallic oxides are sprinkled onto the surface before firing. Richer and durable colors can be produced by "dry dipping" fired bricks into a mixture of clay slip and crushed flint. Then, after drying, the bricks receive a second layer of the same mixture plus a metallic oxide colorant. After drying, a third final layer containing sodium, potassium, and calcium carbonate is applied and the bricks are fired for a second time. A "wet-dipping" method is simpler but produces a less durable glaze surface; the three layers are applied to unfired bricks, and only a single firing is then needed (Trench 2000).

Mineralogical studies of clays and bricks from ancient Mesopotamia (Matson 1984) found nonplastics such as quartz, chert, opal, chlorite, muscovite, calcite, iron oxides, epidote, amphiboles, and pyroxenes. Grain-size analysis showed that over 90% of the material from a Babylonian brick was finer than 0.15 mm. With such a fine-textured clay, the brick-makers had to temper the material heavily with straw. Matson studied clays at Babylon in thin section, confirming that the dominant mineral size was 0.03 mm, except for larger muscovite and chlorite flakes. Mafic minerals were relatively abundant, including minute grains of magnetite.

Childs (1988) incorporated thin-section analysis, along with a variety of other analytical techniques, to characterize the raw material used for bricks in the Kagara region of Tanzania. She tested the hypothesis that clay resources in the region were specialized and their access controlled in ancient Tanzania; she was especially interested in whether or not iron-smelters seeking suitable refractory clays for their furnaces and tuyeres would have to compete for the same resources with potters making vessels and house bricks. Field survey and ethnographic research helped her identify potential clay sources. These were sampled and made into briquettes for laboratory analyses. Thin sections were then made from brick fragments, potsherds, and tuyere sherds (from ancient iron smelting) excavated at archaeological sites in the region.

Four types of clays were available: soils containing kaolinitic clay and abundant sand particles; a more plastic clay rich in kaolinite, illite, or montmorillonite clay minerals; an even more plastic kaolinitic clay found in swamp deposits; and subsurface clay and sand grains reworked and brought to the surface at termite mounds. The ancient vessels were made with a more fine-grained clay than that used for bricks. Nonplastic inclusions added to decrease stickiness and improve drying and firing properties included grog. None of the modern clays were a good match, but Childs presumed that the clays were collected in a stream bed or erosion gully on one of the many hillsides in the region.

Some of the tuyeres also had a grog temper, added to a fine-grained kaolinitic swamp clay of high quality (which would impart the necessary refractory qualities). This raw material was matched to the modern kaolinitic swamp clays. Supporting that finding, the ancient tuyere sherds sometimes had white lumps, which were also seen in macroscopic studies in the field as small, dense lumps of white clay. These clay lumps also suggested that the tuyeres were made from wet, freshly gathered clay rather than from dried, crushed, processed, and rewetted clays. Tuyeres from one site were more coarse grained with a greater number of impurities, more like the pottery clays, and were clearly less successful for tuyere use. Two other sites also had coarse-grained and more impure clays, but they were not similar to the pottery clays.

The bricks analyzed by Childs were used in the construction of the iron-smelting furnaces. They were made with a sandy, iron oxide-rich clay that had no artificial temper component. The material would not have been very plastic, but was still suitable for forming bricks; lack of refractory qualities might have been compensated for by the thickness of the bricks. She concluded that the sandy clays found on the nearby hillsides were used for making these bricks. Differences from site to site, coupled with within-site similarities, suggested that very local clay sources were utilized.

The clays used for the furnace bricks would not be suitable for either pottery or tuyeres. Some of the other clays available would have sufficed, but would have needed tempering and transportation to the smelting sites. The swamp clays used for tuyeres would not be ideal for pottery, and the pottery clays would not be ideal for tuyere use. Thus, at most, competition for resources would have been restricted to the swamp clays. The ironworkers would have been motivated to control access to their best clay sources.

Helmi (1990) studied mud brick (adobe or sun-dried brick) from two archaeological sites in Egypt (Abu-Sir and Mataria) using thin-section petrography as one of a wide range of analytical techniques applied. She noted that mud brick was used extensively in ancient Egypt. Clay was easily available, so this material was a good choice for building in the dry Egyptian climate. Eventually, stone became the most important material for tombs and temples, whereas houses and palaces continued to be built from mud brick. Unfortunately, unfired brick was also a less durable material than stone, so few of the ancient buildings have survived. The basic raw material consisted of the usual clay, silt, sand, and fibrous organic materials such as straw. Helmi's work was aimed primarily at identifying causes of deterioration, as at the two archaeological sites under study, the bricks were found to be in very poor condition.

A thin section of an Abu-Sir brick showed quartz and orthoclase feldspar within a matrix of fine-grained silt, clay, and organic materials. A brick from Mataria showed both fine and coarse quartz and plagioclase feldspar grains within a matrix of silt and clay. Helmi concluded that the main cause of deterioration was the loose packing and the presence of angular, poorly sorted mineral grains. The quartz and feldspars appeared in a loosely packed matrix of silt, clay, and organics, thus the cementing matrix was not evenly distributed between sand grains. Large areas of fine silt and clay were seen, while other areas had concentrations of angular quartz. The overall percentage of silt and clay was also rather high. The high water table at the sites allowed leaching and erosion of the clay, as well as shrinking, cracking, and breaking of the bricks along with dissolution and migration of soluble salts. Other environmental contributors included fluctuations in temperature and humidity and windborne sand, with deleterious effects accelerated by the loose packing and poorly sorted texture of the brick.

Jerome (1993) focused on the materials analysis of sun-dried and partially fired brick from Bronze Age partitions excavated at the Minoan site of Palaikastro, Crete. The site was considered especially important because the town represented one of the best-preserved examples of Minoan town planning and architectural features, was second in size only to Knossos, and the harbor town was occupied continuously for over 700 years during the Bronze Age. Some of the brick partitions under study dissolved after their original excavation, upon exposure to the elements; others are currently being protected by shelters. A wide variety of analytical techniques was employed on samples taken, including thin-section petrography, as a first step toward the development of conservation approaches.

The clay/soil used for the bricks was identified as *terra rossa*, typical of the arid Mediterranean regions, with limestone as the parent rock. The bricks were very porous, containing small pebble inclusions (including the local limestone). Deposits of calcite and other salt crystals were visible on the surface. Jerome noted that calcite, often found in Mediterranean brick-making soils, acted as a consolidant. Thin sections revealed that the pebbles were mainly chert, marble, and "beach rock." Grog was also added as a temper material—the finer the grog particles mixed in with the clay, the stronger the resulting material. Thin sections also showed the presence of a variety of accessory minerals, including muscovite mica, calcite, hematite, kyanite, and staurolite. A lightening of the clay matrix near edges of the brick, observed macroscopically, was found to be a result of migration of soluble salts outwards. Ultimately, bricks from the remaining walls were found to be relatively stable and only light conservation intervention was recommended.

Moropoulou et al. (1993) included thin-section analysis in a study of bricks from the Corfu Venetian fortress (which actually spans from the Venetian period of the fifteenth century to the English period of the seventeenth century). Bricks of differing types were examined, as part of research into the observed varying responses to environmental conditions. There were two major visual categories of brick: red and yellow. The red bricks had a high calcite content and a medium-grained texture embedded with anorthite and quartz crystals. Iron oxides were released during firing and dispersed throughout the matrix, imparting a red color to the bricks. In thin section, microcrystalline calcite could be seen dispersed within a vitrified matrix. There were scattered large pyroxene grains (especially augite). Veins of primary calcite, secondary calcite, and small quantities of quartz were also visible. XRD analysis of a variety of red brick samples also identified the presence of anorthite, alkali feldspars, dolomite, gypsum, muscovite, iron spinel, and wollastonite in red bricks from various locations and periods. The yellow bricks, in contrast, had a lower calcite content. The less vitrified, more porous matrix was embedded with quartz crystals, and had a low concentration of iron oxides in the form of hematite crystals surrounded by hydrolyzed limonite. The illustrated thin section showed a large chunk of hematite surrounded by limonite, with a fine-grained and homogeneous matrix embedded with quartz grains of various sizes. Other minerals identified by XRD analysis included augite, calcite, dolomite, anorthite, albite, other plagioclase feldspars, alkali feldspars, muscovite, biotite, and gypsum. The yellow bricks were more susceptible to chemical and physical deterioration than the red bricks, with the exception of a group of red bricks with small pores, which were more susceptible to salt degradation (cracking). However, overall, the bricks of the fortress were relatively resistant to environmental degradation.

Pavía Santamaría et al. (1993) studied the mineralogy, texture, and physical properties of brick masonry on the San Bartolome church tower in Logoño, Spain, the only Moorish tower (sixteenth century) in the city. The bricks varied in shape, color, mineralogy, texture, hardness, and porosity. The characterization study of the bricks included thin-section analysis, along with a wide variety of analytical techniques. Minerals identified included quartz, calcite, diopside, gehlenite, wollasonite, plagioclase feldspar, and iron-titanium oxides. Seven different textures were seen, some heterogeneous and some homogeneous. Macropores were mainly absent, although mercury porosimetry found a total porosity of 30–38%, with most pores having a radius of 0.3–0.5 µm. These were visible in the illustrated thin sections, with rounded or irregular shapes. The thin sections also revealed many grain overgrowths, a result of partial transition to high-temperature phases. Also visible were gypsum crystals filling fissures and cracks, and secondary calcite replacing grains and filling pores and fractures (both are evidence of deterioration). Other analytical methods helped to identify the additional presence of some halite, aluminum salts, and potassium nitrate in some of these fissures.

Espinosa Gaitán et al. (1996) examined bricks from the Alcazaba, an eleventh–fifteenth century building in Malaga, Spain (a military fortification from the Islamic period). Deterioration in some areas was extensive, so the goal was to determine causes and mechanisms of the observed weathering. Research was also intended to shed light on the constructive chronology of the building. For that purpose, mineralogical and petrographic characterization of bricks of known date from the most important building and reconstruction periods was carried out, for comparison to bricks of uncertain date. Finally, the source of the raw materials used to make the bricks was also of interest.

Thin-section analysis contributed by providing a means of studying the overall texture of the bricks, the nature of the temper, the vitrification state of the matrix, and identification of the primary and secondary minerals present. A group of similar bricks was identified. In plane polarized light, all had a very dark red (almost black) matrix, indicating a high degree of vitrification. In some bricks, under crossed polarized light, phyllosilicates could be seen in microcrystalline areas of the matrix. The temper was composed mainly of quartz and metamorphic lithics (phyllites and slates) that were rich in phyllosilicates. In some cases, feldspars and grog (with a more vitreous matrix and finer grain size) were also present. Porosity of the bricks in this group was high, and the thin-section analysis showed that it consisted of very rounded macropores, cracks, and fissures.

The other bricks examined were found to be texturally very different from the identified group, with most having a poorly vitrified matrix, which under crossed polarizers was very birefringent and rich in phyllosilicates. Patches of more vitrified matrix appeared, especially near the edges of the bricks. The temper additive was similar in composition to that used in the identified group, but differed in size and homogeneity. One brick was quite different from all others, containing abundant foraminifer remains and calcite crystals. Porosity of all of the other bricks was also high, although the individual pores were not as large as in the identified group.

All the bricks studied were found to have gypsum as a deterioration product filling pores and fissures, and in a few cases, gypsum was also identified as part of the matrix. Secondary calcite was also found as a deterioration product in some pores and fissures. Additional work by SEM with elemental analysis identified soluble salts (halite) in the matrix, probably from marine aerosols, as well as some rare accessory minerals (pyrite, zircon, and monazite).

It was concluded that most of the bricks analyzed were made of the same raw material, with a few exceptions. Since gypsum was found in all the bricks, and was sometimes clearly a component of the matrix, gypsum deterioration products could not be attributed exclusively to the mortar. In many cases the mortar was lime, and the gypsum may in fact have originated as a component of the raw materials used for the bricks. Clays found near the building were consistent with the primary brick material, and did contain gypsum. The nonplastics in the brick were probably a natural rather than an added temper, except for the grog. Firing was probably done at 800–850 °C, based on the mineral content and degree of vitrification, although most of the earliest (eleventh-century) bricks appeared to have had higher firing temperatures than the later ones. The state of preservation of the bricks was relatively good, which the authors attributed to the preponderance of macropores over micropores, decreasing the risk of decrystallization of salts.

Pearce et al. (1996) analyzed Roman bricks that had been reused in the Torre Civica (a medieval tower) in Pavia, northwestern Italy, and recovered after its collapse in 1989. A variety of techniques, including thin-section petrography, was employed. In addition to contributing to a better understanding of Roman brick production in Pavia, goals included comparing and assessing the type of information provided by different analytical approaches. The authors began with XRD analysis to identify the minerals present. This step was followed by thin-section analysis for more detailed study of the minerals identified by XRD, and to better study the temper and the reactions that took place during firing of the bricks. The initial XRD work indicated that many of the mineralogical differences between bricks might be attributable to differing firing regimes of similar raw materials. The most abundant minerals found were quartz and feldspars (alkali and plagioclase), followed by gehlenite and hematite; other minerals were rare and sometimes absent (micas, calcite, amphiboles, and diopside). The authors pointed out that the amount of feldspar probably increased during firing, as a result of the transformation of clay minerals, carbonates, and quartz. Similarly, quartz would then have been reduced in abundance. Gehlenite would have formed by reaction of clay minerals and calcite. Calcite might be original to the paste (if firing temperatures had not exceeded 850 °C), or as coarse inclusions, but could also be a deterioration product (secondary precipitation). Clues as to firing temperature included the absence of clay minerals, scarcity or absence of calcite, and the presence of gehlenite (with occasional diopside), suggesting a temperature of between 800 and 950 °C. Gehlenite without diopside indicated a maximum of 850 °C, while the presence of both indicated a range of 900–950 °C. The absence of wollastonite and anorthite, which form as a reaction of gehlenite and quartz above 1050 °C, provided a ceiling temperature.

The thin sections showed a reddish, fine-grained matrix embedding coarser temper grains. Relic micas from the original clay material were visible. Some of the temper was identified as a deliberate addition; both the silicates (minerals and rock fragments) and carbonates varied in grain size, and had rough edges and an uneven distribution. In contrast, some tempers were identified as probably natural constituents of the clay, on the basis of grain size (finer), having smoother edges, and a more even distribution. The temper was primarily comprised of quartz, feldspar, argillaceous rock fragments, and clay pellets. There was also some calcite, but it was suggested that its prevalence was probably reduced during firing reactions. Some secondary precipitation of calcite was also seen. Additional feldspar, plus gehlenite and sometimes diopside, formed in the matrix as a result of reactions of the clay, quartz, original feldspars, and carbonates.

Three groupings were identified on the basis of texture, composition, and qualitative assessment of temper abundance. The first had a fine-grained matrix with an abundance of low-temperature components (quartz, feldspar, clay pellets and relic carbonate grains or calcite "ghosts"). The second group was similar, but with a more abundant temper which often included lithic fragments, while the third group had a fine-grained matrix varying in color from dark red to gray, and an uneven distribution of temper (comprised of quartz, feldspar, and mica). It appeared to have been fired in a less oxidizing atmosphere than the first two groups.

Among the interesting conclusions was the fact that the fabric groups identified by thin-section petrography did not correspond to groupings made by traditional archaeological criteria (which involved visual macroscopic assessments of fabric, texture, color, presence of vacuoles, and the presence, distribution, and type of inclusions). The authors proposed that some of the parameters of archaeological classification may reflect local within-kiln firing conditions, resulting in products of a single kiln firing being assigned to different fabrics. They also noted that the thin-section analysis provided important technological information regarding clay processing, tempering practices, and firing temperature. Instrumental neutron activation analysis suggested that all the bricks were made with a single clay source. Thus, the differences seen in thin section may reflect different tempering practices at different brick kilns. They may have varied in the amount and type (such as quartz-rich river sand or crushed river cobbles) of temper that was added to the clay.

In a study of Central American earthen materials, Mueller and Hansen (2001) examined porosity to characterize water transport capability, as a way of predicting problems from uprising damp and fluctuations in relative humidity. Thin sections were employed to quantify pore size distribution and to create a porosity profile from surface to interior, using the Image-Pro Plus software package. Color differences between matrix grains (less than 30 μm in size), sand grains, and macropores helped in separating and quantifying the components.

Wolf (2002) studied large medieval bricks from St. Urban, a monastery in the canton of Lucerne, Switzerland. Since medieval buildings in Switzerland were more often constructed with stone, this brick monastic complex was of special interest. The primary goal of the project was to determine the brick production technology. As is typical, a variety of analytical techniques was employed, involving mineralogical, microstructural, and open porosity analysis as well as color and fabric studies. Raw materials, fabrication methods, firing temperatures, soak times (length of time when the maximum firing temperature was maintained), range of temperatures occurring in a single firing, distribution of temperatures within the kiln, firing atmosphere, and the fuel consumption and time needed for drying and firing of these very large bricks were all of interest. Briquettes of known firing temperature were used as standards; the archaeological bricks were then analyzed in light of that data. A significant feature of this study was that analytical results were tested with a replication experiment using an imitation of a medieval kiln, and it was this experiment that elicited much of the information beyond that of firing temperatures and soak times.

Thin-section petrography was used to a somewhat limited extent in this study, primarily for fabric analysis, looking at density and opaqueness to the matrix. Two fabric types were identified, reflecting different temperature regimes: one type was characterized by a porous, birefringent matrix, representing firing below 900 °C; in contrast, the other type was characterized by a dense and more or less optically isotropic matrix, representing firing at 900 °C or above. Mineralogical analysis (XRD was used in this project) allowed these larger ranges to be refined. It should be noted that, except for identification of clay minerals, many of the minerals identified through XRD could potentially be identified in thin section as well, unless they appear as very tiny particles.

Wolf observed significant mineralogical changes that included: the breakdown of chlorite and montmorillonite at firing temperatures below 600 °C; crystallization of hematite between 600 and 700 °C; the disappearance of illite between 800 and 900 °C; the development of spinel above 900 °C; breakdown of plagioclase feldspar between 900 and 1000 °C; and the progressive disappearance of potassium feldspar above 1100 °C. He expected mullite to appear in these calcium oxide-poor clays above 1050 °C, but it was not detected. Taken together, six groups of mineral associations were developed to identify firing temperature. It should be noted that the use of these mineralogical appearances to determine temperature depends upon knowing something about the original raw materials. For example, if chlorite or plagioclase feldspar were never present to begin with, or if the raw material was particularly hematite rich, interpretations would differ. Nonetheless, the presence of abundant chlorite, plagioclase feldspar, or mullite can provide certain temperature parameters.

Wolf concluded that fabric analysis was rapid, and provided temperature estimates that were sufficient for the determination of the range of firing temperatures. Mineralogical analysis and study of vitrification structures by SEM were more time consuming, but did give estimates that were precise enough to calculate soak times. Mercury porosimetry proved to be problematic for estimating firing temperatures of these bricks, and it was emphasized that the results highlighted the importance of carefully considering the purpose of estimating firing temperatures in order to choose the appropriate method.

Pavía (2006) included thin-section petrography as part of an analytical study of Irish brick. The goal was to better understand the firing technology, provenance, and sources of raw materials of bricks used to build Rathfarnham Castle in Dublin. The castle was built around 1618 or earlier, originally with limestone rubble. Clay brick was added in 1771, during remodeling and augmentation of the building, making the structure one of the earliest uses of brick in Ireland. The bricks were compared with blocks of fired local clay. Thin-section analysis focused on a study of temper grains (larger than 0.015 mm in diameter), whether natural or artificial, to identify local or foreign sources, and on quantifying sintering and vitrification, including matrix birefringence and the amount of pores and cements. Additional work was done by SEM and XRD analysis.

Pavía concluded that the brick was hand-molded with a silica-based, noncalcareous clay of glacial origin (inferred by the mineralogy and by the presence of pebbles and a coarse matrix) that had been gathered locally. The clay included fluxes and a high percentage of nonplastic inclusions. Many of the inclusions were rock fragments of a wide variety of types such as chert, quartzite, granite, limestone, sandstone, and shale. Minerals included quartz, feldspars, and muscovite. The many pebbles and the color inhomogeneity suggested little processing of the raw material. Fragments of fossiliferous limestones contained sponge spicules, crinoids, and small foraminiferae, typically found in the limestones of the Dublin district.

Bricks were fired at relatively low temperatures ranging from 750 to 900 °C. Quartz, feldspar, and mica grains appeared in a flaky matrix with a significant amount of intergranular porosity. About one-third of the bricks showed no vitrification, while others had localized melting and welding of particles, with areas of glassing and bloating pores (indicating strong vitrification). During firing, the calcite of limestone inclusions began to break down and form calcium silicates (diopside and wollastonite). In thin section, transformations could also be seen in quartz, feldspar (overgrowths and segregated phases), and muscovite grains, and newly formed feldspar, diopside, hornblende, and andalusite were visible. High-temperature silica forms, euhedral hematite, and spinel were also formed during firing. There was much variation—some bricks had abundant sand-sized quartz, feldspar, mica, and rock fragments in a birefringent, porous matrix, suggesting low firing temperatures; others were less porous, had opaque matrices, exhibited transformed and newly formed minerals, and a matrix that was beginning to sinter and vitrify, suggesting higher firing temperatures.

9.3 Tiles

Terracotta tiles (Fig. 9.3.1) have long been used as decorative and functional architectural materials. The clay is often pressed into a rectangular frame to form a slab, or formed into more complex designs with a mold. Curved tiles are made by draping a slab formed in a rectangular mold over

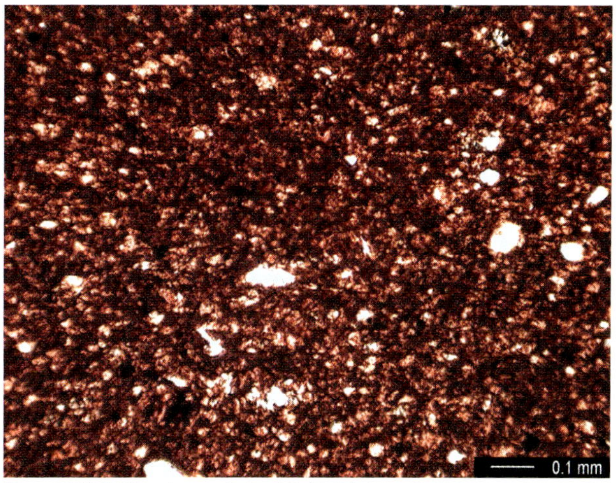

Figure 9.3.1 Tile, plane polarized light. This twentieth-century terracotta tile is very hematite rich, fired in oxidizing conditions. There is also a fine-grained sand component (mainly quartz).

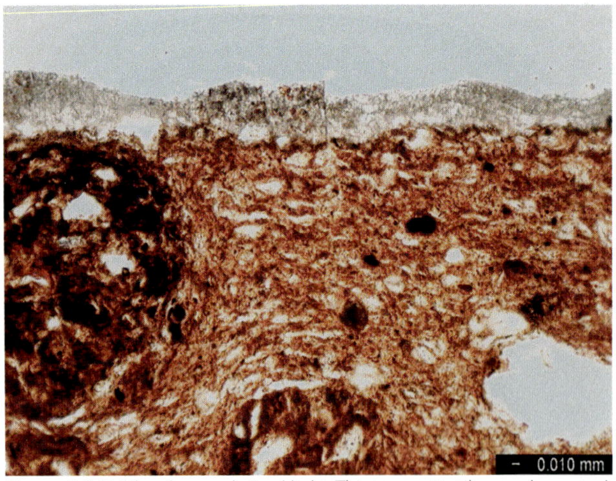

Figure 9.3.2 Tile, plane polarized light. This terracotta tile was decorated with a glaze layer. The tile body is red and hematite rich, fired in oxidizing conditions. The quartz sand ranges from coarse to fine in size, and the tile is rather porous.

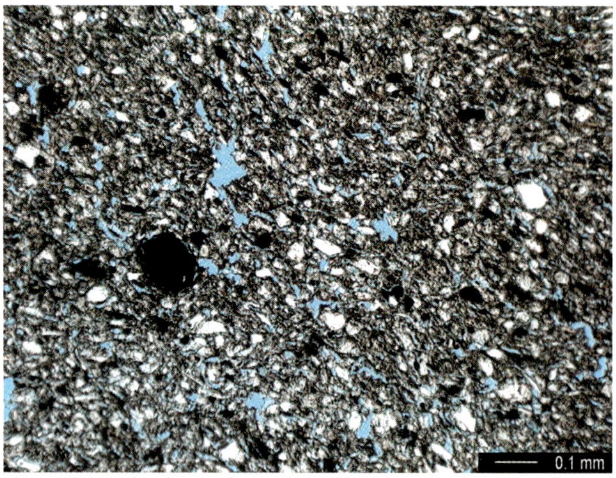

Figure 9.3.3 Tile, plane polarized light. This Chinese architectural tile was fired under reducing conditions. Reduced iron oxides are visible, and the clay is colored black by scattered carbonized organic matter. The tile is very quartz rich, as well as porous.

a rounded form. After thorough drying, terracotta tiles are fired to obtain a durable product that may be covered with paint or enamel, slipped, or glazed (Fig. 9.3.2). In addition to roofing materials, terracotta tiles are incorporated into decorative areas along the sides of buildings (Whitbread 1995).

The clay used for tiles is usually more plastic than that used for bricks; tiles may also have a hollow rather than solid construction, unlike structural bricks. The clay is usually fine grained, so that decorative details can be formed, and sand or grog tempers reduce shrinkage and distortion during drying and firing. Ceramic or plaster molds help to slowly remove moisture from the clay during long drying periods. The surface of the clay is compressed and smooth, to reduce cracking during firing, and give greater definition to the molded elements.

Firing regimes are intended to create a vitrified "fire skin" on the surface with a closed-pore structure that will protect the more permeable underbody from pollutants, rainfall, and environmental fluctuations. Adequate firing to fuse the surface is crucial because most tiles will be used in an outdoor environment. Some tiles are fired under oxidizing conditions to produce reds, oranges, and pinks (as in Figs. 9.3.1 and 9.3.2); others are fired under reducing conditions to produce deep blues, grays, and shades of black (Fig. 9.3.3).

In China, both bricks and roof tiles are often made by a hydrogen reduction process, which probably dates to the Han dynasty (206 B.C.–A.D. 220) or even earlier (Kerr and Wood 2004). The kiln is first charged with extra fuel, then the firebox and flue openings are closed. Water is introduced into the kiln from a pool on the roof, without subsequent introduction of air, and the water in the kiln then dissociates. This firing regime results in tiles with a distinctive gray color, and with a microstructure that shows significant sintering.

A variety of recipes and fabrication processes for tile production can be found. Some of these include mixtures of clay, sand, ground glass or frit, feldspar, flint, and mineral colorants. Other formulations and manufacturing processes involved attempts to imitate stone (especially in Europe and America during the mid-eighteenth to mid-nineteenth century) (Freestone et al. 1984; Kelly 1990; Fidler 1994; Kelly 1994; Stratton 1994; Henrickson and Blackman 1999; Trench 2000). In addition to roofing tiles, tiles can be used for architectural embellishment (either indoors or outdoors), for flooring material, or for specialized uses such as flue tiles (in the form of a hollow box, used in the underfloor systems that distributed hot air around Roman villas and bath houses) (González-García et al. 1990; Middleton 1997; Fawcett 1998).

Betts (1982) conducted a textural analysis of Roman brick and tile in Britain. Grain sizes were measured by using a calibrated eyepiece graticule split into 0.015 mm divisions, the size interval used for the grain size classes that were constructed. Size frequency distribution was determined using the "ribbon method" of point counting. Starting from an arbitrary point, 150 grains were measured, selecting those grains whose centers fell in a 0.15 mm band across the thin section while ensuring that 50 grains each were chosen from the top, middle, and bottom of the thin section. Size was measured using the long axis of each grain. Muscovite micas were excluded, in the belief that including tabular grains along with the other irregularly shaped ones (most of the other grains were quartz) would introduce inaccuracies.

As part of this study, an assessment was made of the variation within individual tiles, using a detailed study of one medieval and two Roman roof tiles from York. Between five and nine thin sections were made from different parts of each individual tile. Grain-size distribution was measured for each thin section, using 150 grains. From that data a mean grain-size distribution for each tile was computed. Betts found that the largest difference between thin sections from the same tile occurred in the smaller size range (< 0.015–0.045 mm), with differences ranging from 1 to 4%. He noted that part of the problem is lack of precision that can appear in measuring very small grains. In the larger size categories the difference between individual thin sections and the mean size values was usually below 1%. He then checked the effect of increasing the number of grains counted, but concluded that 150 grains was sufficient to define the fabric of the tiles under study.

Using this method, Betts analyzed tiles from a Roman kiln site at Grimescar and then compared these with tiles from an excavation at Castleford. A chi-square test of grain-size categories found that the fabrics were significantly different, with the main differences seen in the larger size categories (above 0.045 mm). As a check, the chi-square analysis was repeated using grain-size data in phi (ϕ) intervals, with the same results.

Darvill and Timby (1982) analyzed 61 stamped ceramic tiles excavated at a site in Britain dated to between the early second century and the mid-third century. The petrographic analysis of the tiles was part of research on the economics of heavy industry in Roman Britain. The first step was to try to identify whether or not the tiles from the site came from multiple production sources. The tiles were first divided into groups on the basis of their stamps and die forms. Mineralogical and grain-size analysis data were used to examine the groups and to compare them with products from known kiln sites. Some of the tiles could be matched to the Minety kiln site in Wiltshire, but one group of tiles had a stamp and die very different from the ones used at Minety, and could also be distinguished petrographically by a much higher quantity of mica. No kiln source for this type of tile has as yet been identified.

Middleton et al. (1991, 1992) developed computer programs for image analysis of ceramic thin sections for a study of Romano-British relief-patterned flue tiles. Relief patterning was applied to the tiles using wooden rollers, not primarily for decoration, but to provide a key for the application of plaster. A variety of patterns, or dies, was used. An unanswered question was whether or not the appearance of the same die pattern at different sites implies that the tiles were made at centralized production sites and then transported to other sites, or whether instead it implies the presence of itinerant tile-makers who traveled with their own toolkit of roller dies. The authors set out to answer this question for two sites in Surrey, England.

Two main fabric groups could easily be distinguished in thin section, but subdivision of the first group was difficult because there were so many similarities and so many samples to analyze (about 60 tiles). They therefore developed a computer-based procedure to establish fabric groupings, writing their own preprocessing programs, and using existing statistical packages for cluster analysis and multidimensional scaling. Images were compared and a similarity index developed using a scale of 1–5. Ultimately, it was concluded that the relief-patterned tiles were made at one of the sites, where kilns were also found, and were used there as well as transported to the other site for use in construction.

Middleton (1997) analyzed petrographic thin sections of samples from Roman flue tiles, looking at a variety of kiln sites excavated at Roman sites in England, and found that the different kilns could be characterized primarily by the amounts and relative proportions of silt and sand inclusions in the tiles. For example, one of the sites had tiles with a high proportion of fine silty inclusions, whereas those from another had only a small amount of silt. Another kiln had tiles that were moderately silty, but were characterized by relatively common flakes of mica; yet another had tiles with fine sand in an almost silt-free clay matrix. These varying fabrics were also distinguishable chemically, giving different trace element signatures.

Renaissance *sgraffito* tiles excavated in Udine, in northeastern Italy, were studied by Fabbri et al. (2000). Made with carbonate clays (containing dolomite and calcite), the mainly incised tiles were covered with a slip topped by a transparent glaze. They were either square, or formed an elongated hexagon. The tiles at Udine were made in the late fifteenth and beginning of the sixteenth century, which was significant because *sgraffito* ceramic tiles were rare in the Renaissance. They were made to cover internal walls, and were probably manufactured locally. Thin sections, along with other analytical techniques, were used to study the slip and glaze. They showed the transparent glaze layer, a clay slip layer, incised lines resulting in direct contact of glaze and body, and the ceramic body. The body was fine grained and porous, with an anisotropic matrix. Nonplastic inclusions ranged up to 100 μm in size, and consisted of angular to rounded quartz grains, altered potassium feldspar, mafic minerals (mainly biotite), plagioclase feldspar, and rare metamorphic lithics. XRD analysis also indicated newly formed pyroxene associated with gehlenite in most samples, as well as traces of hematite, calcite, and illite. Both the slip and the glaze layers were of variable thickness—the glaze varied from 50 to 160 μm, and the thicker slip layer varied from 100 to 250 μm. The authors found that the tendency was for the thicker slips to be coupled with the thinner transparent glazes, while the thinner slips were coupled with thicker transparent glaze. It was proposed that the different combinations of slip and glaze thickness might indicate different factories.

In thin section, the white slip showed a very fine-grained crystalline texture, comprised of clay and quartz. The transparent glaze layer over it often had cracks, variable-sized bubbles, and deglazing textures. Some of the tile samples had an alteration deposit (about 50 μm thick) and infilling in cracks. Sharp boundaries indicated poor reactions between layers during firing. This lack of reaction suggested a double firing process, with pre-firing of the body at a higher temperature and a second firing at a lower temperature to apply the glaze. Such a procedure would match that known to have been used in production of *sgraffito* pottery in the Renaissance.

Brownish-red to yellowish-red streaks were visible in many tiles macroscopically. It was concluded that the streaking indicated heterogeneity, and thus suggested that the raw materials were not pre-worked to obtain homogeneous ceramic bodies. The heterogeneity was probably inherent in the quarried clay, rather than being an artificial mixture of different clays intended to produce a certain aesthetic effect, because surfaces were completely covered with slip and glaze.

Some of the most famous and spectacular tiles known are the decorative architectural tiles made in Islamic lands, of glazed stonepaste/fritware (Porter 1995). The materials, production methods, and glazing techniques were intricately tied in with stonepaste ceramic vessels of the same traditions (see section 6.1). Thus, this tile material consisted of crushed quartz combined with a small amount of very plastic clay (such as white montmorillonite) and an alkaline crushed glass frit. The clay served as a flux to help induce melting of the crushed glass (or prepared frit) during firing. The result was a fused glassy network surrounding the many quartz crystals (and often a few feldspar grains) forming a very porous ceramic body. In the microstructure, the crushed quartz particles (often with some accompanying feldspars) are held together by a glass formed from the clay-frit mixture (Kingery and Smith 1984; Canby 1997). Both coarse-grained and fine-grained structures occur, ranging from white to yellow or pale brown, with varying amounts of clay versus quartz (Keblow Bernsted 2003). Most of the tiles were glazed, and examples are found throughout the Islamic world, especially in Persia, Egypt, Syria, Iraq, Turkey, and Central Asia (Figs. 9.3.4 and 9.3.5). Decorative techniques include luster, underglaze painting, and overglaze painting (Porter 1995).

The Ottoman Iznik tiles of sixteenth-century Turkey are considered extraordinary features of the mosques and other buildings of Istanbul where they can be seen in abundance. They are characterized by the usual body formed by ground quartz, with some white clay and glass frit added to form the glassy network, a layer of more finely ground quartz slip, underglaze pigments, and, finally, a colorless, transparent glaze. Kingery and Vandiver (1986) studied examples of these tiles in detail. They pointed out that large numbers of tiles were needed for decorating buildings, so it would have been advantageous to use inexpensive and easily prepared materials for the bulk of the tile body. Thus sand still containing iron impurities, giving a yellowish-gray color to the tile body, could be offset by adding a thin layer of slip made from more pure, carefully selected raw materials. This fine, white layer (quartz-clay-frit) served as a surface for application of the glaze layer.

Kingery and Vandiver illustrated tiles composed of a body of about 80% angular quartz grains from sand, held together with a fused, glassy mixture of clay and frit. The white slip on top of that incorporated finer sized crushed quartz pebbles that were freer of iron than the sand. They proposed that the raw material for the slip consisted of about 80% crushed quartz with about 10% white clay and 10% glass frit. The top layer was a glassy glaze, with a clear area of glaze penetration into the clay slip region. Underglaze painting was covered with a lead-alkali silicate glaze, giving brilliant colors with sharp outlines. The glaze was applied by pouring a slurry over the surface, giving a uniform thickness. Firing was accomplished at about 900 °C in a muffle kiln where the tiles were protected from flue gases.

Henderson and Raby (1989) noted that sixteenth-century Iznik tiles used an alkaline-lead frit rather than the usual alkaline-lime one. They found that the lead-rich frits used for Iznik tiles tended to envelop the silica fraction more completely than lime-rich frits, since they remained fluid for longer and at lower temperatures during firing. Alkali-lime frits produced a less developed network "glue" with the glassy phase only partly coating the silica crystals. In thin sections, some stonepaste tiles showed a thick layer consisting of silica (mainly quartz) grains surrounded by frit and glass, a thin slip layer, an interaction layer containing both silica grains (larger than seen in the slip) and glaze, and a surface glaze (Fig. 9.3.6). The fine slip layer provided an opaque white background for the polychrome underglaze decoration (Henderson and Raby 1989).

Fabbri et al. (2002a) included thin-section petrography in their study of decorative architectural tiles used as wall facings in the Timurid necropolis of Shahi Zinda in Samarkand, Uzbekistan. This monumental cemetery included buildings erected at different times, and the materials studied were produced at the end of the fourteenth century and throughout the fifteenth century, including richly decorated ceramic wall facings both inside and outside the buildings. The thin-section analysis showed that two different ceramic bodies could be distinguished based on grain size, color,

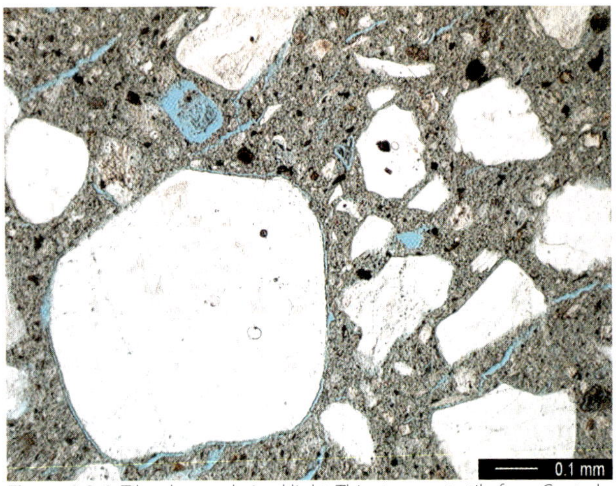

Figure 9.3.4 Tile, plane polarized light. This stonepaste tile from Central Asia has more clay content than is seen in some Islamic tiles. Along with scattered bits of frit, there are two populations of sand size: a fine-grained one within the clay, and a larger grained sand that was added.

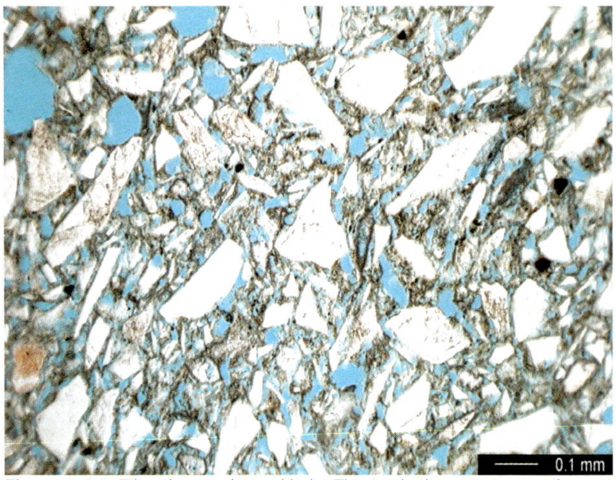

Figure 9.3.5 Tile, plane polarized light. This Iznik tile contains much less clay than the tile in Figure 9.3.4. It is composed primarily of crushed, angular quartz, surrounded by a fine network of glass. The tile is very porous.

and temper composition. Within those two groups, samples were very uniform. The first group had a very fine to fine grain size (average size of temper grains was below 100 μm), and with a temper proportion of between 5 and 10%. The clay matrix was brown-green in color, semi-isotropic, and heterogeneous due to clay aggregates and bohnerz. It was relatively porous with pores of varying shape and size (up to 1 mm). The temper included angular quartz, biotite and muscovite micas, plagioclase and alkali feldspars, opaque minerals, and rare larger inclusions of polycrystalline or metamorphic quartz with rounded edges. Calcium minerals formed during firing included pyroxene and gehlenite. The applied coating for this ceramic body was, with one exception, an opaque glaze, with a thickness averaging around 400 μm. The glazes usually had a thick dark dappling and bubbles that could appear over the entire layer. A structure was usually evident, consisting of a clearer part in contact with the ceramic body and a line of separation between two glassy areas, with a wavy pattern at the intersection. The bottom support layer directly over the ceramic material was always relatively narrow in width, generally between 20 and 50 μm.

The second group differed in having a light-colored (pale yellow) clay body and large temper size. The matrix was often very porous, with large pores (about 1 mm in size). The temper material was usually angular detrital quartz grains that could range up to 500 μm in size. There were also smaller quantities of feldspars (often partly altered and perthitic), opaque minerals, and relicts of biotite. Traces of pyroxene and cristobalite were found in only one sample. The coating for this group was usually a transparent glaze (with two exceptions), with an average thickness of around 400 μm. These glazes had more clarity and uniformity than the glazes of the first group, although there were some bubbles near the point of contact with the ceramic support. One black transparent glaze included some undissolved quartz crystals. Two other glaze samples had a thick network of fractures that could extend as far as the ceramic body interface area. Secondary material (probably calcium carbonate) was visible in the fractures. One of those samples (with a violet glaze) also had a crystalline interface layer of about 100 μm, probably quartz. One sample had a layer of white slip on top of the ceramic body, which was then covered by a transparent glaze. This technique was used to produce white relief lettering, with a thickness that could extend up to 1.2 mm. Other samples of this group exhibited essentially no body–glaze interface layer. It was proposed that this material was made of stonepaste.

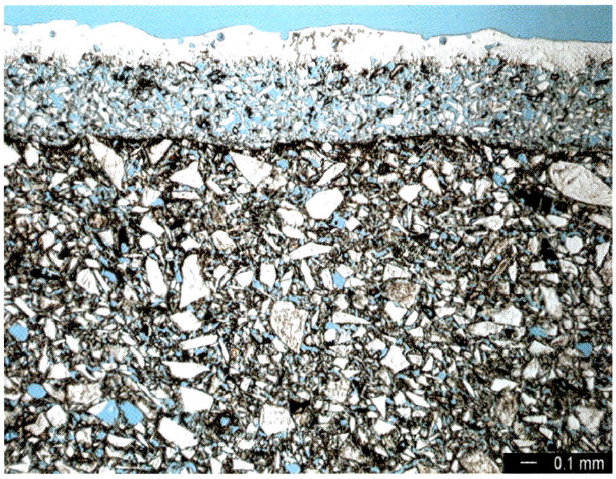

Figure 9.3.6 Tile, plane polarized light. This Iznik tile (Maintai, Dome of the Rock) shows a typical layer structure: a thick body of angular quartz grains surrounded by clay, frit, and glass; a thin fine-grained clay slip layer; a thick interaction layer containing silica grains larger than are found in the slip and glaze; and a glassy surface glaze.

For both groups, the presence of pyroxene was taken to indicate a firing temperature range between 900 and 950 °C—the one instance of cristobalite pointed to temperatures above 1000 °C. The first group was used for the older buildings, and the tiles were modeled in molds or manually. The second group was used for more recent buildings (fifteenth century), when the walls were covered with a ceramic mosaic technique that differed from the earlier technique in that the surface was always flat. Thus a plastic body was not needed, and the siliceous mixture could be used to produce a material that was whiter in color.

9.4 Casting Core Materials

When a bronze is cast by the lost-wax method, a fundamental technological choice that must be made is whether or not to cast a solid object or a hollow one. If the choice is a hollow casting, then a wax model is either built up around a core or a pre-assembled hollow wax model is filled in with core material. Iron rods or wires may be used as armatures to support the core material. Runners for pouring in metal and risers for release of gases are usually attached to the wax model, which is then covered by a fine clay (the investment) which forms an imprint of the wax image on its inner surface. The assembly is heated and the molten wax poured out of the investment, leaving a cavity for the metal. Molten metal is poured in through the runners while the displaced air and gases from the metal escape through the risers. As it cools, the metal takes the shape of the imprint formed on the inner surface of the investment by the wax model. The investment is then broken away, revealing the metal sculpture. For hollow-cast objects, the resulting sculpture is a thin shell of metal surrounding the core.

Hollow castings use considerably less metal than solid ones, but another advantage of clay cores lies in the process of casting itself. Large solid metal sculptures sometimes solidify with a flawed, spongy surface due to gas bubbles. As one solution, some of the gas from molten metal can be directed through a porous clay core, improving the initial appearance of the metal surface and thus requiring less finishing work. For ancient and historic metal sculptures, core materials based on clay were used, with sand and/or organic material incorporated (Reedy 1991). Both sand and organic materials are advantageous because they increase porosity of the clay and thus facilitate escape of gases during casting (Figs. 9.4.1–9.4.4). They also help to prevent shrinking and cracking of the core so it can better withstand the high casting temperatures (Gajurel and Vaidya 1984). Typical organic material that might be added includes straw, rice husks, or textile clippings. These materials tend to carbonize partially or completely during casting. Since fine, well-processed clays are not necessary or useful, an artisan can obtain material near the workshop, rather than having to rely on special clay deposits. Therefore, clay cores are potentially useful for characterizing sculptures from a specific region or workshop site (Reedy 1997b).

The first reported thin-section analysis of clay core materials was that of Gettens (1969) in a technical study of Chinese bronze vessels. He removed core material from a leg, handle, knob, or other appendage of 25 different vessels. Thin-section analysis of this material included identifying the mineral constituents, and, for quartz, recording grain size and grain shape. He found that all the samples were loess, similar to a loess material reported by geologists to cover wide areas of China. Therefore, he felt that analysis of core materials from ancient Chinese bronzes was unlikely to ever be useful for distinguishing between specific regions or workshop sites within China. However, later work showed that Chinese loess deposits are not in fact uniform in mineralogy or particle size, and elemental analyses of core materials from Chinese bronzes has demonstrated that regional groups can be discerned (Holmes and Harbottle 1991).

Matson (1981) conducted a petrographic analysis of thin sections of a clay core from a Classical Greek bronze horse in the Metropolitan Museum of Art, New York that was undergoing an authenticity study. He found the core to be a sandy material, mainly composed of quartz and calcite (with foraminafera). Accessories were iron-oxide minerals, sandstone fragments, feldspars, zircon, rutile, and chalcedony. He concluded that a binder of fine-grained calcite paste had been deliberately added to quartz sand to form the core material.

Schneider (1989b) and Formigli and Schneider (1993) analyzed thin sections of clay core materials from a variety of European bronzes (ancient Greek, Etruscan, Classical Greek, Hellenistic-Roman, and Renaissance), and found the materials to range from fine calcareous clays (sometimes

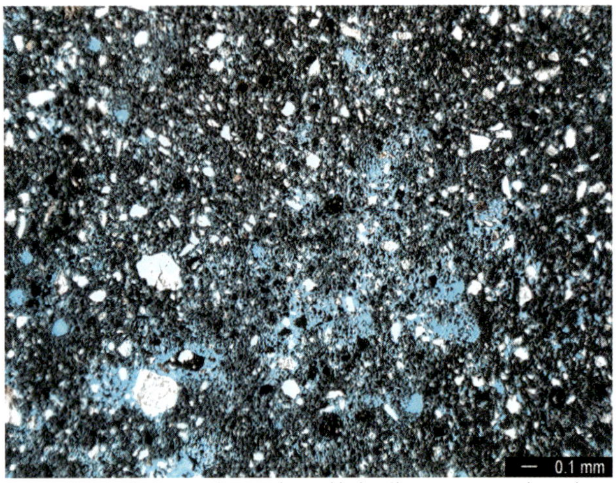

Figure 9.4.1 Clay core, plane polarized light. Clay core materials, such as this one from a Khmer bronze (a loose grain mount in blue epoxy), tend to be very sandy.

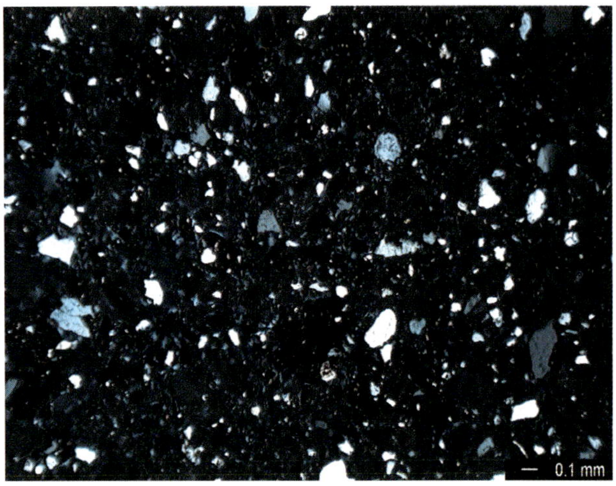

Figure 9.4.2 Clay core, crossed polarized light. The sand in this clay core (from a Khmer bronze) has a univariate distribution, and was probably all a natural component of a sandy clay (rather than having a deliberate additive).

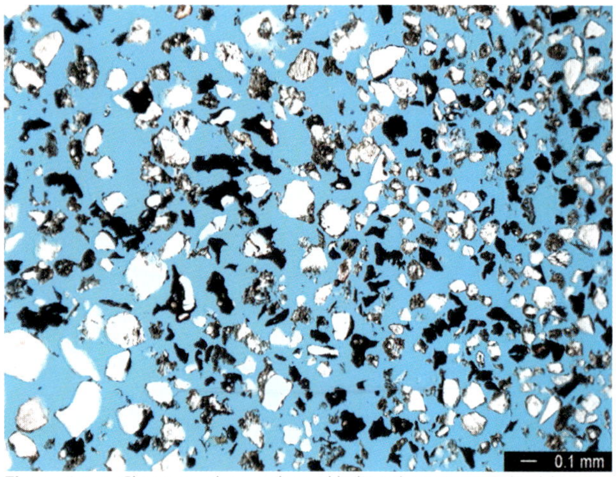

Figure 9.4.3 Clay core, plane polarized light. When mounted in blue-dyed epoxy, the sand fraction as well as the clay and carbonized organic matter of a clay core are visible. This example is from a Khmer bronze.

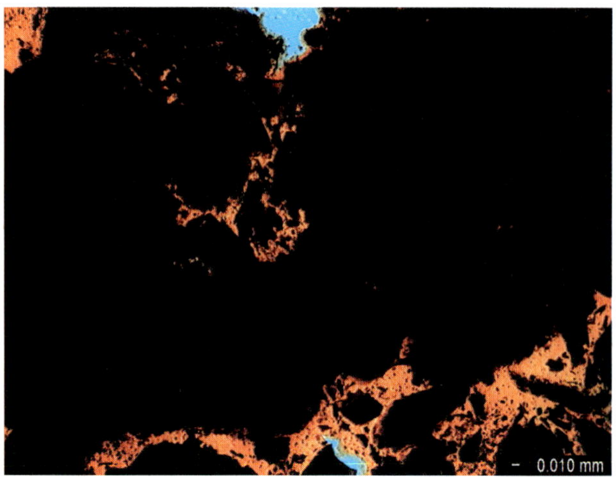

Figure 9.4.4 Clay core, plane polarized light. Organic material such as straw in a clay core tends to carbonize, because the core is enclosed within the reducing atmosphere of the bronze during the intense heat of casting.

containing microfossils) to loams with high percentages of rock fragments or coarse angular sand. Cut hair or other organic materials had been added to most of the core materials. Among the conclusions that they were able to draw from preliminary studies was that the two rare fifth-century B.C. Greek original bronzes found near Riace, Italy, were probably not cast in the same workshop.

The Riace casting cores were analyzed again, in more detail, by Lombardi and Vidale (1998). The interiors of the bronzes were completely cleaned out in the late 1980s, resulting in many kilograms of clay core material being available for analysis. Studies done at the time found that the leg and chest cores were separately formed by rolling and applying around iron bars concentric slabs of silty material mixed with animal hair, a technique known to have been used for other large Classical Greek bronzes. In addition to thin-section petrography, other methods of analysis were completed on the material. Clay cores from both had a significant amount of animal hair, as well as some vegetal fibers, added to the clay. In areas farther away from the hot metal, organic materials remained uncarbonized.

One statue had a fine-grained (less than 20–30 μm), clay-carbonate matrix with larger inclusions of an added lithic temper of two classes (100–200 μm and 250–1200 μm) —about 50 different components were identified in the material. The second statue did not have the added lithic temper; instead, it had a fine-grained (less than 10–20 μm) matrix with only a few coarse inclusions that did not exceed 100 μm. The petrographic information obtained from the clay core study was used to compare the material with the geology of southern Italy and of Greece, to try to identify the provenance of the statues. The authors concluded that the objects were probably cast in Greece, possibly on the plain of Argos, in the northeastern Peloponnesus.

Lombardi (2002) sampled casting core material from the Lupa Capitolina (Capitoline She-Wolf), the famous bronze that serves as the symbol of Rome. About 2 kg of core material was removed during conservation and restoration work on the piece. Thirty thin sections were studied, and a variety of other analytical techniques was also employed. The core was built up with layers of sandy clay-silt material, with vegetal fibers added, probably to improve the binding of the sand-silt-clay mixture and to mitigate shrinkage of the core. Since much of the organic material carbonized during casting, porosity of the core material was high (25–35%). A variety of rocks and minerals was identified (such as volcanic lithics and carbonate lithics) as well as other components such as grog, bone flakes, and unburnt wood. Based on the petrographic composition of the core material, a provenance for the sculpture was identified in the bronze foundry areas of the Lower Tiber valley, close to Rome.

Clay core materials from Renaissance bronzes were analyzed in thin section, both to investigate whether these data might be helpful in provenance and attribution studies (Reedy 1991) and to compare the results with written accounts of production methods given by three Renaissance-period artist-scholars: Biringuccio (ca. 1540), Vasari (ca. 1550), and Cellini (ca. 1568) (Weisman and Reedy 2002). Forty of the sculptures studied were well-documented pieces made for the Hapsburg Kunstkammer or known collectors, now in the collection of the Kunsthistorisches Museum in Vienna, and included pieces attributed to various regions of Italy, France, Germany, Prague, or to Flemish artists, ranging in date from 1450 to 1750. Twenty-four additional Renaissance bronzes from the J. Paul Getty Museum were also studied.

Cellini (Ashbee 1967) described a typical lost wax casting, which he used to produce a large bronze for King Francis. He first made a model in clay of the size he wanted the final piece to be, estimating the amount of shrinkage as about the thickness of a finger. He then sufficiently baked the clay core, drawing out the moisture. An even coat of wax (a finger's thickness) was spread over the core for carving. After the carving of the sculpture, the investment placed over the wax was done in two distinct materials: the inner layers, closest to the poured metal after the wax was melted out, were porous and rich in additives (such as burnt oxhorn cores, iron filings, and dung), while the outer layers were simply clay.

According to Vasari (Brown 1907), the artists usually made a clay core as large as the finished sculpture, and then perfected it as much as possible. He stated that sometimes a plaster mold would be made corresponding to the relief of a model, rather than the artist simply sculpting the wax freehand. The plaster mold was then lined with wax. Separately, an organic-rich clay core was formed over an iron rod, and, as Cellini had also described, was baked in successive layers to remove all moisture. The moisture was removed, in Vasari's words, so that during casting the metal "does not spurt nor do injury, as has happened many times, involving the death of the masters and the ruin of the work." After a final firing, the core and the wax linings from the mold pieces were joined. Vasari described the sculpture itself as having a thin skin of wax pieces smoothed over at the joins. An outer investment would be applied to that wax layer, the wax melted out, and the metal poured in.

Biringuccio (Smith and Gnudi 1959) further explained that plaster or clay molds were used if one was working from an existing bronze, marble, or terracotta model. He said the wax pattern then could be cast by pouring molten wax inside this plaster mold. If the sculpture was to be hollow rather than solid, a baked clay core was cut to the exact size needed, and the wax and plaster mold placed over it. He also described an alternative procedure in which the hollow, assembled outer mold was filled with a liquid clay core mixed with cloth clippings, dung, ashes, and a little plaster of Paris (a powdery and slightly hydrated calcium sulfate).

All the Renaissance core materials analyzed were found to be made of a sandy clay. The percentage of quartz (the major sand component) varied considerably, ranging from less than 5% up to greater than 50% of the core material, indicating no real standardization in the sandiness of the material selected by Renaissance artists. The size of quartz grains ranged from less than 0.1 mm up to 0.4 mm, which is generally considered fine or medium grained, rather than coarse.

Cellini (Ashbee 1967) clearly recognized the usefulness of the sand component of clay cores. He suggested that the clay could be collected in many places, but was preferably obtained near rivers, "for there it has a certain sandiness." Biringuccio (Smith and Gnudi 1959), on the other hand, believed that in selecting clay for core material "aside from actual trial, I believe there is little that can help you, since the clay itself has no color or visible sign that I know of to show how satisfactory it is." He suggested trying clays collected in fields that had not been cultivated very long and were not well manured, from kilns where house tiles were made, or from river banks. He also advised against using very pure, unsandy, clays.

Many other accessory minerals were found in the core materials, but most appeared to be natural constituents of sand or of a sandy clay rather than deliberate additions. The most common accessories included plagioclase feldspar, alkali feldspars, micas (mainly muscovite and chlorite), clinopyroxenes, zircon, apatite, epidote, glauconite, lithic fragments, hematite, and calcite. In the case of hematite, which frequently comprised more than 5% of a specimen, the implication is that iron-rich clays were being selected by some artists, either deliberately or as an accident of workshop location. In contrast, calcite rarely comprised more than 5% of a specimen, and thus appears to be an accidental rather than deliberate component.

Organic material was found in the majority of the specimens. However, it often appeared as carbon that was present at such a low percentage (less than 5%) that it possibly derived solely from natural organic matter in the soil rather than from intentional organic additions. Less than one-third of the samples had enough carbon to conclude it might have originated from added organic matter.

Cellini (Ashbee 1967) provided an interesting description of how he prepared an organic-rich clay for his core material. After collecting a sandy clay, he dried it and put it through a coarse sieve to remove pebbles, roots, glass particles, and other contaminants. He then mixed the sieved clay with cloth frayings (about half as much cloth as clay). The mixture was hydrated to the consistency of dough, beaten with a thick iron rod, and left to sit for about five months so the cloth frayings would begin to decompose. He believed this procedure helped the core material to hold together better during casting.

A few of the core samples were found to contain gypsum (hydrous calcium sulfate) or anhydrite (calcium sulfate), the primary constituents of plaster of Paris, often as minor constituents of the sandy clay core. Because they are somewhat soluble in water, it is possible that some of this component in cores could have dissolved during core and investment production, when the material was wet. (It is important to use non-aqueous methods of preparing petrographic specimens in order to ensure that the gypsum component is not affected.) As is common in plaster of Paris, the gypsum found in Renaissance core materials was the beta hemihydrate phase (Green 1984).

These results, coupled with other studies of individual bronzes and/or the works of specific Renaissance artists (Stone 1982; Pichler and Vendl 1987; Sauer et al. 1988; Schneider 1989b; Formigli and Schneider 1993; Bewer 1994) indicate that, although a plaster-based core was not unknown during the Renaissance, it was used infrequently and probably within specific workshops. For example, Stone (1982) found that the artist Antico (ca. 1460–1528) produced bronzes with plaster piece molds lined with a thin shell of wax into which a core, consisting of a liquid plaster of Paris mixed with fine sand, was poured. The piece mold would subsequently have been removed, leaving a hardened plaster-sand core with a wax surface. Samples from three bronzes by Antico showed that the core material contains anhydrite, gypsum, and quartz. Stone noted that Cellini, in 1565, described bronze cores composed primarily of clay, while mentioning briefly an alternative type composed of a mixture of plaster, ash, and powdered brick. Cellini went on to caution that the Tuscany gypsum is not as good for that purpose as the gypsum of France, Mantua, and Milan. Stone discussed other mid-sixteenth-century texts concerning casting, one of which suggests the use of a damp clay with cloth clippings, horse dung, bone ash, and a little plaster of Paris.

Multivariate statistical analyses (discriminant analysis) were carried out using petrographic data for 64 Renaissance bronzes. The results indicated that it may be possible to use such data to distinguish between products of different workshops (Reedy 1991). However, group sizes representing objects of secure attribution as to place of production were still very small, so these results remain preliminary.

Thin-section analysis of clay core materials from Himalayan bronzes (Afghanistan, northwest Pakistan, northern India, Nepal, and Tibet) was helpful in supporting or refuting provenance attributions for many of the sculptures (Reedy 1991, 1997b). Data were recorded for 45 variables. Since many of the core materials were rich in carbonized organic material, the percentage of carbon was assessed by comparison to visual estimation charts. The core materials were comprised of very sandy clay, so the percentage of quartz was also estimated; average size was categorized as fine, medium, or coarse, and average shape was recorded as angular/subangular or rounded/subrounded. Quartz textures were noted as present or absent (such as inclusions, polygonized, polycrystalline, microcrystalline, vacuoles, authigenic overgrowths, undulatory extinction, or embayments), and the presence or absence of other minerals was recorded. Calcite and hematite sometimes occurred abundantly versus scarcely, but most other non-quartz minerals occurred in minor amounts, if at all. The most commonly occurring minerals in Himalayan core materials were plagioclase feldspar, microcline, other alkaline feldspars, chlorite, muscovite, biotite, hypersthene, hornblende, zircon, apatite, metamorphic and sedimentary lithics, epidote, clinozoisite, sillimanite, and olivine.

All the presence/absence variables were coded as 1/0 so that these data could be included in statistical analyses along with the percentage variables. A series of discriminant analyses was performed, using sculptures of more secure provenance as the group of "knowns" for regions of origin. The multivariate combinations of variables that best discriminated between those groups were calculated, then used to construct a mathematical classification function to classify each of the "uncertain" objects to the regional group it best fit. Using only the thin-section data for clay cores, 100% of known objects could be correctly classified for some of the regional comparisons. The method was most successful when comparing fewer groups of related regions, rather than trying to compare many groups or vast regions. However, improvement was possible if the data were combined in a discriminant analysis with other datasets (such as metal composition and/or casting and decorating variables).

Clay core materials from 69 Southeast Asian bronzes were analyzed by the same method as that employed for the Himalayan bronzes, and a statistical analysis of the data used to identify where art-historical attributions might be supported or otherwise (Reedy 1997c; Woodward 1997). As with Himalayan bronzes, these cores tended to be made of a coarse and sandy clay with added organic material (straw and rice husks) that fully or partially carbonized during casting. The presence of quartz-based sand component varied greatly. A large number of minerals in addition to quartz were identified (micas, feldspars, amphiboles, clinopyroxenes, hematite, epidote, tourmaline, and metamorphic lithics such as schist and quartzite).

Preliminary multivariate statistical analyses (using discriminant analyses) of the thin-section data showed that characterizing clay core materials was very promising as a method for supporting or refuting attributions of bronzes to specific regions or workshops (Ayutthaya, provincial Auytthaya, Sukothai, Lan Na, central Thailand, and Khmer). Although group sizes were too small to make definitive conclusions, it was clear that thin-section data had potential for helping to clarify attributions of Southeast Asian bronzes.

Digital image analysis with Image-Pro Plus was used in the study of clay core materials from Khmer sculptures (in the collections of the National Museum of Phnom Penh, the National Museum of Bangkok, and the Los Angeles County Museum of Art) (Reedy and Meyers 2007). Quantitative data for a wide variety of parameters were available to supplement mineral identifications and the presence/absence data for nine quartz textures and 15 non-quartz minerals. Samples were double vacuum impregnated in blue embedding resin, which helped to preserve sample material during grinding processes while also facilitating areal percentage calculations. Four minerals occurred frequently: chlorite, plagioclase feldspar, muscovite, and hematite. Nine other components occurred often enough to have potential for characterizing core materials for specific workshop locations: biotite, epidote, zircon, rutile, pyroxene, microcline feldspar, other alkali feldspars, apatite, and metamorphic lithics. Among the variables related to silica minerals and textures, some occur so often they are unlikely to ever be useful in distinguishing workshops: quartz with undulose extinction, polycrystalline quartz, microcrystalline quartz, polygonized quartz, and inclusions in quartz grains. Less common were quartz grains containing vacuoles, cristobalite, silica glass, and fibrous (cryptocrystalline) quartz. Most core materials also contained moderate amounts of carbonized organic matter.

The quantitative data produced by image analysis aided interpretations of fabrication technology. The analysis found that the technology of manufacture of Khmer clay core materials left traces of various kinds, and there is the potential to detect those traces in the laboratory through thin-section petrography. Khmer bronze-casters appeared to have used a coarse clay with a natural sand content to build their cores. The amount of sand and the variety of mineral grain types in the sand may be an indication of where along the floodplain they collected the clay, and therefore those values may prove to be as useful in characterizing workshops as identification of individual minerals present. Non-quartz minerals and various silica minerals and textures that occurred often enough to have the potential to characterize a workshop—but not so often as to appear in all workshops—may also be useful. Total porosity and plasticity, achieved through the sandiness of the material selected combined with the amount of organic material present or added, may be a marker of technological style. Forgers working in a modern city may purchase refined clay, thus the criteria of processed as opposed to unprocessed material may also contribute to resolving some authenticity questions. Exploratory stepwise discriminant analyses were promising, indicating that it should be possible to differentiate among some Khmer bronze-casting workshops on the basis of data such as percentage of sand, quartz size characteristics, and presence or absence of some mineral types and textures.

References

Abbate-Zohar, M. A., U. Zezza, F. Ventale, and M. Setti. 1985. Decay stage and causes of the terracotta ornaments (cloister of the Carthusian monastery, Pavia, Italy). In *Proceedings of the 5th International Congress on Deterioration and Conservation of Stone*, ed. G. Félix, 443–51. Lausanne: Presses Polytechniques Romandes.

Abd El-Hady, M. M. 1994. Durability of monumental marble in Egypt. In *Proceedings of the 3rd International Symposium on the Conservation of Monuments in the Mediterranean Basin*, eds. V. Vassina, H. Ott, and F. Zezza, 631–4. Venice: Soprintendenza ai Beni Artistici e Storici di Venezia.

Adams, A. E. and W. S. MacKenzie. 1998. *A color atlas of carbonate sediments and rocks under the microscope*. New York: Wiley.

Adams, A. E., W. S. MacKenzie, and C. Guilford. 1997. *Atlas of sedimentary rocks under the microscope*. [Reprint of 1984 edition.] New York: Longman.

Adan-Bayewitz, D. and M. Wider. 1992. Ceramics from Roman Galilee: A comparison of several techniques for fabric characterization. *Journal of Field Archaeology* 19(2): 189–205.

Agarwal, P. K. 1991. Study of lithic materials in the Victoria Memorial Building. *Conservation of Cultural Property in India* 24: 47–52.

Agresti, A. 2002. *Categorical data analysis*. New York: Wiley- Interscience.

Ahmed, W. 2000. Petrographic examination methods. *Buehler Tech-Notes* 3(5): 1–6.

al-Saa'd, Z. and A. Roussan. 1999. Determination of the initial firing conditions of Ayyubid/Mamluk painted pottery excavated from northern Jordan. *Geoarchaeology* 14(4): 333–49.

Alaimo, R., R. Giarrusso, L. Lazzarini, F. Mannuccia, and P. Meli. 1996. The conservation problems of the theatre of Eraclea Minoa (Sicily). In *Proceedings of the 8th International Congress on Deterioration and Conservation of Stone*, vol. 2, ed. J. Riederer, 1085–95. Berlin: Möller Druck und Verlag.

Alberti, S. A., F. Antonelli, S. Cancelliere, L. Lazzarini, F. Mannuccia, and F. Santalucia. 2000. Castlemaniace, Syacuse (Sicily): The deterioration of the marble of the monumental portal and window. In *Proceedings of the 9th International Congress on Deterioration and Conservation of Stone*, vol. 2, ed. V. Fassina, 649–60. Amsterdam: Elsevier.

Allen, C. S. M. 1991. Thin sections of Bronze Age pottery from the East Midlands of England. In *Recent developments in ceramic petrology*, eds. A. Middleton and I. C. Freestone, 1–15. London: British Museum.

Allen, J. B. and T. J. Charsley. 1968. *Nepheline-syenite and phonolite*. London: HMSO.

Allessandrini, G., R. Peruzzi and L. Di Capitani. 1976. Investigations on the decay of Candoglia marble used in the Milan Duomo. In *The conservation of stone*, ed. R. Rossi-Manaresi, 137–67. Bologna: Centro per la Conservazione delle Sculpture All'Aperto.

Allison, P. 1968. *African stone sculpture*. New York: Frederick A. Praeger.

Allman, M. and D. F. Lawrence. 1972. *Geological laboratory techniques*. New York: Arco.

Alonso, F. J., R. M. Esbert, J. Alonso, and J. Ordaz. 1994. Saline spray action on a treated dolomitic stone. In *The conservation of monuments in the Mediterranean basin*, eds. V. Fassina, H. Ott, and F. Zezza, 867–70. Venice: Soprintendenza ai Beni Artistici e Storici di Venezia.

Alvarez Arza, R., M. Catapotis, M. A. Cau Ontiveros, P. M. Day, and A. M. Rauret i Dalmau. 2002. Pottery production in Bronze Age Catalonia: The case of Pixarelles Cave. In *Modern trends in scientific studies on ancient ceramics*, eds. V. Kilikoglou, A. Hein, and Y. Manniatis, 265–76. Oxford: British Archaeological Reports International Series 1011.

Amadori, M. L., B. Fabbri, and M. Pacciarelli. 2002. The interpretation of the complex fabrics of Bronze Age potteries from Imola (Italy). In *Modern trends in scientific studies on ancient ceramics*, eds. V. Kilikoglou, A. Hein, and Y. Manniatis, 185–93. Oxford: British Archaeological Reports International Series 1011.

Antonelli, F., F. Bernardini, S. Capedri, L. Lazzarini, and E. Montagnari Kokeli. 2004. Archaeometric study of protohistoric grinding tools of volcanic rocks found in the Karst (Italy-Slovenia) and Istria (Croatia). *Archaeometry* 46(4): 537–52.

Ardeleanu-Jansen, A. 2002. The terracotta figurines from Mohenjo Daro: Considerations on tradition, craft and ideology in the Harappan civilization (*c.* 2400–1800 BC). In *Indian archaeology in retrospect.* Volume II: *Protohistory, archaeology of the Harappan civilization*, eds. S. Settar and R. Korisettar, 205–22. New Delhi: Manohar.

Ariño, N., A. Gomez-Bolea, and C. Saiz-Jimenez. 1996. *Rocella Phycopsis* Ach.: Colonization and weathering processes on granite. In *Degradation and conservation of granitic rocks on monuments*, 400–404. Brussels: European Commission, Directorate-General XII, Science, Research and Development.

Arnold, D. E. 1971. Ethnomineralogy of Ticul, Yucatan potters: Etics and emics. *American Antiquity* 36: 20–40.

—. 1972. Mineralogical analyses of ceramic materials from Quinua, department of Ayacucho, Peru. *Archaeometry* 14(1): 93–102.

Arnold, D. E., H. Neff, and M. D. Glascock. 2000. Testing assumptions of neutron activation analysis: Communities, workshops, and paste preparation in Yucatán, Mexico. *Archaeometry* 42(2): 301–16.

Ashbee, C. R. trans. 1967. *The treatises of Benvenuto Cellini on goldsmithing and sculpture*. New York: Dover Publications.

Asher, F. M. 1987. Eastern Indian sculptures: A preliminary analysis of stone materials. *Journal of Field Archaeology* 14(2): 235–41.

Ashurst, J. and F. G. Dimes. 1990. *Conservation of building and decorative stone*, vol. 1. London: Butterworth-Heinemann.

Atterbury, P. 1982. *The history of porcelain*. New York: William Morrow and Company.

Baker, P. E. 1993. Archaeological stone of Easter Island. *Geoarchaeology* 8(2): 127–39.

Banchelli, A., F. Fratini, M. Germani, P. Malesani, and C. Manganelli Del Fa. 1997. The sandstones of Florentine historic buildings: Individuation of the marker and determination of the supply quarries of the rocks used in some Florentine monuments. *Science and Technology for Cultural Heritage* 6: 13–22.

Banerji, P. K. 1982. The khondalites of Orissa, India: A case history of confusing terminology. *Journal of the Geological Society of India* 23: 155–59.

Barbin, V., K. Ramseyer, D. Decrouez, and R. Herb. 1989. Marbres blancs: Caractérisation par cathodoluminescence. *Comptes Rendus de l'Académie des Sciences: Série II, Mécanique, Physique, Chimie, Sciences de l'Univers, Sciences de la Terre* 308: 861–66.

Barbin, V., U. Zezza, and E. S. Pardo. 1995. The cathodomicrofacies of Macael white marbles (Almeria, Spain). In *The study of marble and other stones used in antiquity*, eds. Y. Maniatis, N. Herz, and Y. Basiakos, 131–35. London: Archetype.

Barker, A. J. 1990. *Introduction to metamorphic textures and microstructures*. New York: Chapman and Hall.

Barlow, J. A. 1991. New light on red polished ware. In *Cypriot ceramics: Reading the prehistoric past*, eds. J. A. Barlow, D. L. Bolger, and B. Kling, 51–57. University Museum Monograph 74. Philadelphia, PA: A. G. Leventis Foundation and the University Museum of Archaeology and Anthropology, University of Pennsylvania.

Barnett, W. K. 1991. The identification of clay collection and modification in prehistoric potting at the early Neolithic site of Balma Margineda, Andorra. In *Recent developments in ceramic petrology*, eds. A. Middleton and I. C. Freestone, 17–37. London: British Museum.

Barone, G., S. Ioppolo, G. Puglisi, and G. Tigano. 2002. Archaeometric results and archaeological problems of pottery from the area of Messina (Sicily). In *Modern trends in scientific studies on ancient ceramics*, eds. V. Kilikoglou, A. Hein, and Y. Manniatis, 219–26. Oxford: British Archaeological Reports International Series 1011.

Bartl, K., G. Schneider, and S. Bohme. 1995. Notes on "brittle wares" in north-eastern Syria. *Levant* 27: 165–77.

Basso, E., D. Binder, B. Messiga, and M. P. Riccardi. 2006. The Neolithic pottery of Abri Pendimoun (Castellar, France): A petro-archaeometric study. In *Geomaterials in cultural heritage*, eds. M. Maggetti and B. Messiga, 33–48. London: Geological Society, Special Publications 257.

Basu, A. 1985a. Influence of climate and relief on composition of sands released at source areas. In *Provenance of arenites*, ed. G. G. Zuffa, 1–18. Dordrecht: D. Reidel.

—. 1985b. Reading provenance from detrital quartz. In *Provenance of arenites*, ed. G. G. Zuffa, 231–47. Dordrecht: D. Reidel.

Bedal, L., S. Fleming, M. De Schauensee, and R. Hancock. 1995. Second millennium B.C. pottery at Hasanlu Tepe and Dinkha Tepe: INAA and petrographic studies. In *Materials issues in art and archaeology IV*, eds. P. B. Vandiver, J. R. Druzik, J. L. Galvan Madrid, I. C. Freestone, and G. S. Wheeler, 453–67. Pittsburgh, PA: Materials Research Society.

Behre, C. H. Jr. 1933. *Slate in Pennsylvania*. Harrisburg, PA: Department of Internal Affairs, Topographic and Geologic Survey Bulletin M16.

Bell, F. G. 1990. Engineering geology and building stones of historical monuments: construction materials; geological origin; quarries. In *Engineering geology of ancient works, monuments and historical sites*, eds. P. G. Marinos and G. C. Koukis, 1867–74. Rotterdam: A. A. Balkema.

Berry, R. M. 1969. Cylinder seal mineralogy and petrology. *American Journal of Archaeology* 73(1): 67–69.

Best, M. G. 1982. *Igneous and metamorphic petrology*. San Francisco, CA: W. H. Freeman.

Betancourt, P. P. 1979. *Vaselike ware: An early Bronze Age pottery style in Crete*. Göteborg: Paul Äströms Forlag.

Betancourt, P. P. and G. H. Myer. 1994. Phyllite fabrics in eastern Crete. In *The Ceramic Cultural Heritage: Proceedings of the International Symposium "The Ceramics Heritage" of the 8th CHMTEC-World Ceramics Congress and Forum on New Materials, Florence, Italy, 28 June–2 July, 1994*, ed. P. Vincenzini, 395–99. Faenza: TECHNA.

Betts, I. M. 1982. Roman brick and tile: A study in fabric variability. In *Current research in ceramics: Thin-section studies*, eds. I. C. Freestone, C. Johns, and T. Potter, 63–71. London: British Museum Occasional Paper No. 32.

Beurdeley, J.-M. 1989. *Majapahit*. Paris: Beurdeley & Cie.

Bewer, F. 1994. Studying the technology of Renaissance bronzes. In *Materials issues in art and archaeology IV*, eds. P. B. Vandiver, J. R. Druzik, J. L. G. Madrid, I. C. Freestone, and G. S. Wheeler, 701–09. Pittsburgh, PA: Materials Research Society.

Beynon, D. W., J. Donahue, R. T. Schaub, and R. A. Johnston. 1986. Tempering types and sources for Early Bronze Age ceramics from Bab edh-Dhra' and Numeira, Jordan. *Journal of Field Archaeology* 13(3): 297–305.

Bianchetti, P. L., G. Lombardi, and G. Meucci. 1982. Study of the degradation of "tuff" blocks used in the Roman temple of Cibele (Rome, Italy). In *Proceedings of the 4th International Congress on Deterioration and Preservation of Stone Objects*, eds. K. L. Gauri and J. A Gwinn, 29–38. Louisville, KY: University of Louisville.

Bianchetti, P. L., G. Lombardi, S. Marini, and C. Meucci. 1994. The volcanic rocks of the monuments of the Forum and Palatine (Rome): Characterization, alterations, and results of chemical treatments. In *Lavas and volcanic tuffs*, eds. A. E. Charola, R. J. Koestler, and G. Lombardi, 83–105. Rome: ICCROM.

Biggs, D. L. 1957. *Petrography and origin of nodular cherts*. Urbana, IL: Illinois State Geological Survey.

Bignot, G. 1985. *Elements of micropalaeontology*. Boston, MA: International Human Resources Development Corporation.

Birks, T. 1998. *The complete potters' companion*. Boston, MA: Little, Brown and Company.

Biscontin, G., C. Felix, P. Maravelaki, and E. Zendri. 1992a. Characteristics, weathering forms and mechanisms of Istria stone in Venice. In *The conservation of monuments in the Mediterranean basin*, eds. D. Decrouez, J. Chamay, and F. Zezza, 141–53. Geneva: Musée d'Histoire Naturelle and Musée d'Art et d'Histoire.

Biscontin, G., P. Maravelaki, and E. Zendra. 1992b. The case of the "Portable di S. Maria Dei Servi" in Venice: Results of physico-chemical inquiries. In *The conservation of monuments in the Mediterranean basin*, eds. D. Decrouez, J. Chamay, and F. Zezza, 155–66. Geneva: Musée d'Histoire Naturelle and Musée d'Art et d'Histoire.

Blackman, M. J., S. Mery, and R. P. Wright. 1989. Production and exchange of ceramics on the Oman peninsula from the perspective of Hili. *Journal of Field Archaeology* 16(1): 61–77.

Blanc, A. 1995. Marbles and decorative stones used in French medieval monuments. In *The study of marble and other stones used in antiquity*, eds. Y. Maniatis, N. Herz, and Y. Basiakos, 57–63. London: Archetype.

Blanc, A., P. Blanc, and D. Magnan 1995. Les marbres blancs du site de la Bauve à Meaux: Un grand absent, le marbre pyrénéen. In *Les marbres blancs des Pyrénées: Approches scientifiques et historiques*, comps. J. Cabanot et al., 119–30. Saint-Bertrand-de-Comminges: Musée Archéologique Départemental.

Blanc, A., L. L. Holmes, and G. Harbottle. 2002. Lutetian limestones in the Paris region: Petrographic and compositional characterization. In *Interdisciplinary studies on ancient stone*, eds. J. J. Herrmann, Jr., N. Herz, and R. Newman, 103–09. London: Archetype.

Blanco Varela, M. T., E. Menendez, and M. Hoyos. 1992. Study of surface decay of the marbles and serpentine from the Descalzas Reales convent at Madrid. In *The conservation of monuments in the Mediterranean basin*, eds. D. Decrouez, J. Chamay, and F. Zezza, 167–75. Geneva: Musée d'Histoire Naturelle and Musée d'Art et d'Histoire.

Bond, A. 1976. Multiple sources of pumice in the Aegean. *Nature* 259(5540): 194–95.

Bouchain, I. and B. Velde. 2001. Grain distribution by image analysis of thin sections in some Gaulo-Roman common ware, St. Marchel (Indre) France. In *Archaeology and clays*, ed. I. C. Druc, 71–80. Oxford: British Archaeological Reports.

Boulter, S. 1992. Back from the grave: the conservation of a slate burial chest. *Orientations* 23(12): 43–45.

Bouquillon, A., B. Barthelemy de Saizieu and A. Duval. 1995. Glazed steatite beads from Merhgarh and Nausharo (Pakistani Balochistan). In *Materials research in art and archaeology IV*, eds. P. B. Vandiver, J. R. Druzik, J. L. Galvan Madrid, I. C. Freestone, and G. S. Wheeler, 527–38. Pittsburgh, PA: Materials Research Society vol. 352.

Boxt, M. A. and C. L. Reedy. 1985. Preliminary thin section studies of chert artifacts from Northern Belize. *Journal of New World Archaeology* 5(4): 13–23.

Bradley, S. M. and A. P. Middleton. 1988. A study of the deterioration of Egyptian limestone sculpture. *Journal of the American Institute for Conservation* 27(2): 64–86.

Braun, D. 1983. Pots as tools. In *Archaeological hammers and theories*, eds. J. A. Moore and A. S. Keene, 107–34. New York: Academic Press.

Bray, T. 1995. The Panzaleo puzzle: Non-local pottery in northern highland Ecuador. *Journal of Field Archaeology* 22(2): 137–56.

Broekmans, T., A. Adriaens, and E. Pantos. 2006. Insights into North Mesopotamian "metallic ware." *Archaeometry* 48(2): 219–27.

Bronitsky, G. 1986. The use of materials science techniques in the study of pottery construction and use. In *Advances in archaeological method and theory*, vol. 9, ed. M. Schiffer, 275–330. New York: Academic Press.

Bronitsky, G. and R. Hamer. 1986. Experiments in ceramic technology: The effects of various tempering materials on impact and thermal-shock resistance. *American Antiquity* 51: 89–101.

Brown, G. B. ed. 1907. *Vasari on technique*. London: J. M. Dent.

Brown, P. W. and J. R. Clifton. 1978. Adobe. I: The properties of adobe. *Studies in Conservation* 23(4): 139–46.

Brown, P. W., C. R. Robbins, and J. R. Clifton. 1979. Adobe. II: Factors affecting the durability of adobe structures. *Studies in Conservation* 24(1): 23–39.

Brown, V. M. and J. A. Harrell. 1995. Topological and petrological survey of ancient Roman quarries in the Eastern Desert of Egypt. In *The study of marble and other stones used in antiquity*, eds. Y. Maniatis, N. Herz, and Y. Basiakos, 221–34. London: Archetype.

Budak, M. 1991. The function of shell temper in pottery. *Minnesota Archaeologist* 50: 53–59.

Bugini, R., L. Folli and F. Rossi. 1995. Stone materials from a Roman Republican temple, Brescia (northern Italy). In *Conservation, preservation and restoration: Traditions, trends and techniques*, eds. G. Kamalakar and V. Pandit Rao, 191–96. Hyderabad: Birla Archaeological and Cultural Research Institute.

Bunch, B. and C. I. Fell. 1949. A stone-axe factory at Pike of Stickle, Great Langdale, Westmorland. *Proceedings of the Prehistoric Society* XV: 1–20.

Busdraghi, P., G. Cruciani-Fabozzi, M. G. Pancani, M. L. Polichetti, and S. Vannucci. 1991. Indagini sulla statuaria commissionata dai Buonaccorsi per le loro dimore di Macerate e Monte Santo. In *Le pietre nell'architettura: Struttura e superfici*, eds. G. Biscontin and D. Mietto, 375–87. Padova: Libreria Progetto Editore Padova.

Butler, S. 1989. Steatite in Norse Shetland. *Hikuin* 15: 193–206.

Buxeda I Garrigós, J., V. Kilikoglou, and P. M. Day. 2001. Chemical and mineralogical alteration of ceramics from a Late Bronze Age kiln at Kommos, Crete: The effect on the formation of a reference group. *Archaeometry* 43(3): 349–71.

Buxeda I Garrigós, J., M. A. Cau Ontiveros, and V. Kilikoglou. 2003. Chemical variability in clays and pottery from a traditional cooking pot production village: Testing assumptions in Pereruela. *Archaeometry* 45(1): 1–17.

Cabanot, J. 1995. Les chapiteaux de marbre antérieurs à l'époque romane de la région Aquitaine. In *Les marbres blancs des Pyrénées: Approches scientifiques et historiques*, comps. Jean Cabanot et al., 223–60. Saint-Bertrand-de-Comminges: Musée Archéologique Départemental.

Calia, A., L. Lazzarini, E. Pellegrino, M. Preite Martinez, and B. Turi. 1992. The identification of the marbles of the great apsidal window of Bari cathedral: A mineralogical, petrographic and isotopic study. *Science and Technology for Cultural Heritage* I: 191–99.

Campbell, J. W. P. and W. Pryce. 2003. *Brick: A world history*. London: Thames & Hudson.

Canby, S. R. 1997. Islamic lustreware. In *Pottery in the making: Ceramic traditions*, eds. I. C. Freestone and D. Gaimster, 110–15. Washington, DC: Smithsonian Institution Press.

Caner-Saltik, E. N., S. Demirci, A. Türkmenoğlu, A. Özgenoğlu, H. Göktürk, A. Özer, H. Böke, and E. Inalpulat. 1995. Examination of surface deterioration of Göreme tuffs for the purpose of conservation. In *The safeguard of the rock-hewn churches of the Göreme valley*, 85–94. Rome: ICCROM.

Capedri, S., R. Grandi, and G. Venturelli. 2003. Trachytes used for paving Roman roads in the Po Plain: Characterization by petrographic and chemical parameters and provenance of flagstones. *Journal of Archaeological Science* 30(4): 491–509.

Capone, P. 2006. Rio Grade glaze ware technology and production: Historic expediency. In *The social life of pots*, eds. J. A. Habicht-Mauche, S. L. Eckert, and D. L. Huntley, 216–31. Tucson, AZ: University of Arizona Press.

Carozzi, A. V. 1993. *Sedimentary petrography*. Englewood Cliffs, NJ: PTR Prentice Hall.

Casadio, R., B. Fabbri, and R. Maldera. 1991. Ancient fire resistant pottery from Emilia Romagna (Italy). In *Archeometry '90*, eds. E. Pernicka and G. A. Wagner, 237–46. Basel: Birkhäuser Verlag.

Casi, G., G. Tampone, and S. Vannucci. 1988. Restoration works on the Ducci-Del Rosso Mansion in Sansepolcro (Italy). In *The Engineering geology of ancient works, monuments and historical sites*, eds. P. G. Marinos and G. C. Koukis, 989–98. Rotterdam: A. A. Balkema.

Cau Ontiveros, M. A., P. M. Day, and G. Montana. 2002. Secondary calcite in archaeological ceramics: Evaluation of alteration and contamination processes by thin section study. In *Modern trends in scientific studies on ancient ceramics*, eds. V. Kilikoglou, A. Hein, and Y. Manniatis, 9–18. Oxford: British Archaeological Reports International Series 1011.

Chandler, G. M. 2001. *Development of a field petrographic analysis system and its application to the study of socioeconomic interaction networks of the early Harappan Northwestern Indus valley of Pakistan*. Oxford: British Archaeological Reports, 995.

Charola, A. E. 1997. *Death of a moai. Easter Island statues: Their nature, deterioration and conservation*. Los Osos, CA: Easter Island Foundation Occasional Paper 4.

Charola, A. E., R. J. Koestler, and G. Lombardi. 1994. *Lavas and volcanic tuffs*. Rome: ICCROM.

Chayes, F. 1954a. *Petrographic modal analysis*. New York: Wiley.

—. 1954b. The theory of thin-section analysis. *Journal of Geology* 62: 92–101.

Chiari, G., A. Invernizzi, and G. Bertolotto. 1993. Investigation and restoration of clay fragmentary statues from Old Nisa, Turkenistan. In *Terra 93: 7th International Conference on the Conservation of Earthen Architecture*, 228–30. Lisbon: Direcção Geral dos Edificios e Monumentos Naçionais.

Childs, S. T. 1988. Clay resource specialization in ancient Tanzania: Implications for cultural process. In *Ceramic ecology revisited 1987: The technology and socioeconomics of pottery*, ed. C. C. Kolb, 1–31. Oxford: British Archaeological Reports International Series 436(I).

Chinn, R. W. 2002. *Ceramography: Preparation and analysis of ceramic microstructures*. Materials Park, OH: ASM International.

Christaras, B. 1990. Weathering of granites and mechanical behaviour changing in northern Greece. In *The conservation of monuments in the Mediterranean basin*, ed. F. Zezza, 335–39. Brescia: Grafo.

Chui-mei, H. and B. Bronson. 1987. The ceramics of Changsha, China: Historical and technological background. *Archeomaterials* 2: 73–81.

Clarke, S. and R. Engelbac. 1990. *Ancient Egyptian construction and architecture*. New York: Dover.

Clayton, C. J. 1983. The chemical environment of flint formation in Upper Cretaceous chalks. In *The scientific study of flint and chert*, eds. G. de G. Sieveking and M. B. Hart, 43–54. Cambridge: Cambridge University Press.

Cohen, B. 2006. *The colors of clay: Special techniques of Athenian vases*. Los Angeles, CA: J. Paul Getty Museum.

Collins, C. 1995. *Care and conservation of palaeontological material*. Oxford: Butterworth-Heinemann.

Collins, G. H. 1977. Appendix 1. Petrology. In *Late medieval monumental sculpture in the West Highlands*, K. A. Steer and J. W. M. Bannerman, 195–200. Edinburgh: Royal Commission on the Ancient and Historical Monuments of Scotland.

Conolly, J. R. 1965. The occurrence of polycrystallinity and undulatory extinction in quartz sandstones. *Journal of Sedimentary Petrology* 35(1): 116–35.

Cooper, F. 1986. The stones of Bassai. In *Science in Archaeology: Proceedings of a Meeting held at the British School at Athens, January 1985*, 21–34. Athens: Fitch Laboratory Occasional Paper 2.

Cooper, J. 1982. The research potential of molluscan shell in shell-tempered pottery. In *Current research in ceramics: Thin-section studies*, eds. I. C. Freestone, C. Johns, and T. Potter, 161–62. London: British Museum Occasional Paper No. 32.

Cotkin, S. J., C. Carr, M. L. Cotkin, A. E. Dittert, and D. T. Kremser. 1999. Analysis of slips and other inorganic surface materials on Woodland and Early Fort Ancient ceramics, south-central Ohio. *American Antiquity* 64(2): 316–42.

Courtois, L. 1962–63. Examen minéralogique de quelques roches de monuments Gréco-Bouddhiques. *Arts Asiatiques* 9: 107–13.

Courtois, L. C. and B. Velde. 1980. Petrographic and electron microprobe studies of Cypriot White Slip ware (Late Bronze Age). *Revue d'Archéometrie* 3: 37–43.

Courty, M. A. and V. Roux. 1995. Identification of wheel throwing on the basis of ceramic surface features and microfabrics. *Journal of Archaeological Science* 22: 17–50.

Crosby, D. D. B. and J. G. Mitchell. 1987. A survey of British metamorphic hone stones of the 9th to 15th centuries AD in the light of potassium-argon and natural remnant magnetization studies. *Journal of Archaeological Science* 14: 483–506.

Cruciani-Fabozzi, G., F. Mariano, and S. Vannucci. 1991. Ricerche sui litotipi carbonatici nell'architettura post classica delle Marche, con particolare riferimento alla pietra d'Istria. In *Le pietre nell'architettura: Struttura e superfici*, eds. G. Biscotin and D. Mietto, 235–46. Padova: Libreria Progetto Editore Padova.

Cuomo Di Caprio, N. and S. J. Vaughan. 1993. An experimental study in distinguishing grog (chamotte) from argillaceous inclusions in ceramic thin sections. *Archeomaterials* 7(1): 21–40.

Curtis, G. H. 1959. The petrology of artifacts and architectural stone at La Venta. *Bureau of American Enthology Bulletin* 170: 284–89.

Dajnowski, A., E. Farrell, and P. Vandiver. 1992. The technical examination of some Neolithic Chinese Liangzhu ceramics in the Harvard University Art Museums collection. In *Materials issues in art and archaeology III*, eds. P. B. Vandiver, J. R. Druzik, G. S. Wheeler, and I. C. Freestone, 609–20. Pittsburgh, PA: Materials Research Society.

D'Albis, A. 1984. Steps in the manufacture of the soft-paste porcelain of Vincennes, according to the books of Hellot. In *Ancient technology to modern science*, ed. W. D. Kingery, 257–72. Columbus, OH: American Ceramic Society.

—. 1985. The history of innovation in European porcelain manufacture and the evolution of style: Are they related? In *Technology and style*, ed. W. D. Kingery, 397–412. Columbus, OH: American Ceramic Society.

D'Amico, C. 2005. Neolithic "greenstone" axe blades from northwestern Italy across Europe: A first petrographic comparison. *Archaeometry* 47(2): 235–52.

D'Amico, C. and E. Starnini. 2006. Prehistoric polished stone artefacts in Italy: A petrographic and archaeological assessment. In *Geomaterials in cultural heritage*, eds. M. Maggetti and B. Messiga, 257–72. London: Geological Society, Special Publications 257.

Darvill, T. and J. Timby. 1982. Textural analysis: A review of potentials and limitations. In *Current research in ceramics: Thin-section studies*, eds. I. C. Freestone, C. Johns, and T. Potter, 73–87. London: British Museum Occasional Paper No. 32.

Day, P. 1989. Technology and ethnography in petrographic studies of ceramics. In *Archaeometry: Proceedings of the 25th International Symposium*, ed. Y. Maniatis, 139–47. Amsterdam: Elsevier.

Day, P. M. and V. Kilikoglou. 2001. Analysis of ceramics from the kiln. *Hesperia Supplements* 30: 111–33.

De Andres, A. M. and J. L. Balcazar. 1989. Study of ceramic sherds of Early Bronze Age shelters (Peña Corva, Guadalajara, Spain). In *Archaeometry: Proceedings of the 25th International Symposium*, ed. Y. Maniatis, 593–601. Amsterdam: Elsevier.

De Benedetto, G. E., P. Acquafredda, M. Masieri, G. Quarta, L. Sabbatini, P. G. Zambonin, M. Tite, and M. Walton. 2004. Investigation of Roman lead glaze from Canosa: Results of chemical analyses. *Archaeometry* 46(4): 615–24.

De Casa, G., G. Giglio, G. Lombardi, and M. Mariottini. 1994. Characterization and state of decay of the volcanic tuff of the Tabularium in the Roman Forum, Italy. In *Lavas and volcanic tuffs*, eds. A. E. Charola, R. J. Koestler, and G. Lombardi, 107–27. Rome: ICCROM.

De Domingo, C. A. and A. W. Johnston. 1997. Trade between Kommos, Crete and East Greece: A petrographic study of Archaic transport amphorae. In *Archaeological Sciences 1995: Proceedings of a Conference on the Application of Scientific Techniques to the Study of Archaeology*, eds. A. Sinclair, E. Slater, and J. Gowlett, 62–68. Oxford: Oxbow Monograph 64.

De Geyter, G. 1985. Petrographical characterization of siliceous natural building stones in northern Belgium. In *Proceedings of the 5th International Congress on Deterioration and Conservation of Stone*, vol. 1, ed. G. Félix, 3–11. Lausanne: Presses Polytechniques Romandes.

De Paepe, P. 1979. A petrological study of coarse wares from Thorikos, S. E. Attica (Greece). *Miscellanea Graeca* 2: 61–88.

De Paepe, P., L. Moens, P. Roos, and M. Waelkens. 1988. Source analysis of the raw materials of four classical marble sculptures using chemical, microscopic and isotopic criteria. In *Classical marble: Geochemistry, technology, trade*, eds. N. Herz and M. Waelkens, 389–93. Dordrecht: Kluwer.

De Witte, E., A. Tervren, R. J. Koestler, and A. E. Charola. 1988. Conservation of the Göreme rock: Preliminary investigations. In *Proceedings of the 6th International Congress on Deterioration and Conservation of Stone*, 346–55. Torun: Nicholas Copernicus University Press.

Dean, N. 1988. Geochemistry and archaeological geology of the Carrara marble, Carrara, Italy. In *Classical marble: Geochemistry, technology, trade*, eds. N. Herz and M. Waelkens, 315–23. Dordrecht: Kluwer.

Dearman, W. R. and T. Y. Irfan. 1978. Assessment of the degree of weathering in granite using petrographic and physical index tests. In *Deterioration and protection of stone monuments*, 2.4. Paris: RILEM.

Deer, W. A., R. A. Howie, and J. Zussman. 1980. *An introduction to the rock-forming minerals*. London: Longman.

DeHayes, S. M. and D. Stark, eds. 1994. *Petrography of cementitious materials*. Philadelphia, PA: American Society of Testing and Materials.

Del Monte, M., C. Sabbioni, and O. Vittori. 1981. Airborne carbon particles and marble deterioration. *Atmospheric Environment* 15(5): 645–52.

Dermentzopoulos, T., T. Mimides, E. Davi, S. Kazas, T. Perraki, and K. Sachpazis. 1988. Building stones of ancient monuments in Attica: An outline. In *Engineering geology of ancient works, monuments and historical sities*, eds. P. G. Marinos and G. C. Koukis, 619–29. Rotterdam: A. A. Balkema.

Dermitzakis, M. D., M. Tsipoura-Vlachou, and M. Stamatakis. 1990. Petrological and geological study of the building material from the underground pit of Gortys area, Crete island. In *The engineering geology of ancient works, monuments and historical sites*, eds. P. G. Marinos and G. C. Koukis, 2049–56. Rotterdam: A. A. Balkema.

Descombes, A. 1986. La poterie tradionanelle au rwanda. *La revue de la céramique et du verre* 30: 33–35.

Di Florio, M. R., S. Lorenzoni, and E. Z. Lorenzoni. 1991. The rocks from the archaeological Buddhist sites of the Middle Swat valley (Pakistan NW). Abstract in *Archaeological Stone: Scientific and Technical Studies, British Museum Conference 14–16 November 1991*. London: British Museum.

Di Florio, M. R., C. Faccenna, S. Lorenzoni, and E. Z. Lorenzoni. 1993a. Lithological analysis of materials used in the buildings of the sacred area of Pānr I (Swat valley, northern Pakistan) and their origins. Appendix D in *Pānr I (Swat, Pakistan)*, by D. Faccenna, A. N. Khan, and I. H. Nadiem, 357–72. Rome: IsMEO.

Di Florio, M. R. S., S. Lorenzoni, L. M. Olivieri, and E. Zanettin Lorenzoni. 1993b. Evidence of ancient quarrying in Lower Swat, N. W. F. P., Pakistan. A geological study. *Science and Technology for Cultural Heritage* 2: 63–74.

Dickinson, W. R. 1985. Interpreting provenance relations from detrital modes of sandstones. In *Provenance of arenites*, ed. G. G. Zuffa, 333–61. Higham, MA: D. Reidel.

Dickinson, W. R. and R. Shutler Jr. 1974. Probable Fijian origin of quartzose temper sands in prehistoric pottery from Tonga and the Marquesas. *Science* 185: 454–57.

—. 1979. Petrography of sand tempers in Pacific Island potsherds: Summary. *Geological Society of America Bulletin* 90(1): 993–95.

—. 2000. Implications of petrographic temper analysis for Oceanian prehistory. *Journal of World Prehistory* 14(3): 203–66.

Dickinson, W. R., J. Takayama, E. A. Snow, and R. Shutler Jr. 1990. Sand temper of probable Fijian origin in prehistoric potsherds from Tuvalu. *Antiquity* 64: 307–12.

Dickinson, W. R., B. M. Butler, D. R. Moore, and M. Swift. 2001. Geologic sources and geographic distributions of sand tempers in prehistoric potsherds from the Mariana Islands. *Geoarchaeology* 16(8): 827–54.

Dietrich, R. V. and B. J. Skinner. 1979. *Rocks and minerals*. New York: Wiley.

Dimes, F. G. 1990a. Igneous rocks. In *Conservation of building and decorative stone*, vol. 1, eds. J. Ashurst and F. G. Dimes, 37–60. London: Butterworth-Heinemann.

—. 1990b. Metamorphic rocks. In *Conservation of building and decorative stone*, vol. 1, eds. J. Ashurst and F. G. Dimes, 135–49. London: Butterworth-Heinemann.

Dimitrijević, R., S. Karamata, A. McPherron, and D. Milovanović. 1974. Mineralogic and petrologic investigations of some Neolithic tool samples from Divostin near Kragujevac (Serbia). *Bulletin Scientifique, Section A, Sciences Naturelles, Techniques et Medicales* 19(5/6): 134–36.

Dingelstadt, C., R. Dreesen, J. Thorez, G. Lorenzi, D. Bossiroy, D. Antenucci, and J. Bainier. 2000. A petrographic atlas as a decision-tool in replacement and substitution of ornamental stones in historical buildings and monuments. In *Proceedings of the 9th International Congress on Deterioration and Conservation of Stone*, vol. 2, ed. V. Fassina, 33–41. Amsterdam: Elsevier.

Dodge, H. 1988. Decorative stones for architecture in the Roman empire. *Oxford Journal of Archaeology* 7(1): 65–80.

Doherty, C. J. and A. L. Maske. 1998. Characterization of the Takatori stoneware from Chikuzen province, Japan. *Archaeometry* 40(1): 71–95.

Dollery, D. 1994. Damage to shale. In *Conservation of geological collections*, ed. R. E. Child, 14–17. London: Archetype.

Donahue, J., D. R. Watters, and S. Millspaugh. 1990. Thin section petrography of Northern Lesser Antilles ceramics. *Geoarchaeology* 5(3): 229–54.

Dooley, K. and N. Herz. 1995. Provenance determinations of early American marbles. In *The study of marble and other stones used in antiquity*, eds. Y. Maniatis, N. Herz, and Y. Basiakos, 243–52. London: Archetype.

Douglas, J. and S. S. Sorensen. 2007. Mineralogical characteristics of Khmer sculpture in the Bayon style. In *Scientific research on the sculptural arts of Asia*, eds. J. G. Douglas, P. Jett, and J. Winter, 115–24. London: Archetype.

Douglass, A. A. and D. M. Schaller. 1993. Sourcing Little Colorado White Ware: A regional approach to the compositional analysis of prehistoric ceramics. *Geoarchaeology* 8(3): 177–201.

Dreesen, R. and M. Dusar. 2004. Historical building stones in the province of Limburg (NE Belgium): Role of petrography in provenance and durability assessment. *Materials Characterization* 53: 273–87.

Dupree, N. H. 1977. *An historical guide to Afghanistan*, 2nd edn. Kabul: Afghan Tourist Organization.

Drucker, P., R. F. Heizer, and R. J. Squier. 1955. *Excavations at La Venta, Tabasco*. Washington, DC: Smithsonian Institution.

Echallier, J.-C. 1983. L'analyse pétrographic des céramiques archéologiques (principe, techniques et limites de la méthode). *Archéologie du Midi Méditerranéen* 21(9): 60–67.

Echallier, J-C. and S. Mery. 1992. L'Evolution mineralogique et physicochimique des pates calcaires au cours de la cuisson: Expérimentation en laboratoire et application archéologique. In *Sciences de la terre et céramiques archéologiques: Expérimentations, applications, documents et travaux No. 16*, ed. S. Mery, 87–120. Cergy: Centre Polytechnique Saint-Louis à Cergy.

Efes, Y. and S. Luckat. 1976. Relations between corrosion of sandstones and uptake rates of air pollutants at the Cologne Cathedral. In *Second International Symposium on the Deterioration of Building Stones*, 193–200. Athens: Ministry of Culture.

Ehlers, E. G. 1987. *Optical mineralogy*. Palo Alto, CA: Blackwell Scientific.

Eliscu, F. 1959. *Sculpture techniques in clay, wax, slate*. Philadelphia and New York: Chilton.

Elssner, G., H. Hoven, G. Kiessler, and P. Wellner. 1999. *Ceramics and ceramic composites: Materialographic preparation*. New York: Elsevier.

Erguvanli, K., I. Eriş, M. Ahunbay, and Z. Ahunbay. 1988. The significance of research on old quarries for the restoration of historic buildings with special reference to the Mamara region, Turkey. In *Engineering geology of ancient works, monuments and historical sites*, eds. P. G. Marinos and G. C. Koukis, 631–38. Rotterdam: A. A. Balkema.

Erlin, B. and D. Stark, eds. 1990. *Petrography applied to concrete and concrete aggregates*. Philadelphia, PA: American Society of Testing and Materials.

Errington, E. and J. Cribb, eds. 1992. *The crossroads of Asia: Transformation in image and symbol in the art of ancient Afghanistan and Pakistan*. Cambridge: Ancient India and Iran Trust.

Esbert, R. M., J. Alonso, F. J. Alonso, and J. Ordaz. 1995. Suggested properties for evaluating the behaviour of a waterproofed limestone under aggressive environmental conditions. In *Methods of evaluating products for the conservation of porous building materials in monuments*, 331–39. Rome: ICCROM.

Esbert, R. M., F. J. Alonso, F. Diaz-Pache, J. Ordaz, and A. Pérez-Ortiz. 1996. Superficial protection of granite face to saline spray. In *Degradation and conservation of granitic rocks in monuments*, 337–42. Brussels: European Commission, Directorate-General XII, Science, Research and Development.

Espinosa Gaitán, J., R. Villegas Sánchez, and E. Sebastián Pardo. 1996. Characterization of the bricks used in the construction of the Alcazaba (Malaga, Spain). In *Proceedings of the 8th International Congress on Deterioration and Conservation of Stone*, vol. 3, ed. J. Riederer, 1679–85. Berlin: Möller Druck und Verlag.

Evens, E. D., L. V. Grinsell, S. Piggott, and F. S. Wallis. 1962. Fourth report of the sub-committee of the south-western group of museums and art galleries on the petrological identification of stone axes. *Proceedings of the Prehistoric Society* N.S. XXVIII: 209–66.

Faber, E. W., V. Kilikoglou, P. M. Day, and D. E. Wilson. 2002. A technological study of Middle Minoan polychrome pottery from Knossos, Crete. In *Modern trends in scientific studies on ancient ceramics*, eds. V. Kilikoglou, A. Hein, and Y. Manniatis, 129–41. Oxford: British Archaeological Reports International Series 1011.

Fabbri, B., S. Gualtieri, C. Mingazzini, P. Spadea, P. Casadio, R. Costantini, and G. Malisani. 2000. Archaeometric investigations of *sgraffito* ceramic tiles (fifteenth–sixteenth centuries) recovered from excavations in Udine (north-east Italy). *Archaeometry* 42(2): 317–24.

Fabbri, B., S. Gualtieri, and C. Mingazzini. 2002a. Materials and techniques of the ceramic wall facings in the Timurid Necropolis of Shahi Zinda (Samarkand, Uzbekistan). In *Modern trends in scientific studies on ancient ceramics*, eds. V. Kilikoglou, A. Hein, and Y. Manniatis, 351–60. Oxford: British Archaeological Reports International Series 1011.

Fabbri, B., S. Gualtieri, and S. Santoro. 2002b. The importance of the firing atmosphere in the production of coarse ceramics with calcite and chamotte inclusions. In *Proceedings of the 31st International Symposium on Archaeometry*, eds. E. Jerem, K. T. Biró, and E. Rudner, 191–95. Oxford: Archaeopress.

Faccenna, D., A. N. Khan, and I. N. Nadrem. 1993. *Pānr I (Swat, Pakistan)*. Rome: IsMEO Reports and Memoires XXVI.

Farnsworth, M. 1964. Greek pottery: A mineralogical study. *American Journal of Archaeology* 68: 221–28.

Fassina, V., M. Rossetti, M. Oddone, S. Mazzocchin, and S. Calogero. 1994. Marine spray and polluted atmosphere as factors of damage to monuments in the Mediterranean coastal environment: An analytical study of some samples from the Sanctuary of Demeter in Eleusis

(Athens). In *Third International Symposium on the Conservation of Monuments in the Mediterranean Basin*, eds. V. Fassina, H. Ott, and F. Zessa, 287–94. Venice: Soprintendenza ai Beni Artistici e Storici di Venezia.

Fawcett, J. 1998. *Historic floors: Their history and conservation.* Oxford: Butterworth-Heinemann.

Feinman, G. M., S. Banker, R. F. Cooper, G. B. Cook, and L. M. Nicholas. 1989. A technological perspective on changes in the the ancient Oaxacan Grayware ceramic tradition: Preliminary results. *Journal of Field Archaeology* 16(3): 331–44.

Felts, W. M. 1942. A petrographic examination of potsherds from ancient Troy. *American Journal of Archaeology* 46: 237–44.

Fenn, D. B., T. Rutenbeck, L. I. Knab, J. R. Clifton, and B. B. G. Nistal-Moret. 1987. Structural monitoring of Castillo de San Marcos National Monument. In *Old Cultures in New Worlds: 8th General Assembly and International Symposium ICOMOS International Council on Monuments and Sites*, 400–407.

Ferreira, C. M. 1992. Building rocks of historical monuments of Ouro Preto, MG - Brazil: Causes and rates of their decay. In *Proceedings of the 7th International Congress on Deterioration and Conservation of Stone*, 77–85. Lisbon: Laboratorio Nacional de Engenharia Civil.

Ferring, C. R. and T. Pertulla. 1987. Defining the provenance of red slipped pottery from Texas and Oklahoma by petrographic methods. *Journal of Archaeological Science* 14: 437–56.

Fidler, J. 1994. Fragile remains: An international review of conservation problems in the decay and treatment of architectural terracotta and faience. In *Architectural ceramics: Their history, manufacture and conservation*, ed. J. M. Teutonico, 7–25. London: James & James.

Fieller, N. R. J. and P. T. Nicholson. 1991. Grain size analysis of archaeological pottery: The use of statistical models. In *Recent developments in ceramic petrology*, eds. A. Middleton and I. C. Freestone, 71–111. London: British Museum.

Fiora, L., G. Chiari, and R. Compagnoni. 1996. Granitic rocks in the historic buildings of Turin (Piedmont, Italy). In *Degradation and conservation of granitic rocks in monuments*, 375–80. Brussels: European Commission, Directorate-General XII, Science, Research and Development.

Fitzner, B. and K. Heinrichs. 1994. Damage diagnosis at monuments carved from bedrocks in Petra, Jordan. In *The conservation of monuments in the Mediterranean basin*, eds. V. Fassina, H. Ott, and F. Zezza, 663–71. Venice: Soprintendenza ai Beni Artistici e Storici di Venezia.

Fitzner, B., K. Heinrichs, M. Volker, and R. Campos Perez. 1993. Damage investigations on natural stone monuments in Minas Gerais, Brazil. In *Conservation of Stone and Other Materials: Proceedings of the International RILEM/UNESCO Congress held at the UNESCO Headquarters, Paris, 29 June–1 July 1993*, ed. M.-J. Thiel, 360–67. London: E & FN Spon.

Fitzpatrick, S. M., W. R. Dickinson, and G. Clark. 2003. Ceramic petrography and cultural interaction in Palau, Micronesia. *Journal of Archaeological Science* 30: 1175–84.

Fitzpatrick, S. M., H. Takamiya, H. Neff, and W. R. Dickinson. 2006. Compositional analysis of Yayoi-Heian period ceramics from Okinawa: Examining the potential for provenance study. *Geoarcheology* 21(8): 803–22.

Flügel, E. 1982. *Microfacies analysis of limestones*, trans. K. Christenson. Berlin: Springer-Verlag.

Fobe, R., H. Sweevers, G. Vleugels, and R. Van Grieken. 1993. Weathering of miocene ferruginous sandstone in ancient buildings in northeastern Belgium. *The Science of the Total Environment* 132(1): 53–70.

Folk, R. L. 1974. *Petrology of sedimentary rocks.* Austin, TX: Hemphill.

Folk, R. L. and J. S. Pittman. 1971. Length-slow chalcedony: A new testament for vanished evaporites. *Journal of Sedimentary Petrology* 41: 1045–58.

Folk, R. L. and W. C. Ward. 1957. Brazos river bar: A study in the significance of grain size parameters. *Journal of Sedimentary Petrology* 27: 3–26.

Formigli, E. and G. Schneider. 1993. Antiche terre di fusione. Indagini archeometriche sulle terre di fusione di bronzi greci, romani e rinascimentali. In *Antiche officine del bronzo: Materali, instrumenti, techniche. Atti del seminario di studi ed esperimenti. Murlo 26–31 luglio 1991*, ed. E. Formigli, 69–102. Siena: New Publishing Image.

Fornaseri, M., L. Malpieri, and L. Tolomeo. 1975. Provenance of pumices in the north coast of Cyprus. *Archaeometry* 17(1): 112–16.

Fornaseri, M., L. Lazarini, P. Pensabene, M. Preite Martinez, and B. Turi. 1995. "Lapis Niger" and other black limestones used in antiquity. In *The study of marble and other stones used in antiquity*, eds. Y. Maniatis, N. Herz, and Y. Basiakos, 235–40. London: Archetype.

Forte, M. 1994. Archaeometric and digital computer analysis on Etruscan bucchero from Marzabott. In *The Ceramics Cultural Heritage: Proceedings of the International Symposium of the 8th CIMTEC World Ceramics Congress and Forum on New Materials*, ed. P. Vincenzini, 529–39. Faenza: TECHNA.

Foster, A. 1981. Slating. In *Handbook of building crafts in conservation*, ed. J. Bowyer, 302–15. New York: Van Nostrand Reinhold.

Fowles, S. M., L. Minc, S. Duwe, and D. V. Hill. 2007. Clay, conflict, and village aggregation: Compositional analyses of Pre-Classic pottery from Taos, New Mexico. *American Antiquity* 72(1): 125–52.

Franchi, R. and U. Gori. 1993. Problems involved in conservation of the structure and stone material of the Ducal Palace in Urbania, Pesaro, Italy. In *Conservation of Stone and Other Materials: Proceedings of the International RILEM/UNESCO Congress held at the UNESCO Headquarters, Paris, 29 June–1 July 1993*, ed. M.-J. Thiel, 485–92. London: E & FN Spon.

Francus, P. 1998. An image-analysis technique to measure grain-size variation in thin-sections of soft clastic sediments. *Sedimentary Geology* 121: 289–98.

Fratini, F., C. Manganelli del Fa, E. Pecchioni, and A. Scala. 1994. The sculptures in Bomarzo Park, Viterbo, Italy: Deterioration and conservation problems of the Peperino. In *Lavas and volcanic tuffs*, eds. A. E. Charola, R. J. Koestler, and G. Lombardi, 129–41. Rome: ICCROM.

Freestone, I. C. 1991. Extending ceramic petrology. In *Recent developments in ceramic petrology*, eds. A. Middleton and I. C. Freestone, 399–410. London: British Museum.

—. 1995. Ceramic petrography. *American Journal of Archaeology* 99(1): 111–17.

—. 2001. Post-depositional changes in archaeological ceramics and glass. In *Handbook of archaeological sciences*, eds. D. R. Brothwell and A. M. Pollard, 615–25. Chichester: Wiley.

Freestone, I. C. and M. J. Hughes. 1989. Appendix III: Examination of ceramics from Qasrij Cliff and Khirbet Qasrij. In *Excavations at Qasrij Cliff and Khirbet Qasrij*, J. Curtis and D. Collon, 61–75. London: Trustees of the British Museum.

Freestone, I. C. and M. S. Tite. 1997. The technology of German stoneware glazes. In *German stoneware 1200–1900*, D. Gaimster, 354–58. London: British Museum.

Freestone, I. C., C. Johns, and T. Potter, eds. 1982. *Current research in ceramics: Thin-section studies.* London: British Museum Occasional Paper No. 32.

Freestone, I. C., M. Bimson, and M. S. Tite. 1984. The constitution of Coade Stone. In *Ancient technology to modern science*, ed. W. D. Kingery, 293–304. Pittsburgh, PA: American Ceramic Society.

Fueten, F. 2006. Petrographic image processing of thin sections using the rotating polarizer stage. *SAS Bulletin* 29(1): 8–12.

Fukang, Z. 1987. Technical studies of Changsha ceramics. *Archeomaterials* 2: 83–92.

Gaimster, D. 1997a. *German stoneware 1200–1900.* London: British Museum.

—. 1997b. Stoneware production in medieval and early modern Germany. In *Pottery in the making: Ceramic traditions*, eds. I. C. Freestone and D. Gaimster, 122–27. Washington, DC: Smithsonian Institution Press.

Gajurel, C. L. and K. K. Vaidya. 1984. *Traditional arts and crafts of Nepal.* New Delhi: S. Chand.

Galanos, A. and Y. Doganis. 2003. The remnants of the epidermis on the Parthenon: A valuable analytical tool for assessing condition. *Studies in Conservation* 48(1): 3–16.

Galetti, G., L. Lazzarini, and M. Maggetti. 1992. A first characterization of the most important granites used in antiquity. In *Ancient stones: Quarrying, trade and provenance*, eds. M. Waelkens, N. Herz, and L. Moens, 167–77. Leuven: Leuven University Press.

Gandin, A., E. Capezzuoli, and A. Ciacci. 2006. The stone of the inscribed Etruscan stelae from the Valdelsa area (Siena, Italy). In *Geomaterials in cultural heritage*, eds. M. Maggetti and B. Messiga, 273–82. London: Geological Society, Special Publications 257.

Garcia del Cura, M. de los A., M. Hoyos, and C. Saiz-Jimenez. 1992. Petrographic characterization and weathering of limestone tesserae from mosaics of Italica (Spain). In *The conservation of monuments in the Mediterranean basin*, eds. D. Decrouez, J. Chamay, and F. Zezza, 469–75. Geneva: Musée d'Histoire Naturelle and Musée d'Art et d'Histoire.

Garcia-Vallès, M., J. Molera, and M. Vendrell-Saz. 1997. A siliceous sandstone in an urban environment: The decay and cleaning of the church of Betlem (Barcelona, Catalonia). *International Journal for Restoration of Buildings and Monuments* 3(5): 469–85.

Garrett, E. 1986. A petrographic analysis of Black Mesa ceramics. In *Spatial organization and exchange*, ed. S. Plog, 114–142. Carbondale, IL: Southern Illinois University Press.

Garza-Valdes, L. A. 1991. Technology and weathering of Mesoamerican jades as guides to authenticity. In *Materials issues in art and archaeology*, eds. P. B. Vandiver, J. Druzik, and G. S. Wheeler, 321–57. Pittsburgh, PA: Materials Research Society.

Gauri, K. L. and J. K. Bandhyopadhyay. 1999. *Carbonate stone: Chemical behavior, durability, and conservation*. New York: Wiley.

Germann, K., G. Gruben, H. Knoll, V. Valis, and F. J. Winkler. 1988. Provenance characteristics of Cycladic (Paros and Naxos) marbles: A multivariate geological approach. In *Classical marble: Geochemistry, technology, trade*, eds. N. Herz and M. Waelkens, 251–62. Dordrecht: Kluwer.

Gettens, R. J. 1969. *The Freer Chinese bronzes*. Volume III: *Technical studies*. Washington, DC: Smithsonian Institution Oriental Studies 7.

Gherdan, K., G. Y. Szakmány, T. Weiszburg, and G. Ilon. 2002. Petrological investigation of Bronze and Iron Age ceramics from West Hungary: Vaskeresztes, Velem, Sé, Gór. In *Modern trends in scientific studies on ancient ceramics*, eds. V. Kilikoglou, A. Hein, and Y. Manniatis, 305–12. Oxford: British Archaeological Reports International Series 1011.

Gibbins, W. A. 1988. Carvingstone, the foundation of a northern economy. *Inuit Art Quarterly* 3(4): 4–8.

Gibson, A. and A. Woods. 1997. *Prehistoric pottery for the archaeologist*. London: Leicester University Press.

Gilead, I. and Y. Goren. 1989. Petrographic analyses of fourth millennium BC pottery and stone vessels from the Northern Negev, Israel. *Bulletin of the American Schools of Oriental Research* 275: 5–14.

Giot, P-R. and G. Querré. 1987. Quelques productions céramiques médiévales de Bretagne: Les incidences des caractéristiques minéralogiques. In *La céramique Ve–XIXe s. – fabrication, commercialisation, utilisation: Actes du premier Congrès international d'archéologie médiévale, Paris, 4–6 oct*, eds. J. Chapelot, H. Galinié, and J. Pilet-Lemière, 149–56. Caen: Université de Caen.

Glover, J. E. 1975. The petrology and probable stratigraphic significance of Aboriginal artifacts from part of south-western Australia. *Journal of the Royal Society of Western Australia* 58(3): 75–85.

—. 1984. The geological sources of stone artefacts in the Perth basin and nearby areas. *Australian Aboriginal Studies* 1: 17–25.

Goedicke, H. 1964. Some remarks on stone quarrying in the Egyptian Middle Kingdom (2060–1786 B.C.). *Journal of the American Research Center in Egypt* 3: 43–50.

Goffer, Z. 1980. *Archaeological chemistry*. New York: Wiley.

Goins, E. and C. L. Reedy. 2000. Digital image analysis in microscopy for objects and architectural conservation. *Objects Specialty Group Postprints*, vol. 7, 122–33. Washington, DC: American Institute for Conservation.

Goldberg, P., B. Gould, A. Killebrew, and J. Yellin. 1986. Comparison of neutron activation and thin-section analyses on Late Bronze Age ceramics from Deir el-Balah. In *Proceedings of the 24th International Archaeometry Symposium*, eds. J. Olin and M. J. Blackman, 341–51. Washington, DC: Smithsonian Institution Press.

Gomez, B. and C. Doherty. 2000. A preliminary petrographic analysis of Cypriot White Slip II ware. *Archaeometry* 42(1): 109–18.

González-García, F., V. Romero-Acosta, G. García-Ramos, and M. González-Rodríguez. 1990. Firing transformations of mixtures of clays containing illite, kaolinite and calcium carbonates used by ornamental tile industries. *Applied Clay Science* 5(4): 361–75.

Gorgoni, C., I. Kokkinakis, L. Lazzarini, and M. Mariottini. 1992. Geochemical and petrographic characterization of *rosso antico* and other white-grey marbles of Mani (Greece). In *Ancient stones: Quarrying, trade and provenance*, eds. M. Waelkens, N. Herz, and L. Moens, 155–65. Leuven: Leuven University Press.

Gorgoni, C., L. Lazzrini, P. Pallante, and B. Turi. 2002. An updated and detailed mineropetrographic and C-O stable isotopic reference database for the main Mediterranean marbles used in antiquity. In *Interdisciplinary studies on ancient stone*, eds. J. J. Herrmann, Jr., N. Herz, and R. Newman, 115–31. London: Archetype.

Gosselain, O. P. and A. Livingstone-Smith. 1995. The ceramics and society project: An ethnographic and experimental approach to technological choice. In *The aim of laboratory analyses of ceramics in archaeology*, eds. A. Lindahl and O. Stilborg, 147–60. Stockholm: Kungl. Vitterhets Historie och Antikvitets Adademien.

Götze, J. and H. Siedel. 2004. Microscopic scale characterization of ancient building sandstones from Saxony (Germany). *Materials Characterization* 53: 209–22.

Gramly, R. M. 1984. Mount Jasper: A direct-access lithic source area in the White Mountains of New Hampshire. In *Prehistoric quarries and lithic production*, eds. J. E. Ericson and B. A. Purdy, 11–21. Cambridge: Cambridge University Press.

Grapes, R. 2006. *Pyrometamorphism*. New York: Springer.

Green, G. W. 1984. Gypsum analysis with the polarizing microscope. In *The chemistry and technology of gypsum*, ed. R. A. Kuntze, 22–47. Philadelphia, PA: American Society of Testing and Materials.

Green, H. S., A. H. V. Smith, B. R. Young and R. K. Harrison. 1980. The Caergwrle bowl: Its composition, geological source and archaeological significance. *Science* 80(1): 26–30.

Greensmith, J. T. 1989. *Petrology of the sedimentary rocks*, 7th edn. London: Unwin Hyman.

Gribble, C. D. and A. J. Hall. 1985. *A practical guide to optical mineralogy*. London: Allen and Unwin.

Grissom, C.A. 1994. The deterioration and treatment of volcanic stone: A review of the literature. In *Lavas and volcanic tuffs*, eds. A. E. Charola, R. J. Koestler, and G. Lombardi, 4–29. Rome: ICCROM.

Gruber, P. and B. Sternad. 1981. The detection and identification of gypsum alteration products in stone by means of ultraviolet fluorescence microscopy. *Studies in Conservation* 26: 161–67.

Gruen, A., F. Remondino, and L. Zhang. 2002. Reconstruction of the Great Buddha of Bamiyan, Afghanistan. *International Archives of Photogrammetry and Remote Sensing* 34(5): 363–68.

Guozhen, L. and G. Lingxiang. 1986. Development of Chinese celadon and its influence. In *High-technology ceramics past, present, and future: The nature of innovation and change in ceramic technology*, ed. W. D. Kingery, 129–51. Westerville, OH: American Ceramic Society.

Guozhen, L. and Z. Xiqiu. 1985. The development of Chinese white porcelain. In *Technology and style*, ed. W. D. Kingery, 217–36. Columbus, OH: American Ceramic Society.

Habicht-Mauche, J. A., S. L. Eckert, and D. L. Huntley. 2006. *The social life of pots: Glaze wares and cultural dynamics in the Southwest, AD 1250–1680*. Tucson, TX: University of Arizona Press.

Hankey, V. 1976. A late bronze age temple at Amman: II. Vases and objects made of stone. *Levant* 6: 160–78.

Harrell, J. A. 1992. Ancient Egyptian limestone quarries: A petrological study. *Archaeometry* 34(2): 195–211.

Harrell, J. A. and T. B. Brown. 1995. An Old Kingdom basalt quarry at Widan el-Faras and the quarry road to Lake Moeris. *Journal of the American Research Center in Egypt* 32: 71–91.

Harris, V. 1997. Ash-glazed stonewares in Japan. In *Pottery in the making: Ceramic traditions*, eds. I. C. Freestone and D. Gaimster, 80–85. Washington, DC: Smithsonian Institution Press.

Harrison-Hall, J. 1997a. Chinese porcelain from Jingdezhen. In *Pottery in the making: Ceramic traditions*, eds. I. C. Freestone and D. Gaimster, 194–99. Washington, DC: Smithsonian Institution Press.

—. 1997b. Ding and other whitewares of Northern China. In *Pottery in the making: Ceramic traditions*, eds. I. C. Freestone and D. Gaimster, 182–87. Washington, DC: Smithsonian Institution Press.

Hart, D. 1988. *The building magnesian limestones of the British Isles*. Garston: Building Research Establishment.

Harvey, R. D. 1977. Descriptive characterization of the grain textures, mineralogy, density and structures of ornamental and building stones as Carrara marble, Istrian stone, Greek marble, Valpolicella stones, San Giuliano marble (Pisa) and their reactivities with sulfur dioxide. In *Art e Ambient*: 3–11. Modena: Editore Poligrafico Artioli.

Hasan, A. S., A. Ahmad, and F. Khan. 1995. Conservation of Kankar stone antiquities: A case study. In *Conservation, preservation and restoration: Traditions, trends and techniques*, eds. G. Kamalakar and V. P. Rao, 185–90. Hyderabad: Birla Archaeological and Cultural Research Institute.

Hays, T. R. and F. A Hassan. 1974. Mineralogical analysis of Sudanese Neolithic ceramics. *Archaeometry* 16(1): 71–79.

Heath, L., P. Russell, and C. Zucker. 1991. Tibetan clay sculpture and the construction of colossal statues. *Chö-yang* 5: 122–30.

Hegde, K. T. M. 1979. Analysis of ancient Indian deluxe wares. In *Archaeophysika 10: Proceedings of the 18th International Symposium on Archaeometry and Archaeological Prospection, Bonn, 14–17 March 1978*, 141–55. Köln: Rheinland-Verlag in Kommission bei Rudolf Habelt Verlag.

—. 1983. The art of the Harappan microbead. *Archaeology* 36(3): 68–72.

Hegmon, M. 1995. Pueblo I ceramic production in southwest Colorado: Analyses of igneous rock temper. *Kiva* 60(3): 371–90.

Heidke, J. M. and E. J. Miksa. 2000. Correspondence and discriminant analysis of sand and sand temper compositions, Tonto basin, Arizona. *Archaeometry* 42(2): 273–99.

Heidke, J. M., E. J. Miksa, and H. D. Wallace. 2002. A petrographic approach to sand-tempered pottery provenance studies: Examples from two Hohokam local systems. In *Ceramic production and circulation in the greater Southwest: Source determination by INAA and complementary mineralogical investigations*, eds. D. M. Glowacki and H. Neff, 152–78. Los Angeles, CA: UCLA Cotsen Institute of Archaeology.

Heiken, G. 1979. Pyroclastic flow deposits. *American Scientist* 67: 564–71.

Heimann, R. B. and M. Maggetti. 1981. Experiments on simulated burial of calcareous Terra Sigillata (mineralogical change). In *Scientific studies in ancient ceramics*, ed. M. J. Hughes, 163–77. London: British Museum Occasional Paper 19.

Heinrich, E. W. 1956. *Microscopic petrography*. New York: McGraw-Hill.

Heizer, R. F., F. Stross, T. R. Hester, A. Albee, I. Perlman, F. Asaro and H. Bowman. 1973. The Colossi of Memnon revisited. *Science* 182(4118): 1219–25.

Heller, D. S. and N. Herz. 1995. Weathering of dolomitic marble and the role of oxalates. In *The study of marble and other stones used in antiquity*, eds. Y. Maniatis, N. Herz, and Y. Basiakos, 267–76. London: Archetype.

Helmi, F. M. 1990. Deterioration and conservation of some mud brick in Egypt. In *Sixth International Conference on the Conservation of Earthen Architecture: Adobe 90*, ed. K. Grimstad, 277–82. Marina del Rey, CA: Getty Conservation Institute.

—. 1992. Weathering of some marble from Egypt. In *The conservation of monuments in the Mediterranean basin*, eds. D. Decrouez, J. Chamay, and F. Zezza, 209–17. Geneva: Musée d'Histoire Naturelle and Musée d'Art et d'Histoire.

—. 1994. Study of the deterioration of volcanic rocks from Egypt. In *Lavas and volcanic tuffs*, eds. A. E. Charola, R. J. Koestler, and G. Lombardi, 53–61. Rome: ICCROM.

Helmi, F. M. and H. R. Attia. 1996. Characterization and conservation of Seti-1 Temple stone, Abydous, Upper Egypt. In *Proceedings of the 8th International Congress on Deterioration and Conservation of Stone*, vol. 2, ed. J. Riederer, 1123–36. Berlin: Möller Druck und Verlag.

Helmold, K. P. 1985. Provenance of feldspathic sandstones – the effect of diagenesis on provenance interpretations: A review. In *Provenance of arenites*, ed. G. G. Zuffa, 139–63. Higham, MA: D. Reidel.

Henderson, J. 1989. Iznik ceramics: A technical examination In *Iznik: The pottery of Ottoman Turkey*, eds. N. Atasoy and J. Raby, ch. 6. London: Alexandria Press.

—, 2000. *The science and archaeology of materials*. London and New York: Routledge.

Henderson, J. and J. Raby. 1989. The technology of fifteenth-century Turkish tiles: An interim statement on the origins of the Iznik industry. *World Archaeology* 21(1): 115–32.

Henrickson, R. C. and M. J. Blackman. 1999. Hellenistic production of terracotta roof tiles among the ceramic industries at Gordion. *Oxford Journal of Archaeology* 18(3): 307–26.

Henson, F. A. 1953. The geology of ornamental stones. *Discovery* 14: 218–21.

Hernández, M. A. V. 1996. Granitic materials and historical monuments: Study of the factors and mechanisms of weathering and application to historical heritage conservation. In *Degradation and conservation of granitic rocks in monuments*, 1–44. Brussels: European Commission, Directorate-General XII, Science, Research and Development.

Herz, N. 1955. Petrofabrics and classical archaeology. *American Journal of Science* 253: 299–305.

Herz, N. and E. G. Garrison. 1998. *Geological methods for archaeology*. New York: Oxford University Press.

Hetzner, D. 1998. Quantitative image analysis. Part I: Principles. *Buehler Tech-Notes* 2(4): 1–5.

Higgins, R. A. 1969. *Catalogue of the terracottas in the Department of Greek and Roman Antiquities, British Museum, 1*. London: British Museum.

Hoard, R. J., M. J. O'Brien, M. G. Khorasgany, and V. S. Gopalaratnam. 1995. A materials-science approach to understanding limestone-tempered pottery from the Midwestern United States. *Journal of Archaeological Science* 22: 823–32.

Hodges, H. W. M. 1963. The examination of ceramic materials in thin section. In *The scientist and the archaeologist*, ed. E. Pyddoke, 101–10. New York: Roy.

Holmes, L. L. and G. Harbottle. 1991. Provenance study of cores from Chinese bronze vessels. *Archeomaterials* 5(2): 165–84.

Holmes, L. L., C. T. Little, and E. V. Sayre. 1986. Elemental characterization of medieval limestone sculpture from Parisian and Burgundian sources. *Journal of Field Archaeology* 13(4): 419–38.

Horowitz, A. S. and P. E. Potter. 1971. *Introductory petrography of fossils*. Berlin: Springer-Verlag.

Hovorka, D. and L. Illášová. 1995. Antigorite serpentinite: rare raw material of Neolithic polished stone industry from Slovakia. *Archeologické Rozhledy* 47(3): 357–62.

Howard, H. 1982. Clay and the archaeologist. In *Current research in ceramics: Thin-section studies*, eds. I. C. Freestone, C. Johns, and T. Potter, 145–58. London: British Museum Occasional Paper No. 32.

Hull, E. 1872. *A treatise on the building and ornamental stones of Great Britain and foreign countries*. London: Macmillan.

Hunt, P. N. and D. R. Griffiths. 1989. Optical petrology in the field. *World Archaeology* 21(1): 165–72.

Huntington, S. L. (with contributions by J. C. Huntington). 1985. *The art of ancient India*. New York and Tokyo: Weatherhill.

Hutchison, C. S. 1974. *Laboratory handbook of petrographic techniques*. New York: Wiley.

Impey, O. 1996. *The early porcelain kilns of Japan: Arita in the first half of the seventeenth century*. Oxford: Clarendon Press.

Isphording, W. C. and E. M. Wilson. 1974. The relationship of "volcanic ash," sak lu'um, and palygorskite in northern Yucatan Maya ceramics. *American Antiquity* 39(3): 483–88.

Isaacson, J. S. and T. F. Aleto. 1989. Petrographic analysis of ceramic thin sections from La Puná Island, Ecuador. *Archeomaterials* 3: 61–67.

Ixer, R. A. 1997. Detailed petrography of the Stonehenge dolerites using reflected light petrography: A return to the light. In *Archaeological Sciences 1995: Proceedings of a Conference on the Application of Scientific Techniques to the Study of Archaeology*, eds. A. Sinclair, E. Slater, and J. Gowlett, 11–17. Oxford: Oxbow Monograph 64.

Ixer, R. A. and S. Lunt. 1991. The petrography of certain pre-Spanish pottery from Peru. In *Recent developments in ceramic petrology*, eds. A. Middleton and I. C. Freestone, 137–64. London: British Museum.

Jackson, M. D., F. Marra, R. L. Hay, C. Cawood, and E. M. Winkler. 2005. The judicious selection and preservation of tuff and travertine building stone in ancient Rome. *Archaeometry* 47(3): 485–510.

Jan, M. Q. and R. A. K. Tahirkheli. 1969. The geology of the lower part of

Indus Kohistan (Swat), West Pakistan. *Geological Bulletin, Peshawar University* 4(1): 1–13.

Jerome, P. S. 1993. Analysis of Bronze Age mudbricks from Palaikastro, Crete. In *Terra 93: 7th International Conference on the Conservation of Earthen Architecture*, 381–86. Lisbon: Direcção Geral dos Edificios e Monumentos Naçionais.

Jester, T. C. 1995. *Twentieth-century building materials: History and conservation*. Washington, DC: McGraw-Hill.

Jiazhi, L. 1984. The evolution of Chinese pottery and porcelain technology. In *Ancient technology to modern science*, ed. W. D. Kingery, 135–62. Columbus, OH: American Ceramic Society.

Jones, L. D. 1986. *Lowland Maya pottery: The place of petrological analysis*. Oxford: British Archaeological Reports International Series 288.

—. 1991. Tempering trends in Lowland Maya pottery. In *Recent developments in ceramic petrology*, eds. A. Middleton and I. C. Freestone, 165–83. London: British Museum.

Jones, M. S. and G. E. Thompson. 1996. The use of stone thin sections in the study of decay of individual components of stone by pollutants. In *Proceedings of the 8th International Congress on Deterioration and Conservation of Stone*, vol. 2, ed. J. Riederer, 771–82. Berlin: Möller Druck und Verlag.

Jones, R. E. (with contributions by J. Boardman, H. W. Catling, C. B. Mee, W. W. Phelps, and A. M. Pollard). 1986. *Greek and Cypriot pottery: A review of scientific studies*. Athens: British School at Athens, Fitch Laboratory Occasional Paper 1.

Jope, E. M. and G. E. Dunning. 1954. The use of blue slate for roofing in mediaeval England. *Antiquities Journal* 34: 209–17.

Jornet, A., M. J. Blackman, and J. S. Olin. 1984. 13th to 18th century ceramics from the Paterna-Manises area (Spain). In *Ancient technology to modern science*, ed. W. D. Kingery, 235–55. Columbus, OH: American Ceramic Society.

Josephs, R. L. 2005. Applying micromorphological terminology to ceramic petrology. *Geoarchaeology* 20(8): 861–65.

Joyner, L. 1997. Byzantine and Frankish cooking wares at Corinth, Greece: Changes in diet, style and raw material exploitation. In *Archaeological Sciences 1995: Proceedings of a Conference on the Application of Scientific Techniques to the Study of Archaeology*, eds. A. Sinclair, E. Slater, and J. Gowlett, 82–87. Oxford: Oxbow Monograph 64.

Kamilli, D. C. and C. C. Lamberg-Karlovsky. 1979. Petrographic and electron microprobe analysis of ceramics from Tepe Yahya, Iran. *Archaeometry* 21(1): 47–59.

Kamilli, D. C. and A. Steinberg. 1985. New approaches to mineral analysis of ancient ceramics. In *Archaeological geology*, eds. G. Rapp, Jr. and J. A. Gifford, 313–30. New Haven and London: Yale University Press.

Kapur, S., N. Sakarya, and E. A. FitzPatrick. 1993. Fibrous minerals in 16th-century Iznik ceramics. *Geoarchaeology* 8(4): 333–37.

Kars, E. A. K., H. Kars, and R. D. McDonnell. 1991. Greenstone axes from eastern central Sweden: A technological-petrological approach. *Archaeometry* 34(2): 213–22.

Katsarou, S., A. Sampson, and E. Dimou. 2002. Obsidian as temper in the Neolithic pottery from Viali, Greece. In *Modern trends in scientific studies on ancient ceramics*, eds. V. Kilikoglou, A. Hein, and Y. Manniatis, 111–20. Oxford: British Archaeological Reports International Series 1011.

Keblow Bernsted, A-M. K. 2003. *Early Islamic pottery materials and techniques*. London: Archetype.

Keith, V. and M. Cooper. 1974. *Nonparametric design and analysis*. Ottowa: University of Ottowa Press.

Keen, L. and J. Radley. 1971. Report on the petrological identification of stone axes from Yorkshire. *Proceedings from the Prehistoric Society* 37: 16–37.

Kelly, A. 1990. *Mrs Coade's stone*. Upton-upon-Severn: Self Publishing Association Ltd. and the Georgian Group.

—. 1994. Coade-Stone: Its character and conservation. In *Architectural ceramics: Their history, manufacture and conservation*, ed. J. M. Teutonico, 32–38. London: James & James.

Kelly, J. G. 1985. Geological aspects of the decay of Irish granitic building materials. In *Fifth International Congress on Deterioration and Conservation of Stone*, vol.1, ed. G. Félix, 431–44. Lausanne: Presses Polytechniques Romandes.

Kelly, J. H. 1979. The Sierra Blanca restudy project. In *Jornada Mogollon Archaeology: Proceedings of the 1st Jornada Conference*, eds. P. H. Beckett and R. N. Wiseman, 107–32. Las Cruces, NM: New Mexico State University.

Kempe, D. R. C. 1982. Nature and source of the Gandhara sculptural schist. *Journal of Archaeological Science* 9: 25–28.

—. 1986. Gandhara sculptural schist: Proposed source. *Journal of Archaeological Science* 13: 79–87.

Kempe, D. R. C. and A. P. Harvey, eds. 1983. *The petrology of archaeological artefacts*. Oxford: Clarendon Press.

Kempe, D. R. C. and J. A. Templeman. 1983. Techniques. In *The petrology of archaeological artefacts*, eds. D. R. C. Kempe and A. P. Harvey, 26–52. Oxford: Clarendon Press.

Kenoyer, J. M. and M. Vidale. 1992. A new look at stone drills of the Indus valley tradition. In *Materials issues in art and archaeology III*, eds. P. B. Vandiver, J. R. Druzik, G. S. Wheeler, and I. C. Freestone, 495–518. Pittsburgh, PA: Materials Research Society.

Kerr, P. F. 1977. *Optical mineralogy*. New York: McGraw Hill.

Kerr, R. and N. Wood. 2004. *Joseph Needham's science and civilization in China*. Volume 5: *Chemistry and chemical technology, part XII: Ceramic technology*. Cambridge: Cambridge University Press.

Kholief, M. M. and M. A. Barakat. 1992. Rock composition of the Sphinx. In *Proceedings of the 1st International Symposium on the Great Sphinx*, ed. F. A. Esmael, 111–18. Cairo: Ministry of Culture, Egyptian Antiquities Organization.

Kilikoglou, V., G. Vekinis, and Y. Maniatis. 1995. Toughening of ceramic earthenwares by quartz inclusions: An ancient art revisited. *Acta Metallurgica et Materialia* 43(8): 2959–65.

Kilka, T. 1992. An example of study where the petrography prevails over the chemistry: The Bronze Age ceramics from Fiave (Italy). In *Sciences de la terre et céramiques archéologiques: Expérimentations, applications, documents et travaux No. 16*, ed. S. Mery, 61–72. Cergy: Centre Polytechnique Saint-Louis á Cergy.

Kilmurry, K. 1982. The manufacture of Stamford ware: An application of thin-sectioning and neutron activation analysis. In *Current research in ceramics: Thin-section studies*, eds. I. C. Freestone, C. Johns, and T. Potter, 105–11. London: British Museum Occasional Paper No. 32.

Kingery, W. D. 1986. The development of European porcelain. In *High-technology ceramics past, present, and future: The nature of innovation and change in ceramic technology*, ed. W. D. Kingery, 153–80. Westerville, OH: American Ceramic Society.

Kingery, W. D. and D. Smith. 1984. The development of European soft-paste (frit) porcelain. In *Ancient technology to modern science*, ed. W. D. Kingery, 273–92. Columbus, OH: American Ceramic Society.

Kingery, W. D. and P. B. Vandiver. 1986. The eighteenth-century change in technology and style from the famille-verte palette to the famille-rose palette. In *Technology and style*, ed. W. D. Kingery, 363–81. Columbus, OH: American Ceramic Society.

—. 1986. *Ceramic masterpieces: Art, structure, and technology*. New York: Free Press.

Kirchner, D., F. Fockenberg, and P. W. Mirwald. 1991. Baumberger Kalksandstein: A mineralogic-petrographic and petrophysical characterization of a prominent building and stone cutting material in NW Germany. In *Science, technology and European cultural heritage*, eds. N. S. Baer, C. Sabioni, and A. I. Sors, 531–34. Oxford: Butterworth-Heinemann.

Knight, D., P. Marsden, and J. Carney. 2003. Local or nonlocal? Prehistoric granodiorite-tempered pottery in the East Midlands. In *Prehistoric pottery: People, pattern and purpose*, ed. A. Gibson, 111–25. Oxford: British Archaeological Reports International Series 1156.

Koh Choo, C. K., S. Kim, H. T. Kang, J. Y. Do, Y. E. Lee, and G. H. Kim. 1999. A comparative scientific study of the earliest kiln sites of Koryŏ celadon. *Archaeometry* 41(1): 51–69.

Koh Choo, C. K., Y. E. Lee, I. W. Shim, W. K. Choo, G. H. Kim, W. Y. Huh, and S. C. Chun. 2004. Compositional and microstructural study of Koryŏ celadon and whiteware excavated from Sŏri kiln in Kyŏnggi province. *Archeometry* 46(2): 247–65.

Kokkorou-Alevras, G., V. Mandi, A. P. Grimanis, and Y. Manniatis. 1995. The traditional archaeological characterisation of marble sculpture and the results of modern scientific techniques. In *The study of marble and other stones used in antiquity*, eds. Y. Maniatis, N. Herz, and Y. Basiakos, 95–102. London: Archetype.

Kolb, C. C. 1977. Technological investigations of Mesoamerican "thin orange" ceramics. *Current Anthropology* 18(3): 534–36.

—. 1989. Kushano-Sasanian storage jars from Central Asia: A study in ceramic ecology and ethnoarchaeology. In *Ceramic ecology 1988: Current research on ceramic materials*, ed. C. C. Kolb, 175–259. Oxford: British Archaeological Reports International Series 513.

Koumantakis, J. and A. Orfanoudakis. 1988. Etude des calcarénites dans lesquelles sont creusés de tombeaux de la période mycénienne tardive à Mazarakata et Metaxata de Céphalonie (Grèce). In *The engineering geology of ancient works, monuments and historical sites*, eds. P. G. Marinos and G. C. Koukis, 691–98. Rotterdam: A. A. Balkema.

Kozlowski, R., J. Magiera, J. Weber, and J. Haber. 1990. Decay and conservation of Pińczów porous limestone. I. Lithology and weathering. *Studies in Conservation* 35: 205–21.

Krauze, J., S. Skibinski, A. Warzenczak, and M. Urbanowski. 1991. Investigation and conservation of the Holy Trinity Memorial in Banská Štiavnica (ČSFR). *Wiener berichte über naturwissenschaft in der kunst* 6/7/8: 381–96.

Krishnan, K. 2002. A survey of ceramic analysis with specific reference to pottery. In *Indian archaeology in retrospect*, vol. 1, eds. S. Settar and R. Korisettar, 377–87. New Delhi: Manohar.

Krishnan, K. and K. Shah. 2005. Beyond wares and shapes: Gaining a petrographic perspective on ancient Indian pottery. In *Science in archaeology and archaeo-materials*, ed. A. K. Biswas, 135–56. New Delhi: D. K. Printworld.

Krumbein, W. C. 1938. Size frequency distributions of sediments and the normal phi curve. *Journal of Sedimentary Petrology* 8: 84–90.

Kularatnam, K. 1979. Ceylon building stones. *Ancient Ceylon* 3: 211–23.

Kuntze, R. A. 1983. *The chemistry and technology of gypsum*. Philadelphia, PA: American Society of Testing and Materials, ASTM Special Technical Publication 861.

Lagrou, D., R. Dereeson, and L. Broothaers. 2004. Comparative quantitative petrological analysis of Cenozoic aquifer sands in Flanders (N. Belgium): Overall trends and quality assessment. *Materials Characterization* 53: 317–26.

Lahanier, C. 1976. Laboratoire de recherche des musées de France examen en laboratorie d'une tête de roi barbare. *Arts Asiatiques* 32: 80–81.

Lal, B. B. 1985. Weathering and disintegration of stone monuments. In *Fifth International Congress on Deterioration and Conservation of Stone*, ed. G. Félix, 213–22. Lausanne: Presses Polytechniques Romandes.

Lapuente, M. P. 1995. Mineralogical, petrographical and geochemical characterization of white marbles from Hispania. In *The study of marble and other stones used in antiquity*, eds. Y. Maniatis, N. Herz, and Y. Basiakos, 151–60. London: Archetype.

Lapuente, P. and B. Turi 1995. Marbles from Portugal: Petrographic and isotopic characterization. *Science and Technology for Cultural Heritage* 4(2): 33–42.

Lapuente, P., M. P. Martinez, B. Turi, and P. Blanc. 2002. Characterization of dolomitic marbles from the Malaga province (Spain). In *Interdisciplinary studies on ancient stone*, eds. J. J. Herrmann, Jr., N. Herz, and R. Newman, 152–62. London: Archetype.

Laue, S., C. B. Böhm, and D. Jeannette. 1996. Saltweathering and porosity: Examples from the crypt of St. Maria in Kapitol, Cologne. In *Proceedings of the 8th International Congress on Deterioration and Conservation of Stone*, vol. 1, ed. J. Riederer, 513–22. Berlin: Möller Druck und Verlag.

Laurenzi, M. T. 1971 Studi e osservazioni sullo stato di conservazione del portale centrale di San Petronio a Bologna. In *La conservazione delle sculture all'Aperto*, comps. R. R. Manaresi and E. Riccòmini, 117–32. Bolgona: Gallerie di Bologna.

Laviano, R. 1992. The rock churches of Madonna degli Angeli and Madonna delle Croci (Sassi di Materia, Basilicata, southern Italy): Mineralogical and geochemical characteristics and decay of calcarenite di Gravina. In *Science and Technology for Cultural Heritage* 1: 87–95.

Laviano, R. and I. M. Muntoni. 2006. Provenance and technology of Apulian Neolithic pottery. In *Geomaterials in cultural heritage*, eds. M. Maggetti and B. Messiga, 49–62. London: Geological Society, Special Publications 257.

Lawal, H. B. 2003. *Categorical data analysis with SAS and SPSS applications*. Mahwah, NJ: Lawrence Erlbaum.

Lazzarini, L. 1990. *Rosso antico* and other red marbles used in antiquity: A characterization study. In *Marble: Art historical and scientific perspectives on ancient sculpture*, eds. M. True and J. Podany, 237–52. Malibu, CA: J. Paul Getty Museum.

—. 2002. The origins and characterization of *breccia nuvolata*, *marmor Sagarium*, and *marmor Triponticum*. In *Interdisciplinary studies on ancient stone*, eds. J. J. Herrmann, Jr., N. Herz, and R. Newman, 58–67. London: Archetype.

Lazzarini, L. and F. Antonelli. 2003. Petrographic and isotopic characterization of the marble of the island of Tinos (Greece). *Archaeometry* 45(4): 541–52.

Lazzarini, L. and S. Calogero. 1989. Early local and imported Byzantine *sgraffito* ware in Venice: A characterization and provenance study. In *Archaeometry: Proceedings of the 25th International Symposium*. ed. Y. Maniatis, 571–84. Amsterdam: Elsevier.

Lazzarini, L. and S. Cancelliere. 2000. Characterization of the white marble of two unpublished ancient Roman quarries on the islands of Fourni and Skyros (Greece). *Periodico di Mineralogia* 69(1): 49–62.

Lazzarini, L., G. Moschini, and B. M. Stievano. 1980. A contribution to the identification of Italian, Greek and Anatolian marbles through a petrological study and the evaluation of Ca/Sr Ratio. *Archaeometry* 22(2): 173–83.

Lazzarini, L., M. Mariottini, M. Pecoraro, and P. Pensabene. 1988. Determination of the provenance of marbles used in some ancient monuments in Rome. In *Classical marble: Geochemistry, technology, trade*, eds. N. Herz and M. Waelkens, 399–409. Dordrecht: Kluwer.

Lazzarini, L., U. Masi, and P. Tucci. 1995. Petrographic and geochemical features of the Carystain marble, "Cipollino verde", from the ancient quarries of southern Euboea (Greece). In *The study of marble and other stones used in antiquity*, eds. Y. Maniatis, N. Herz, and Y. Basiakos, 161–69. London: Archetype.

Lazzarini, L., G. Lombardi, F. Marconi, and C. Meucci.1996. New data on the characterisation and conservation of the Easter Island's pyroclastics used for the Moais. In *Proceedings of the 8th International Congress on Deterioration and Conservation of Stone*, vol. 2, ed. J. Riederer, 1147–57. Berlin: Möller Druck und Verlag.

Lazzarini, L., C. Gorgoni, P. Pallante, and B. Turi. 1997. Identification of ancient white marbles in Rome. II: The Portico of Octavia. *Science and Technology for Cultural Heritage* 6(2): 185 98.

Le Maitre, R. W., ed. 1989. *A classification of igneous rocks and glossary of terms*. Oxford: Blackwell Scientific Publications.

Leavengood, P., J. Twilley, and T. van Halm. 1996. Isamu Noguchi's "Black Sun": Investigation of a cracking phenomenon. In *Proceedings of the 8th International Congress on Deterioration and Conservation of Stone*, vol. 1, ed. J. Riederer, 239–51. Berlin: Möller Druck und Verlag.

Leese, M. N. 1983. The statistical treatment of grain size data from pottery. *Proceedings of the 22nd Symposium on Archaeometry*, eds. A. Aspinall and S. E. Warren, 47–55. Bradford: University of Bradford.

Lepsius, G. R. 1890. *Griechische marmonstudien*. Berlin: Konige.

Letsch, J. and W. Noll. 1983. Phase formation in several ceramic subsystems at 600 °C–1000 °C as a function of oxygen fugacity. *Ceramic Forum International CFI Berichte der DKG* 60: 259–67.

Linné, S. 1965. The ethnologist and the American Indian potter. In *Ceramics and man*, ed.. F. R. Matson, 20–42. Chicago, IL: Aldine Publishing Company.

Liu, D. 2000. *Ca ca: Zang chuan fo jiao mo zhi ni fo Xiang (tsha tsha: Tibetan clay molded Buddhist images)*. Tianjin Shi: Tianjin ren min mei shu chu ban she.

Lombard, J. P. 1987. Provenance of sand temper in Hohokam ceramics, Arizona. *Geoarchaeology* 2(2): 91–119.

Lombardi, G. 2002. A petrographic study of the casting core of the Lupa Capitolina bronze sculpture (Rome, Italy) and identification of its provenance. *Archaeometry* 44: 601–12.

Lombardi, G. and M. Vidale. 1998. From the shell to its content: The casting cores of the two bronze statues from Riace (Calabria, Italy). *Journal of Archaeological Science* 25: 1055–66.

London, G. 1981. Dung-tempered clay. *Journal of Field Archaeology* 8(2): 189–95.

Lopez Azcona, M. C., F. Mingarro, M. A. Garcia del Cura, S. Ordoñez, J. P. Calvo, R. Fort, M. Bastillo, M. Peinado, and A. M. Alonso Zarza. 1991. Degradation of building materials of the Toledo Cathedral (Spain). In *Science, technology and European cultural heritage*, eds. N. S. Baer, C. Sabbioni, and A. I. Sors, 925–29. Oxford: Butterworth-Heinemann.

Lowenstein, J. G. 1985. Westerwald salt-glazed stoneware. In *Technology and style*, ed. W. D. Kingery, 383–96. Columbus, OH: American Ceramic Society.

Lucchi, F. R. 1985. Influence of transport processes and basin geometry on sand composition. In *Provenance of arenites*, ed. G. G. Zuffa 19–45. Dordrecht: D. Reidel.

Luczanits, C. 2004. *Buddhist sculpture in clay*. Chicago, IL: Serindia.

Luedtke, B. E. 1992. *An archaeologist's guide to chert and flint*. Los Angeles, CA: Institute of Archaeology, University of California, Los Angeles, Archaeological Research Tools 7.

MacKenna, S. A. 1978. Report on the thin-sectioning of an amphora-sherd. Appendix in *Amphorae from Brochley Hill*, by S. A. Castel. *Britannia* 9: 389–90.

MacKenzie, W. S. and A. E. Adams. 1994. *A color atlas of rocks and minerals in thin section*. London: Manson Publishing.

MacKenzie, W. S. and C. Guilford. 1981. *Atlas of rock-forming minerals in thin section*. Harlow: Longman.

MacKenzie, W. S., C. H. Donaldson and C. Guilford. 1982. *Atlas of igneous rocks and their textures*. New York: Wiley.

MacSween, A. 1995. Fabric analysis of coarse ware assemblages: Examples from three sites in the Northern Isles of Scotland. In *The aim of laboratory analyses of ceramics in archaeology*, eds. A. Lindahl and O. Stilborg, 127–38. Stockholm: Kungl. Vitterhets Historie och Antikvitets Ademien.

MacSween, A., J. R. Hunter, and S. E. Warren. 1987. Analysis of coarse wares from the Orkney islands. In *Science and archaeology*, eds. E. A. Slater and J. O. Tate, 95–106. Glasgow: British Archaeological Reports British Series 196(I).

Maggetti, M. 1981. Composition of Roman pottery from Lousonna (Switzerland). In *Scientific studies in ancient ceramics*, ed. M. J. Hughes, 33–49. London: British Museum.

—. 1982. Phase analysis and its significance for technology and origin. In *Archaeological ceramics*, eds. J. S. Olin and A. D. Franklin, 121–33. Washington, DC: Smithsonian Institution Press.

—. 2005. The Alps: A barrier or a passage for ceramic trade. *Archaeometry* 47(2): 389–401.

Mahmoud, A-M. A. and K. A. Bard. 1993. Sources of the Predynastic grinding stones in the Hu-Semaineh region, Upper Egypt, and their cultural context. *Geoarchaeology* 8(3): 241–45.

Mainman, A. 1982. Studies of Anglo-Saxon pottery from Canterbury. In *Current research in ceramics: Thin-section studies*, eds. I. C. Freestone, C. Johns, and T. Potter, 93–100. London: British Museum Occasional Paper No. 32.

Maish, J., M. Svoboda, and S. Lansing-Maish. 2006. Technical studies of some Attic vases in the J. Paul Getty Musem. In *The colors of clay: Special techniques in Athenian vases*, ed. B. Cohen, 8–16. Los Angeles, CA: J. Paul Getty Museum.

Malaga-Starzec, K., T. Sahlin, and O. Lindqvist. 2000. Laboratory investigation of weathering behaviour of fresh and impregnated limestone and sandstone from central Sweden. In *Proceedings of the 9th International Congess on Deterioration and Conservation of Stone*, vol. 1, ed. V. Fassina, 349–56. Amsterdam: Elsevier.

Mallory-Greenough, L. M., J. D. Greenough, and J. V. Owen. 1998. Provenance of temper in a New Kingdom Egyptian pottery sherd: Evidence from the petrology and mineralogy of basalt fragments. *Geoarchaeology* 13(4): 391–410.

Mandal, S. 1997. Striking the balance: The roles of petrography and geochemistry in stone axe studies in Ireland. *Archaeometry* 39(2): 2898–308.

Maniatis, Y. and M. Tite. 1981. Technical examination of Neolithic-Bronze Age pottery from central and southeast Europe and from the Near East. *Journal of Archaeological Science* 8: 59–76.

Maniatis, Y., R. W. Jones, I. K. Whitbread, A. Kostikas, A. Simopoulos, Ch. Karakalos, and C. K. Williams, II. 1984. Punic amphoras found at Corinth, Greece: An investigation of their origin and technology. *Journal of Field Archaeology* 11: 205–22.

Mann, J. C. 1986. Composition and origin of material in pre-Columbian pottery, San Salvador Island, Bahamas. *Geoarchaeology* 1(2): 183–95.

Marcus, E. S. 1995. A petrographic analysis of storage jars from Tel Nami, Israel: Ceramic production and trade networks of the Middle Bronze IIA period (ca. 2000–1750 BC). In *Materials issues in art and archaeology IV*, eds. P. B. Vandiver, J. R. Druzik, J. L. Galvan Madrid, I. C. Freestone, and G. S. Wheeler, 597–603. Pittsburgh, PA: Materials Research Society.

Margolis, S. V. 1989. Authenticating ancient marble sculpture. *Scientific American* 260(6): 104–14.

Margolis, S. V. and W. Showers. 1988. Weathering characteristics, age, and provenance determinations on ancient Greek and Roman marble artifacts. In *Classical marble: Geochemistry, technology, trade*, eds. N. Herz and M. Waelkens, 233–42. Dordrecht: Kluwer.

—. 1990. Ancient Greek and Roman marble sculpture: Authentication, weathering, and provenance determinations. In *Marble: Art historical and scientific perspectives on ancient sculpture*, eds. M. True and J. Podany, 283–99. Malibu, CA: J. Paul Getty Museum.

Marinos, G. P. 1948. Notes on the structure of Greek marbles. *American Journal of Archaeology* 246: 386–89.

Maritan, L. and C. Mazzoli. 2004. Phosphates in archaeological finds: Implications for environmental conditions of burial. *Archaeometry* 46(4): 673–83.

Maritan, L., C. Mazzoli, V. Michielin, D. Morandi Bonacossi, M. Luciani, and G. Molin. 2005. The provenance and production technology of Bronze Age and Iron Age pottery from Tell Mishrifeh/Qatna (Syria). *Archaeometry* 47(4): 723–44.

Marshall, J. 1951. *Taxila*. Cambridge: Cambridge University Press.

Martin-Socas, D., M. D. Camalish-Massieu, M. L. Tejedor-Salguero, and A. Rodriguez-Rodriguez. 1989. Mineralogical composition and evaluation of firing temperatures of the Purchena pottery (Almeria, Spain). In *Archaeometry: Proceedings of the 25th International Symposium*, ed. Y. Maniatis, 149–57. Amsterdam: Elsevier.

Martinez, G. M. and E. N. Martinez. 1991. Characterization of stone from the Metropolitan Cathedral and from the facade of the National Museum at Tepotzotlan, Mexico. *Studies in Conservation* 36: 99–110.

Maskelyne, N. S. 1878. Stonehenge: The petrology of its stones. *Wiltshire Archaeological and Natural History Magazine* 17: 147–60.

Mason, R. B. 1991. Petrography of Islamic ceramics. In *Recent developments in ceramic petrology*, eds. A. Middleton and I. C. Freestone, 185–209. London: British Museum.

Mason, R. B. and L. B. Golombek. 1991. Differentiating early Chinese-influenced blue and white ceramics of Egypt, Syria, and Iran. In *Archaeometry '90*, eds. E. Pernicka and G. A. Wagner, 465–74. Basel: Birkhäuser Verlag.

—. 2003. The petrography of Iranian Safavid ceramics. *Journal of Archaeological Science* 30: 251–61.

Mason, R. B. and E. J. Keall. 1988. Provenance of local ceramic industry and the characterization of imports: Petrography of pottery from medieval Yemen. *Antiquity* 62: 452–63.

Mason, R. B., J. R. Hallett, and E. J. Keall. 1989. Provenance studies of Islamic pottery from Yeman: INAA and petrographic analysis. In *Archaeometry: Proceedings of the 25th International Symposium*, ed. Y. Maniatis, 543–50. Amsterdam: Elsevier.

Matson, F. R. 1970. Some aspects of ceramic technology. In *Science in archaeology: A survey of progress and research*, eds. D. Brothwell and E. Higgs, 592–602. New York: Praeger.

—. 1981. Appendix V: Analyses of the core sand found in the horse. *In Technical examination of the Classical bronze horse from the Metropolitan Museum of Art*, by K. C. Lefferts, L. J. Majewski, E. V. Sayre, and P. Meyers. *Journal of the American Institute for Conservation* 21(1): 39–42.

—. 1984. The brickmakers of Babylon. In *Ancient technology to modern science*, ed. W. D. Kingery, 61–76. Columbus, OH: American Ceramic Society.

—. 1985. Glazed brick from Babylon: Historical setting and microprobe analyses. In *Technology and style*, ed. W. D. Kingery, 133–56. Columbus, OH: American Ceramic Society.

Matthew, A. J., A. J. Woods, and C. Oliver. 1991. Spots before the eyes: New comparison charts for visual percentage estimating in archaeological material. In *Recent developments in ceramic petrology*, eds. A. Middleton and I. C. Freestone, 211–63. London: British Museum.

Mausfield, S. A. and G. Grassegger. 1994. The changing environment of pore solutions in natural building stones during immission accelerated weathering processes. In *La conservazione dei monumenti nel bacino del Mediterraneo. Materiali lapidei e monumenti: Metodologie per l'analisi del degrado e la conservazione*, eds. V. Fassina, H. Ott, and F. Zezza, 129–35. Venice: Soprintendenza ai Beni Artistici e Storici di Venezia.

McBride, E. F. 1985. Diagenetic processes that affect provenance determinations in sandstone. In *Provenance of arenites*, ed. G. G. Zuffa, 95–113. Hingham, MA: D. Reidel.

McGee, E. S. 1999. *Colorado yule marble: Building stone of the Lincoln Memorial*. Washington, DC: U.S. Geological Survey Bulletin 2162, National Park Service and U.S. Department of the Interior, U.S. Geological Survey.

McKee, H. J. 1971. Early ways of quarrying and working stone in the United States. *Bulletin of the Association for Preservation Technology* 3(1): 44–58.

—. 1973. *Introduction to early American masonry: Stone, brick, mortar and plaster*. Washington, DC: National Trust for Historic Preservation.

McPherron, A. and D. Milovanović. 1974. A note on the petrologic characteristics of a tool from a Neolithic site near Divostin (Kragujevac, Serbia). *Bulletin Scientific, Section A, Sciences Naturelles, Techniques et Medicales* 19(5–6): 134.

Merrill, G. P. 1891. *Stones for building and decoration*. New York: Wiley.

Merriman, M. 1892. The strength and weathering qualities of roofing slates. *Perspectives* 27: 331–48.

Meyers, P. and L. van Zelst. 1977. Neutron activation analysis of limestone objects: A pilot study. *Radiochimica Acta* 24: 197–204.

Michelaki, K., L. Minc, and J. O'Shea. 2002. Integrating typological and physico-chemical approaches to examine the potter's choices: A case from Bronze Age Hungary. In *Modern trends in scientific studies on ancient ceramics*, eds. V. Kilikoglou, A. Hein, and Y. Manniatis, 313–22. Oxford: British Archaeological Reports International Series 1011.

Middleton, A. 1991. Ceramics: Materials for all reasons. In *Science and the past*, ed. S. Bowman, 16–36. Toronto: University of Toronto Press.

Middleton, A. and I. C. Freestone. 1991. *Recent developments in ceramic petrology*. London: British Museum.

—. 1997. Tiles in Roman-Britain. In *Pottery in the making*, eds. I. C. Freestone and D. Gaimster, 158–63. Washington, DC: Smithsonian Institution Press.

Middleton, A., I. C. Freestone and M. N. Leese. 1985. Textural analysis of ceramic thin sections: Evaluation of grain sampling procedures. *Archaeometry* 27(1): 64–74.

Middleton, A. P., M. N. Leese, and M. R. Cowell. 1991. Computer-assisted approaches to the grouping of ceramic fabrics. In *Recent developments in ceramic petrology*, eds. A. Middleton and I. C. Freestone, 265–76. London: British Museum Occasional Paper No. 81.

Middleton, A. P., M. R. Cowell, and E. W. Black. 1992. Romano-British relief-patterned flue tiles: A study of provenance using petrography and neutron activation analysis. In *Sciences de la terre et céramiques archéologiques: Expérimentations, applications, documents et travaux No. 16*, ed. S. Mery, 49–59. Cergy: Centre Polytechnique Saint-Louis à Cergy.

Miksa, E. J. 2001. Criteria for evaluating multiple components in pottery paste. In *Archaeology and clays*, ed. I. C. Druc, 81–93. Oxford: British Archaeological Reports.

Miller, B. L.1939. *Northampton County Pennsylvania: Geology and geography*. Harrisburg, PA: Department of Internal Affairs, Topographic and Geologic Survey, Fourth Series, Bulletin C48.

Miller, S, F. M. McGibbon, D. H. Caldwell, and N. A. Ruckley. 2006. Geological tools to interpret Scottish medieval carved sculpture: Combined petrological and magnetic susceptibility analysis. In *Geomaterials in cultural heritage*, eds. M. Maggetti and B. Messiga, 283–305. London: Geological Society, Special Publications 257.

Mirwald, P. W., K. Kraus, and A. Wolff. 1988. Stone deterioration on the cathedral of Cologne. *Durability of Building Materials* 5(3/4): 549–70.

Mitchell, J. G., H. Askvik, and H. G. Resi. 1984. Potassium-argon ages of schist honestones from the Viking Age sites at Kaupang (Norway), Aggersborg (Denmark), Hedeby (West Germany) and Wolin (Poland), and their archaeological implications. *Journal of Archaeological Science* 11: 171–76.

Moens, L. 1987. Empreintes d'une pierre. In *Marbres Helléniques*: 90–107. Bruxelles: Credit Communal.

Moens, L., P. Roos, J. De Rudder, J. Hoste, P. De Paepe, J. Van Hende, R. Marechal, and M. Waelkens. 1988a. White marble from Italy and Turkey: An archaeometric study based on minor- and trace-element analysis and petrography. *Journal of Radioanalytical and Nuclear Chemistry* 123(1): 333–48.

—. 1988b. A multi-method approach to the identification of white marbles used in antique artifacts. In *Classical marble: Geochemistry, technology, trade*, eds. N. Herz and M. Waelkens, 243–50. Dordrecht: Kluwer.

Moens, L., P. Roos, J. De Rudder and J. Hoste. 1989. Chemical and petrographical identification of white marbles from the Mediterranean area. I. Comparison between Carrara and Marmara marbles. In *Archaeometry: Proceedings of the 25th International Symposium*, ed. Y. Maniatis, 613–24. Amsterdam: Elsevier.

Moens, L. P. Roos, J. De Rudder, J. Hoste, P. De Paepe, J. Van Hende, R. Marechal and M. Waelkens.1990. Scientific provenance determination of ancient white marble sculptures using petrographic, chemical, and isotopic data. In *Marble: Art historical and scientific perspectives on ancient sculpture*, eds. M. True and J. Podany, 111–24. Malibu, CA: J. Paul Getty Museum.

Moens, L., P. De Paepe, and M. Waelkens. 1992. Multidisciplinary research and cooperation: Keys to a successful provenance determination of white marble. In *Ancient stones: Quarrying, trade and provenance*, eds. M. Waelkens, N. Herz, and L. Moens, 247–52. Leuven: Leuven University Press.

Moens, L., P. De Paepe, and M. Waelkens. 1995. Une approache multidisciplinaire permettant de déterminer l'origine des marbres blancs utlisés pendant l'Antiquité classique. In *Les marbres blancs des Pyrénées: Approches scientifiques et historiques*, comps. J. Cabanot et al., 33–47. Saint-Bertrand-de-Comminges: Musée Archéologique Départemental.

Molina, E., J. García-Talegón, and M. A. Vicente. 1996. Granitic facies originated by palaeoweathering processes: The quarries of Avila (Spain). In *Degradation and conservation of granitic rocks in monuments*, 45–51. Brussels: European Commission, Directorate-General XII, Science, Research and Development.

Montana, G., H. Mommsen, I. Iliopoulos, A. Schwedt, and M. Denaro. 2003. The petrography and chemistry of thin-walled ware from an Hellenistic-Roman site at Segesta (Sicily). *Archeometry* 45(3): 375–89.

Moore, D. T. 1978. The petrography and archaeology of English honestones. *Journal of Archaeological Science* 5: 61–73.

Moore, D. T. and W. A. Oddy. 1985. Touchstones: Some aspects of their nomenclature, petrography and provenance. *Journal of Archaeological Science* 12(1): 59–80.

Moreland, G. C. 1968. Preparation of polished thin sections. *American Mineralogist* 53: 2070–74.

Morey, J. E. and K. C. Dunham. 1953. A petrographical study of medieval hones from Yorkshire. *Proceedings of the Yorkshire Geological Society* 29(2): 141–48.

Morgenstein, M. 2006. Geochemical and petrographic approaches to chert tool provenance studies: Evidence from two western USA Holocene archaeological sites. In *Geomaterials in cultural heritage*, eds. M. Maggetti and B. Messiga, 307–21. London: Geological Society, Special Publications 257.

Moroni, B., G. Poli, and D. Querci. 2002. Provenance of basaltina and travertine employed in the construction of Orvieto Cathedral. In *Interdisciplinary studies on ancient stone*, eds. J. J. Herrmann, Jr., N. Herz, and R. Newman, 169–75. London: Archetype.

Moropoulou, A., P. Theoulakis, K. Bisbikou, G. Biscontin, E. Zendri, A. Bakolas, and P. Maravelaki. 1993. Behaviour of the bricks in the

Corfu Venetian fortress. In *Conservation of stone and other materials*, ed. M. -J. Thiel, 402–10. London: E & FN Spon.

Moropoulou, A., Chr. Kourteli, A. Bisbikou, Th. Tsiourva, M. Beazi, and F. Zezza. 1995. Environmental impact assessment on the porous stone masonries of the Rethymnon Fortress. In *Architectural studies, materials and analysis*, eds. C. A. Brebbia and B. Leftheris, 137–50. Southhampton and Boston, MA: Computational Mechanics Publication, Structural Studies of Historical Buildings IV, Volume 1.

Morris, E. L. 1982. Iron Age pottery from western Britain: Another petrological study. In *Current research in ceramics: Thin-section studies*, eds. I. C. Freestone, C. Johns and T. Potter, 15–25. London: British Museum Occasional Paper 32.

Morter, J. and H. Iceland. 1995. Notes on an eastern Calabrian assemblage in the Stentinello tradition. In *The ceramics cultural heritage*, ed. P. Vincenzini, 241–49. Faenza: Techna.

Morton, A. C. 1985. Heavy minerals in provenance studies. In *Provenance of arenites*, ed. G. G. Zuffa, 249–77. Dordrecht: D. Reidel.

Morzadec, H. 1992. La céramique "proto-onctueuse" de Bretagne occidentale: Approche pétrographique et géochimique. In *Sciences de la terre et céramiques archéologiques: Expérimentations, applications, documents et travaux No. 16*, ed. S. Mery, 73–86. Cergy: Centre Polytechnique Saint-Louis à Cergy.

Mueller, U. and E. F. Hansen. 2001. Use of digital image analysis in conservation of building materials. In *Proceedings of the 8th Euroseminar on Microscopy Applied to Building Materials*, eds. B. Georgali and E. E. Toumbakari, 603–10. Athens: Cosmosware.

Myer, G. H. and P. P. Betancourt. 1981. The composition of Vasilike Ware and the production of the mottled colours of the slip. In *Scientific studies in ancient ceramics*, ed. M. J. Hughes, 51–55. London: British Museum.

National Materials Advisory Board. 1982. *Conservation of historic stone buildings and monuments: Report of the Committee on Conservation of Historic Stone Buildings and Monuments*. Washington DC: National Academy Press.

Newman, R. 1984. *The stone sculpture of India*. Cambridge, MA: Center for Conservation and Technical Studies, Harvard University Art Museums.

—. 1990. Weathering layers and the authentication of marble objects. In *Marble: Art historical and scientific perspectives on ancient sculpture*, eds. M. True and J. Podany, 263–82. Malibu, CA: J. Paul Getty Museum.

—. 1992. Application of petrography and electron microprobe analysis to the study of Indian stone sculpture. *Archaeometry* 34(2): 163–74.

—.1997. Appendix A: Petrographic analysis. In *The sacred sculpture of Thailand*, by H. W. Woodward, Jr., 261–65. Baltimore, MD: Walters Art Gallery.

Nijs, R., G. De Geyter, and B. Fobe. 1988. Behaviour of the historic Balegem stone and its substitute, the Gobertange stone in Flemish monuments. In *The engineering geology of ancient works, monuments and historic sites*, eds. P. G. Marinos and G. C. Koukis, 843–48. Rotterdam: A. A. Balkema.

Noble, J. V. 1988. *The techniques of painted Attic pottery*. [Reprint of 1965 edition.] New York: Thames & Hudson.

Noll, W. 1981. Mineralogy and technology of the painted ceramics of ancient Egypt. In *Scientific studies in ancient ceramics*, ed. M. J. Hughes, 143–54. London: British Museum.

O'Brien, P. F. and T. P. Cooper. 1994. Comparative study of granite cleaning techniques. In *Granite weathering and conservation*, eds. E. Bell and T. P. Cooper, 65–70. Dublin: Director of Buildings' Office, University of Dublin, Trinity College.

O'Brien, N. R. and R. M. Slatt. 1990. *Argillaceous rock atlas*. New York: Springer-Verlag.

Ohri, V.C. ed. 1975. *Arts of Himachal*. Simla: State Museum.

Olcese, G. 1991. Roman coarse ceramics from Albintimilium (Ventimiglia, Italy): An example of archaeometric and archaeological studies. In *Archaeometry '90*, eds. E. Pernicka and G. A. Wagner, 495–504. Basel: Birkhäuser Verlag.

Ordaz, J. and R. M. Esbert. 1985. Porosity and capillarity in some sandstone and dolomite monumental stones. In *Proceedings of the 5th International Congress on Deterioration and Conservation of Stone*, vol. 1, ed. G. Félix, 93–102. Lausanne: Presses Polytechniques Romandes.

Ordaz, J., R. M. Esbert and L. M. Suarez de Rio. 1978. A proposed petrographical index to define mineral and rock deterioration in granite rocks. In *Deterioration and protection of stone monuments*, 2.6. Paris: RILEM.

Ordaz, J., R. M. Esbert, F. J. Alonso, V. G. Ruiz de Argandoña, and A. Rodriguez Rey. 1989. Petrophysical characterization and deterioration of the stones of "La Pedrera" (Barcelona, Spain). In *The conservation of monuments in the Mediterranean basin*, ed. F. Zezza, 271–74. Brescia: Grafo.

Orton, C., P. Tyers, and A. Vince. 1993. *Pottery in archaeology*. Cambridge: Cambridge University Press.

Outwater, J. O., Jr. 1957. The Pre-Columbian stonecutting techniques of the Mexican plateau. *American Antiquity* 22: 258–64.

Owen, J. V. 1997. Quantification of early Worcester porcelain recipes and the distinction between Dr Wall- and Flight-period wares. *Journal of Archaeological Science* 24(4): 301–11.

—. 1998. On the earliest products (ca 1751–1752) of the Worcester porcelain manufactory. *Historical Archaeology* 32(4): 63–75.

—. 2001. Provenance of eighteenth-century British porcelain sherds from sites 3B and 4E, fortress of Louisbourg, Nova Scotia: Constraints from mineralogy, bulk paste, and glaze compositions. *Historical Archaeology* 35(2): 106–19.

Owen, J. V. and T. E. Day. 1998. Assessing and correcting the effects of chemical weathering of potsherds: A case study using soft-paste porcelain wasters from the Longton Hall (Staffordshire) factory site. *Geoarchaeology* 13(3): 265–86.

Owen, J. V. and M. L. Morrison. 1999. Sagged phosphatic Nantgarw porcelain (ca. 1813–1820): Casualty of overfiring or a fertile paste? *Geoarchaeology* 14(4): 313–32.

Owen, J. V. and J. Sandon. 1998. Petrology of Gilbody, Pennington and Christian/Pennington (18th century Liverpool) porcelains and their distinction from some contemporary phosphatic and magnesian/plombian British wares. *Journal of Archaeological Science* 25: 1131–47.

Owen, J. V., J. O. Wilstead, R. W. Williams, and T. E. Day. 1998. A tale of two cities: Compositional characteristics of some Nantgarw and Swansea porcelains and their implications for kiln wastage. *Journal of Archaeological Science* 25: 359–75.

Owen, J. V., B. Adams, and R. Stephenson. 2000. Nicholas Crisp's "Porecellien": A petrological comparison of sherds from the Vauxhall (London, ca. 1751–1764) and Indeo pottery (Bovey Tracey, Devonshire; ca. 1767–1774) factory sites. *Geoarchaeology* 15(1): 43–78.

Page, B. M. 1951. *Talc deposits of steatite grade, Inyo county, California*. San Francisco, CA: State of California Division of Mines Special Report 8.

Page, H. and J. Greenwood. 1986. The treatment of waterlogged "shale" artefacts by using freeze drying. *Conservation News* 29: 12.

Pal. P. 1978. *The sensuous immortals*. Los Angeles, CA: Los Angeles County Museum of Art.

—. 1986. *Indian sculpture*. Volume 1: *A catalogue of the Los Angeles County Museum of Art collection*. Los Angeles, Berkeley and London: Los Angeles County Museum of Art and University of California.

—. 1988. *Indian sculpture*. Volume 2: *A catalogue of the Los Angeles County Museum of Art collection*. Los Angeles, Berkeley and London: Los Angeles County Museum of Art and University of California.

Palmiter, T. V. and P. F. Johnson. 1988. Techniques for archaeothermometry. In *Materials issues in art and archaeology*, eds. E. V. Sayre, P. Vandiver, J. Druzik, and C. Stevenson, 135–40. Pittsburgh, PA: Materials Research Society.

Panjwani, P. A. 1989. Petrography of Harappan pottery from Lothal. *Man & Environment* 13: 83–86.

Papageorgakis, J. and E. Kolaiti. 1992. The ancient limestone quarries of Profitis Elias near Delfi (Greece). In *Ancient stones: Quarrying, trade and provenance*, eds. M. Waelkens, N. Herz, and L. Moens, 37–41. Leuven: Leuven University Press.

Papageorgakia, J. and E. Mposkos. 1988. Building stones of the Minoan Palace of Knossos. In *The engineering geology of ancient works, monuments and historical sites*, eds. P. G. Marinos and G. Koukis, 649–59. Rotterdam: A. A. Balkema.

Papageorgakis, J., N. Mourtzas, and A. Orfanoudaki. 1992. Bronze Age quarries on the eastern coastal zone of Crete (Greece). In *Ancient stones: Quarrying, trade and provenance*, eds. M. Waelkens, N. Herz, and L. Moens, 21–27. Leuven: Leuven University Press.

Pavía, S. 2006. The determination of brick provenance and technology using analytical techniques from the physical sciences. *Archaeometry* 48(2): 201–18.

Pavía Santamaría, S., S. Caro Calatayud, L. Garcia Viguera, and A. Rodriguez Bea. 1993. Mineralogy, texture and physical properties of brick masonry: The San Bartolome church tower in Logroño, La Rioja, Spain. In *Conservation of stone and other materials*, ed. M. -J. Thiel, 351–59. London: E & FN Spon.

Pavía Santamaría, S., P. O'Brien, and T. P. Cooper. 1996a. Evaluation of cleaning methods for granite based on petrographic examinations. *Materials and Structure* 29(187): 185–89.

—. 1996b. Petrographic examinations of cleaning methods for granite. In *Degradation and conservation of granitic rocks in monuments*, 423–27. Brussels: European Commission, Directorate-General XII, Science, Research and Development.

Paynter, S., F. Okyar, S. Wolf, and M. S. Tite. 2004. The production technology of Iznik pottery: A reassessment. *Archaeometry* 46(3): 421 37.

Peacock, D. P. S. 1968. A petrological study of certain Iron Age pottery from western England. *Proceedings of the Prehistoric Society* 34: 414–27.

—. 1970. The scientific analysis of ancient ceramics: A review. *World Archaeology* 1: 375–89.

—. 1971. Petrography of certain coarse pottery. In *Excavations at Fishbourne*, ed. B. Cunliffe, 225–29. London: Society of Antiquaries Research Report 27.

—. 1977. Pompeian Red Ware. In *Pottery and early commerce: Characterization and trade in Roman and later ceramics*, ed. D. P. S. Peacock, 147–62. London: Academic Press.

—. 1985. Petrology and ceramics: A review of work on Medieval potteries. In *The archaeologist and the laboratory*, ed. P. Phillips, 5–8. London: Council for British Archaeology Research Report 58.

—. 1994. The amphorae: Typology, fabric and chronology. In *Excavations at Carthage: The British mission.* Volume 2: *The circular harbour, north side, the pottery*, eds. M. G. Fulford and D. P. S. Peacock, 42–52. Oxford: Oxford University Press for the British Academy.

Peacock, D. P. S. and R. Tomber. 1991. Roman amphora kilns in the Sahel of Tunisia: Petrographic investigation of kiln material from a sedimentary envrironment. In *Recent developments in ceramic petrology*, eds. A. Middleton and I. C. Freestone, 289–302. London: British Museum.

Peacock, D. P. S. and D. R. Williams. 1986. *Amphorae and the Roman economy*. London: Longman.

Peacock, D. P. S., Olwen Williams-Thorpe, R. S. Thorpe, and A. G. Tindle. 1994. Mons Claudianus and the problem of the "granito del foro": A geological and geochemical approach. *Antiquity* 68(259): 209–30.

Pearce, M., S. Meloni, M. Setti, N. Genova, M. Oddone, and A. Muggia. 1996. A multidisciplinary investigation of Roman bricks from the Torre Civica, Pavia (NW Italy). *Science and Technology for Cultural Heritage* 5(2): 19–27.

Pearson, G. T. 1992. *Conservation of clay and chalk buildings*. London: Donhead.

Peña, J. T. 1990. Internal red-slip cookware (Pompeian Red Ware) from Cetamura del Chianti, Italy: Mineralogical composition and provenience. *American Journal of Archaeology* 94: 647–61.

—. 1991. The reconstruction of ancient ceramic technology through the comparative analysis of ancient and contemporary potters' materials: A case study for the Central Tiber valley of Italy. In *Materials issues in art and archeology II*, eds. P. B. Vandiver, J. Druzik, and G. S. Wheeler, 511–21. Pittsburgh, PA: Materials Research Society.

—. 1992. Raw material use among nucleated industry potters: The case of Vasanello, Italy. *Archeomaterials* 6(2): 93–122.

Penny, N. 1993. *The materials of sculpture*. New Haven, CT: Yale University Press.

Perath, I. 1984. *Stone building and building stone in Israel: A historical review*. Jerusalem: Geological Survey of Israel, Environmental Geology Division.

Pérez-Ortiz, A., J. Ordaz, R. M. Esbert, and F. J. Alonso. 1994. Microfissuring evolution of the granite from the Axeitos dolmen along the salt crystallization test. In *The conservation of monuments in the Mediterranean basin*, eds. V. Fassina, H. Ott, and F. Zezza, 115–19. Venice: Soprintendenza ai Beni Artistici e Storici di Venezia.

Peters, T. and R. Iberg. 1978. Mineralogical changes during firing of calcium-rich brick clays. *Ceramic Bulletin* 57(5): 503–09.

Pettijohn, F. J. 1975. *Sedimentary rocks*. New York: Harper and Row.

Pettijohn, F. J., P. E. Potter, and R. Siever. 1987. *Sand and sandstone*, 2nd edn. New York: Springer-Verlag.

Phillips, W. R. 1971. *Mineral optics: Principles and techniques*, San Francisco, CA: W. H. Freeman.

Phillips, W. R. and D. T. Griffen. 1981. *Optical mineralogy: The nonopaque minerals*. San Francisco, CA: W. H. Freeman.

Philpotts, A. R. and N. Wilson. 1994. Application of petrofabric and phase equilibria analysis to the study of a potsherd. *Journal of Archaeological Science* 21(5): 607–18.

Pichler, B. and A. Vendl. 1987. Authentication of a Renaissance bronze statue. In *Recent advances in the conservation and analysis of artifacts*, ed. J. Black, 169–71. London: Summer Schools Press.

Picot, P. and J.-P. Ragot. 1972. Étude de la détérioration en surface des matériaux du temple de Borobudur (Indonésie). In *Proceedings of the 1st International Symposium on the Deterioration of Building Stones*, 77–88. Chambéry: Les Imprimeries Réunies de Chambéry.

Pierpont, R. N. 1987. Slate roofing. *APT Bulletin* 19(2): 10–23.

Plehwe-Leisen, E. von, H. D. Castello Branco, E. Wendler, D. D. Klemm, and R. Snethlage. 1992. The Prophets of Aleijadinho in Congonhas, Brazil. Considerations on a conservation program for soapstone. In *Proceedings of the 7th International Congress on Deterioration and Conservation of Stone*, eds. J. Delgado Rodrigues, F. Henriques and F. Telmo Jeremias, 1415–25. Lisbon: Laboratorio Nacional de Engenharia Civil.

Poli, G., C. Manganelli Del Fá, B. Moroni, D. Faraone, and F. Fratini. 1992. Lithoid materials from the apse of Orvieto Cathedral (Terni, Italy): Deterioration characteristics and provenance. *Science and Technology for Cultural Heritage* 1: 15–31.

Porat, N., J. Yellin, and L. Heller-Kallai. 1991. Correlation between petrography, NAA, and ICP analyses: Application to Early Bronze Egyptian pottery from Canaan. *Geoarchaeology* 6(2): 133–49.

Porter, J. W. 1962. Temper in Bluff pottery from the Cahokia area. *Southern Illinois University Museum Lithic Laboratory, Research Report 1.*

—. 1964. Comment on Weaver's *Technological Analysis of Lower Mississippi Ceramic Materials. American Antiquity* 29(4): 520–21.

Porter, V. 1995. *Islamic tiles*. New York: Interlink Books.

Poster, A. 1986. *From Indian earth: 4,000 years of terracotta art*. Brooklyn, NY: Brooklyn Museum of Art.

Prag, A. J. N. W, F. Schweizer, and J. Li W. Williams. 1974. Hellenistic glazed wares from Athens and southern Italy: Analytical techniques and implications. *Archaeometry* 16(2): 153–87.

Prothero, D. R. 1983. Petrographic analysis of slides. In *Patterns of chert acquisition among woodland groups within the Delaware watershed: A lithologic approach*, L. Lavin. New York: Ph.D. dissertation, Department of Anthropology, New York University.

Protz, R. and A. J. VandenBygaart. 1998. Towards systematic image analysis of soil micromorphology. *Sciences of Soils* 3: 4. http://192.129.24.144/licensed_materials/10112/tocs/t8003001.htm [accessed January 6, 2008]

Pruesser, F. 1993. Scientific examination of the Getty Kouros. In *The Getty Kouros Colloquium*, 63–64. Athens: Kapon.

Purdy, B. A. 1981. *Florida's prehistoric stone technology*. Gainesville, FL: University of Florida.

Purdy, B. A. and F. N. Blanchard. 1973. Petrography as a means of tracing stone tools from Florida. *Florida Anthropologist* 26(1): 121–25.

Pyddoke, E. ed. 1963. *The scientist and archaeology*. London: Phoenix House.

Qingzheng, W. 2002. *A dictionary of Chinese ceramics*. Singapore: Sun Tree Publishing Limited.

Quaresima, R., C. Giampaolo, F. Spernanzoni, and R. Volpe. 2006. Identification, characterization and weathering of the stones used in medieval religious architecture in L'Aquila (Italy). In *Geomaterials in cultural heritage*, eds. M. Maggetti and B. Messiga, 323–35. London: Geological Society, Special Publications 257.

Quinn, P. S. and P. M. Day. 2007. Calcareous microfossils in Bronze Age Aegean ceramics: Illuminating technology and provenance. *Archaeometry* 49(4): 775–93.

Racine, L. A. 1997. *Re-examination after excavation: The problems of attributing wares to three New Jersey stoneware potteries*. Newark, NJ: M.A. thesis. University of Delaware.

Rainey, M. M. and W. B. Whalley. 1994. The importance of microfractures in granitic rock breakdown with reference to building stone decay. In *Granite weathering and conservation*, eds. E. Bell and T. P. Cooper, 36–45. Dublin: Director of Buildings' Office, University of Dublin, Trinity College.

Rapp, G., Jr. 2002. *Archaeomineralogy*. Berlin: Springer.

Rapp, G., Jr. and C. L. Hill. 1998. *Geoarchaeology: The earth-science approach to archaeological interpretation*. New Haven and London: Yale University Press.

Rathgen, F. 1901. Konservierung von altertumsfunden aus thon. *Prometheus* 12: 77–79.

—. 1908. Mitteilungen aus dem laboratorium der Königlichen Museum zu Berlin: I. Über konservierung babylonischer tonafeln: II. Konservierung eines tonsarges. *Museumskunde* 4: 12–24.

Reedy, C. L. 1991. Petrographic analysis of casting core materials for provenance studies of copper alloy sculptures. *Archeomaterials* 6(2): 121–63.

—. 1992a. Gandhara and related terracotta technologies. In *The crossroads of Asia: Transformation in image and symbol*, eds. E. Errington and J. Cribb, 278–80. Cambridge: Ancient India and Iran Trust.

—. 1992b. Petrographic analysis of Gandhara stone sculptures. In *The crossroads of Asia: Transformation in image and symbol*, eds. E. Errington and J. Cribb, 264–77. Cambridge: Ancient India and Iran Trust.

—. 1994. Thin-section petrography in studies of cultural materials. *Journal of the American Institute for Conservation* 33: 115–29.

—, 1996, Optical mineralogy. In *Excavations at Anshan (Tal-e Malyan: The Middle Elamite Period)*, ed. E. Carter, 109–12. Philadelphia, PA: University of Pennsylvania Museum.

—. 1997a. Technical studies of Gandharan art. In *Gandharan art in context: East–West exchanges at the crossroads of Asia*, eds. R. Allchin, B. Allchin, N. Kreitman, and E. Errington, 267–83. New Delhi: Regency.

—. 1997b. *Himalayan bronzes: Technology, style, and choices*. London and Newark, NJ: University of Delaware Press.

—. 1997c. Analysis of the clay cores. In *The sacred sculpture of Thailand: The Alexander B. Griswold collection, the Walters Art Gallery*, ed. H. W. Woodward, Jr., 266–77. Baltimore, MD: Walters Art Gallery.

—. 2006a. Review of digital image analysis of petrographic thin sections in conservation research. *Journal of the American Institute for Conservation* 45(2): 127–46.

—. 2006b. Multiple functions and multiple histories of Tibetan *tsha-tshas* (votive clay images). In *The object in context: Crossing conservation boundaries*, eds. D. Saunders, J. H. Townsend, and S. Woodcock, 144–50. London: International Institute for Conservation of Historic and Artistic Works.

Reedy, C. L. and P. Meyers. 2007. New methods for analyzing thin sections of casting core materials: A case study with Southeast Asian bronzes. In *Scientific research on the sculptural arts of Asia*, eds. J. Douglas, P. Jett, and J. Winter, 101–12. London: Archetype.

Rehman, J.-U. and A. Zeb. 1970. The geology of the Shah Dheri-Kabul area, Swat. *Geological Bulletin University of Peshawar* 5: 96–110.

Renfrew, C. 1971. Obsidian and pumice: The use of recent igneous rocks in Aegean prehistory. *Acta of the First International Scientific Congress on the Volcano of Thera* 1, ed. A. Kalogeropoulou, 430–36. Athens: Archaeological Services of Greece, General Direction of Antiquities and Restoration.

Renfrew, C. and J. S. Peacey. 1968. Aegean marble: A petrological study. *Annual of the British School at Athens* 63: 45–66.

Rice, P. 1977. Whiteware pottery production in the valley of Guatemala: Specialization and resource utilization. *Journal of Field Archaeology* 4: 221–33.

—. 1987. *Pottery analysis: A sourcebook*. Chicago, IL: University of Chicago Press.

—. 1996. Recent ceramic analysis: 2. Composition, production, and theory. *Journal of Archaeological Research* 4: 165–202.

Ricq-de Bouard, M. and F. G. Fedele. 1993. Neolithic rock resources across the Western Alps: Circulation data and models. *Geoarchaeology* 8(1): 1–22.

Riederer, J. and C. Prasartset. 1995. Conservation and preservation of sandstone: A case study on historical monuments in Thailand. In *Conservation, preservation and restoration traditions, trends and techniques*, eds. G. Kamalakar and V. Pandit Rao, 205–12. Hyderabad: Birla Archaeological and Cultural Research Institute.

Rigby, V., A. P. Middleton, and I. C. Freestone. 1989. The Pruny workshop: Technical examination of La Tène bichrome painted pottery from Champagne. *World Archaeology* 21(1): 1–16.

Riley, J. A. 1982. The petrological analysis of Aegean ceramics. In *Current research in ceramics: Thin-section studies*, eds. I. C. Freestone, C. Johns, and T. Potter, 1–7. London: British Museum Occasional Paper No. 32.

Riley, J. A., P. Hopke, R. Martin, and J. W. Porter. 1994. The diffusion of technological knowledge: A case study in North American ceramic analysis. In *Ancient technologies and archaeological materials*, eds. S. U. Wisseman and W. S. Williams, 41–58. Amsterdam: Gordon and Breach Science Publishers.

Ritchie, Carson I. A. 1973. *Soft stone carving*. London and New York: St. Martin's Press.

Roa, H. M. 1994. Geological outline of Easter Island and petrographico-structural features of its lithic monuments. In *Lavas and volcanic tuffs*, eds. A. E. Charola, R. J. Koestler, and G. Lombardi, 253–58. Rome: ICCROM.

Röder, G. 1992. Felsberg granite as a substitute for Claudianus granite. In *Ancient stones: Quarrying, trade and provenance*, eds. M. Waelkens, N. Herz, and L. Moens, 131–35. Leuven: Leuven University Press.

Rodrigues, J. D. 1996. Conservation of granitic rocks with application to the megalithic monuments. In *Degradation and conservation of granitic rocks in monuments*, 161–242. Brussels: European Commission, Directorate-General XII, Science, Research and Development.

Rodriguez-Navarro, C., E. Sabastian-Pardo, and U. Zezza. 1994. Petrophysical-mechanical parameters for decay evaluation of building stones. A case study: Jaen Cathedral (Andalusia, Spain). In *The conservation of monuments in the Mediterranean basin*, eds. V. Fassina, H. Ott, and F. Zezza, 595–603. Venice: Soprintendenza ai Beni Artistici e Storici di Venezia.

Rodriguez-Rey, A. and M. Montoto. 1978. Insolation weathering phenomena in crystalline rocks. In *Deterioration and protection of stone monuments*, 2.7. Paris: RILEM.

Roos, P., L. Moens, J. De Rudder, P. De Paepe, J. Van Hende, and M. Waelkens. 1988. Chemical and petrographical characterization of Greek marbles from Pentelikon, Naxos, Paros and Thasos. In *Classical marble: Geochemistry, technology, trade*, eds. N. Herz and M. Waelkens, 263–72. Dordrecht: Kluwer.

Rosen, A. M. 1986. *Cities of clay*. Chicago, IL: University of Chicago Press.

Rossi-Manaresi, R. 1976. Treatments for sandstone consolidation. In *The conservation of stone*, ed. R. Rossi-Manaresi, 547–71. Bologna: Centro per la Conservazione delle Sculture all'aperto.

Rossi-Manaresi, R. and G. Chiari. 1980. Effectiveness of conservation treatments of a volcanic tuff very similar to adobe. In *Proceedings of the 3rd International Symposium on Mudbrick (Adobe) Preservation*, 29–38. Ankara: ICOM.

Rossi-Manaresi, R. and A. Tucci. 1989. Pore structure and salt crystallization: "Salt decay" of Agrigento biocalcarenite and "case hardening" in sandstone. In *The conservation of monuments in the Mediterranean basin*, ed. F. Zezza, 97–100. Brescia: Grafo.

Roux, V. and M. A. Courty. 1998. Identification of wheel-fashioning methods: Technological analysis of 4th–3rd millennium BC Oriental ceramics. *Journal of Archaeological Science* 25: 747–63.

Russell, A. D. 1982. Domestic Early Saxon pottery from East Anglia: An interim report. In *Current Research in ceramics: Thin-section studies*, eds. I. C. Freestone, C. Johns, and T. Potter, 101–03. London: British Museum Occasional Paper No. 32.

Ryan, W. 1978. *Properties of ceramic raw materials.* Oxford: Pergamon Press.

Rye, O. S. 1976. Keeping your temper under control: Materials and the manufacture of Papuan pottery. *Archaeology and Physical Anthropology in Oceania* IX: 106–37.

—. 1981. *Pottery technology: Principles and reconstruction.* Washington, DC: Taraxacum.

Sabine, P. A. 1958. Black mica in the pottery fabric. Appendix in Threlfall, *Excavation of a medieval settlement at Beere, North Tawton, Devon. Medieval Archeology* 2: 140.

Sakarya, N., S. Kapur, and E. A. Fitzpatrick. 1990. Preliminary study of the microstructure and mineralogy of 12th and 13th century ceramics, Samsat, southeastern Turkey. *Geoarchaeology* 5(3): 275–81.

Saleh, S. A., F. M. Helmi, M. M. Kamal, and A-F. E. El-Banna. 1992. Study and consolidation of sandstone: Temple of Karnak, Luxor, Egypt. *Studies in Conservation* 37: 93–104.

Salvadori, O., D. Pinna, and G. C. Grillini. 1994. Lichen-induced deterioration on an ignimbrite of the Vulsini Complex (Central Italy). In *Lavas and volcanic tuffs*, ed. A. E. Charola, 143–53. Rome: ICCROM.

Santallier, D., J. P. Fillion, and A. Mignot. 1998. A propos de ce que l'on appelle les "roches vertes" an archéologie. *Revue d'Archéométrie* 22: 44–55.

Santamaría, P., S. Caro Calatayud, F. S. Pérez Lorente, and A. Duffy. 1994. More about gypsum crusts on historic buildings (Gypsum crusts on calcareous sandstone from La Rioja, northern Spain). In *The conservation of monuments in the Mediterranean basin*, eds. V. Fassina, H. Ott, and F. Zezza, 585–93. Venice: Soprintendenza ai Beni Artistici e Storici di Venezia.

Santi, P., F. Antonelli, and A. Renzulli. 2005. Provenance of medieval *Pietra Ollare* artefacts found in archaeological sites of central-eastern Italy: Insights into the Alpine soapstone trade. *Archaeometry* 47(2): 253–64.

Sappa, G., G. Giglio, and G. De Casa. 1995. Mechanical characteristics of some volcanic tuffs used in the buildings of ancient Rome. In *Materials issues in art and archaeology IV*, eds. P. B. Vandiver, J. R. Druzik, J. L. Galvan Madrid, I. C. Freestone, and G. S. Wheeler, 733–43. Pittsburgh, PA: Materials Research Society.

Sauer, R., B. Pichler, and J. Weber. 1988. Untersuchungen am Kernmaterial. In *Wiener Berichte über Naturwissenschaft in der Kunst*, eds. A. Vendl, B. Pichler, and J. Weber, 318–31. Vienna: VWGÖ.

Schäfer, A. and T. Teyssen. 1987. Size, shape and orientation of grains in sand and sandstones: Image analysis applied to rock thin sections. *Sedimentary Geology* 52: 251–71.

Schiffer, M. B. 1990. The influence of surface treatment on heating effectiveness of ceramic vessels. *Journal of Archaeological Science* 17: 373–81.

Schmid, J., M. Ambühl, D. Decrouez, S. Müller, and K. Ramseyer. 1999. A quantitative fabric analysis approach to the discrimination of white marbles. *Archaeometry* 41(2): 239–52.

Schmitt, A. 1998. Amphorae from Lyons: Petrographic and chemical arguments. *Archaeometry* 40(2): 293–310.

Schneider, G. 1989a. A technological study of North-Mesopotamian stone ware. *World Archaeology* 21(1): 30–50.

—. 1989b. Investigation of crucibles and moulds from bronze foundries in Olympia and Athens and determination of provenance of bronze statues. In *Archaeometry: Proceedings of the 25th International Symposium*, ed. Y. Maniatis, 305–10. Amsterdam: Elsevier.

Schneider, J. S. 2002. Stone textures and function: A relationship between milling tools and subsistence as derived from western American quarries data. In *Interdisciplinary studies on ancient stone*, eds. J. J. Herrmann, Jr., N. Herz, and R. Newman, 381–93. London: Archetype.

Schneider, H., K. Okada, and J. Pask. 1994. *Mullite and mullite ceramics.* Chichester: John Wiley & Sons.

Scholle, P. A. 1979. *A color illustrated guide to constituents, textures, cements, and porosities of sandstones and associated rocks.* Tulsa, OK: American Association of Petroleum Geologists.

Scholle, P. A. and D. S. Ulmer-Scholle. 2003. *A color guide to the petrography of carbonate rocks: Grains, textures, porosity, diagenesis.* Tulsa, OK: American Association of Petroleum Geologists.

Schwarz-Mackensen, G. and W. Schneider. 1983a. Fernbeziehungen in frühneolithikum - rohstonvensorgung am beispiel des actinolit-hornblende schist. *Archäologisches Mitteilungen aus Norwestdeutschland* 1: 165–76.

—. 1983b. Wo liegen die hauptliefergebiete für das rohmaterial donauländischer steinbelle und -äxte in Mitteleuropa? *Archäologisches Korrespondenzblatt* 13: 305–14.

—. 1986. Petrographie und Herkunft des Rohmaterials neolithischer Steinbeile und-äxte im nördlichen Harzvorland. *Archäologisches Korrespondenzblatt* 16: 29–44.

Scoffin, T. P. 1987. *An introduction to carbonate sediments and rocks.* Glasgow and London: Blackie.

Sengupta, R. 2002. Preservation and restoration of Buddha Colossi at Bamiyan in Afghanistan by India. In *Bamiyan: Challenge to world heritage*, ed. K. Warikoo, 150–55. New Delhi: Bhavana Books & Prints.

Sestier, C. 2005. Utilisation du degraissant vegetal en contexte Neolithique: Hypotheses technologiques et experimentation. In *International Congress of Prehistory: Section 2, Archaeometry; Colloque/Symposium 2.1, Pottery Manufacturing Processes; Reconstitution and Interpretation*, eds. A. Livingstone-Smith, D. Bosquet, and R. Martineau, 81–94. Oxford: Archaeopress.

Shanmugam, G. 1985. Types of porosity in sandstones and their significance in interpreting provenance. In *Provenance of arenites*, ed. G. G. Zuffa, 115–37. Hingham, MA: D. Reidel.

Shareq, A., V. M. Chmyriov, K. F. Stazhilo-Alekseev, V. I. Drunov, P. J. G. Gannon, L. N. Rossovskiv, A. Kh. Kafarskiy, and E.P. Malyarov. 1977. *Mineral resources of Afghanistan.* Kabul: United Nations Development Programme.

Shelley, D. 1993. *Igneous and metamorphic rocks under the microscope.* London: Chapman & Hall.

Shepard, A. O. 1939. Appendix A: Technology of La Plata pottery. In *Archaeological studies in the La Plata district*, ed. E. Morris, 249–87. Washington, DC: Carnegie Institution of Washington Publication 519.

—. 1942. Rio Grande glaze paint ware. A study illustrating the place of ceramic technological analysis in archaeological research. *Carnegie Institution of Washington Publication* 528: 129–262.

—. 1953. Notes on color and paste composition. *Archaeological Studies in the Petrified Forest National Monument* 27: 177–93.

—. 1965. Red Grande glaze-paint pottery: A test of petrographic analysis. In *Ceramics and man*, ed. F. R. Matson, 62–87. Chicago, IL: Aldine Publishing Company.

—. 1968. *Ceramics for the archaeologist.* [Reprint of 1956 edition.] Washington, DC: Carnegie Institute Institution of Washington.

Sheperd, W. 1972. *Flint.* London: Faber and Faber.

Shimron, A. E. 1985. The petrography and origin of two whetstones from the tomb near Enan. *Atiquot* 17: 83–89.

Shipley, W. E. III and E. Graham. 1987. Petrographic analysis and preliminary source identification of selected stone artifacts from the Maya sites of Seibal and Uaxactun, Guatemala. *Journal of Archaeological Science* 14: 367–83.

Shotton, F. W. 1959. New petrological groups based on axes from the West Midlands. *Proceedings of the Prehistoric Society* 25: 135–43.

—. 1969. Petrological examination. In *Science in archaeology*, eds. D. Brothwell and E. Higgs, 571–77. New York: Praeger Publishers.

Shoval, S., P. Beck, and E. Yadin. 2006. The ceramic technology used in the manufacture of Iron Age pottery from Galilee. In *Geomaterials in cultural heritage*, eds. M. Maggetti and B. Messiga, 101–17. London: Geological Society, Special Publications 257.

Simpkins, D. L. and D. J. Allard. 1986. Isolation and identification of Spanish moss fiber from a sample of Stallings and Orange Series ceramics. *American Antiquity* 51(1): 102–17.

Singh, M. 1993. Studies on weathering of Kailansanatha Temple, Kancheepuram. *Current Science* 64(8): 559–65.

Singh, M., K. K. Jain, T. Singh, and L. Lazzarini. 1995. Evaluation of organosilanes for the consolidation of khondalite stone. In *Conservation, preservation and restoration: Traditions, trends and techniques*, eds. G. Kamalakar and V. P. Rao, 213–16. Hyderabad: Birla Archaeological and Cultural Research Institute.

Skibo, J. M. 1992. *Pottery function: A use alteration perspective.* New York: Plenum Press.

Skibo, J. M., M. B. Schiffer, and K. C. Reid. 1989. Organic-tempered pottery: An experimental study. *American Antiquity* 54(1): 122–46.

Smith, C. S. and M. T. Gnudi. trans. 1959. *The Pirotechnia of Vannoccio Biringuccio*. Boston, MA: MIT.

Smith, M. S. and M. B. Trinkley. 2006. Fibre-tempered pottery of the Stallings Island Culture from the Crescent site, Beaufort County, South Carolina: A mineralogical and petrographical study. In *Geomaterials in cultural heritage*, eds. M. Maggetti and B. Messiga, 119–25. London: Geological Society, Special Publications 257.

Sondi, I. and D. Slovenec. 2003. The mineralogical characteristics of the *Lamboglia 2* Roman-age amphorae from the Central Adriatic (Croatia). *Archaeometry* 45(2): 251–62.

Southward, J. A. 1979. A summary of ceramic technology, plant remains and shell identification analyses from LA 4921, Three Rivers, New Mexico. In *Jornada Mogollon Archaeology: Proceedings of the 1st Jornada Conference*, eds. P. H. Beckett and R. N. Wiseman, 91–102. Las Cruces, NM: New Mexico State University.

Spadea, P., T. Perusini, and A. Frangipane. 1996. Dolostones used in Middle Age in Friuli (NE Italy). In *Proceedings of the 8th International Congress on Deterioration and Conservation of Stone*, vol. 1, ed. J. Riederer, 155–58. Berlin: Möller Druck und Verlag.

Stanier, P. 1985. *Quarries and quarrying*. Princes Risborough: Shire Publications, Shire album 134.

Steponaitis, V. 1984. Technological studies of prehistoric pottery from Alabama: Physical properties and vessel function. In *The many dimensions of pottery*, eds. S. Van der Leeuw and A. C. Pritchard, 79–122. Amsterdam: Universiteit van Amsterdam.

Sterflinger, K., F. Blazquez, M. Farcia-Valles, W. E. Krumbein, and M. Vendrell-Saz. 1996. Patina, microstromatolites and black spots as related to biodeterioration processes of granite. In *Degradation and conservation of granitic rocks in monuments*, 391–97. Brussels: European Commission, Directorate-General XII, Science, Research and Development.

Sterner, J. 1989. Who is signaling whom? Ceramic style, ethnicity and taphonomy among the Sirak Bulahey. *Antiquity* 63: 451–60.

Steward, J. D., R. Pieper, M. J. Devonshire, C. Langham, D.A. Rosen, and M. A. Simon. 1981. *The Schermerhorn row block: A study in nineteenth-century building technology in New York city*. Waterford, NY: New York State Parks, Recreation and Historic Preservation, Bureau of Historic Sites.

Steward, J.D., P. Fralick, R. G. V. Hancock, J. H. Kelley, and E. M. Garrett. 1990. Petrographic analysis and INAA geochemistry of prehistoric ceramics from Robinson Pueblo, New Mexico. *Journal of Archaeological Science* 17(6): 601–25.

Stoltman, J. B. 1989. A quantitative approach to the petrographic analysis of ceramic thin sections. *American Antiquity* 54: 147–60.

—. 1991. Ceramic petrography as technique for documenting cultural interaction: An example from the Upper Mississippi valley. *American Antiquity* 56: 103–20.

—. 2001. The role of petrography in the study of archaeological ceramics. In *Earth sciences and archaeology*, eds. P. Goldberg, V. T. Holliday, and C. R. Ferring, 297–326. New York: Kluwer/Plenum.

Stoltman, J. B., J. H. Burton, and J. Haas. 1992. Chemical and petrographic characterizations of ceramic pastes: Two perspectives on a single data set. In *Chemical characterization of ceramic pastes in archaeology*, ed. H. Neff, 85–92. Madison, WI: Prehistory Press.

Stone, R. 1982. Antico and the development of bronze casting in Italy at the end of the Quattrocento. *Metropolitan Museum Journal* 16: 87–116.

Stone, R. W. 1932. *Building stones of Pennsylvania*. Harrisburg, PA: Pennsylvania Geological Survey Fourth Series Bulletin M 15.

Storemyr, P. 1997. *The stones of Nidaros: An applied weathering study of Europe's Northernmost medieval cathedral*. Trondheim: Ph.D. dissertation, Department of Architectural History, Norwegian University of Science and Technology.

—. 2004. Weathering of soapstone in a historical perspective. *Materials Characterization* 53: 191–207.

Storey-Maskelyne, N. 1878. Stonehenge: The petrology of its stones. *Wiltshire Archaeological and Natural History Magazine*: 17: 147–60.

Stratton, M. 1994. The nature of terracotta and faience. In *Architectural ceramics: Their history, manufacture and conservation*, ed. J. M. Teutonico, 45–54. London: James & James.

Streeten, A. D. F. 1982. Textural analysis: An approach to the characterization of sand-tempered ceramics. In *Current research in ceramics: Thin-section studies*, eds. I. C. Freestone, T. Potter, and C. Johns, 123–34. London: British Museum Occasional Paper No. 32.

Stross, F. H., R. L. Hay, F. Asaro, H. R. Bowman, and H. V. Michel. 1988. Sources of the quartzite of some ancient Egyptian sculptures. *Archaeometry* 30(1): 109–19.

Takács-Biró, C. 1983. Sources of raw materials used for the manufacture of chipped stone implements in Hungary. In *The scientific study of flint and chert*, eds. G. de G. Sieveking and M. B. Hart, 121–32. Cambridge: Cambridge University Press.

Tarzi, Z. 1985. Hadda. In *Encyclopedia universalis*, 64–66. Paris.

Thomachot, C. and D. Jeannette. 2000. Petrophysical properties modifications of Strasburg's cathedral sandstone by black crusts. In *Proceedings of the 9th International Congress on Deterioration and Conservation of Stone*, vol. 1, ed. V. Fassina, 265–73. Amsterdam: Elsevier.

Thomas, H. H. 1923. The source of the stones of Stonehenge. *Antiquaries Journal* 3: 239–60.

Thornton, D. 1997. Maiolica production in Renaissance Italy. In *Pottery in the making: Ceramic traditions*, eds. I. C. Freestone and D. Gaimster, 116–21. Washington, DC: Smithsonian Institution Press.

Tite, M. 1989. Iznik pottery: An investigation of the methods of production. *Archaeometry* 31(2): 115–32.

—. 1995. Firing temperature determinations: How and why? In *The aim of laboratory analyses of ceramics in archaeology*, eds. A. Lindahl and O. Stilborg, 47–42. Stockholm: Kungl. Vitterhets Historie och Antikvitets Adademien.

Tite, M. and V. Kilikoglou. 2002. Do we understand cooking pots and is there an ideal cooking pot? In *Modern trends in scientific studies on ancient ceramics*, eds. V. Kilikoglou, A. Hein, and Y. Manniatis, 1–6. Oxford: British Archaeological Reports International Series 1011.

Tomber, R. and J. Dore. 2002. *The national Roman fabric reference collection: A handbook*. London: Museum of London Archaeology Service.

Tomkins, P. and P. M. Day. 2001. Production and exchange of the earliest ceramic vessels in the Aegean: A view from Early Neolithic Knossos, Crete. *Antiquity* 75: 259–60.

Topal, T. and V. Doyuran. 1996. The physico-chemical properties of the weathered Cappadocian tuff and their significance on conservation studies. In *Proceedings of the 8th International Congress on Deterioration and Conservation of Stone*, vol. 1, ed. J. Riederer, 223–31. Berlin: Möller Druck und Verlag.

Török, A. 2004. Comparison of the process of decay of two limestones in a polluted urban environment. In *Stone deterioration in polluted urban environments*, eds. D. J. Mitchell and D. E. Searle, 74–92. Enfield, NH: Science Publishers.

Toth, N. 1987. The first technology. *Science* 256(4): 104–13.

Trench, L. ed. 2000. *Materials and techniques in the decorative arts: An illustrated dictionary*. Chicago, IL: University of Chicago Press.

Tröger, W. E. 1979. *Optical determination of rock-forming minerals*. Stuttgart: E. Schweizerbart'sche Verlagsbuchhandlung.

Truncer, J., M. D. Glascock and H. Neff. 1998. Steatite source characterization in eastern North America: New results using instrumental neutron activation analysis. *Archaeometry* 40(1): 23–44.

Tsirtsoni, Z. and P. Yiouni. 2002. Neolithic cooking vessels from Dikili Tash (eastern Macedonia, Greece): A technological and functional approach. In *Modern trends in scientific studies on ancient ceramics*, eds. V. Kilikoglou, A. Hein, and Y. Manniatis, 103–10. Oxford: British Archaeological Reports International Series 1011.

Tucci, G. 1988. *Indo-Tibetica*. Volume 1: *Stupa, art, architectonics, and symbolism (Mchod-rten and tsha-tsha in Indian and Western Tibet)* [Reprint of 1938 Italian edition.] New Delhi: Aditya Prakashan.

Türkmenoğlu, A. G., E. H. Göktürk, and E. N. Caner. 1991. The deterioration of tuffs from the Cappadocia region of Turkey. *Archaeometry* 33(2): 231–38.

Twilley, J. and J. C. Podany. 1986. Deterioration of Arizona red sandstone and refinement of methods for its treatment. In *Case studies of the conservation of stone and wall paintings*, eds. N. S. Brommelle and P. Smith, 174–79. London: International Institute for Conservation of Historic and Artistic Works.

Tykot, R. H., J. J. Herrmann, Jr., N. J. Van der Merwe, R. Newman, and K. O. Allegretto. 2002. Thasian marble sculpture. In *Interdisciplinary studies on ancient stone*, eds. J. J. Herrmann, Jr., N. Herz, and R. Newman, 188–95. London: Archetype.

Ubac, R. 1956. Slate engravings. *Graphis* 12: 346–48.

Unterwurzacher, M., J. Polleres, and P. Mirwald. 2005. Provenance study of marble artefacts from the Roman burial area of Faschendorf (Carinthia, Austria). *Archaeometry* 47(2): 265–73.

Upadhyay, M. K. 1991. Preservation of Ashokan Pillars in Bihar. *Conservation of Cultural Property in India* 24: 79–81.

Valenstein, S. G. 1989. *A handbook of Chinese ceramics.* New York: Metropolitan Museum of Art.

Van de Mieroop, M. 1999. *Cuneiform texts and the writing of history.* London and New York: Routledge.

Van der Ark, L. A., N. A. Croon, and K. Sijtsma, eds. 2005. *New developments in categorical data analysis for the social and behavioral sciences.* Mahwah, N J: Lawrence Erlbaum.

Van Gemert, D., E. Ulrix, and W. Viaene. 1988. Physico-chemical erosion of limestone: A micro-macro approach. In *The engineering geology of ancient works, monuments and historical sites*, eds. P. G. Marinos and G. C. Koukis, 805–08. Rotterdam: A. A. Balkema.

Vandiver, P. B. and W. D. Kingery. 1986. Song Dynasty celadon glazes from Dayo near Longquan. In *Scientific and Technological Insights on Ancient Chinese Pottery and Porcelain: Proceedings of the International Conference on Ancient Chinese Pottery and Porcelain*, ed. S. Chao, 187–91. Beijing: Shanghai Institute of Ceramics.

Vandiver, P. B., L. A. Cort, and C. A. Handwerker. 1989a. Variations in the practice of ceramic technology in different cultures: A comparison of Korean and Chinese celadon glazes. In *Cross-craft and cross-cultural interactions in ceramics*, eds. P. E. McGovern and M. D. Notis, 347–88. Westerville, OH: American Ceramic Society.

Vandiver, P. B., O. Soffer, B. Klima, and J. Svoboda. 1989b. The origins of ceramic technology at Dolni Věstonice, Czechoslovakia. *Science* 246: 1002–08.

Vandiver, P. B., O. Soffer, B. Klima, and J. Svoboda. 1990. Venuses and wolverines: The origins of ceramic technology, ca. 26,000 B.P. In *The changing roles of ceramics in society: 26,000 B.P. to the present*, ed. W. D. Kingery, 13–81. Westerville, OH: American Ceramic Society.

Vandiver, P. B., A. Underhill, L. Fenghsi, Y. Haiguang, and F. Hui. 2002. Longshan pottery: Its manufacture at Liangchengzhen in southeast Shandong province. In *International Symposium on Ancient Ceramics*, Shanghai Institute of Ceramics, 574–86. Shanghai: Chinese Academy of Sciences.

Varti-Matarangas, M., N. Beloyannis, V. Mitsaki-Hafner, J. Katsikis and E. Panelias. 1995. Lithofacies determination, origin and decay of the building stones in the monuments of the ancient Olympia area. In *The study of marble and other stones used in antiquity*, eds. Y. Maniatis, N. Herz, and Y. Basiakos, 295–302. London: Archetype.

Vaughn, D. M. 1993. An analysis of prehistoric greenstone artifacts in northeast Alabama. *Geoarchaeology* 8(6): 515–29.

Vaughan, S. J. 1990. Petrographic analysis of the early Cycladic wares from Akrotiri, Thera. In *Thera and the Aegean World III*, eds. D. A. Hardy, C. G. Doumas, J. A. Sakellarakis, and P. M. Warren, 470–87. London: Thera Foundation.

—. 1991a. Late Cypriot Base Ring Ware: Studies in raw materials and technology, In *Recent developments in ceramic petrology*, eds. A. Middleton and I. C. Freestone, 337–68. London: British Museum.

—. 1991b. Material and technical characterization of base ring ware: A new fabric typology. In *Cypriot ceramics: Reading the prehistoric record*, eds. J. A. Barlow, D. L. Bolger, and B. Kling, 119–30. University Museum Monograph 74. Philadelphia, PA: A. G. Levantis Foundation and the University Museum of Archaeology and Anthropology, University of Pennsylvania.

—. 1995. Ceramic petrology and petrography in the Aegean. *American Journal of Archaeology* 99(1): 115–17.

Vaughan, S. J., V. Kilikoglou, and A. Papagiannopoulou. 1995. An interdisciplinary study of Middle Cycladic white wares from Akotiri on Thera. In *Materials issues in art and archaeology IV*, eds. P. B. Vandiver, J. R. Druzik, J. L. Galvan Madrid, I. C. Freestone, and G. S. Wheeler, 445–52. Pittsburgh, PA: Materials Research Society.

Velde, B. and I. C. Druc. 1999. *Archaeological ceramic materials: Origin and utilization.* Berlin: Springer.

Vendrell-Saz, M., W. E. Krumbein, C. Urzi, and M. Garcia-Vallès. 1996. Are patinas of Mediterranean monuments really related to the rock substrate? In *Proceedings of the 8th International Congress on Deterioration and Conservation of Stone*, vol. 2, ed. J. Riederer, 609–24. Berlin: Möller Druck und Verlag.

Vendrall-Saz, M., J. Molera, J. Roqué, and J. Pérez-Arantegui. 2006. Islamic and Hispano-Moresque (múdejar) lead glazes in Spain: A technical approach. In *Geomaterials in cultural heritage*, eds. M. Maggetti and B. Messiga, 163–73. London: Geological Society, Special Publications 257.

Venkatachalapathy, R. and K. Ramaswamy. 1992. Petrographic studies on artifacts of Narayanakuppam Sambhuvarayar district, Tamil Nadu. In *Role of Chemistry in Archaeology: 1st International Colloquium, 15–18 November 1991*, eds. M. C. Ganorkar and N. R. Rao, 41–46. Hyderabad: Birla Institute of Scientific Research.

Verges-Belmin, V. 1994. Pseudomorphism of gypsum after calcite, a new textural feature accounting for the marble sulphation mechanism. *Atmospheric Environment* 28(2): 295–304.

Vince, A. 1982. Medieval and post-medieval Spanish pottery from the city of London. In *Current research in ceramics: Thin-section studies*, eds. I. C. Freestone, C. Johns, and T. Potter, 135–44. London: British Museum Occasional Paper No. 32.

—. 1989. The petrography of Saxon and early medieval pottery in the Thames valley. In *Scientific analysis in archaeology and its interpretation*, ed. J. Henderson, 163–77. Los Angeles, CA: UCLA Institute of Archaeology.

Vincent, B. 1988. *Prehistoric ceramics of northeastern Thailand with special reference to Ban Na Di.* Oxford: British Archaeological Reports International Series 461.

—. 1991. Ceramic technology in Thailand. *Indo-Pacific Prehistory Association Bulletin* 10: 341–48.

Visser, H., A. Gervais, K. Gervais, H.-H. Neumann, H. Rösch, and E. Wendler. 1996. Recognition of stone provenance and case histories of deterioration of two ancient Elbe sandstone tombstones from Hamburg and Hanover (northern Germany). In *Proceedings of the 8th International Congress on Deterioration and Conservation of Stone*, vol. 1, ed. J. Riederer, 1–10. Berlin: Möller Druck und Verlag.

Wagner, E. and C. Schubert. 1972. Pre-Hispanic workshop of serpentinite artifacts, Venezuelan Andes, and possible raw material source. *Science* 175(4024): 888–90.

Wainwright, I. N. M. and J. L. Down. 1986. Slatechuck Creek argillite: Its structure, composition and dimensional stability. *Journal of the International Institute for Conservation–Canadian Group* 10/11: 8–19.

Wallis, F. S. 1963. Petrological examination. In *The scientist and archaeology*, ed. E. Pyddoke, 80–100. London: Phoenix House.

Wang, I. 1995. Automatic identification of rocks in thin sections using texture analysis. *Mathematical Geology* 27(7): 847–65.

Wang, Y. X. and Zhu, H. R. 1988. The mechanism of the corrosion of black marble and its preventive measure. In *The engineering geology of ancient works, monuments and historical sites*, vol. 2, eds. P. G. Marinos and G. C. Kourkis, 963–68. Rotterdam: A. A. Balkema.

Warren, J. 1999. *Conservation of brick.* Oxford: Butterworth-Heinemann.

Washington, H. S. 1898. The identification of marbles used in Greek sculpture. *American Journal of Archaeology* 2: 1–18.

Watson, O. 2004. *Ceramics from Islamic lands.* London: Thames & Hudson.

Weinstein, M. 1958. *The world of jewel stones.* New York: Sheridan House.

Weisman, B. M. and C. L. Reedy. 2002. Technical studies on Renaissance bronzes. In *Materials issues in art and archaeology VI*, eds. P. Vandiver, M. Goodway, and J. L. Mass, 483–95. Pittsburgh, PA: Materials Research Society.

Weiss, L. E. 1954. Fabric analysis of some Greek marbles and its application to archaeology. *American Journal of Science* 252: 641–62.

Weiss, T., B. Leiss, H. Oppermann, and S. Siegesmund. 1999. Microfabric of fresh and weathered marbles: Implications and consequences for the reconstruction of the Marmorpalais Potsdam. *Zeitschrift der Deutschen Geologischen Gesellschaft* 150(2): 313–32.

Whitbread, I. K. 1986a. A microscopic view of Greek transport amphorae. In *Science in archaeology*, eds. R. E. Jones and H. W. Catling, 49–52. Athens: British School at Athens, Fitch Laboratory Occasional Paper 2.

—. 1986b. The characterisation of argillaceous inclusions in ceramic thin sections. *Archaeometry* 28(1): 79–88.

—. 1989. A proposal for the systematic description of thin sections towards the study of ancient ceramic technology. In *Archaeometry: Proceedings of the 25th International Symposium*, ed. Y. Maniatis, 127–38. Amsterdam: Elsevier.

—. 1991. Image and data processing in ceramic petrology. In *Recent developments in ceramic petrology*, eds. A. Middleton and I. C. Freestone, 369–86. London: British Museum.

—. 1995. *Greek transport amphorae: A petrological and archaeological study*. Athens: The British School at Athens, Fitch Laboratory Occasional Paper 4.

—. 1996. Detection and interpretation of preferred orientation in ceramic thin sections. In *Imaging the Past*, eds. T. Higgins, P. Main and J. Long, 173–81. London: British Museum Occasional Paper No. 114.

—. 2001. Ceramic petrology, clay geochemistry and ceramic production: From technology to the mind of the potter. In *Handbook of archaeological sciences*, eds. D. R. Brothwell and A. M. Pollard, 449–59. Chichester: Wiley.

Williams, C. L. 1983. The cherts of the Upper Greensand (Cretaceous) of south-east Devon. In *The scientific study of flint and chert*, eds. G. de G. Sieveking and M. B. Hart, 63–70. Cambridge: Cambridge University Press.

Williams, D. F. 1982. Aspects of prehistoric pottery-making in Orkney. In *Current research in ceramics: Thin-section studies*, eds. I. C. Freestone, C. Johns, and T. Potter, 9–13. London: British Museum Occasional Paper No. 32.

—. 1983. Petrology of ceramics. In *The petrography of archaeological artefacts*, eds. D. R. C. Kempe and A. P. Harvey, 301–29. Oxford: Clarendon Press.

Williams, H. and R. F. Heizer. 1965a. Sources of rocks used in Olmec monuments. *Contributions of the University of California Archaeological Research Facility* 1(1): 1–40.

—. 1965b. Geological notes on the ruins of Mitla and other Oaxacan sites, Mexico. *Contributions of the University of California Archaeological Research Facility* 1(1): 41–54.

—. 1965c. Stones used for colossal sculpture at or near Teotihuacan. *Contributions of the University of California Archaeological Research Facility* 1(1): 55–70.

Williams, H., F. J. Turner, and C. M. Gilbert. 1982. *Petrography: An introduction to the study of rocks in thin sections*. San Francisco, CA: W. H. Freeman.

Williams, J. and D. Jenkins. 1997. Provenancing of prehistoric and Roman pottery using petrographic and mineralogical analysis: Case studies from North Wales. In *Archaeological Sciences 1995: Proceedings of a Conference on the Application of Scientific Techniques to the Study of Archaeology*, eds. A. Sinclair, E. Slater, and J. Gowlett, 92–99. Oxford: Oxbow Monograph 64.

Williams-Ellis, C. 1919. *Cottage building in cob, pisé, chalk and clay*. New York: Charles Scribner's Sons.

Williams-Thorpe, O. 1988. Provenance and archaeology of Roman millstones from the Mediterranean area. *Journal of Archaeological Science* 15: 253–305.

Williams-Thorpe, O. and R. S. Thorpe. 1990. Millstone provenancing used in tracing the route of a fourth-century BC Greek merchant ship. *Archaeometry* 32(2): 115–37.

Williams-Thorpe, O., D. Aldiss, I. J. Rigby, and R. S. Thorpe. 1999. Geochemical provenancing of igneous glacial erratics from southern Britain, and implications for prehistoric stone implement distributions. *Geoarchaeology* 14(3): 209–46.

Winkler, E. M. 1994. *Stone in architecture: Properties, durability*. Berlin: Springer-Verlag.

Wolf, S. 2002. Estimation of the production parameters of very large medieval bricks from St. Urban, Switzerland. *Archaeometry* 44(1): 37–65.

Woodforde, J. 1976. *Bricks to build a house*. London: Routledge & Kegan Paul.

Woods, A. J. 1982. Thin sections and ceramic technology: An introduction. In *Microscopy in archaeological conservation*, eds. M. Corfield and K. Foley, 13–18. London: British Museum.

—. 1985a. Form, fabric, and function: Some observations on the cooking pot in antiquity. In *Technology and style*, ed. W. D. Kingery, 157–72. Columbus, OH: American Ceramic Society.

—. 1985b. An introductory note on the use of tangential thin sections for distinguishing between wheel-thrown and coil/ring-built vessels. *Bulletin of the Experimental Firing Group* 3: 100–114.

Woodward, H. W., Jr. 1997. *The sacred sculpture of Thailand: The Alexander B. Griswold collection*. Baltimore, MD: Walters Art Gallery.

Worthing, M. A. 1982. A petrographic method of sourcing potsherds from the Port Moresby area of Papua New Guinea. *Archaeology in Oceania*. 17: 79–82.

Wright, A. 1991. *Craft techniques for traditional buildings*. London: Batsford.

Xidakis, G. S. and I. S. Samaras. 1989. Engineering properties of some building stones used in ancient cities of the Aegean Thrace. In *The conservation of monuments in the Mediterranean basin*, ed. F. Zezza, 133–38. Brescia: Grafo.

Yardley, B. W. D., W. S. MacKenzie and C. Guilford. 1990. *Atlas of metamorphic rocks and their textures*. New York: Longman Scientific & Technical.

Young, W. J. and B. Ashmole. 1968. The Boston relief and the Ludovisi throne. *BMFA* 66: 124–66.

Zantedeschi, C. and A. Zanco. 1993. Distinctive characteristics of euganean trachytes for their identification in ancient monuments. *Science and Technology for Cultural Heritage* 2: 1–10.

Zezza, F., F. Macri, D. Decrouez, V. Barbin, K. Ramseyer, J. Garcia-Rowe, and C. Saiz-Jimenez. 1992. Origin, weathering and biological colonization of the marble terminal column on the Via Appia, Brindisi, Italy. In *The conservation of monuments in the Mediterranean basin*, eds. D. Decrouez, J. Chamay, and F. Zezza, 99–112. Geneva: Musée d'Histoire Naturelle and Musée d'Art et d'Histoire.

Zhongtang, Y. 1992. Study on gas composition of ancient Yaozhou celadon glaze and its archaeological significance. In *Science and Technology of Ancient Ceramics: Proceedings of the 1992 International Symposium*, 201–10. Beijing: Science Press.

Zubrow, E., J. Fountain, A. Pelcin, K. Allen, P. Bush, K. Bartolotta, and L. Hunt. 1988. Prehistoric distribution networks of domestic ceramics in Sorsogon, Phillippines: An archaeometric study. In *Ceramic ecology revisited 1987: The technology and socioeconomics of pottery*, ed. C. C. Kolb, 33–56. Oxford: British Archaeological Reports International Series 436(i).

Zuixiong, L., N. Agnew, and P-M. Lin. 1997. Chemical consolidation of conglomerate and sand at the Magao grottoes. In *Conservation of ancient sites on the silk road*, ed. N. Agnew, 194–212. Los Angeles, CA: J. Paul Getty Trust.

Subject index

Note: Page references in bold type indicate major coverage. Page references in italics indicate illustrations.